Fungi
Multifaceted Microbes

Fungi

Multifaceted Microbes

Editors
B.N. Ganguli
S.K. Deshmukh

CRC Taylor & Francis
Taylor & Francis Group
Boca Raton London New York

CRC is an imprint of the Taylor & Francis Group,
an informa business

CONSISTING OF KNOWLEDGE

Anamaya Publishers
New Delhi

Editors
B.N. Ganguli M.S., Ph.D. (Illinois, USA)
Chair Professor, Agharkar Research Institute, Pune
Distinguished Professor, Bharati Vidyapeeth (Deemed University), Pune
Visiting Faculty, Mumbai University, Institute of Chemical Technology, Mumbai, India

S.K. Deshmukh Ph.D. (Sagar University)
Senior Research Scientist
Nicholas Piramal Research Centre, Mumbai, India

Copublished by CRC Press
Taylor and Francis Group
6000 Broken Sound Pkwy, NW
Suite 300, Boca Raton FL 33487, USA
with Anamaya Publishers, New Delhi, India

ISBN 1-4200-4328-5

Sold and distributed exclusively in all the countries, except India, by CRC Press
6000 Broken Sound Pkwy, NW, Suite 300, Boca Raton FL 33487, USA

In India, sold and distributed by Anamaya Publishers
F-154/2, Lado Sarai, New Delhi-110 030, India

Printed in India.

Preface

Fungi play a crucial role in many ecological processes. In plants, animals and geological sites that range from the aquatic to the terrestrial and urban structures such as historic monuments. They have also been harnessed to yield some of man's most useful products in medicines, in nutrition and in a variety of chemicals.

This book has attempted to cover some of these aspects from the wide range of topics that give fungi their multifaceted properties. Topics that range from fungal biochemistry and biotechnology and targets for drug discovery (Chapters 1, 2 and 12) to classical and molecular taxonomy (Chapters 2, 7 and 10), evolutionary aspects (Chapter 8), secondary metabolites from marine resources (Chapter 13), applications as mycoinsecticides (Chapters 14 and 15) and immunodiagnostics (Chapter 16) to name just a few. A special chapter on nanotechnology using fungi (Chapter 17) is an area that is of high interest today.

It is, of course, impossible in any single book to address every aspect of fungal biochemistry and biotechnology, given its enormous scope and dimensions.

But this volume contains articles from as wide a spectrum of international scientists as possible so that their viewpoints and their science is reflected and available to many others in the field. Thus we have authors from Australia, Brazil, Canada, Czechoslovakia, Estonia, Finland, Hungary, Russia, U.S.A. and India. The extensive literature survey will be extremely useful for scientists in the field.

The editors acknowledge with thanks the efforts and support of all the contributors to this book and are grateful for their patience. Dr. B.N. Ganguli thanks Professor Janos Berdy, Dr. T. Anke and the Journal of Antibiotics, Japan and specially the team of Microbiologists at the Hoechst Centre for Basic Research at Mulund, India who were with him for more than two decades.

Dr. S.K. Deshmukh thanks Dr. (Mrs.) Swati Piramal, Director, Dr. Somesh Sharma, CSO and Dr. H. Sivaramakrishnan, Nicholas Piramal Research Centre, for their encouragement and assistance.

EDITORS

Dr. M.J. Thirumalachar (1914-1999)

It is a honor and a pleasure to dedicate this book to Dr. M.J. Thirumalachar, a mycologist of international repute whose contributions on several new genera and species of fungi are well known.

My association with him goes back to the years between 1955 and 1965, during the formative years of my career, soon after graduation, when I joined the Biochemistry Department of the Research Laboratories of the Hindustan Antibiotics at Pimpri near Pune. He was head of Mycology and later became the Director of that center.

Through his encouragement and support I was able to continue my studies and research that ultimately changed my career and life.

A man of few words, his dedication to research and meticulous attention to the details of every experiment were remarkable. There were no holidays or hours for him. I saw him in the laboratory every day without fail till late in the night.

Dr. Thirumalachar published extensively in many areas. There are several monographs on rusts, description of new genera of smuts such as *Mundkurella*, a new genus of downy mildew, *Sclerophthora*, to give only a few examples. He reported the actinomycete genus *Chainia* in Nature (1955) and from a new strain *Streptomyces pimprina* he isolated an antifungal antibiotic named Hamycin. I worked on that project at that time. Interestingly he believed in the isolation of microorganisms from the areas close to the laboratory by careful observation. He reported several other new molecules such as Dermostatin, Aureofungin, Antiamoebin.

He was elected Fellow of the National Science Academy in 1966, received the S.S. Bhatnagar Award in 1967 and the Sunder Lal Hora medal in 1969. He was President of the Indian Phytological Society and the first President of the Mycological Society of India when it was formed in 1973.

I lost touch with him when I moved to the United States for higher studies. He also retired and migrated to the U.S.A. where he set up the Jeersanidhi-Anderson Institute at Walnut Creek, California with his son M.J. Narasimhan, Jr. From there he published work on new formulations of antifungals with reduced toxicity such as Jaimycin 1 but with the same potency. He patented Phyton 27 as a non-phytotoxic systemic fungicide for the control of Dutch elm disease and other tree wilts.

I met him again in 1980 and thereafter almost every year at the Interscience Congress on Antimicrobial Agents and Chemotherapy held annually in the U.S.A. His pleasure at meeting me and appreciation of my progress in both life and science are moments that I will cherish forever.

I thank him.

B.N. Ganguli

Contents

1. Fungal Secondary Metabolites: A Pandora's Box of Treasures

B.N. Ganguli
Agharkar Chair Professor, Agharkar Research Institute, Pune-411004, India
Former Deputy Director and Head of Microbiology, Hoechst Centre for Basic Research, Mulund, Mumbai-400080, India

SUMMARY Fungi are ubiquitous in nature and have, for centuries, provided us with a variety of useful products from cheese to alcoholic beverages, organic acids, pigments, enzymes, food and lately antibiotics and small molecular weight compounds with a variety of pharmacological activities. Various scientists have estimated that a million or even more fungi exist. The actual figure could be several millions due to the present day exploration of a variety of eco-sites, greater intensity of the efforts, newer methods of isolation and molecular taxonomic identification techniques. This leads to the assumption that fungi have the greatest potential for producing bioactive metabolites? A Pandora's box with unlimited potential? How many and how useful? This potential can be realized only by sophisticated fermentation techniques, screens using innovative but relevant targets and using miniaturized High Thru-put Screening Methods. This has been illustrated herein by a number of examples of metabolites with activities in innovative screens. They must be combined with "state of the art" chromatographic separations and early characterization of the metabolites through advanced instrumentation methods, followed by dereplication before structure elucidation is attempted. Secondary metabolite libraries and "in house" data bases are often of great value to industries in the search for new lead molecules for drug discovery programs.

Introduction

There are at least 1.5 to 1.75 million fungi going by literature estimates. Out of these approximately 74,000 species are described, i.e. ca 5% only (Hawksworth, 1991; Suryanarayanan et al. 2005; Reddy, 2005). There is the whole area of endophytic fungi that has not really been explored or exploited in any great detail. There are at least 750,000 species of plants on the earth and if 2 to 5 endophytes exist in each, we are reaching figures up to approximately 3 million (Personal Communication, Katz). Strobel and Daisy (2003) commented that endophytes could be a goldmine of secondary metabolites. *Pestalotiopsis* spp. can be considered as "the *E. coli* of the rain forests" and *P. microspora* (Speg) a "microbial factory" of bioactive secondary metabolites. According to them a variety of chemical structures are seen such as taxol, torreyanic acid, ambuic acid, cryptocandin, subglutinol A and B and others.

Obviously, the secondary metabolites described today are produced by a select few of the above, using conventional methods in most cases and therefore constitute a very small fraction of the numbers that may exist—a real Pandora's box of treasures still awaits the resourceful scientist.

In spite of this extremely limited exploitation, the capacity of fungi to produce a wide variety of useful molecules is well documented in numerous articles and books. Fungi have found extensive utilization in industrial enzyme production, as biofertilizers and biopesticides in agriculture, in bioremediation processes, in transformations of a variety of organic compounds (Moo-Young, 1985).

Bioactives from Fungi

This article restricts itself to a few selected bioactives and specifically to the human health area. Berdy (2005), in a personal review, estimates that 200,000 to 250,000 bioactive natural products are known. Out of these, approximately 22,000 to 23,000 are bioactive microbial products. Another 20,000 to 25,000 microbial metabolites, so-called "inactives" are known.

From fungi, 8600 bioactives are described, including 4900 antibiotics and 3700 with other activities. Among these, species of *Aspergillus, Penicillium* and *Fusarium* have produced approximately 2200 compounds. Other filamentous and endophytic species, such as *Trichoderma, Phoma, Alternaria, Acremonium* and *Stachybotrys* have yielded several hundred bioactives. Basidiomycetes such as *Ganoderma, Lactarius* or *Aureobasidium* have produced another 2000. Yeasts and Myxomycetes are poor producers (Berdy, 2005). But how many are used in human therapy? Just 15 or so!

Mycophenolic acid is one of the earliest metabolites to be described but is still useful today. The discoveries of Penicillin, Cephalosporin, Cyclosporin, Rapamycin, all fungal products, are landmarks in the history of human therapeutics. They ushered in the "Golden Age" of antibiotics that has lasted almost to the end of the 20th century. But gradually in the late eighties the emphasis shifted to pharmacologically active compounds that are aimed at improving the quality of our lives, for example pain killers. Cyclooxygenase-2 inhibitors such as the "coxcibs" dominated the market as billion dollar blockbusters, until toxicity problems surfaced. Other areas have been cancer therapy, tackling arthritis in the elderly, diabetes and obesity problems that lead to cardiovascular diseases. Several cholesterol lowering compounds, the majority produced by fungi—the "statins"—are known.

Have we exhausted the metabolite producing capacity of fungi?—in the few we have exploited to produce new molecules? Certainly not! We have barely touched the tip of the iceberg.

Production, Detection, Purification, Characterization and Dereplication of Secondary Metabolites

It is a combination of all of the above processes that increases the chances to find novel or new lead molecules (Berdy, 1989, 2005; Ganguli, 1989, 2004, 2005; Yarborough et al. 1993; Monaghan et al. 1995). New molecules will come from exploitation of neglected genera, from those that have eluded cultivation and/or are difficult to cultivate. What is true for the actinomycetes is equally true for fungi. New molecules will come from innovative fermentation methods and manipulation of fermentation parameters, using both known and new species (Monaghan et al. 1995; Cragg et al. 1997; Biabani and Laatsch, 1998; Knight et al. 2003). Selective detection methods using novel, biologically relevant, disease based targets can provide novel lead molecules (Demain, 1999; Hill et al. 1998; Donadio et al. 2002). The whole area of genetic manipulations in fungi, using the tools of molecular biology, is leading to "expression based cloning systems" and to the production of homologous and heterologous proteins in fungi (Broekhuijsen et al. 1993; Dalboge, 1997; Maras et al. 1999; Hamer et al. 2001; Punt et al. 2002).

Metagenomics uses community extracted DNA and cloning to express metabolite libraries from uncultivated organisms. A technique used extensively today.

The simple bioassay to detect anti-infectives is still extremely useful provided it is made very sensitive and reproducible and uses a wide selection of sensitive and resistant test strains. Its limitations are the inability to detect non antibacterials. High Thru-put Screens (HTS) are now routine for pharmacologically active molecules. Active in receptor binding assays, enzyme inhibition, protein-protein interactions, DNA-protein interactions, transcription factor regulation, most of which are carried out in cell line based assays and/or microtitre plate based robotic assays. The limitations are a need to pre-process the complex fermentation broth and remove interfering substances from the media or the crude extract. Schmid et al. (1999) described a multi-stage, automated solid phase extraction system to make the extracts compatible with HTS methods.

Rapid methods (one minute) have been described for taxonomy as well as qualitative and quantitative secondary metabolite profiling in crude fungal extracts using HPLC, diode array detectors and flow injection analyses together with electrospray ionization mass spectrometry. Direct infusion ES-MS has also been used (Smedsgard and Frisvad, 1996; Julian et al. 1998; Zahn et al. 2001; Higgs et al. 2001). The only requirement is that the compound must be detectable by either positive or negative ion electrospray ionization.

Dereplication is an important aspect in the search for new compounds from the already existing thousands that have been reported (Cordell and Shin, 1999). Databases play an important role in the elimination of known molecules (Laatsch, 2000). Several others exist.

Activities of Fungal Metabolites Against Some Selected Targets

1. Serine/Threonine Phosphatase 2A (PP2A) is an intracellular protein phosphatase that dephosphorylates many substrates. Specific inhibitors of PP2A enhance the activity of natural killer cells which leads to prevention of tumor metastasis. Example: Harzianic acid (Kawada et al. 2004).
2. Inhibitors of the extracellular signal related kinase (ERK) activity in the mitogen activated protein kinase-MAPK/ERK-signaling pathway in cells. Example: Terpestacin (Jung et al. 2003) an anti-angiogenesis compound.
3. Interleukin 1-beta Converting Enzyme (ICE) inhibitors: ICE is a cysteine protease that cleaves the biologically inactive 31 KDa precursor of IL-1 B to the active molecule that is a key mediator of inflammation and septic shock. Example: EI-1941-1 and -2. (Koizumi et al. 2003a).
4. Sphingosine kinase inhibitors: The kinase is a key enzyme that regulates the levels of sphingosine-1-phosphate (SP) which is involved in a variety of cellular functions such as angiogenesis, thrombosis, inflammation. Example: S-15183a and b (Kono et al. 2001).
5. Cholesteryl Ester Transfer Protein (CETP). Inhibitors of CETP are effective in prevention of atherosclerosis. Low concentrations of CETP result in high HDL and low LDL cholesterol that resolves the condition. Example: Erbulenols (Tomoda et al. 1998). Similiarly, Acyl-CoA: Cholesterol Acyltransferase (ACAT) inhibitors are treatments for hypercholesterolemia and atherosclerosis. Example: Phenylpropene C (Rho et al. 1998).
6. Low density lipoprotein receptor gene inducers. Example: CJ-12,950, CJ-13,357-Novel lactones (Dekker et al. 1998).
7. Glucose-6-phosphate Translocase T1 Inhibitors: Useful in Non Insulin Dependant Diabetes Mellitus. Example: Kodaistatin (Vertesy et al. 2000).
8. Inhibitors of Insulin like Growth Factor (IGF) transduction: Anticancer agents against IGF dependant tumors. Example: Tyroscherin (Hayakawa et al. 2004).
9. Reversal of Multi-Drug Resistance: P-Glycoprotein binding and inhibition of efflux mechanisms of antitumor agents. Example: Androstan A (Rho et al. 1998).
10. Prolyl Endopeptidase (PEP) inhibitors and Alzeihmer's disease: PEP is a serine protease that cleaves the C terminal side of proline residues and it could be involved in the processing of the amyloid precursor protein along with the endoproteolytic enzymes, the secretases. Example: Kynapcins 13 and 28 (Kim et al. 2002).
11. Inhibitors of Eukaryotic Signal Transduction: Specifically the nuclear transcription factor-NF-kappa-beta and its activation pathway. NF-kB is involved as an immediate early transcriptional activator of genes encoding pro-inflammatory cytokines, chemokines that selectively attract inflammatory cells, inflammatory enzymes (iNOS, Cox-2) and adhesion molecules. Inhibitors of the NF-kB

activation pathway may, therefore, have broad applications as novel therapeutics for a variety of diseases including cancer. Example: Cycloepoxydon and related compounds (Gehrt et al. 1998).

12 Cell wall active antifungals can be detected in a plate assay using a combination of glucan and chitin synthesis inhibitors in *Neurospora crassa* (Taft et al., 1994). Example: Mulundocandin, Deoxymulundocandin (Roy et al. 1987; Mukhopadhyay et al. 1992), (Fig. 1).

13 A simple bioassay for antifungals gave Fumifungin (Mukhopadhyay et al. 1987), Arthrichitin (Vijayakumar et al. 1996), Orbuticin (Roy et al. 1996) (Fig. 5), Aranorosin and related compounds (Roy et al.1987, 1992; Mukhopadhyay et al. 1998; Vijayakumar et al. 1998) (Fig. 2A).

Serendipity in Screening

Several thousands of active metabolites are known from a few select detection methods. Even more exist whose activities we do not know as yet. It may be of interest to test old and established structures in the new target assays. New activities in old compounds are being seen and similiarly, old activities are very often seen in new molecules. The first example is Wortmannin, a very old molecule from *Penicillium wortmanni* which we also isolated in our laboratories in the seventies and discarded. Today, it is found to be an inhibitor of Phosphoinositide-3-kinase (PI3K) that blocks morphological transition and migration of SiHa cells in wound closure assays (Mishra, G.C., Reports of the National Center for Cell Sciences, Pune, 2004).

In 1987, we isolated the lipopeptide Mulundocandin from *Aspergillus sydowii* as an anti-candida compound (Roy et al. 1987) (Fig. 1). Today, with the spread of AIDS and immunocompromised patients, *Pneumocystis carinii* infections are diagnosed against which the lipopeptides are effective.

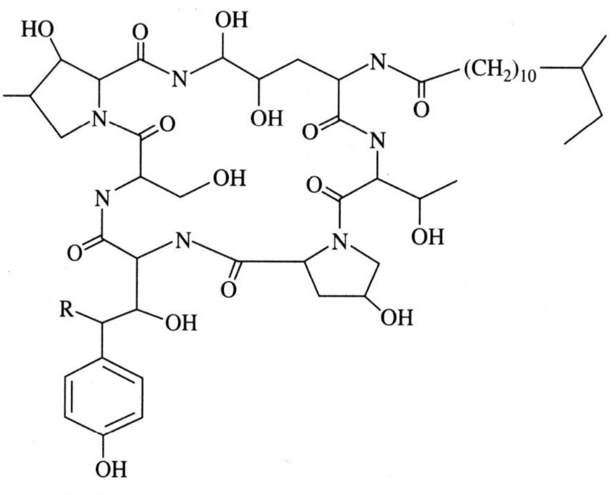

Mulundocandin: R = OH
Deoxymulundocandin: R = H

Fig. 1. Mulundocandin and Deoxymulundocandin lipopeptide active against *Candida* sp. from *Aspergillus sydowii*, isolated at the Hoechst Centre for Basic Research, Mumbai, India.

In 1988, from *Pseudoarachniotus roseus*, we discovered a novel skeleton Aranorosin (Fig. 2a) and several related compounds viz. Aranorosinols, Aranochlors (Roy et al. 1988, 1992; Mukhopadhyay et al. 1998; Vijayakumar et al. 1998) but only as antifungals and cytotoxic compounds. They were never tested for ICE inhibition or as NF-kB pathway inhibitors.

Gehrt et al. (1998) reported three hexaketides-cycloepoxydon and two related compounds that inhibit

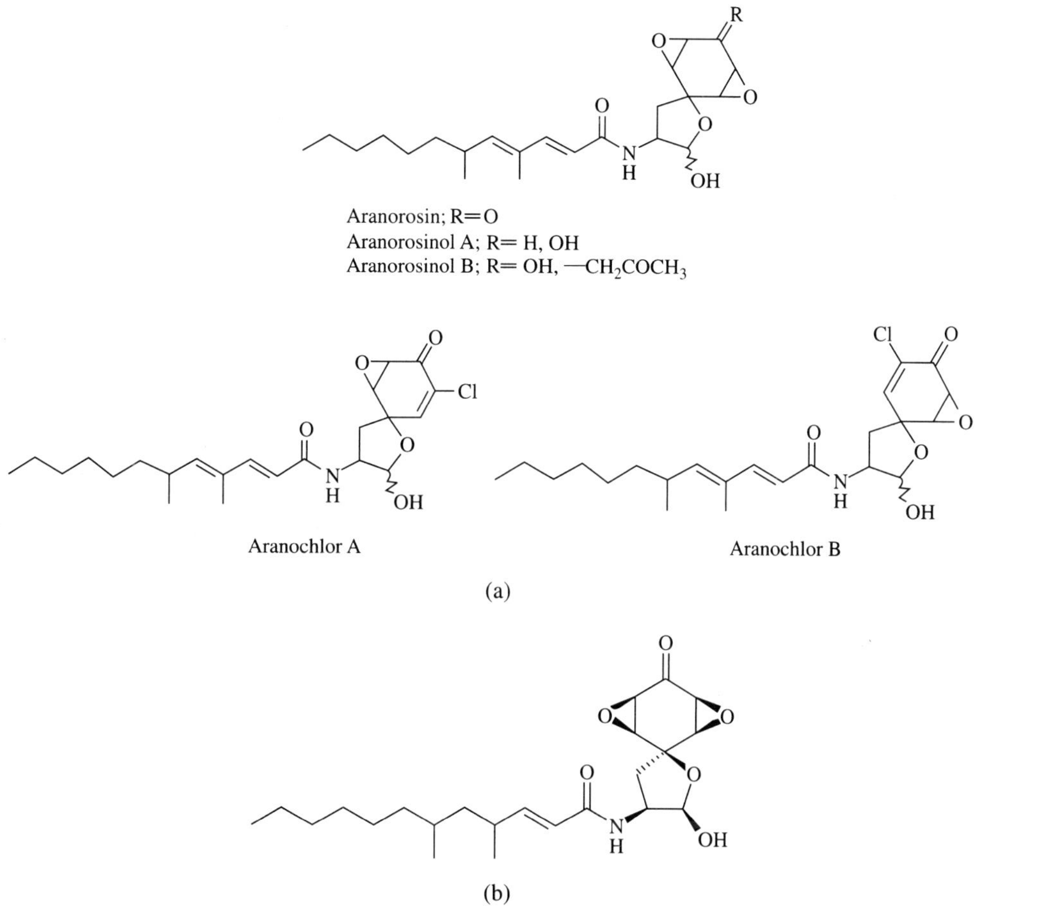

Fig. 2. (a) Aranorosin, Aranorosinol A, B, Aranochlor A and B from *Pseudoarachniotus roseus*. Antifungal compounds isolated at the Hoechst Centre for Basic Research, Mumbai, India, and (b) EI-2128-1, from a *Penicillium* sp. Interleukin-1beta converting enzyme (ICE) inhibitor (reproduced from Koizumi et al. 2003a, b).

the activation of the NF-kB and AP-1 mediated signal transduction pathways in COS-7 cells (Fig. 3a). Compounds bearing similar structural features as Aranorosins in a different target assay.

Koizumi et al. (2003a, b) reported EI-2128-1 (Fig. 2b), from a *Penicillium* sp. as an ICE inhibitor. Its structure is 4′, 5′-dihydroaranorosin. They also reported other compounds with related structural features, EI-1941-1 and EI-1941-2 both ICE inhibitors (Fig. 3b).

Ichikawa et al. (2001a, b) found different structures in a fungus (diterpenes) and a basidiomycete (substituted pyridine-2-carboxylates) inhibiting the production of the cytokines-IL 1B and TNF alpha (Fig. 4a, b).

It appears that the inhibition of the NF-kB pathway, inhibition of the enzyme ICE and production of the cytokines IL-1B and TNF alpha are linked and inhibited by compounds with an epoxy structure. But it does not explain why others, without such a structural feature, still show activity. It is necessary to check the quantitative inhibitory concentrations to find the true inhibitors. What is not covered here are several metabolites from actinomycetes with the same activity profile.

Jung et al. (2003) identified Terpestacin from *Embellisia chlamydospora* as an anti-angiogenesis molecule. It was originally isolated from an *Arthrinium* sp. as a synctium formation inhibitor in HIV infections (Oka

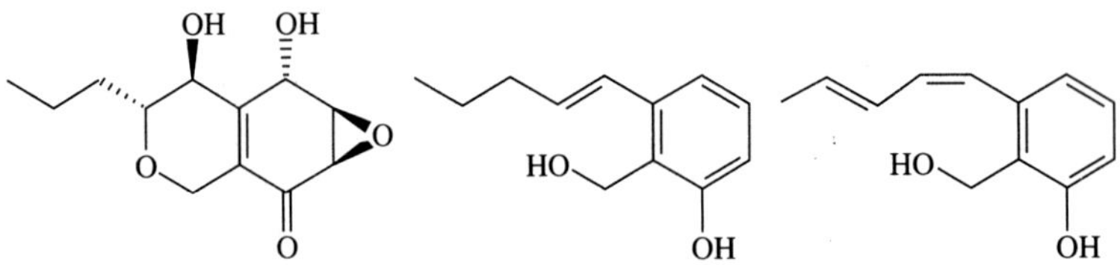

(a)

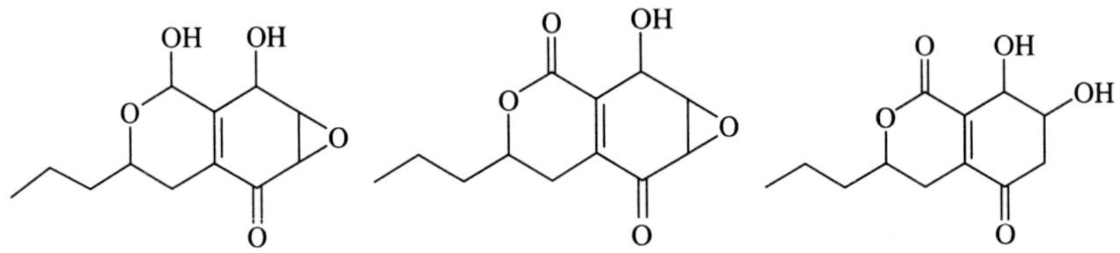

(1) (2) (3)

(b)

Fig. 3. (a) Cycloepoxydon and related hexaketides from a *Xylaria* sp. NF-kB and AP-1 pathway inhibitors. Reproduced from Gehrt et al. (1998), and (b) EI-1941-1, EI-1941-2, EI-1941-3 from *Farrowia* sp. (1) and (2) are ICE inhibitors and (3) is inactive (reproduced from Koizumi et al. 2003a, b).

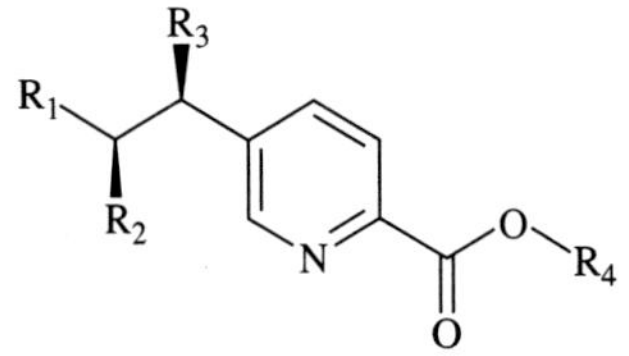

	R_1	R_2	R_3	R_4
CJ 14877 (I)	Me	OH	OH	Me
CJ 14897 (II)	Me	OAc	OH	Me

(a)

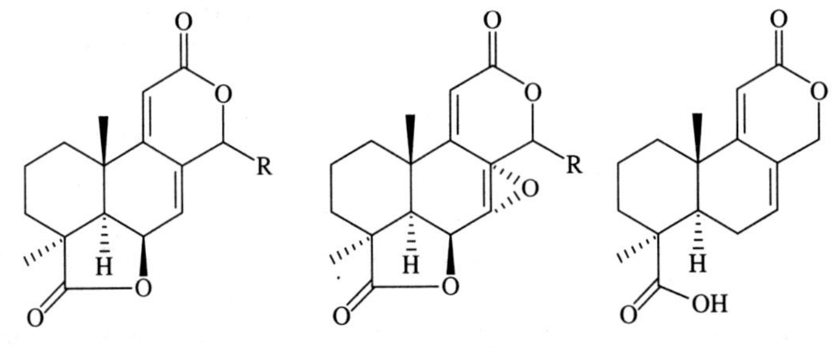

	R
LL-Z1271α	·····OMe
LL-Z1271γ	·····OH
CJ-14445	····H

	R
PR 1388	·····OMe
CJ 14515	∿∿OH

CJ 14604

(b)

Fig. 4. (a) Methyl 5 substituted pyridine-2-carboxylates from a basidiomycete, *Marasmiellus* sp. IL-1B and TNF alpha production inhibitors (reproduced from Ichikawa et al. (2001b) and (b) Diterpenes from a fungus *Oidiodendron griseum* CL37215. IL-1B and TNF alpha production inhibitors (reproduced from Ichikawa et al. 2001a).

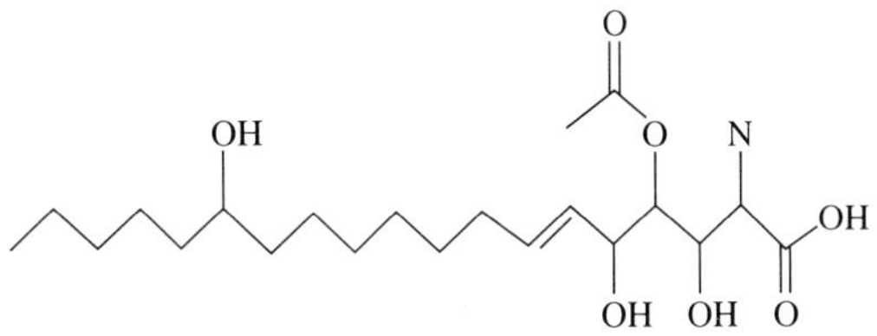

Fumifungin

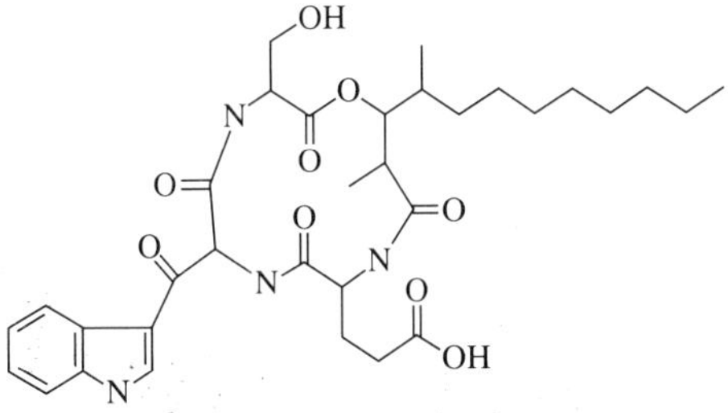

Arthrichitin

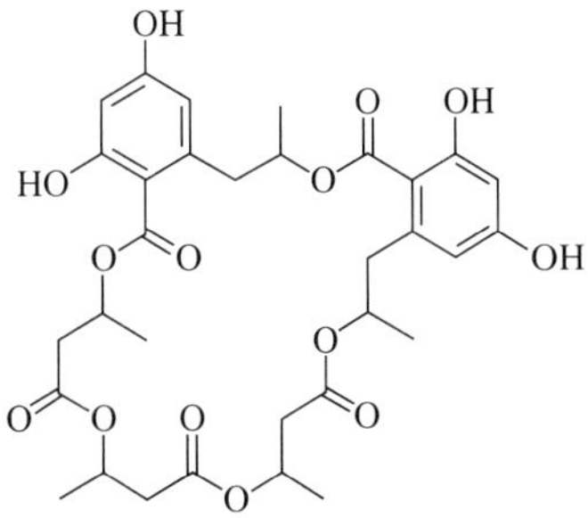

Orbuticin

Fig. 5. Antifungal compounds from simple bioassays, isolated at the Hoechst Centre for Basic Research, Mumbai, India.

et al. 1993) and also from *Bipolaris cynodontis* as a phytotoxin (Sugawara et al. 1985). It has a structure similar to Radicol which is an inhibitor of the molecular chaperone heat shock protein, HSP 90.

Kim et al. (2004) found the known compound hydroxy methyl benzoic acid (HMBA) in *Curvularia* sp. KF119, with the new activity of a cell cycle inhibitor. Earlier, in 1990 and 1991, other investigators had found the compound to be a phytotoxin.

Mycophenolic acid is a well known inhibitor of inosine monophosphate dehydrogenase with antiviral and antitumor activities (Williams, 1968; Carter, 1969). Ui et al. (2005) have reported that it is a synctium formation inhibitor and gp 120 expression reducer in a HIV entry assay.

From the above examples it is evident that as different target assays are being developed, new activities are seen, in both, old and new compounds. Activities in different structural moieties have sometimes appeared to be similar. But it must be realised that complex biochemical pathways are inter- and intradependant on several key enzymes. The whole area of isozymes, for instance, cannot be ignored. For example, there are 17 histone deacetylases (HDAC) in humans which strongly suggests that different isozymes have different functions in different cells, tissues, organs (Johnstone, 2002). Different activities of tetrapeptides reported by various investigators suggest that specificity differences in isozymes may exist (Mori et al. 2003).

A few representative examples of novel bioactives from fungi, reported in the Journal of Antibiotics (2003 to 2005) are given in Table 1 to give an idea of the varied spectrum of activities seen.

Conclusion

This article is a brief summary of the potential of fungi to produce a wide range of compounds with highly interesting activities. In order to realise that potential and bring new therapeutics to the common man, an enormous amount of investigation of the compounds needs to be carried out. The hurdles are enormous and the expenses equally high. The whole area of toxicity versus therapeutic efficacy is still crucial in *in vivo*

Table 1. Bioactives from Fungi Reported in the Journal of Antibiotics, Japan (2003-2005)

S. No.	Source	Compound	Target/Inhibitor	Reference
1	*Penicillium* sp. FK-1-1436	Atrovenetinone	HIV-Integrase	Shiomi et al. (2005)
2	*Exophiala pisciphila*	Exophillic Acid	HIV-Integrase	Ondeyka et al. (2003)
3	*Myrothecium cunctum* No. 002	FR227244 Triterpene glycoside	1,3-beta-glucan synthase	Kobayashi et al. (2004)
4	Pycnidial fungus-Not identified	Spinosine Sulfate A	Neuraminidase	Shibazaki et al. (2004)
5	Strain MT 51005-Not identified	CRM 51006 related to Anguilosporal	Phospholipase C	Oh et al. (2004)
6	*Periconia byssoides*	Peribysins	Cell adhesion	Yamada et al. (2005)
7	*Penicillium* sp.	Deoxyverrucosidin	GRP 78/BiP Down regulator	Choo et al. (2005)
8	*Penicillium algidum (Psychrotolerant)*	Cyclic Peptides Psychrophilin D	Anti-cancer, Anti-plasmodium	Dalsgaard et al. (2005)
9	*Penicillium terrestre* Marine derived	Benzoquinones, Bisorbicillinoids	Cytotoxic compounds	Liu et al. (2005)
10	*Phoma* sp. No. 00144	FR225654 stemphylorin type	Gluconeogenesis	Ohtsu et al. (2005)
11	*Aspergillus terreus*	Terreulactones A, B, C, D	Acetylcholinesterase inhibitors	Cho et al. (2003)
12	*Penicillium* sp.	S14-95, Terpenoid	Interferon gamma mediated signal transduction pathway inhibitor	Erkel et al. (2003)
13	*Cunninghamella* sp. F-1490	ICM 0201 Morpholine carboxylic acid derivative	Osteoclastogenesis inhibitor	Inoue et al. (2003)
14	*Acremonium* sp. 27082	FR235222 Cyclic Tetrapeptide	Mammalian histone deacetylase inhibitor, immunosuppressant	Mori et al. (2003)
15	*Aspergillus* sp. CFR1046	Asperaldin naphthoquinone	Inhibitor of rat lens aldose reductase	Rao et al. (2003)
16	Unidentified Fungus FK-0929	Epohelmins A, B Epoxyheptamethylenamine type compounds	Lanosterol synthase inhibitors	Sakano et al. (2004)
17	*Aspergillus niveus* L49575	Aspochalamins, A-D and Aspochalsin Z	Weak antibacterial, cytotoxic activity	Gebhardt et al. (2004)
18	*Panaeolus reterugis (Fr)* Gill poisonous mushroom	Panepolic acid, Paneolilludinic acid, Iludane sesquiterpenes	Antibacterial agent against Gram positives	Ma et al. (2004)
19	*Bipolaris brizae* *Stachybotrys chartarum*	Cochlioquinones and Epi-cochlioquinones	Antagonists of human chemokine receptor, CCR-5:anti-HIV	Yoganathan et al. (2004)
20	*Verticillium hemipterigenum* BCC 2370 Insect pathogen	Ascochlorin glycoside	Antiviral, antitumor, Ras-farnesyl protein transferase inhibitor	Seephonkai et al. (2004)
21	*Chaetomium subspirale*	Oxaspirodon, spiro compound	Inhibitor of inducible TNF-alpha expression	Jan Rether et al. (2004)
22	*Micromucor ramannianus* var *anguilisporus*	Dimeric depsipeptides K S-506a, 506m	Inhibitors of protein Tyrosine Phosphatase I B (PTP1B)	Oh et al. (2004)

23	*Metarrhizium* sp. FK-1079	Fungerin. Isoprenylated urokanic acid derivative	Inhibits microtubule polymerization and arrests cell cycle in M phase	Koizumi et al. (2004)
24	Marine *Aspergillus* sp.	Dihydroxyisoechinulin A	Radical scavenging activity and UV-A protectant	Li et al. (2004)
25	*Sporormiella minima*	FR 171456: structure to be reported	Cholesterol biosynthesis inhibitor in Hep G-2 cells and up regulator of cell surface LDL receptor	Hatori et al. (2004)
26	*Acremonium kiliense* Grutz TC-1703	TMC-260 Pyrrolidinone derivative, analog of Pramanicin	Inhibitor of IL-4 signal transduction and supresses IgE production	Sakurai et al. (2003)
27	*Byssochlamys nivea*	Byssochlamysol, ergostane steroid	Inhibitor of Insulin like Growth factor IGF dependent cell growth using MCF cell	Mori et al. (2003)
28	*Ustilago maydis* and *Geotrichum candidum*	Glycolipid mixtures: Ustilipids	Neurotensin receptor antagonists. Effective in Schizophrenia, Parkinson's, Alzeihmer's	Kurz et al. (2003)

studies in animals and then in humans. Chronic toxicity studies of numerous kinds in different organs, pharmacokinetics and finally the effective concentration for treatment takes several years (ca 8 to 12 years) and costs $ 1 to 1.5 billion for one compound to reach the market and a great deal of technology has to be developed for economic viability.

Compounds fail because surrogate toxicity markers and animal tests cannot really replace tests in humans.

Which is why only a few compounds (may be 15 or so) are used in humans.

Recombinant DNA technology, fungal expression systems and the possibilities of gene cloning into heterologous systems are not described. That is a book by itself! It is mainly addressed to enzymes and proteins so far.

Fungal diseases, which are highly problematic and systemic fungal infections are difficult to treat and hence not covered. Antifungals from fungi are known, but non-toxic, orally effective compounds are badly required.

Then there is the whole area of applications of fungi—producers of so many useful enzymes in every day life. Produced by innovative solid state fermentations, they have found extensive use in many bioremediation systems-removal of toxic metals, carcinogenic dyes from effluents. They are wonderful fertilizers and biopesticides. Fungi and their enzymes carry out numerous biotransformations many of which are indispensible to Industry.

No attempt is made to give creativity indices in fungi nor specify the best producers for the simple reason that microorganisms can be manipulated to give a variety of different compounds. Their genetic capacities are limited by our limited detection techniques, metabolic engineering methods and conventional fermentation methods.

Then there remains the question: how different are fungal metabolites, chemically? Are their structures basically related? Are they from the same metabolic pathways? It is difficult to say and there is not enough information. Many similar structures are seen in diverse organisms as shown in some examples. But significant differences in activities are seen, which, if carefully analyzed, can lead to therapeutically effective compounds. It is not the level of activity that becomes important but the specificity and the therapeutic window. The example can be given of the interesting immunosupressant FR 235222 reported by Mori et al. (2003). The compound is a cyclic tetrapeptide, a class of compounds seen fairly regularly. But it has unique structural features such as three residues not usually found in proteins: a methylproline, an isovaline and a 2-amino-9-hydroxy-8-oxodecanoic acid moiety. It may be these residues that give the molecule its high activity and oral availability.

New skeletons do appear in the literature especially from neglected genera and neglected ecosystems such as the marine environment, with greater frequency than earlier. Very often they appear from minor components that are identified much more easily due to the sensitive separation and detection methods, coupled with sophisticated spectroscopic techniques and use of numerous databases.

Ackowledgements

The author thanks Dr. Janos Berdy, Prof. T. Anke and the Journal of Antibiotics, Japan for their kind permission to use some of the data shown.

References

Bérdy, J. (1989). The discovery of new bioactive microbial metabolites. Screening and identification. In: Bioactive Metabolites from Microorganisms, *Progress in Industrial Microbiology*, Vol. 27 (Eds.) Bushell, M.E. and Grafe, U., Elsevier, Amsterdam, pp. 2–26.

Bérdy, J. (2005) Bioactive Microbial Metabolites: A Personal View. Journal of Antibiotics 58(1): 1–26.

Biabani, M.A.F. and Laatsch, H. (1998) Advances in chemical studies on low-molecular weight metabolites of marine fungi. Journal für Praktische Chemie 340: 589–607.

Broekhuijsen, M.P., Mattern, I.E., Contreras, R., Kinghorn, J.R. and van den Hondel, C.A.M. J. J. (1993) Secretion of heterologous proteins by *Aspergillus niger*: production of active human interleukin-6 in a protease-deficient mutant by KEX2-like processing of a glucoamylase–hIL6 fusion protein. Journal of Biotechnology 31: 135–145.

Carter, S.B., Franklin, T.J., Jones, D. F., Leonard, B. J., Mills, S. F., Turner, R. W. and Turner, W. B. (1969) Mycophenolic acid: an anti-cancer compound with unusual properties. Nature 223: 848–850.

Cho, K.M., Kim, W.G., Lee, C.K., and Yoo, I.D. (2003) Terreulactones A, B, C, and D: novel acetylcholinesterase inhibitors produced by *Aspergillus terreus*. I. Taxonomy, fermentation, isolation and biological activities. Journal of Antibiotics **56**(4): 344–350

Choo, S.J., Park, H.R., Ryoo, I.J., Kim, J.P., Yun, B.S., Kim, C.J., Shin-ya, K. and Yoo, I. D. (2005) Deoxyverrucosidin, a novel GRP78/BiP down-regulator, produced by *Penicillium* sp. Journal of Antibiotics 58 (3): 210–213.

Cordell, G.A., and Shin, Y.G. (1999) Finding the needle in the haystack. The dereplication of natural product extracts. Pure and Applied Chemistry 71: 1089–1094.

Cragg, G.M., Newman, D.J. and Snader, K.M. (1997) Natural products in drug discovery and development. Journal of Natural Products (Lloydia) 60: 52–60.

Dalboge, H. (1997) Expression cloning of fungal enzyme genes: a novel approach for efficient isolation of enzyme genes of industrial relevance. FEMS Microbiology Reviews 21(1): 29–42.

Dalsgaard, P.W., Larsen, T.O. and, Christophersen, C. (2005) Bioactive cyclic peptides from the psychrotolerant fungus *Penicillium algidum*. Journal of Antibiotics 58(2): 141–144.

Dekker, K.A., Aiello, R., Hirai, H., Inagaki, T., Sakakibara, T., Suzuki, Y., Thompson, J.F., Yamauchi, Y. and Kojima, N. (1998) Novel lactone compounds from *Mortierella verticillata* that induce the human low density lipoprotein receptor gene: Fermentation, isolation, structural elucidation and biological activities. Journal of Antibiotics 51: 14–20.

Demain, A.L. (1999) Pharmaceutically active secondary metabolites of microorganisms. Applied Microbiology and Biotechnology 52: 455–463.

Donadio, S., Carrano, L., Brandi, L., Serina, S., Soffientini, A., Raimondi, E., Montanini, N., Sosio, M. and Gualerzi, C.O. (2002) Targets and assays for discovering novel antibacterial agents. Journal of Biotechnology 99(3):175–185.

Erkel, G., Rether, J., Anke, T. and Sterner, O. (2003) S14-95, a novel inhibitor of the JAK/STAT pathway from a *Penicillium* species. Journal of Antibiotics 56(4): 337–343.

Ganguli, B.N. (1989) A Microbial Screening Programme: Scope and limitations. In Bioactive Metabolites from Microorganisms, *Progress in Industrial Microbiology*, Vol. 27 (Eds.) Bushell, M.E. and Grafe , U., Elsevier Amsterdam, pp. 27–37.

Ganguli, B.N. (2004). Microbial Diversity - The Challenge. In Microbiology and Biotechnology in Sustainable Development (Ed) Jain, P.C., CBS Publishers & Distributors, New Delhi, India, pp. 1–11.

Ganguli, B.N. (2005) Old Wine in New Bottles, New Wine in Old Bottles, New Wine in New Bottles: A Study of Microbial Metabolite Diversity. In Abstract of the International Conference on Microbial Diversity, Current Perspective and Potential Applications. April 16–18, New Delhi, pp. 19.

Gebhardt, K., Schimana, J., Holtzel, A., Dettner, K., Draeger, S., Beil W., Rheinheimer, J. and Fiedler, H.P. (2004) Aspochalamins A-D and Aspochalasin Z produced by the endosymbiotic fungus *Aspergillus niveus* LU 9575. I. Taxonomy, fermentation, isolation and biological activities. Journal of Antibiotics 57(11): 707–714.

Gehrt, A., Erkel, G., Anke, T. and Sterner, O. (1998) Cycloepoxydon, 1-hydroxy-2-hydroxymethyl-3-pent-1-enylbenzene and 1-hydroxy-2-hydroxymethyl-3-pent-1,3-dienylbenzene, new inhibitors of eukaryotic signal transduction. Journal of Antibiotics 51(5): 455–463.

Hamer, L., Adachi, K., Montenegro-Chamorro, M.V., Tanzer, M.M., Mahanty, S.K., Lo, C., Tarpey, R.W., Skalchunes, A.R., Heiniger, R.W., Frank, S.A., Darveaux, B.A., Lampe, D.J., Slater, T.M., Ramamurthy, L., DeZwaan, T.M., Nelson, G.H., Shuster, J.R., Woessner, J. and Hamer, J. E. (2001) Gene discovery and gene function assignment in filamentous fungi. Proceedings of the National Academy of Sciences, USA, 98(9): 5110–5115.

Hawksworth, D.L. (1991) The fungal dimension of biodiversity: magnitude, significance, and conservation. Mycological Research 95: 641–655.

Hayakawa, Y., Yamashita, T., Mori, T., Nagai, K., Shin-Ya, K. and Watanabe, H. (2004) Structure of tyroscherin, an antitumor antibiotic against IGF-1-dependent cells from *Pseudallescheria* sp. Journal of Antibiotics 57(10): 634–638.

Hatori, H., Shibata, T., Tsurumi, Y., Nakanishi, T., Katsuoka, M., Ohtsu, Y., Sakamoto, K., Takase, S., Ueda, H., Hino, M. and Fujii, T. (2004) FR171456, a novel cholesterol synthesis inhibitor produced by *Sporormiella minima* No. 15604. I. Taxonomy, fermentation, isolation, physico-chemical properties. Journal of Antibiotics 57(4): 253–259.

Higgs, R.E., Zahn, J.A., Gygi, J.D. and Hilton, M.D. (2001) Rapid Method to Estimate the Presence of Secondary Metabolites in Microbial Extracts. Applied and Environmental Microbiology 67(1): 371–376.

Hill, D.C., Wrigley, S.K. and Nisbet, L.J. (1998) Novel screen methodologies for identification of new microbial metabolites with pharmacological activity. Advances in Biochemical Engineering/Biotechnology 59: 73–121.

Ichikawa, K., Hirai, H., Ishiguro, M., Kambara, T., Kato, Y., Kim, Y.J., Kojima, Y., Matsunaga, Y., Nishida, H., Shiomi, Y., Yoshikawa, N., Huang, L.H., and Kojima, N. (2001a) Cytokine production inhibitors produced by a fungus, *Oidiodendron griseum*. Journal of Antibiotics 54(9): 697–702.

Ichikawa K., Hirai H., Ishiguro M., Kambara T., Kato Y., Kim Y.J., Kojima Y., Matsunaga Y., Nishida H., Shiomi Y., Yoshikawa N. and Kojima N. (2001b) Novel cytokine production inhibitors produced by a basidiomycete, *Marasmiellus* sp. Journal of Antibiotics 54(9): 703–709.

Inokoshi, J., Shiomi, K., Masuma, R., Tanaka, H., Yamada, H. and Omura, S. (1999) Funalenone, a novel collagenase inhibitor produced by *Aspergillus niger*. Journal of Antibiotics 52(12): 1095–1100.

Inoue, H., Kumagai, H., Osono, M., Matsufuji, M., Sameshima, T., Kawamura, N., Someno, T., Ishizuka M. and Takeuchi, T. (2003) ICM0201, a new inhibitor of osteoclastogenesis from *Cunninghamella* sp. F-1490. I. Taxonomy, fermentation, isolation and biological activities. Journal of Antibiotics 56(3): 209–213.

Julian, R.K., Higgs, R.E., Gygi, J.D. and Hilton, M.D. (1998) A method for quantitatively differentiating crude natural extracts using high-performance liquid chromatography-electrospray mass spectrometry. Analytical Chemistry 70: 3249–3254.

Jung, H.J., Lee, H.B., Kim, C.J., Rho, J.R., Shin, J., and Kwon, H.J. (2003) Anti-angiogenic Activity of Terpestacin, a Bicyclo Sesterterpene from *Embellisia chlamydospora*. Journal of Antibiotics 56: 492–496.

Johnstone, R.W. (2002) Histone-deacetylase inhibitors: novel drugs for the treatment of Cancer. Nature Reviews Drug Discovery 1(4): 287–299.

Kawada, M. Yoshimoto, Y. Kumagai, H. Someno, T. Momose, I. and Kawamura, N. (2004) PP2A inhibitors, Harzianic acid and related compounds produced by fungus strain F–1531. Journal of Antibiotics 57: 235–237.

Kim, S.I., Park, I.H. and Song, K. S. (2002) Kynapcin-13 and -28, New Benzofuran Prolyl Endopeptidase inhibitors from *Polyozellus multiplex*. Journal of Antibiotics 55(7): 623–628.

Kim, S.N., Kim, W.G., Jung, H.J., Shim, J.S., Kim, H.J., Lee, H.B., Han, J.T., Kim, C.J., Baek, N.I., Yoo, I.D. and Kwon, H.J. (2004) Cell cycle inhibitory activity of 4-hydroxy-3-(3'-methyl-2'-butenyl)-benzoic acid from *Curvularia* sp. KF119. Journal of Antibiotics 57(9): 605–608.

Knight, V., Sanglier, J. J., DiTullio, D., Braccili, S., Bonner, P., Waters, J., Hughes, D., and Zhang, L. (2003) Diversifying microbial natural products for drug discovery. Applied Microbiology and Biotechnology 62: 446–458.

Kobayashi, M., Kanasaki, R., Ezaki, M., Sakamoto, K., Takase, S., Fujie, A., Hino, M. and Hori, Y. (2004) FR227244, a novel antifungal antibiotic from *Myrothecium cinctum* No. 002. Taxonomy, fermentation, isolation and physico-chemical properties. Journal of Antibiotics 57(12): 780–787.

Koizumi, F., Agatsuma, T., Ando, K., Kondo, H., Saitoh, Y, Matsuda, Y., and Nakanishi, S. (2003a) EI-2128-1, a novel interleukin-1beta converting enzyme inhibitor produced by *Penicillium* sp. E-2128. Journal of Antibiotics 56(11): 891–898.

Koizumi, F., Matsuda, Y., and Nakanishi, S. (2003b) EI-1941-1 and -2, novel interleukin-1beta converting enzyme inhibitors produced by *Farrowia* sp. E-1941. I. Biochemical characterization of EI-1941-1 and -2. Journal of Antibiotics 56(5):464–469.

Koizumi, Y., Arai, M., Tomoda, H. and Omura, S. (2004) Fungerin, a fungal alkaloid, arrests the cell cycle in M phase by inhibition of microtubule polymerization. Journal of Antibiotics 57 (7): 415–420.

Kono, K., Tanaka, M., Ono, Y., Hosoya, T., Ogita, T. and Kohama, T. (2001) S–15183a and b, new sphingosine kinase inhibitors, produced by a fungus. Journal of Antibiotics 54(5): 415–420.

Kurz, M., Eder, C., Isert, D., Li, Z., Paulus, E.F., Schiell, M., Toti, L., Vertesy, L., Wink, J., Seibert, G. (2003)

Ustilipids, acylated beta-D-mannopyranosyl D-erythritols from *Ustilago maydis* and *Geotrichum candidum*. Journal of Antibiotics 56(2): 91–101.

Laatsch, H. (2000) Antibase 2000, A Database for Rapid Identification of Microbial Natural Products. Chemical Concepts, Weinheim.

Li, Y., Li, X., Kang, J.S., Choi, H.D., and Son, B.W. (2004) New radical scavenging and ultraviolet-A protecting prenylated dioxopiperazine alkaloid related to isoechinulin A from a marine isolate of the fungus Aspergillus. Journal of Antibiotics 57(5):337–340.

Liu W., Gu Q., Zhu W., Cui C., Fan G. (2005) Two new benzoquinone derivatives and two new bisorbicillinoids were isolated from a marine-derived fungus *Penicillium terrestre*. Journal of Antibiotics 58(7): 441–446.

Ma, W., Huang, Y., Lin, L., Zhu, X., Chen, Y., Xu, H. and Wei, X. (2004) Two new biologically active illudane sesquiterpenes from the mycelial cultures of *Panaeolus retirugis*. Journal of Antibiotics 57(11): 721–725.

Manning, A.M. and Anderson, D.C. (1996) Transcription factor: A New Frontier in Drug Discovery. Annu. Rep. Med. Chem 29: 235–244.

Maras, M., van Die I., Contreras R. and van den Hondel CA. (1999) Filamentous fungi as production organisms for glycoproteins of bio-medical interest. Glycoconjugate Journal 16(2): 99–107.

Monaghan, R. L., Polishook, J. D., Pecore, V. J., Bills, G. F., Nallin-Omstead, M. and Streicher, S. L. (1995) Discovery of novel secondary metabolites from fungi: Is it really a random walk through a random forest? Canadian Journal of Botany 73 (Suppl. 1): S925–S931.

Moo-Young, M. (ed.) (1985) Comprehensive Biotechnology: Vol 3: The Practice of Biotechnology: Current Commodity Products. Pergamon Press. Oxford, England.

Mori, T., Shin-ya, K., Aihara, M., Takatori, K. and Hayakawa, Y. (2003) Byssochlamysol, a new antitumor steroid against IGF-1-dependent cells from *Byssochlamys nivea*. I. Taxonomy, fermentation, isolation and biological activity. Journal of Antibiotics 56(1): 1–5.

Mori, H. Urano, Y., Abe, F., Furukawa, S., Furukawa, S., Tsurumi, Y., Sakamoto, K., Hashimoto, M., Takase, S., Hino. M. and Fujii, T. (2003) FR235222, a fungal metabolite, is a novel immunosuppressant that inhibits mammalian histone deacetylase (HDAC). I. Taxonomy, fermentation, isolation and biological activities. Journal of Antibiotics 56(2): 72–79.

Mukhopadhyay, T., Bhat, R.G., Roy, K., Vijayakumar, E.K.S. and Ganguli, B.N. (1998) Aranochlor A and aranochlor B, two new metabolites from *Pseudoarachniotus roseus*: production, isolation, structure elucidation and biological properties. Journal of Antibiotics 51(4): 439–441.

Mukhopadhyay, T., Roy, K., Bhat, R. G., Sawant, S. N., Bumbach, J. and Ganguli, B. N.(1992) Deoxymulundocandin-A new echinocandin type antifungal antibiotic. Journal of Antibiotics 45: 618–623.

Mukhopadhyay, T., Roy, K., Coutinho, L., Rupp, R.H., Ganguli, B.N. and Fehlhaber, H. W. (1987) Fumifungin, a new antifungal antibiotic from *Aspergillus fumigatus* Fresenius 1863. Journal of Antibiotics 40(7): 1050–1052.

Oh, H., Kim, B.S., Bae, E.Y., Kim, M.S., Kim, B.Y., Lee, H.B., Kim, C.J. and Ahn, J.S. (2004) Inhibition of PTP1B by metabolites from *Micromucor ramannianus* var. *angulisporus* CRM000232. Journal of Antibiotics 57(8): 528–531.

Oh, W.K., Oh, H., Kim, B.Y., Kim, B.S. and Ahn, J.S. (2004) CRM-51006, a new phospholipase C (PLC) inhibitor, produced by unidentified fungal strain MT51005. Journal of Antibiotics 57(12): 808–811.

Ohtsu, Y., Sasamura, H., Tanaka, M., Tsurumi, Y., Yoshimura, S., Takase, S., Shibata, T., Hino, M. and Nakajima H. (2005) The novel gluconeogenesis inhibitor FR225654 that originates from *Phoma* sp. no. 00144. I. Taxonomy, fermentation, isolation and physico-chemical properties. Journal of Antibiotics 58(7): 447–51.

Oka, M., Iimura, S., Tenmyo, O., Sawada, Y., Sugawara, M., Ohkusa, N., Yamamoto, H., Kawano, K., Hu, S.L., Fukagawa, Y., et al. (1993). Terpestacin, a new synctium formation inhibitor from *Arthrinium* sp. Journal of Antibiotics 46(3): 367–373.

Ondeyka, J.G., Zink, D.L., Dombrowski, A.W., Polishook, J.D., Felock, P.J., Hazuda, D.J. and Singh, S.B. (2003) Isolation, structure and HIV-1 integrase inhibitory activity of exophillic acid, a novel fungal metabolite from *Exophiala pisciphila*. Journal of Antibiotics 56(12): 1018–1023.

Punt, P. J., van Biezen, N., Conesa, A., Albers, A., Mangnus J. and van den Hondel, C. (2002) Filamentous fungi as cell factories for heterologous protein production. Trends in Biotechnology 20: 200–206.

Reddy, B.N. (2005) Fungal Diversity: Present Perspective and Future Challenges. In Abstract of the International

Conference on Microbial Diversity, Current Prospective and Potential Applications, April 16–18, 2005, New Delhi, pp. 18.

Rao, K.C., Divakar, S., Srinivas, M., Babu, K.N., Karanth, N.G. and Sattur, A.P. (2003) Asperaldin, a new aldose reductase inhibitor from *Aspergilus niger* CFR-1046. I. Fermentation, isolation and characterization. Journal of Antibiotics 56(2): 173–176.

Rether. J., Erkel, G., Anke, T., and Sterner, O. (2004) Oxaspirodion, a new inhibitor of inducible TNF-alpha expression from the Ascomycete *Chaetomium subspirale*. Production, isolation and structure elucidation. Journal of Antibiotics 57(8): 493–495.

Rho, M.C., Toyoshima, M., Hayashi, M., Uchida, R., Shiomi, K., Komiyama, K. and Omura, S. (1998) Enhancement of drug accumulation by andrastin A produced by *Penicillium* sp. FO-3929 in vincristine-resistant KB cells. Journal of Antibiotics 51(1): 68–72.

Roy, K., Mukhopadhyay, T., Reddy, G.C.S., Desikan, K.R. and Ganguli, B.N. (1987) Mulundocandin: A new lipopeptide antibiotic I Taxonomy, Fermentation Isolation and Characterization. Journal of Antibiotics 40: 275–280.

Roy, K., Mukhopadhyay, T., Reddy, G.C.S., Desikan, K.R. Rupp, R.H. and Ganguli, B.N. (1988) Aranorosin, A novel antibiotic from *Pseudoarachniotus roseus* I Taxonomy, Fermentation, Isolation, Chemical and Biological properties. Journal of Antibiotics 41: 1780–1784.

Roy, K., Vijayakumar, E.K.S., Mukhopadhyay, T., Chatterjee, S., Bhat, R.G., Blumbach, J. and Ganguli, B. N. (1992) Aranorosinol A and Aranorosinol B two new metabolites from *Pseudoarachniotus roseus*. Production, isolation, structure elucidation and biological properties. Journal of Antibiotics 45: 1592–1598.

Roy, K., Chatterjee, S., Deshmukh, S.K., Vijayakumar, E.K.S., Ganguli, B.N. and Fehlhaber, H. W. (1996) Orbuticin, a new secondary metabolite from *Acremonium butyri*. Journal of Antibiotics 49(11): 1186–1187.

Sakano, Y., Shibuya, M., Yamaguchi, Y., Masuma, R., Tomoda, H., Omura, S. and Ebizuka, Y. (2004) Epohelmins A and B, novel lanosterol synthase inhibitors from a fungal strain FKI-0929. Journal of Antibiotics 57(9):564–8.

Sakurai, M., Kohno, J., Yamamoto, K., Okuda, T., Nishio, M., Kawano, K. and Ohnuki, T. (2002) TMC-256A1 and C1, new inhibitors of IL-4 signal transduction produced by *Aspergillus niger* var *niger* TC 1629. Journal of Antibiotics 55(8): 685–692.

Sakurai, M., Hoshino, H., Kohno, J., Nishio, M., Kishi, N., Okuda, T., Kawano, K. and Ohnuki, T. (2003) TMC-260, a new inhibitor of IL-4 signal transduction produced by *Acremonium kiliense* Grutz TC 1703. Journal of Antibiotics 56(9): 787–791.

Schmid, I., Sattler, I., Grabley, S. and Thiericke, R. (1999) Natural products in high throughput screening: automated high-quality sample preparation. Journal of Biomolecular Screening 4: 15–25.

Seephonkai, P., Isaka, M., Kittakoop, P., Boonudomlap, U. and Thebtaranonth, Y. (2004) A novel ascochlorin glycoside from the insect pathogenic fungus *Verticillium hemipterigenum* BCC 2370. Journal of Antibiotics 57(1): 10–16.

Shibazaki, M., Tanaka, K., Nagai, K., Watanabe, M., Fujita, S., Suzuki, K., Okada, G. and Yamamoto, T. (2004) YM-92447 (spinosulfate A), a neuraminidase inhibitor produced by an unidentified pycnidial fungus. Journal of Antibiotics 57(12): 812–815.

Shiomi, K., Matsui, R., Isozaki, M., Chiba, H., Sugai, T., Yamaguchi, Y., Masuma, R., Tomoda, H., Chiba, T., Yan, H., Kitamura, Y., Sugiura, W., Omura, S. and Tanaka, H. (2005) Fungal phenalenones inhibit HIV-1 integrase. Journal of Antibiotics 58(1): 65–8.

Smedsgard, J. and Frisvad, J.C. (1996) Using direct electrospray mass spectrometry in taxonomy and secondary metabolite profiling of crude fungal extracts. Journal of Microbiological Methods 25: 5–17.

Strobel, G. and Daisy, B. (2003) Bioprospecting for Microbial Endophytes and Their Natural Products. Microbiology and Molecular Biology Reviews 67: 491–502.

Sugawara, F., Strobel, G., Fisher, L.E., Van Duyne, G.D. and Clardy, J. (1985) Bipolaroxin, a selective phytotoxin produced by *Bipolaris cynodontis*. Proceedings of the National Academy of Sciences, USA, 82(24): 8291–8294.

Suryanarayanan, T.S., Johnson, J.A. and Murali, T.S.(2005) Fungal endophytes of tropical forests: A review in Abstracts of the International Conference on Microbial Diversity, Current Prospective and Potential Applications, April 16-18, 2005, New Delhi, p. 19.

Taft, C.S., Enderlin, C.S. and Selitrennikoff, C.P. (1994) A high throughput in vitro assay for fungal (1, 3) beta-glucan synthase inhibitors. Journal of Antibiotics 47(9): 1001–1009.

Tomoda, H., Tabata, N., Masuma, R., Si, S.Y., and Omura, S. (1998) Erabulenols, inhibitors of cholesteryl ester transfer protein produced by *Penicillium* sp. FO-5637. I. Production, isolation and biological properties. Journal of Antibiotics 51(7): 618–623.

Ui, H., Asanuma, S., Chiba, H., Takahashi, A., Yamaguchi, Y., Masuma, R., Omura, S. and Tanaka, H. (2005) Mycophenolic Acid Inhibits Syncytium Formation Accompanied by Reduction of gp120 Expression. Journal of Antibiotics 58: 514–518.

Yamada, T., Doi, M., Miura, A., Harada, W., Hiramura, M., Minoura, K., Tanaka, R., Numata, A. (2005) Absolute stereostructures of cell-adhesion inhibitors, peribysins A, E, F and G, produced by a sea hare-derived *Periconia* sp. Journal of Antibiotics 58(3): 185–191.

Yoganathan, K., Yang, L.K., Rossant, C., Huang, Y., Ng, S., Butler, M.S. and Buss, A.D. (2004) Cochlioquinones and epi-cochlioquinones: antagonists of the human chemokine receptor CCR5 from *Bipolaris brizae* and *Stachybotrys chartarum*. Journal of Antibiotics 57(1): 59–63.

Vertesy, L., Burger, H.J., Kenja, J., Knauf, M., Kogler, H., Paulus, E.F., Ramakrishna, N.V., Swamy, K.H., Vijayakumar, E.K., and Hammann, P. (2000) Kodaistatins, novel inhibitors of glucose-6-phosphate translocase T1 from *Aspergillus terreus* thom DSM 11247. Isolation and structural elucidation. Journal of Antibiotics 53(7): 677–686.

Vijayakumar, E.K.S., Roy, K., Chatterjee, S., Deshmukh, S.K., Ganguli, B.N., Kogler, H and Fehlhaber, H.W. (1996) Arthrichitin, a new cell wall active metabolite from *Arthrinium phaeospermum*. Journal of Organic Chemistry 61: 6591–6593.

Vijayakumar, E.K., Roy, K., Chatterjee, S., Mukhopadhyay, T., Bhat, R.G. and Ganguli, B.N. (1998) Structure-activity relationships of new aranorosin analogs to antifungal activity. Journal of Antibiotics 51(5): 522–524.

Williams, R.H., Lively, D.H., DeLong, D.C., Cline, J.C. and Sweeny, M.J. (1968) Mycophenolic acid: antiviral and antitumor properties. Journal of Antibiotics 7: 463–464.

Yarbrough, G.G, Taylor, D.P., Rowlands, R.T., Crawford, M.S. and Lasure, L.L. (1993) Screening microbial metabolites for new drugs: theoretical and practical issues. Journal of Antibiotics 46: 535–544.

Yoganathan, K., Yang, L.K., Rossant, C., Huang, Y., Ng, S., Butler, M.S. and Buss, A.D. (2004) Cochlioquinones and epi-cochlioquinones: antagonists of the human chemokine receptor CCR5 from *Bipolaris brizae* and *Stachybotrys chartarum*. Journal of Antibiotics 57(1): 59–63.

Zahn, J.A., Higgs, R.E. and Hilton, M.D. (2001) Use of direct-infusion electrospray mass spectrometry to guide empirical development of improved conditions for expression of secondary metabolites from actinomycetes. Applied and Environmental Microbiology 67: 377–386.

2. Biology and Biotechnology Potential of Saprophytic Entomophthorales

M.C. Srinivasan
Former Head, Microbial Technology and Biochemical Sciences Division,
National Chemical Laboratory, Pune-411008, India

SUMMARY The Entomophthorales are terrestrial fungi, relatively neglected, which are pathogenic to insects and exist as saprophytic forms associated with plant detritus and intestinal contents of amphibians and reptiles. Six families and 23 genera are recognized. This article focuses on the genera *Conidiobolus* and *Basidiobolus*. Their ecology and distribution, taxonomy, physiology of vegetative growth and reproduction including pathogenesis is reviewed. Fermentation of such fungi has given several alkaline proteases and lipases with applications potential.

Introduction

"As genomics and biomedicine are to human health, so ecology and conservation biology are to the planet's health … Their growth is hampered by a seldom acknowledged deficiency; our ignorance of most of the world's biodiversity particularly at the level of individual species, where knowledge is foundational to all the other studies of biodiversity and hence of the whole living environment … . Bacteria, mites, nematodes, worms, fungi, beetles, and other major environmental players are necessarily ignored or at best given "morphospecies" code numbers. Even among the small minority of species diagnosed and named, fewer than 1% have been subjected to the kind of careful biological studies needed to undergird ecology and conservation biology"—Edward O. Wilson (Science 289: 2279, 2000).

Biodiversity and conservation of bio-resources have been topics of global attention in recent times and efforts at the national and international levels to conserve endangered species from extinction have been on the increase. Endangered species of plants and animals have been identified and practical steps initiated for safeguarding their numbers. Microbial diversity on the other hand is a much neglected subject, partly because of the lack of appreciation of their role in sustaining the ecosystem and also because of the efforts needed to undertake extensive microscopic and cultural studies before identification is possible. Fungi are distinguishable by the morphology of their spore forms and mycologists have been active in collecting and describing fungal taxa in different parts of the world. However in most instances the technical description and herbarium specimens are the only available material while living cultures for these novel taxa are not available. This makes it impossible to carry out biochemical and biotechnological studies on these new taxa. In the present context of global appreciation of exploring biodiversity for biotechnology application and discovery of novel bioactive molecules, pure culture isolation and effective conservation in germplasm banks assume great importance. Currently, bioactive molecules and enzymes of industrial interest are being screened from microbes belonging to diverse taxonomic groups including those which were hitherto not explored.

This article presents a review of the biology and biotechnology potential of the saprophytic Entomophthorales (Zygomycotina).

The Entomophthorales comprise terrestrial fungi, which are pathogenic to insects and also saprophytic forms associated with plant detritus and intestinal contents of amphibians and reptilia. By virtue of their formation of zygospores through conjugating gametangia, these fungi are classified under the Zygomycotina along with Mucorales. However, despite their relatedness as Zygomycetes, the organisms belonging to Mucorales and Entomophthorales have glaringly different morphological and pathogenic features. While the Mucorales produce colonies with floccose aerial mycelium, the Entomophthorales produce dense, waxy or deeply furrowed colonies. Unlike Mucorales which produce their spores in multispored sporangia or sporangioles, the Entomophthorales are characterized by simple phototropic conidiophores from which conidia (or single celled sporangioles) are discharged with force.

Under the Entomophthorales six families and 23 genera are recognized.

Ancylistaceae comprises of *Ancylistes* parasitic on *Closterium* and *Conidiobolus* occurring saprophytic in plant detritus.

Basidiobolaceae with one genus *Basidiobolus* is widely distributed in plant detritus and in association with the mycoflora of amphibian and reptile intestines.

Completoriacae with one genus *Completoria* parasitic on fern prothalli.

Entomophthoraceae with 11 genera (Humber, 1989) which are pathogenic to insects of which the genus *Entomophthora* is most widely distributed and investigated.

Neozygitaceae with two genera *Neozygites* and *Thaxterosporium,* obligatory pathogenic to insects (specially *Homoptera*) and mites.

Meristacraceae with 3 genera, *Meristacrum, Ballocephala* and *Zygnemomyces* obligately pathogenic to not only soil invertebrates, particularly nematodes and tardigrades, but also possibly insects.

From a biotechnological point of view the most readily culturable and fast growing members are significant and the discussion is restricted to the genera *Conidiobolus* and *Basidiobolus*. There has been considerable confusion in the differentiation of *Entomophthora* from *Conidiobolus*. Actually the genus *Conidiobolus* was established by Brefeld (1884) for a fungus that overgrew cultures of basidiomycetous fungi *Hirneola* and *Exidia* to distinguish the saprophytic *Entomophthora*-like fungus since at that time species of *Entomophthora* Fresenius (1856), pathogenic to insects, had been regarded as obligate pathogens and unculturable. Subsequently, Sawyer (1929) successfully established cultures of *Entomophthora sphaerosperma* on natural media such as sterilized pork, swordfish and potatoes, and hence the point of separating *Conidiobolus* from *Entomophthora* on the basis of pathogenicity became questionable. The type species of *Conidiobolus, C. utriculosus* with very large conidia and zygospores has not been identified again and we have only Brefeld's description for its validation.

A fast growing Entomopathoraceous fungus capable of readily growing on mycological media containing simple organic nutrients and with the formation of microconidia on radial sterigmata and villose appendages on its conidia has been repeatedly isolated and named variously as *Delacroixia coronata* (Costantin, 1897), *Conidiobolus villosus* (Martin, 1925) and *Entomophthora coronata* (Kevorkian, 1937). The insect pathogenic *Entomophthora* species such as *Entomophthora muscae* on house flies are relatively slow growing and often difficult to maintain in prolonged subcultures on laboratory media. Srinivasan et al. (1964) developed a simple medium containing wheat grain extract supplemented with yeast extract, peptone and glycerin and successfully cultured a strain of *Entomophthora muscae* collected from infected houseflies in India. In a comparison of the morphogenesis of asexual spore formation, they observed that the conidiophores of *E. muscae* were determinate, prismatic and easily differentiated from the vegetative mycelium, whereas in cultures of *Conidiobolus*, the conidiophores were non-determinate and identical with the vegetative mycelium.

They suggested that the conidiophore is macronematous in *Entomophthora* while micronematous in *Conidiobolus* drawing analogy to the term used in the case of Fungi Imperfecti. In a careful study of several strains of rapidly growing *Delacroixia coronata* in culture, Srinivasan and Thirumalachar (1964) observed the micronematous conidiophore development and designated it as *Conidiobolus coronatus* (Cost.) Srinivasan and Thirumalachar. However, in the same year Batko (1964) published the same binomial for this fungus and got precedence due to prior publication. The fungus is now designated as *C. coronatus* (Cost.) Batko.

The genus *Basidiobolus* was established by Eidam (1886) with *B. ranarum* as the type species, which was isolated from the intestinal contents of frogs. The fungus grows readily in culture and is characterized by conidiophores which enlarge apically before developing conidia, which are discharged forcibly. Formation of zygospores in large numbers is a distinguishing feature of all the species of *Basidiobolus* known so far. The genus is placed in a separate family Basidiobolaceae and in a study of phylogenetic divergence of entomophthoralean fungi, Nagahama et al. (1995) determined the nuclear 18S ribosomal RNA gene sequences of *Basidiobolus ranarum* and compared it with the sequence of *Conidiobolus coronatus* and *Entomophtora muscae* besides other species. Based on their data it is evident that *Basidiobolus ranarum* diverted from the other entomophthoralean fungi and belongs to the cluster closer to the anaerobic rumen chytrids like *Spizellomyces* and *Neocallimastix*. *Conidiobolus*, *Entomophthora* and *Zoophthora* clustered with *Mucor*. Their analysis suggests loss of flagella on several lineages in the phylogenetic tree of divergent chytrids. The authors have postulated that the possibility exists that the entomophthoralean fungi are not a monophyletic group and have indicated the possibility of excluding the Basidiobolaceae from the Entomophthorales, and these findings are somewhat contradictory to the classification traditionally followed and accepted at present. McKerracher and Heath (1985) found an organelle that looked like the base of a chytrid flagellum (basal body or kinetosome also acting as a centriole at nuclear division) in the cytoplasm of *Basidiobolus*. No study of any other zygomycetes has uncovered any organelle similar to a basal body or centriole. Berbee and Taylor (1999) have stated that it seems likely that *Basidiobolus* has had a relatively recent loss of its flagellum.

Ecology and Distribution of Saprophytic Entomophthorales

Species of *Conidiobolus* are found worldwide in soils and plant detritus and *C. coronatus* is the most common and abundant species encountered. Several species of *Conidiobolus* have been isolated and studied from plant litter in the United States by Charles Drechsler and from India by Srinivasan and Thirumalachar in a series of papers that describe many *Conidiobolus* isolates including species new to science.

Drechsler (1952) described a technique for isolation of saprophytic Entomophthorales from plant litter by overlaying moistened detritus particles on agar media and isolating pure cultures from conidia forcibly discharged and germinating on the agar medium and the same technique with minor modifications was successfully applied by Srinivasan and Thirumalachar in their studies on Indian species. The fungus is believed to be more prevalent in the warm wet tropical climates like countries of West Africa and their numbers peak up during periods of heavy rainfall. Germination of conidia is observed under very high humidity conditions while below 95% humidity levels poor germination is seen. *Conidiobolus* species have also been isolated in association with fruit bodies of fleshy fungi and as already stated the type species *C. utriculosus* was isolated by Brefeld (1884) as a contaminant while studying spore germination in *Hirneola* and *Exidia*. Couch (1939) isolated *C. brefeldianus* Couch while studying *Septobasidium*. Dring (1958) found *C. rhysosporus* in association with the apothecia of *Ascobolus* species. *Conidiobolus micophilus* and *Conidiobolus mycophagus* were described from India by Srinivasan and Thirumalachar (1965) from decaying mushrooms as two new species.

Basidiobolus species have been identified as saprobes in association with the intestinal mycoflora of

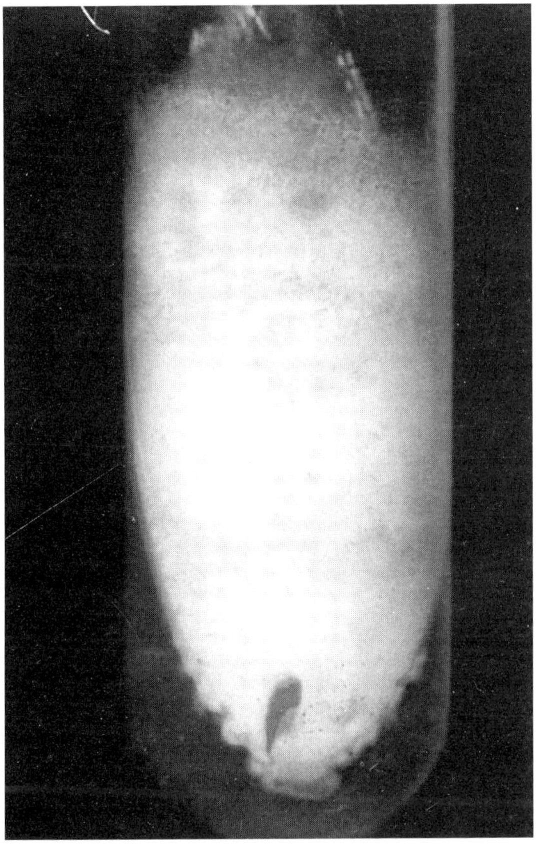

Fig. 1. *Conidiobolus coronatus*: Discharged conidia forming visible cloud on glass above growing colony.

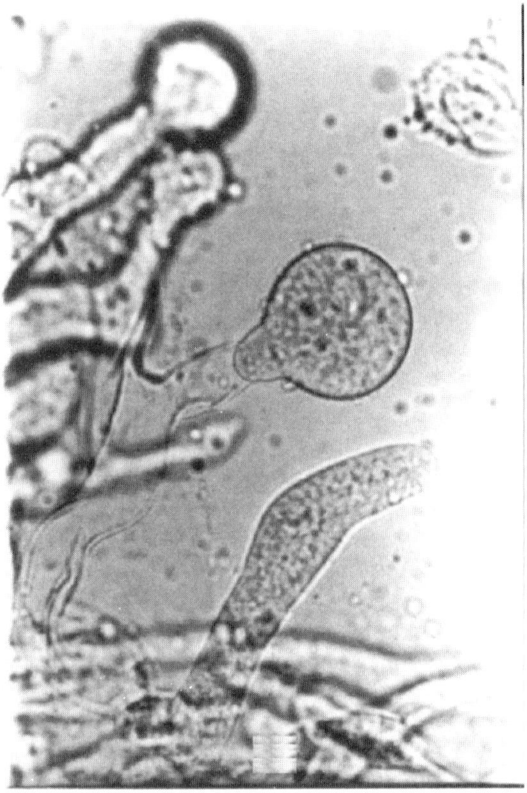

Fig. 2. *Conidiobolus coronatus*: Micronematous conidiophore with mature conidium just prior to its forcible discharge.

amphibians and reptiles and also are frequently isolated from plant detritus by the agar plate overlay technique. Drechsler isolated *Basidiobolus haptosporus, B. meristosporus, B. magnus* and *B. ranarum* from plant litter in the United States while Srinivasan and Thirumalachar (1967) described a new species *B. heterosporus* besides *B. haptosporus* from Indian detritus collection. Benjamin (1962) described *B. microsporus* from dung in the USA.

Both *Conidiobolus* and *Basidiobolus* have been identified as "opportunistic" infections in man and animals where they cause recognizable symptoms. *Conidiobolus* infections appear as chronic sinusitis involving facial and subcutaneous tissues and the paranasal sinuses. *Basidiobolus* causes chronic infections of the peripheral subcutaneous tissues usually on the arms, thighs etc., as hardened nodules spreading laterally.

Taxonomy and Classification

Species differentiation in *Conidiobolus* is primarily based on morphological characteristics of the asexual and sexual spore forms. Characters which are taxonomically significant, include the presence or absence of the zygospores and the characteristics of the zygospore wall architecture, formation of microconidia

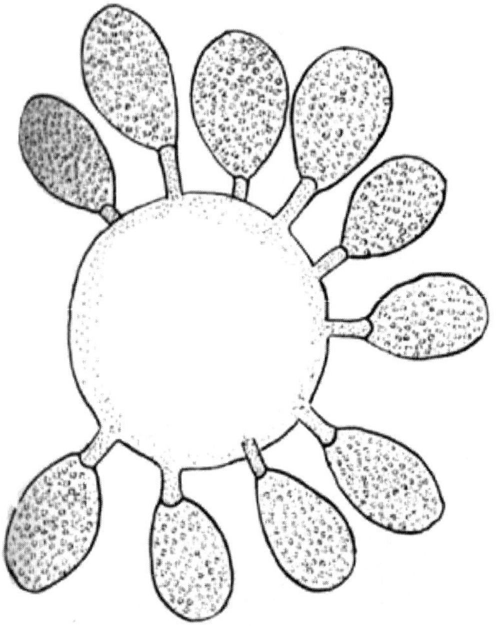

Fig. 3. *Conidiobolus coronatus*: Microconidia formation on radial sterigmata.

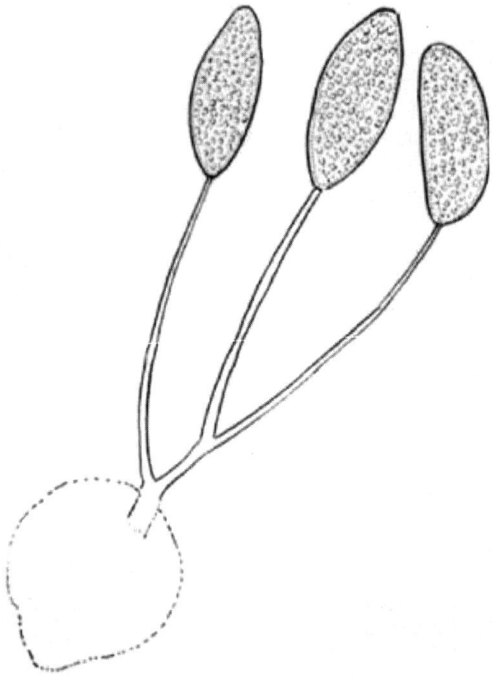

Fig. 4. *Conidiobolus heterosporus*: Elongated conidia formation on branched capillary conidiophores.

formation of secondary fusiform conidia, which are forcibly ejected or released passively from capillary conidiophores, formation of chlamydospores etc. Srinivasan and Thirumalachar (1967) presented an evaluation of the taxonomic characters in the genus and gave the key to identify species. King (1976, 1976a, 1977) in a detailed investigation of *Conidiobolus* using numerical taxonomy studies, biology and cluster analysis and taxonomic considerations gave a description of recognized species. There were certain obvious differences in the conclusions drawn from those arrived at by Srinivasan and Thirumalachar. Both the taxonomic keys are based exclusively on morphological basis and published over 25 years ago. It is obvious that newer insights into speciation of this interesting genus incorporating newer techniques and molecular data is warranted and possibly the taxonomy and classification may show significant variation from both these schemes.

Basidiobolus has fewer recognized species and the primary basis for distinguishing the species is

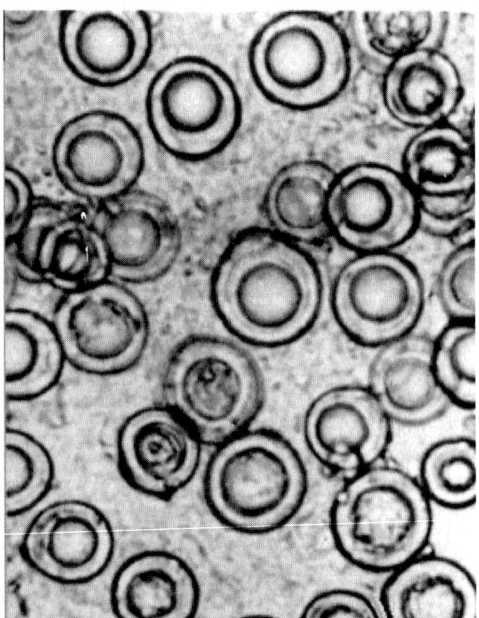

Fig. 5. *Conidiobolus brefeldianus*: Mature zygospores with smooth thick wall and large reserve globules.

the zygospore wall architechture, i.e. whether it is smooth or undulate. *B. ranarum* isolated from frogs or plant detritus has warty and undulate walls and the culture also has a strong musty odour like benzene hexachloride. Drechsler (1947) described *Basidiobolus haptosporus* and subsequently *B. meristosporus* (Drechsler, 1955), which differ in having smooth zygospores. *B. magnus* with larger undulate zygospores was described as a new species by Drechsler (1964). Formation of adhesive secondary conidia on capillary conidiophores appears to be a character common to *B. ranarum* and all the species described by Drechsler. Benjamin (1962) isolated, from the dung of a small animal in California, a *Basidiobolus* with undulate zygospore wall and primary conidia giving rise to microspores radially, which are ovoid at the base and with an attenuated apical portion. No adhesive conidia were formed and the culture also did not have the musty odour. The species was named *B. microsporus* Benjamin. Srinivasan and Thirumalachar (1967a) described a new species characterized by having undulated as well as smooth zygospores and designated it as *B. haptosporus*. This species also has the elongate adhesive conidium like the species described by Drechsler. In evaluating the taxonomic basis for speciation, they considered that the segmentation of the hyphal cells internally to form meristospores was not taxonomically significant since such a state can be readily induced by incorporation of ammonium salt in the medium. Thus, the name *B. haptosporus* was considered valid for all smooth walled zygospore forming isolates and the species *B. meristosporus* was considered as its synonym.

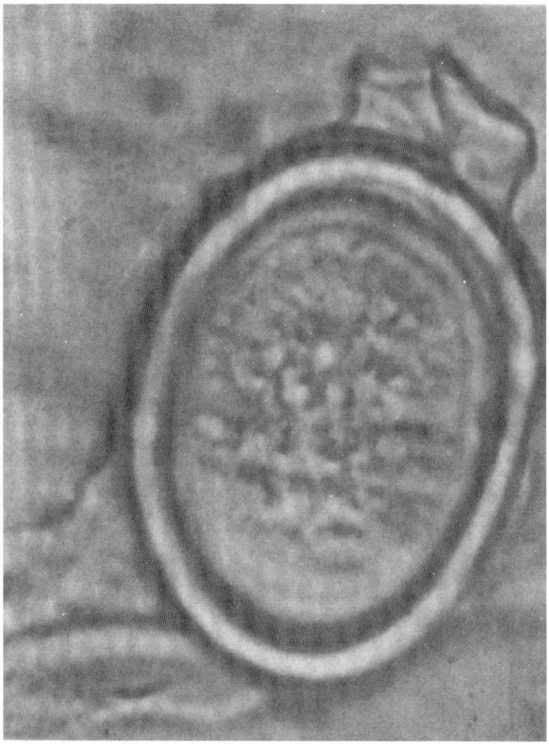

Fig. 6. *Basidiobolus heterosporus*: Smooth thick walled mature zygospore showing persistent gametangial papilla.

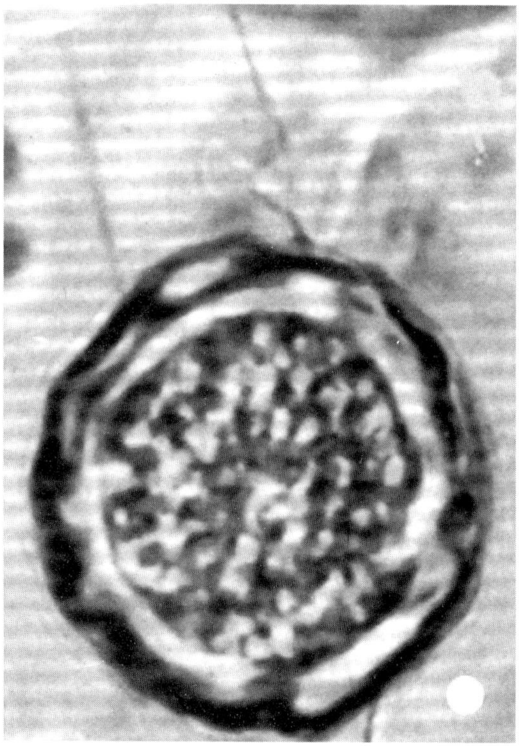

Fig. 7. *Basidiobolus ranarum*: Zygospore with thick undulated wall.

All the pathogenic isolates studied so far have shown only smooth zygospores and accordingly all of them have been referred to as *Basidiobolus haptosporus* (Srinivasan and Thirumalachar (1965a).

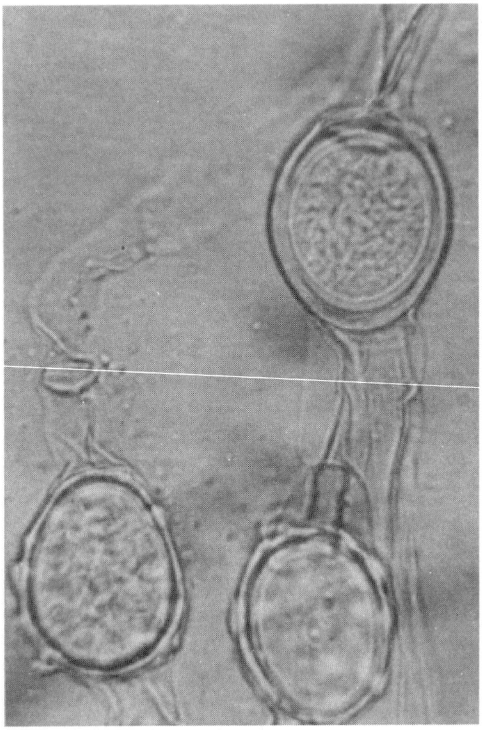

Fig. 8. *Basidiobolus heterosporus*: Showing smooth and undulated walled zygospores developing on the same hyphal segments.

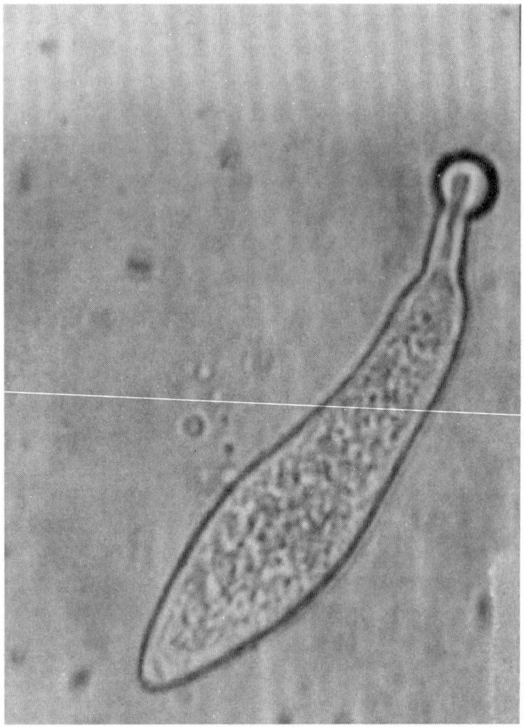

Fig. 9. *Basidiobolus haptosporus*: Elongated secondary conidium, boat shaped with adhesive material on its beak.

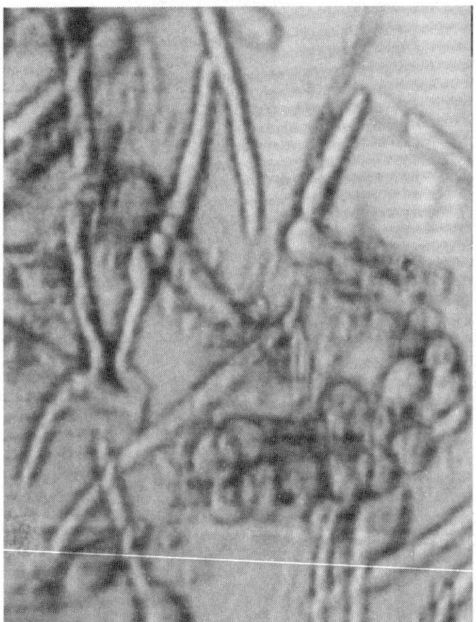

Fig. 10. *Basidiobolus haptosporus:* Developmental stages and rounding up of "darmform" cells inside the mycelia in media containing ammonium chloride.

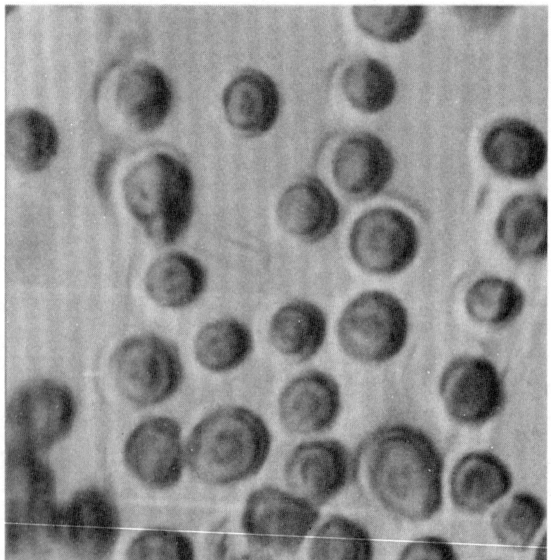

Fig. 11. *Basidiobolus haptosporus:* "Darmform" cells after release from hyphal segments.

Physiology of Vegetative Growth and Reproduction

Colonies of *Conidiobolus* growing on agar media such as glucose, yeast extract, malt extract, agar are flat with little aerial mycelium but show a powdery surface due to the development of conidia on unbranched conidiophores from which they are actively discharged. Conidia are often discharged beyond the growing edge of the parent colony and germinate within a short time by multiple germ tubes and readily establish satellite colonies, which in turn start developing conidia rapidly. In *C. coronatus* and some other species, formation of microconidia on radial sterigmata from the globose conidia is observed, especially on conidia which are forcibly discharged and adhere to the glass surface above the culture in test tubes or among conidia falling back on the parental colony surface. Formation of villose hair from conidia especially in the older cultures is a unique feature of *C. coronatus*. In three species of *Conidiobolus*, viz. *C. eurymitus*, *C. couchii* and *C. iuxtagenitus*, formation of heteromorphic secondary conidia which are fusiform and forcibly discharged are observed. In *C. heterosporus* and some other species, filamentous capillary conidiophores bearing ellipsoid to ovate conidia which are passively liberated are formed, especially in fresh isolates. In cultures, which have been maintained *in vitro* through several subcultures, the formation of these capillary conidia may decline or even be lost. Mycelia undergoing septation in the region beyond the growing edge of colony readily forms hyphal segments sometimes rounding up into tubular or somewhat dilated bodies. In some cases like *C. chlamydosporus* and *C. globuliferus*, well defined chamydospores of definite size and shape may be formed in older cultures. In species with sexual reproduction, conjugation of anisogamous gametangia differentiated from the hyphal segments fuse into a globose zygospore which at maturity appears thick walled, colorless, smooth and with a well developed globose reserve globule in its center. The zygospore wall is warty or undulate in *C. utriculosus and C. rhysosporus*. *Conidiobolus* species are incapable of growing on synthetic media with nitrate or nitrite and require an organic nitrogen source for good growth and reproduction. Peptone in small concentrations and yeast extract added to the medium are favorable for zygospore development.

Basidiobolus colonies appear flat with deeply furrowed surface without aerial mycelium but in some strains a slight fuzzy appearance may be seen due to the large number of conidiophores developing along with sterile mycelial outgrowths. The growing hyphae readily differentiate hyphal segments and in most cases adjacent segments put forth beaklike outgrowths and function like gametangia. Confluence of contents and the ready formation of zygospores is observed and the mature zygospores bear the beak-like protuberances on them, which is characteristic of the genus. Zygospores are smooth walled, thick and with a prominent nucleus in the center visible even in unstained preparations in the case of *B. haptosporus*. In *B. ranarum, B. magnus* and *B. microsporus* the zygospores are undulate while in *B. heterosporus* both smooth and undulate zygospores are developed. The formation of conidia takes place on unbranched conidiophores, which dilate into vesicles before the conidia develop and a sudden rupture of the dilated vesicle forcibly discharges the conidia and the ruptured part is carried by the conidium as a triangular piece during its flight. Formation of capillary conidiophores with elongate to boat shaped secondary conidia with a characteristic beak and adhesive matrix at their tips have been reported in all the species except *B. microsporus*. Characteristic microspores with ovate bodies and elongated apices are formed from globose conidia in *B. microsporus*.

Ready segmentation of mycelia takes place in media containing ammonium salts and in liquid media containing 1% ammonium chloride incubated as stationary cultures, the hyphal segments undergo internal cleavage and the contents round up to release large number of yeast like bodies which settle to the bottom of the tube. These bodies referred to as "Darmform" cells represent an interesting morphogenetic behaviour observed in all *Basidiobolus* species.

Pathogenesis

In recent years several cases of fungal infections in humans and animals caused by widely distributed saprophytic fungi have been reported. These are referred to as "opportunistic" infections, which take place under conditions of low immunity in normal individuals or in patients undergoing therapy for AIDS or immunosuppressed following organ transplants. These fungal infections are different from regular pathogenesis since inoculation of the isolates into healthy hosts fail to reproduce the symptoms and fulfil the well known Koch's postulates. The disease zygomycosis has been named for fungal invasion due to zygomycetes which causes serious symptoms in man and animals. The topic has been reviewed in detail by Ribes et al. (2000).

Species of *Basidiobolus* and *Conidiobolus* have been isolated from cases of zygomycosis of man and animals like horses from different parts of the world. Under the warm, wet climatic conditions of Africa and Asia, these infections appear to be more prevalent although these fungi are widespread around the world. Subcutaneous creeping granuloma in the nasal region and face caused by *Basidiobolus haptosporus* is reported from Uganda (Wilson, 1961). In India Kamalam and Thambiah (1975, 1982, 1984) reported mycotic infection due to *Basidiobolus* with lymph node involvement and muscle invasion. *Conidiobolus coronatus* is the most widespread agent of zygomycosis in man and horses. The infection appears as chronic sinusitis beginning as a swelling of the inferior nasal turbinate and extends to nearly all adjacent tissues involving facial and subcutaneous and paranasal sinuses. Lymph node invasion by *C. coronatus* was reported from India by Kamalam and Thambiah (1978). Phycomycosis of horses by *Entomophthora coronata* (= *Conidiobolus coronatus*) was reported by Emmons and Bridges (1961).

Infection by entomophthoralean fungi is predisposed by low immunity of individuals and often the transmission of fungal spores is through abrasions or insect bites. These fungi are highly proteolytic in nature and Gugnani (1992) discussing Entomophthoromycosis due to *Conidiobolus* suggests that since proteases are secreted rapidly, they may be secreted into infected tissues at the initiation of infection leading to degradation of protein and the subsequent lipase activity hydrolyzes the fatty materials in the subcutaneous tissues. Muscle invasion is facilitated by collagenase activity present in the fungus. Physiological adaptation to grow rapidly at 37°C is an additional factor and most of the strains isolated from the environment grow rather feebly or not at all at 37°C which is the human body temperature. Many things about the etiology of the disease and the predisposing factors which make these commonly occurring species to cause infection are not clear. It is surprising that despite their world wide distribution in abundance there are relatively few cases reported and most of them are from the warm and humid tropical environment of the developing countries of Africa and Asia.

The infection by entomophthoralean fungi share the same situation as seen in the other opportunistic fungi and among the zygomycetes many of the fungi which are common saprophytes like *Mucor* or *Cunninghamella* or those which have been extensively used in the fermentation industry like *Rhizomucor miehei* and *R. pusillus* have also been isolated from cases of Phycomycosis. There is an urgent need to develop reliable methods to diagnose pathogenic strains so that they may not be taken up for use in the fermentation industry leading to large scale multiplication. As an example of efforts in this direction may be mentioned the studies by Cutler and Swatek (1969) who found that human pathogenic *Basidiobolus* strains produce a black pigment on tyrosine agar plates whereas the non pathogenic strains produce pinkpigment or no pigment at all. In the absence of successful infection experiments through artificial inoculation, it will be very difficult to draw meaningful conclusion with regard to not only the Entomophthoralean species but all the opportunistic fungi which cause troublesome infections whose etiology is little understood.

Fermentation Metabolites

Investigations carried out at the National Chemical Laboratory, Pune, India have led to identification of strains of *Conidiobolus coronatus* and *C. macrosporus* capable of secreting high activity alkaline protease in submerged culture. High activity enzyme secretion was reported from *C. coronatus* (Srinivasan et al.

1983) which was compatible with commercial detergents (Phadatare et al. 1993) and capable of being used in animal cell culture as a substitute for conventional trypsin (Chiplunkar et al. 1985). The enzyme was stable to high alkaline pH conditions, similar to Subtilisin and was a serine protease. Tanksale et al. (2001) studied the alkaline proteases from *C. macrosporus* after immobilization on polyamide using glutaraldehyde as a bifunctional agent. The immobilized enzyme exhibited much higher thermal stability at 60°C and was fully active for over twenty cycles of reuse. Lipases with activity and stability at pH 4-8 and upto 60°C were reported from *C. coronatus* (Okafor and Gugnani, 1990).

Ingale et al. (2002) reported alkaline protease secretion in *Basidiobolus haptosporus* when the growth morphogenesis was in the form of fine and dispersed pellets developing from inoculum grown to produce free floating "darmform" cells. Poor to no enzyme secretion occurred when the growth in shake flasks was lumpy and as large pellets. The enzyme was compatible with commercial detergents and also could be applied for the recovery of silver from spent photographic films.

Apart from enzyme technology, *Conidiobolus* strains have been shown to be useful in the production of polyunsaturated fatty acids. In a series of patents Idemitsu-Petrocham, Japan have reported the production of bishomogammalinolenic acid, Dihomogammalinolenic acid and arachidonic acid from diverse *Conidiobolus* species. For example in a medium containing sesame oil and olive oil, *Conidiobolus nanodes* accumulated lipids with significant bis-homogammalinolenic acid (Idemitsu-Petrochem, 1990). In a medium containing alpha linolenic acid or derivatives, omega-3-fatty acid accumulation was observed by *Conidiobolus lamprauges.* The feasibility of using 1-5% of linseed oil as the source of alpha linolenic acid has been patented (Idemitsu-Petrochem, 1990a).

Discussion and Conclusion

There is an urgent need for more critical investigations on the interesting but relatively neglected group of saprophytic Entomophthorales for a better understanding of their biology, taxonomic relationships and biotechnology potential. Their usefulness in fermentation industry is just beginning to emerge and it is possible that many interesting bioactive metabolites remain to be discovered from these fungi if greater attention to screen them is received from biotechnologists. There is also an immediate necessity for a better understanding of the factors which predispose humans and animals to opportunistic invasion by these widespread fungi and reliable methods are required for *in vitro* differentiation of strains with pathogenic potential. Mycologists and biotechnologists, particularly from the tropical regions may greatly benefit by undertaking serious exploration of the saprophytic Entomophthorales and contribute significantly to a better understanding of this group of fungi from the biological as well as the biotechnological point of view.

Post-Script

I am very happy that this volume is being dedicated to Dr. M.J. Thirumalachar whose contributions to the study of Indian fungi and their utilization for the identification of novel metabolites is recognized nationally and internationally. It has been a great pleasure writing this article, reminiscing the experience I had in working with Dr. Thirumalachar and contributing to the knowledge of Indian species of *Conidiobolus* and *Basidiobolus* and also giving the first taxonomic keys to both these genera. I thank the Editors for giving me this opportunity to contribute to this volume.

References

Batko, A. (1964) Notes on entomophthoraceous fungi in Poland. Entomophaga Mémoire Hors Série 2: 129–131.

Benjamin, R.K. (1962) A new *Basidiobolus* that forms microspores. Aliso 5: 223–233.

Berbee, M.L. and Taylor, J.W. (1999) Fungal Phylogeny. In: Molecular Fungal biology, (eds), Oliver, R. and Schweizer, M. Cambridge University Press, New York, 21–77.

Brefeld, O. (1884) *Conidiobolus utriculosus* and minor Untersuch Gesammt Mykol 6: 35-72.

Chiplunkar, J.M. et al.(1985) Applications of alkaline protease from *Conidiobolus* in animal cell culture. Biotechnology Letters. 7: 665–668.

Costantin, J. (1897) Sur une Entomophthore'e nouvelle. *Bull Soc. Mycol. Bot.* 26: 119–130.

Couch, J.N. (1939) A new *Conidiobolus* with sexual reproduction. American Journal of Botany. 26: 119–130.

Cutler, J.E. and Swatek, F.E. (1969) Pigment production by *Basidiobolus* in the presence of tyrosine. Mycologia 61: 130–135.

Drechsler, C. (1947) A *Basidiobolus* producing elongated secondary conidia with adhesive beaks. Bulletin of the Torrey Botanical Club 74: 403–413.

Drechsler, C. (1952) Widespread distribution of *Delacroixia coronata* and other saprophytic entomophthoraceae in plant detritus. Science. 115: 575–576.

Drechsler, C. (1955) A southern *Basidiobolus* forming many sporangia from globose and from elongated adhesive conidia. Journal of the Washington Academy of Sciences 45: 49–56.

Drechsler, C. (1964) An odorous *Basidiobolus* often producing conidia plurally and forming some diclinous sexual apparatus. American Journal of Botany 51: 770–777.

Dring, V.J. (1958) *Conidiobolus rhysosporus* Drechsler in Great Britain. Transactions of British Mycological Society 41: 52–54.

Eidam, E. (1886) *Basidiobolus* eine neue Gattung der Entomophthoraceen. Beitr Biol Pflanz 4: 181–251.

Emmons, C.W. and Bridges, C.H. (1961) *Entomophthora coronata* the etiologic agent of phycomycosis of horses. Mycologia 53: 307–312.

Gugnani, H.C. (1992) Entomophthoromycosis due to *Conidiobolus*. European Journal of Epidemiology. 8: 391–396.

Humber, R.A. (1989) Synopsis of a revised classification for the Entomophthorales (Zygomycotina). Mycotaxon 34: 441–460.

Idemitsu-Petrochem (1990) Preparation of lipid containing bishomogammalinolenic acid by culturing *Conidiobolus* sp. in sesame oil culture medium JP-089982 (Biotechnology Abstracts 91:02817).

Idemitsu-Petrochem (1990a) Preparation of lipid containing omega-3-unsaturated fatty acid in *Conidiobolus lamprauges*. JP 019739 (Biotechnology Abstracts 90:14878).

Ingale, S.S., Rele, M.V. and Srinivasan, M.C. (2002) Alkaline Protease production by *Basidobolus* (NCL 97.1.1): Effect of "darmform" morphogenesis and cultural conditions on enzyme production and preliminary enzyme characterization. World Journal of Microbiology and Biotechnology 18: 403–408.

Kamalam, A. and Thambiah, A.S. (1975) Basidiobolomycosis with lymph node involvements. Sabouraudia 13: 44–48.

Kamalam, A. and Thambiah, A.S. (1978) Lymph node invasion by *Conidiobolus coronatus* and its spore formation "*in vitro*". Sabouraudia 16: 175–184.

Kamalam, A. and Thambiah, A.S. (1982) Basidiobolomycosis following injection injury. Mykosen 25: 512–516.

Kamalam, A. and Thambiah, A.S. (1984) Muscle invasion by *Basidiobolus haptosporus*. Sabouraudia 22: 273–277.

Kevorkian, A.G. (1937) Studies in Entomophthoraceae I Observation on the genus *Conidiobolus*. Journal of Agriculture of The University of Puerto Rico. 21: 191–200.

King, D.S. (1976) Systematics of *Conidiobolus* using numerical taxonomy, I: Biology and cluster analysis. Canadian Journal of Botany 54: 54–65.

King, D.S. (1976a) Systematics of *Conidiobolus* using numerical taxonomy, II: Taxonomic considerations. Canadian Journal of Botany 54: 1285–1296.

King, D.S. (1977) Systematics of *Conidiobolus* using numerical taxonomy, III: Description of recognized species. Canadian Journal of Botany 55: 718–729.

Martin, G.W. (1925) Morphology of *Conidiobolus villosus*. Botanical Gazette 80: 311–318.

McKerracher, L.J. and Heath, L.B. (1985) The structure and life cycle of nucleus associated organelles in two species of *Basidiobolus*. Mycologia 77: 412–417.

Nagahama, T., Sato, H., Shimuzu, M. and Sugiyama, J. (1995) Phylogenetic divergence of the entomorphthoralean fungi: evidence from nuclear 18S ribosomal RNA gene sequences. Mycologia 87: 203–209.

Okafor, J.I. and Gugnani, H.C. (1990) Lipase activity of *Basidiobolus* and *Conidiobolus* species. Mycoses 33(2): 81–85.

Phadatare, S.U., Deshpande, V.V. and Srinivasan, M.C. (1993) High activity alkaline protease from *Conidiobolus coronatus* (NCL 86–8–20): Enzyme production and compatibility with commercial detergents. Enzyme and Microbial Technology 15: 72–76.

Ribes, J.A., Vanover-Sams, C.L. and Baker, D.J. (2000) Zygomycetes in human disease. Clinical Microbiology Reviews 13(2): 236–301.

Sawyer, W.H. (1929) Observations on some entomogenous members of the Entomophthoraceae in artifical culture. American Journal of Botany. 16: 87–121.

Srinivasan, M.C. and Thirumalachar, M.J. (1964) On the identity of *Entomophthora coronata*. Mycopathologia et Mycologia Applicata 24: 391–396.

Srinivasan, M.C. and Thirumalachar, M.J. (1965) Studies on species of *Conidiobolus* from India. IV Sydowia (Annales of Mycology) 19: 86–91.

Srinivasan, M.C. and Thirumalachar, M.J. (1965a) *Basidiobolus* species pathogenic for man. Sabouraudia 4 (1): 32–34.

Srinivasan, M.C. and Thirumalachar, M.J. (1967) Evaluation of taxonomic characters in the genus *Conidiobolus* with key to known species. Mycologia. 59: 698–713.

Srinivasan, M.C. and Thirumalachar, M.J. (1967a) Studies on *Basidiobolus* species from India with discussion on some of the characters used in the speciation of the genus. Mycopathologia et Mycologia Applicata 33: 56–64.

Srinivasan, M.C., Narasimhan, M.J. and Thirumalachar, M.J. (1964) Artificial culture of *Entomophthora muscae* with a discussion on the morphological aspects for differentiation of the genus *Entomophthora* and *Conidiobolus*. Mycologia 56: 683–689.

Srinivasan, M.C. et al. (1983) High activity alkaline protease produced by a *Conidiobolus*. Biotechnology Letters 5: 285–288.

Tanksale, A., Chandra, P.M., Rao, M. and Deshpande, V. (2001) Immobilization of alkaline protease from *Conidiobolus macrosporus* for reuse and improved thermal stability. Biotechnology Letters 23: 51–54.

Wilson, A.M.M. (1961) Two phycomycete isolates from subcutaneous phycomycosis in Uganda. Nature (London) 191: 612–613.

3. Hymenochaetoid Fungi (Hymenochaetales, Hymenomycetes) in India

Erast Parmasto

Institute of Zoology and Botany of the Estonian Agricultural University, 181 Riia St., 51014, Tartu, Estonia

SUMMARY As a result of a study of specimens in Indian, American and European herbaria and author's collections (1978-1979), 20 species of hymenochaetoid fungi (genera *Dichochaete* and *Hymenochaete)* have been found in India. Eight are new for India. Only four states of India had more than five species; hymenochaetoid fungi is a group almost neglected in biodiversity studies in this country. Data on distribution of the species in India are summarized, and a key to the species found or likely to be found in future is presented.

Introduction

The genus *Hymenochaete* Lév. s.l. (Hymenochaetales, Hymenomycetes, Basidiomycota) includes about 120 species distributed throughout the World, but mainly in tropical and subtropical countries. Three species of *Hymenochaete* have been described using types from India: *H. cacao* (= *H. adusta), H. fuscobadia* (= *H. depallens)* and *H. rheicolor.* Scattered data on occurrence of *Hymenochaete* species in India have been published in more than 25 papers; most of these are from the list of Indian fungi by Bilgrami et al. (1979). There are four publications where this group of fungi has been dealt with in detail: Banerjee's (1935) paper on some Thelephoraceae in Bengal (5 species), Rattan's (1977) book on resupinate Aphyllophorales of the North Western Himalayas (7 species), Sharma's (1995) book on Indian Hymenochaetaceae (9 species), and Natarajan and Kolandavelu's (1998) book on resupinate Aphyllophorales of Tamil Nadu State (5 species).

In his world monograph of the genus *Hymenochaete,* Léger (1998) mentioned three species from India *(H. rubiginosa,* and types of *H. fuscobadia* and *H. rheicolor)* that he had studied. In the same paper, he indicated eight additional species citing literature data but omitted several others, possibly having some doubts about the correctness of identification of these *(H. aspera* = *Dichochaete setosa, H. depallens, H. leonina, H. tabacina, H. villosa).*

Altogether, 23 species names of the genus *Hymenochaete* have been mentioned in literature from India until this study. 12 species are accepted here (including synonyms of correct names), 1 is a synonym of a species not yet found in India, 5 are doubtful records or misapplied names, and 5 are unknown species (possibly misprints) or species of other systematic affinities (see the list at the end).

Materials and Methods

This article is based on a study mainly of those specimens filed under *Hymenochaete* in the herbaria BPI, DD, K, NY, O, PAN and TU; several type specimens were also studied in S. Unhappily, attempts to get, for an additional study, Sharma's specimens used for his monograph (1995) were unsuccessful. The author also used his own collections from Himachal Pradesh, Uttar Pradesh and Tamil Nadu states of India (1978-

1979). Microscopic study was carried out by making free-hand sections of basidiomata or as squash mounts in 2% aqueous solution of KOH. For identification of specimens, measurements of spores have been made using a Sony CCD Video Camera attached to a Nikon Labophot 2 microscope and analysed by Global Lab Image (Data Translation Inc.) software. For statistics, 25 or 30 spores were measured in each specimen. Herbarium acronyms are after Holmgren et al. (1990). Species that are new for India are denoted with *.

Results

According to the herbarium specimens studied and reliable literature data, 20 species have been found in 12 states and territories of India. Only in four states more than five species have been found (Tamil Nadu 13, Uttar Pradesh 10, Himachal Pradesh 8, West Bengal 6). Of the seven species that are new for India, three are new taxa (*Hymenochaete gigasetosa, H. indica* and *H. legeri* which, until now, has been found in Hawaii).

The number of species found is low when compared with the number in some better studied, mostly much smaller, territories (for example, in New Zealand it is 30, Brazil 29, Hawaii islands 19, Réunion 16). There are two reasons for it: general rarity of most species of this group of fungi and insufficient attention paid to these fungi (as well as to the other resupinate Aphyllophoroid fungi) in India. Hopefully the *key to species* published below will give some help for the further studies; five taxa not yet found in India but possibly inhabiting here are included in the key, too. Terms used in the key are explained and illustrated in the papers by Sharma (1995), Léger (1998) and Parmasto (2001b).

Key to the Species

Names of the species not yet found in India are in *italics*.

1. Basidiome pileate or effuse-reflexed 2
   Basidiome resupinate, sometimes with slightly elevated margins 10
2. Hymenophore coarsely granulose or colliculose, then warted or irregularly hydnoid; structure fibrillose with elongated cavities; in context dichotomously branched dichohyphae **Dichochaete setosa**
   Hymenophore smooth or lowly tuberculate; dichotomously branched hyphae absent 3
3. Hard when dry (sometimes brittle); in section with a dark line (cortex) made of densely agglutinated hyphae 4
   Soft or coriaceous, usually pliable; cortex absent 8
4. In hymenium very numerous brownish slightly conical hyphidia 2-3 μm diam with thickened or thick walls, sometimes encrusted with small brown granules; spores 4-5(-5.5) × 1.8-2.5 μm **Hymenochaete luteobadia**
   Hyphidia absent 5
5. In hyphal layer scattered or rare setal hyphae (embedded setae) 120-250 × 7-13 μm; spores cylindrical, slightly curved, 4.5-7 × 1.2-2.2(-2.5) μm **H. tabacina**
   Setal hyphae absent; spores elongated ellipsoid, 3-5 × 1.5-3 μm 6
6. Setae crowded or rare, 30-50 × 4.5-7(-8) μm, projecting upto 30 μm 7
   Setae crowded (very numerous), 40-80(-100) × (6-)8-10(-12) μm, projecting usually 40–60 μm **H. rubiginosa**
7. Setae not numerous (rare); in the hymenium oblique submonilioid hyphal tips originating in the hyphal layer **H. villosa**
   Setae very numerous (crowded); in the hymenium submonilioid hyphae absent ***H. ochromarginata*** Talbot (found in Thailand and South Africa)
8. Setae 45-120 μm long; spores ellipsoid or cylindric, 3.8-6 μm long 9

Setae 25-35 μm long; spores broadly ellipsoid or subspherical, 2.7-3.3 × 2.1-2.5 μm **H. adusta**

9. Setae 45-80 μm long; spores 3.8-4.8 × 1.5-2.3 μm; lower part of the hyphal layer (above the hymenium) composed of densely parallel hyphae **H. attenuata**
Setae 60-95(-120) μm long; spores 4.2-6 × 1.7-2.5 μm; hyphal layer composed of subparallel or interwined hyphae **H. rheicolor**

10. In the hymenium branched dendrohyphidia or numerous subhyaline to brown hyphidia (distinct in sterile hymenium, sometimes rare or absent among well-developed basidia) 11
Hyphidia and dendrohyphidia absent.. 18

11. Dendrohyphidia or acanthohyphidia (with numerous short denticles in upper part) present 12
Simple hyphidia present 15

12. Acanthohyphidia present, dendrohyphidia absent; setae broad, 35-50(-60) × 7-14 μm; spores cylindric, 4.2-6 × 1.8-2.4 μm **H. murina**
Acanthohyphidia absent; in hymenium numerous or scattered dendrohyphidia 13

13. Dendrohyphidia scattered, with few long branches; setae 45-100 × 6-15 μm, only when old encrusted with granules; spores cylindric, 6-8.5 μm long 14
Dendrohyphidia very numerous, with short branches forming a dense net above and in the hymenium; setae 60-80(-90) × 6-9(-10) μm, encrusted with granules or crystals; spores short-cylindrical, 6-7 × 2.5-2.8 μm ***H. floridea*** Berk. & Broome (found in Sri Lanka and Australasia)

14. Hymenium dark red or blood colour, without a violet tint; setal hyphae absent; width of spores 2.7-4.2 μm. On angiospermic trees **H. sphaericola**
Hymenium dark vinaceous or light blood colour; setal hyphae present (but sometimes difficult to find); width of spores 1.5-2.8 μm. On conifers ***H. cruenta***

15. Many hyphae and hyphidia monilioid and/or with thickenings on walls; setae (40-)45-70(-80) × (6-)7-9.5(-12) μm; spores 3.2-4.2 × 1.7-2.2 μm **H. semistupposa**
Hyphae never monilioid; hyphidia subcylindrical or with almost conical tip; setae 50-85 × 7-13(-15) μm 16

16. Hyphal layer composed of loosely interwoven hyphae well developed, cortex usually present as a dark line in the middle of the hyphal layer or near the substrate; spores 3.8-5.5 × 2-3 μm **H. leonina**
Hyphal layer absent, rarely very thin and hardly distinguishable; cortex absent........................ 17

17. Spores ellipsoid, with one side flattened, 4.5-6.3(-7) × 2-3 μm **H. innexa**
Spores elongated-ellipsoid, 6-8(-8.5) × 3-4 μm **H. cervina**

18. Hyphal layer absent 19
Hyphal layer present.. 23

19. Setae 45-65 × 5-7 μm, in upper part with 1-3 teeth (protuberances) upto 5 μm long; spores ellipsoid, 4.5-5.5 × 2.5-3.5 μm ***H. pellicula*** Berk. & Broome (found in Sri Lanka and Philippines)
Setae without teeth (protuberances)... 20

20. Setae 25-45 × 5-8 μm.. 21
Setae 35-100(-110) × (6-)7-15(-17) μm 22

21. Spores 6-7.2(-8) × 3.5-4.5 μm; basidiome light-coloured (vinaceous buff) **H. indica**
Spores 3.5-4.5 × 1.5-2 μm; basidiome dark (umber) **H. vallata**

22. Setae (60-)75-100(-110) μm long, tip heavily encrusted with amorphous granules or a subglobose conglomerate of crystals; spores 5.5-8.2 × 2.2-3.2 μm; hymenium Mouse Grey, Vinaceous Buff or whitish **H. legeri**
Setae 35-80(-100) μm long, upper part encrusted with granules or rugose; spores 4.5-6.8(-7) × 1.5-2.3 μm; hymenium brownish or reddish grey, later dark brown ***H. corrugata***

23. Setae (100-)120-200(-250) × 13-23 µm, with blunt or rounded tip, encrusted with small granules **H. gigasetosa**
    Setae (60-)70-150 × 5-13(-17) µm, with acute tip without incrustation 24
24. Spores narrow, 4.5-6.5(-7) × 1.8-2.8(-3.2) µm; setae numerous, slender, upto 9(-10) µm diam. Hymenium smooth, never pelliculose **H. cinnamomea**
    Spores broad, (5.5-)6-7.5 × (3.5-)3.7-4.5 µm; setae not numerous, 9-13(-17) µm diam. Hymenium pelliculous, later deeply cracked and lacerate **H. depallens**

List of Species and Collection Localities

Dichochaete setosa (Sw.: Fr.) Parmasto. -Syn.: *Hymenochaete aspera* Berk. & M.A. Curtis, *Hydnochaete resupinata* (Sw.) Ryvarden
West Bengal: "Bengal, Calcutta ... Oct 1930" (Banerjee, 1935: 17). Himachal Pradesh, Uttar Pradesh: on the distribution map on p. 57 in the Sharma's book (1995) there are two localities indicated in these states.

These are the only localities of this species outside the Americas. Except for the size of setae given as 22.5-49.5 × 9-11, Banerjee's description fits very well in the unique characters of this species.

Hymenochaete adusta (Lév.) Har. & Pat., J. Bot. (Morot) 17: 7, 1903. -Syn.: *H. cacao* (Berk.) Berk.
Arunachal Pradesh and Nagaland (Sharma, 1995, distribution map on p. 61). Meghalaya: Khasie (= Khasi Hills), 7 Jul 1850 Dr. Hooker (Berkeley, 1854: 169; K, holotype; BPI 277664, 277665 and Lloyd Herb. 29713, isotypes of *H. cacao)*. West Bengal: Calcutta (Banerjee, 1935: 23); northern part of West Bengal (Sharma, 1995, distribution map on p. 61); Distr. Darjeeling, Pashok forest, 2.500', 7 Jul 1974 S.K. Gupta 279 (BPI 1100810).

Hymenochaete attenuata (Lév.) Lév., Ann. Sci. Nat. Bot. III 5: 152, 1846
Himachal Pradesh: Simla, S.S. Rattan (BPI 348662); Manipur, Ukhrul, on angiospermic twigs, G.S. Dhingra 80a, 87a (PAN). Uttar Pradesh: Dehra Dun District, Mussoorie, alt. 2000 m, on a fallen twig of *Rhododendron arboreum,* 3 Apr 1979 E. Parmasto (TAA 102061). Tamil Nadu: Tirunelveli Distr., Kalakkadu Sanctuary, 14 Feb 1979 (Natarajan & Kolandavelu, 1998: 14); Kalakad Sanctuary, 14 Feb 1979 E. Parmasto (TAA 103213).

Mentioned also from Sikkim, Gangtok, 1750 m, 14 Jun 1960 by Imazeki et al. (1966: 621), but the measurements of setae (65-90 × 8-10 µm) and spores (5.5-6 × 2.5-3 µm) indicated by them fit better to the closely related species *H. rheicolor*. - *H. attenuata* and *H. rheicolor* are two very closely related species if not conspecific. According to Léger (1998) they have different spore size (up to 4.5 vs 5-6 µm long) and measurements of setae (45-75 × 6-9(-11) vs 60-105(-120) × (8-)9-12(-13) µm); however, most of the specimens collected in Asia or Americas are without spores, and there are specimens with setal length 60-80 or 65-85 µm. "Typical" but not all specimens may be distinguished using the characters indicated in the section "Key to the species."

***Hymenochaete cervina** Berk. & M.A. Curtis, J. Linn. Soc. Bot. 10 (46): 334, 1868
Tamil Nadu: Tirunelveli Distr., Mundanthurai Sanctuary, Kodimudi plantation area, on a fallen twig of an angiospermic tree, 18 Feb 1979 E. Parmasto (TAA 103441).

***Hymenochaete cinnamomea** (Pers.: Fr.) Bres., Atti Accad. Sci., Lett. Arti Agiati III 3 (1): 110, 1897. Described as *6. Hymenochaete* sp. by Rattan, 1977: 25
Himachal Pradesh: Mahasu, Narkanda, on twigs of *Berberis* sp., 28 Jun 1971 S.S. Rattan 5524 (BPI

348657, UT 38575, det. E. Parmasto) and 1 Jul 1971 S.S. Rattan 5541 (K; PAN 5541; BPI 348659; UT 38570). Tamil Nadu: Tirunelveli Distr., Kalakadu Sanctionary, 900 m, 15 Feb 1979 E. Parmasto (TAA 103314). Uttar Pradesh: Dehra Dun, on bark of *Shorea robusta,* 21 Oct 1971 S.S. Rattan 5759 (BPI 348658, UT 39465, det. E. Parmasto).

See also the note at *H. semistupposa.* -Specimens seen are 1-2-layered, only one (Rattan 5541) has up to 10 layers.

Hymenochaete depallens Berk. & Broome, J. Linn. Soc. Bot. 14: 68 (1873). Syn.: *H. fuscobadia* Thind & Adlaka
Himachal Pradesh: Simla, Chadwick Falls, 3 Aug 1971 S.S. Rattan 5571 (K, identified by him as *H. leonina* Berk. & M.A. Curtis; BPI 348656). Punjab: Gopalpur, Bashahar Division, 7,000 ft, 24 Oct 1941 K. Bagchee 9 (K). Uttar Pradesh: Mussoorie, Dhobi Ghat, on *Quercus incana,* 23 Aug 1954 K.S. Thind (K, holotype of *H. fuscobadia).*

***Hymenochaete gigasetosa** Parmasto, Folia Cryptog. Estonica 40: 42. 2004
Uttar Pradesh: Mundali, Chakrata Div., dead branches of *Picea morinda,* 8 Nov 1957 P.S. Rehill (K ex DD 94/Ch57; LY 4479 ex DD 7204, det. as *H. serialis* spec. nov.); same locality, 5 Nov 1957 P.S. Rehill (K ex DD 90/Ch.57); same locality, on dead branches of an unidentified broad-leaved tree, 5 Nov 1957 P.L. Rehill (LY 4399 ex DD 7203, det. as *H. pulvinata* spec. nov.).

Hymenochaete pulvinata and *H. serialis* are herbarium names not published by Rehill.

***Hymenochaete indica** Parmasto, Folia Cryptog. Estonica 40: 42. 2004
Tamil Nadu: Tirunelveli Distr., Mundanthurai Sanctuary, 16 Feb 1979 E. Parmasto (TAA 103329).

Hymenochaete innexa G. Cunn., Trans. Roy. Soc. New Zeal. 85 (1): 47, 1957
Tamil Nadu: Tirunelveli Distr., Mundanthurai Sanctuary, on angiospermic substrates 16 Feb 1979 (Natarajan & Kolandavelu, 1998: 17); Tirunelveli Distr., Kalakad Sanctuary, 900 m, on a dead twig of *Swietenia mahagoni,* 15 Feb 1979 E. Parmasto (TAA 103316); same place, on fallen trunks and twigs of angiospermic trees, 12 - 14 Feb 1979 G. Sekar (TAA 103126, 103220, det. E. Parmasto) and E. Parmasto (TAA 103187, 103273).

***Hymenochaete legeri** Parmasto, *Mycotaxon* (in press)
Tamil Nadu: Tirunelveli Distr., Kalakadu Sanctuary, 900 m, 13 and 14 Feb 1979 E. Parmasto (TAA 103250, 103296).

Spores of the Indian specimen are somewhat smaller than usual in this species, but nevertheless they are bigger than in the closely related *H. corrugata.*

Hymenochaete leonina Berk. & M.A. Curtis, J. Linn. Soc. Bot. 10 (46): 334, 1868
Tamil Nadu: Tirunelveli Distr., Kalakad Sanctuary, 13 Feb 1979 E. Parmasto (TAA 103202); Tirunelveli Distr., Mundanthurai Sanctuary, 16 Feb 1979 E. Parmasto (TAA 103310).

Mentioned also from Uttar Pradesh by Rattan (1977: 24), but the only collection mentioned by him (S.S. Rattan 5571) belongs to *H. depallens* (see above). Sharma (1995: 65) indicated localities in Himachal Pradesh, Uttar Pradesh, West Bengal and Assam on his distribution maps; however, he described the spores of the specimens studied by him as "ellipsoid, 5-6 (7) $\times$ 3-3.5 μm", i.e., as much bigger than the spores of *H. leonina.* In North and Mesoamerican specimens, mean spore size (of 25 spores measured in a specimen) is 4.08-4.57 $\times$ 1.61-1.94 μm in Brazil and in Venezuela 4.60-5.26 $\times$ 2.15-2.60 μm; the measurements given by Sharma indicate another species of *Hymenochaete*. Locality in Sri Lanka mentioned by Massee (1890: 107) belongs to some other fungus (see Petch, 1925). *H. leonina* is mainly distributed in Americas; in Africa it has been only found in Kenya, in Asia only in India.

Hymenochaete luteobadia (Fr.) Höhn. & Litsch., Sitzungsber. K. Akad. Wiss. Wien, Math.-nat. Kl. I 116: 750, 1907
TAMIL NADU: Tirunelveli Distr., Kalakadu Sanctuary, 900 m, 14 Feb 1979 E. Parmasto (TAA 103263). WEST BENGAL: Distr. Jalpaiguri, Kuntimari, 350', 10 Oct 1975 H.S. Chakraborty 1357 (BPI 1103223; this is an otherwise typical specimen with numerous hyphidia with thickened walls, but setae are very numerous, crowded in a way typical for *H. cacao*).

Mentioned also by Rattan (1977: 29-30) from three localities in Himachal Pradesh, but the description given by him (he did not notice the encrusted with brown granules hyphidia with thickened walls characteristic for *H. luteobadia)* fits better to *H. villosa*, not dealt with in his book. The same may be said of numerous localities indicated on the distribution map by Sharma (1995, fig. 25 on p. 66): in his description (p. 67), hymenium is said to be "brown to olive brown" and "paraphysoid hyphae absent".

Hymenochaete murina Bres., Ann. Mycol. 18: 46, 1920
TAMIL NADU: Tirunelveli Distr., Kalakadu Sanctuary, 900 m, 13 Feb 1979 E. Parmasto (TAA 103202); Mundanthurai Sanctuary, Kodimudi plantation area, on a fallen rotten angiospermic twig, 18 Feb 1979 G. Sekar (TAA 103432, det. E. Parmasto).

Hymenochaete rheicolor (Mont.) Lév., Ann. Sci. Nat. Bot. III 5: 151, 1846 -Syn.: *H. sallei* Berk. & M.A. Curtis, *H. tenuissima* Berk.
INDIA: C. Torrend (BPI, Lloyd Herb. 33079, 33743, det. C.G. Lloyd). HIMACHAL PRADESH: Simla, Chail, on fallen branches of *Rhododendron arboreum,* 5 Aug 1971 S.S. Rattan (PAN 5584, UT 38978, det. E. Parmasto). Southwestern MAHARASHTRA (Sharma, 1995, distribution map on p. 68). TAMIL NADU: "Neilgheris P.S.B." (= Nilgiri Hills); according to the original description by Montagne, 1842: 23, the locality is near Goodaloor (= Gudalur) (K and S, isotypes of *Telephora* (*Stereum*) *rheicolor*); southern Tamil Nadu (Sharma, 1995, distribution map on p. 68). WEST BENGAL: Darjeeling (Banerjee, 1935: 19); Darjeeling, Palmajua, 2300 m, 4 May 1960 (Imazeki *et al.*, 1966: 621); Darjeeling Govt. College, 6.500', H.S. Chakraborti 98 (BPI 1100811); Distr. Darjeeling, Singla-Coke forest, 19 Oct 1975 S.K. Gupta 926 (BPI 1103221); Darjeeling, G.H. Cave (BPI 330077, det. C.G. Lloyd).

Hymenochaete rubiginosa (Dicks.: Fr.) Lév., Ann. Sci. Nat. Bot. III 5: 151, 1846
HIMACHAL PRADESH: Chakrata, on *Quercus* sp. and *Q. semecarpifolia,* Oct 1943, and Bashahr, Daran, Oct 1941 (Bagchee, Puri & Bakshi, 1954: 24). TAMIL NADU: Nilgiris (Bagchee, Puri & Bakshi, 1954: 24). UTTAR PRADESH: Rishikesh, on a stump of *Shorea robusta,* 10 Sep 1973 (Rattan, 1977: 32); Dehra Dun, on *Shorea robusta,* 31 Aug 1952 B.K. Bakshi (Léger, 1998: 245); Dehra Dun, Thatra Forest, on a stump of *Shorea robusta,* 26 Mar 1979 E. Parmasto (TAA 103476, 103497); Timli, on a stump of *Shorea robusta,* 26 Mar 1979 E. Parmasto (TAA 103483). WEST BENGAL: Calcutta, Behala and Ballygunj, Aug to Nov 1930-1932 (Banerjee, 1935: 20); Darjeeling, Ramam, 2300 m, May 1960 (Imazeki et al., 1966: 621).

According to the distribution map on fig. 27 of Sharma's (1995) book, also found in PUNJAB, HARYANA, ARUNACHAL PRADESH and MANIPUR. However, *H. rubiginosa* is very similar to *H. villosa,* and that is why these data may be partly doubtful.

Hymenochaete semistupposa Petch, Ann. Roy. Bot. Gard. Perad. 9: 278, 1925
UTTAR PRADESH: Mussorie Hills, Dhobi Ghat on *Quercus incana* 6 Sep 1953 (Thind & Adlakha, 1956: 60). TAMIL NADU (Natarajan & Kolandavelu, 1998).

Data on occurrence in Himachal Pradesh (Rattan, 1977) are incorrect: the only collection indicated by him belongs to *H. cinnamomea.* The descriptions presented by some authors give rise to some doubt. Sharma (1995: 71) did not mention the very characteristic for this species monilioid hyphae; his measurements of setae and description of the structure of stratified basidiomata fit much better to *H. cinnamomea.* Thind and Adlakha (1956: 60) reported presence of light brown subglobose or ellipsoid spores 4-8 × 3-5 μm which obviously are alien, and described hyphal sheaths surrounding setae as 'hyphal pegs'. Nevertheless, the hyphae shown by them on fig. 1, *j* are typical for *H. semistupposa.*

Hymenochaete sphaericola* Lloyd, Mycol. Notes 7 (9): 1338, 1925. Syn.: *H. murashkinskyi* Pilát
JAMMU and KASHMIR, HIMACHAL PRADESH, SIKKIM and UTTAR PRADESH: widely distributed in Indian Himalayas on *Quercus incana, Quercus* sp. and *Rhododendron* sp., also on *Betula utilis, Betula* sp., *Prunus padus, Pyrus baccata* and other angiospermic trees. See details on taxonomy and distribution of this species in the recently published special paper by Parmasto (2001a).

The species has been described under the name of a closely related species *H. cruenta* (Pers.: Fr.) Donk by Sharma (1995: 62-63), and its synonym *H. mougeotii* (Fr.) Cooke by Rattan (1977: 29).

Hymenochaete tabacina (Sowerby: Fr.) Lév., Ann. Sci. Nat. Bot. III 5: 145, 1846
ARUNACHAL PRADESH, JAMMU AND KASHMIR and eastern Himalayas (Sharma, 1995, distribution map on p. 72). MADHYA PRADESH: Madhya Bharat, Sangar (Saksena & Vyas, 1964). UTTAR PRADESH: Ranikhet, Chaubattia, on *Quercus incana,* May 1944; Chakrata, Molta, Jun 1947; Chakrata, Mundali, on *Quercus semicarpifolia,* Jun 1947 (Bagchee, Puri & Bakshi, 1954: 24); Mundali, Chakrata Div., dead twigs of *Picea morinda,* 6 Nov 1957 P.S. Rehill (K ex DD 92/Ch.57).

Hymenochaete vallata* G.H. Cunn., Trans. Roy. Soc. New Zeal. 85 (1): 46 (1957)
TAMIL NADU: Tirunelveli Distr., Kalakadu Sanctuary, on a fallen twig of an angiospermic tree, 14 Feb 1979 E. Parmasto (TAA 103297).

Hymenochaete villosa (Lév.) Bres., Ann. Mycol. 8: 588, 1910. Syn.: *H. nigricans* (Lév.) Pat.
INDIA: leg. D.M. Rajan (BPI 329855, Lloyd Herb. 32983, det. Lloyd as *H. cacao).* HIMACHAL PRADESH: Kulu, on *Quercus incana*, Oct 1940 (Bagchee, Puri & Bakshi, 1954: 24); Kulu, Naggar, on wooden pole, 23 Sep 1971 S.S. Rattan (PAN 5723, UT 39438; det. as *H. luteobadia);* Dalhousie, Ranikhet, on a stump, 21 Sep 1972 S.S. Rattan (UT 39446). MAHARASHTRA: Bombay; Pune Distr., Khandala (Theissen, 1911). SIKKIM: Gangtok, 1750 m, 14 Jun 1960 (Imazeki et al., 1966: 622). TAMIL NADU: Tirunelveli Distr., Mundanthurai Sanctuary, on a fallen twig of an angiospermic tree, 16 Feb 1979 E. Parmasto (TAA 103327). UTTAR PRADESH: Nainital, Bhowali, Oct 1932 (Bagchee, Puri & Bakshi, 1954: 24); Nainital, Bhowali, 27 Aug 1934 E. Parkinson (O); Ranikhet, Chaubatia, Gov. Fruit Garden, 24 May 1944 K. Baghee 3729 (LY 3086, det. as *H. tabacina).* WEST BENGAL: Calcutta and Darjeeling Districts (Banerjee, 1935: 21-22); Darjeeling, Senchal, 2400 m, 16 Apr 1960 (Imazeki *et al.,* 1966: 622); Distr. Darjeeling, Kynanuk, on a fallen log of *Dalbergia* sp., S.K. Gupta 530 (BPI 1100816); Distr. Darjeeling, Samsing Forest, on a dead log of *Quercus* sp., 18 Aug 1974 H.S. Chakraborti 374 (BPI 1100812); Distr. Jalpaiguri, Jainti forest, 19 Oct 1975 S.K. Gupta 1399 (BPI 1103222).

Cf. also *H. luteobadia.*

Species Excluded or Dubious

Hymenochaete carteri Berk., nom. herb. (K)
Found in Bombay; a species without setae, i.e., a taxon outside *Hymenochaete;* included into the list of excluded species of *Hymenochaete* by Cooke (1879: 149) and Saccardo (1888: 603), nevertheless indicated in the list of Indian fungi by Bilgrami, Jamaluddin & Rizwi (1979: 123).

Hymenochaete corrugata (Fr.: Fr.) Lév., Ann. Sci. Nat. Bot. III 5: 152, 1846
Found in Tamil Nadu, Tirunelveli Distr., Kalakadu Sanctuary, on dead twigs of angiospermic trees 12 Feb 1979 (Natarajan & Kolandavelu, 1998: 17). – In southern countries (including India) this species is substituted by the closely related allopatric species *H. legeri.* Differences between the two taxa are indicated above, in the key to species.

Hymenochaete cruenta (Pers.: Fr.) Donk, Persoonia 1 (1): 51, 1959. - See *H. mougeotii* below.

Hymenochaete damicornis (Link) Lév., Ann. Sci. Nat. Bot. III 5: 151, 1846
Data on occurrence of this species on *Dalbergia sissoo* in Uttar Pradesh (Bose and Mehta, 1951) are possibly erroneous; reliable data on its distribution are only from Central and South America.

Hymenochaete fuliginosa (Pers.) Bres. in Burt, 1918 non (Pers.) Lév. sensu Natarajan & Kolandavelu, 1998. *H. fuliginosa* (Pers.) Lév., Ann. Sci. Nat. Bot. III 5: 152, 1846 has been found in boreal zone of Eurasia and North America, doubtful data are also from the Carribbean region and New Zealand. Description of the Indian species by Natarajan & Kolandavelu (setae 10-70 × 2-10 μm, spores 5-7 × 2-4) does not fit well enough to *H. fuliginosa* (Pers.) Bres. *H. fuliginosa* Lév. *sensu* Berk. is a quite different tropical species (cf. Bresadola, 1903: 93), but its identity is unknown; it has been found in Nilgiris (Tamil Nadu) (Massee, 1890: 109).

Hymenochaete fuscoalbida Reid, Thind and Adlakha
A misprint of *H. fuscobadia* in the list of Indian fungi by Bilgrami et al. (1979: 123).

Hymenochaete fuscobadia Thind & Adlakha, Trans. Brit. Mycol. Soc. 41: 133, 1958
Description of this species was published in Reid, Trans. Brit. Mycol. Soc. 41 (1): 133 (1958); later (re)described by Rattan (1977), and Sharma (1995) who used for this type and one more collection from Western Himalayas. On the distribution map by Sharma (1995, fig. 23), four localities are indicated in that region. Rattan (1977: 24) said that *H. fuscobadia* "is very close to *H. leonina*"; however, his *H. leonina* is actually *H. depallens.* I have studied the types of *H. fuscobadia* and *H. depallens;* these are conspecific.

Hymenochaete lemiforma Berk.
Indicated by Bilgrami, Jamaluddin & Rizwi (1979: 123) as if found by Bagchee (1953). No such name in the Bagchee's paper; an unknown name.

Hymenochaete mougeotii (Fr.: Fr.) Cooke, Grevillea 8: 147, 1880 [= *H. cruenta* (Pers.: Fr.) Donk]
Mentioned from India by Berkeley (1854: 170, "Nov 29. Singalelah, Sikkim, 1848. 10,000 feet. (Dr. Hooker.)") and by Imazeki et al. (1966: 621, Sikkim). All specimens collected in India, named *H. mougeotii* and seen by me belong to *H. sphaericola* (see above). Occurrence of *H. cruenta* in India is possible and may be expected on *Abies pindrow* or other firs in Himalayas.

Hymenochaete subpurpurascens (Berk. & Broome) Massee, J. Linn. Soc. Bot. 27: 101, 1890
Found on *Shorea robusta* by Bagchee (1953, 1960; cited by Bilgrami et al. 1979: 123). – According to the published descriptions, basidiomata of this insufficiently known species (data on spores are dubious!) are similar to those of *H. rubiginosa* and *H. villosa;* the collections from India have to be restudied.

Hymenochaete subpurpureus Berk. & Broome
Possibly a misprint for *H. subpurpurascens* (Bilgrami *et al.,* 1979: 123).

Hymenochaete tristicula (Berk. & Broome) Massee, J. Linn. Soc. Bot. 27: 111, 1890 ("*H. tristelia* Massee" in Bilgrami et al., 1979: 123)
Correct name for this species common in India is *Duportella tristicula* (Berk. & Broome) Pat. in Reink., Philipp. J. Sci. 17: 364, 1920; it belongs to the family Corticiaceae s.l.

Acknowledgements

I am much indebted to the Indian University Grant Commission, Drs. K. Natarajan, P.S. Rehill, Prof. C.V. Subramanian and late Prof. K.S. Thind for their generous help during my visit to India in 1978-1979. Curators of the herbaria BPI, DD, K, NY, O, PAN, S and TU kindly helped me in my studies. The Royal Society, London supported my studies of *Hymenochaete* in the Kew herbarium; the work was partly granted by the Estonian Science Foundation (Grant no. 2145). I am grateful to Prof. James Ginns (Canada) for critically reviewing the manuscript.

References

Bagchee, K., Puri, Y.N. and Bakshi, B.K. (1954) Principal diseases and decays of oaks and other hardwoods in India-II. Indian Phytopathology 7 (1):18–42.

Banerjee, S. (1935) Thelephoraceae of Bengal-I. Journal of Indian Botanical Society 14 (1): 13–48, pl. 1–9.

Berkeley, M.J. (1854) Decades of fungi. Decades XLIV.-XLVI. *Hooker's Journal of Botany* 6: 169–170.

Bilgrami, K.S., Jamaluddin and Rizwi, M.A. (1979) *Fungi of India. Part I. List and references.* Today and Tomorrow's Printers and Publishers, New Delhi, pp. 467.

Bose, S.K. and Mehta, P.R. (1951) Record of disease occurrence: Some hitherto unrecorded fungus diseases of Kumaon and Garhwal region of U.P. Plant Protection Bulletin 3: 46–48.

Bresadola, G. ["J."] (1903) Fungi polonici. Annales Mycologici 1 (1): 65–131.

Cooke, M.C. (1879) On Hymenochaete and its allies. Grevillea 8, 145–150.

Holmgren, P.K., Holmgren, N.H. and Barnett, L.C. (eds.) (1990) *Index Herbariorum. Part I:* The herbaria of the World. 8th ed. New York Botanical Garden, Bronx, NY. 693 pp.

Imazeki, R., Kobayasi, Y. and Aoshima, K. (1966) Fungi. *In:* Hara, H. (comp.), *The flora of Eastern Himalaya. Results of the botanical expeditions to Eastern Himalaya organized by the University of Tokyo 1960 and 1963.* The University of Tokyo, Tokyo, p. 611–626.

Léger, J.-C. (1998) Le genre Hymenochaete Léveille. *Biblioth. Mycol.* 171. J. Cramer, Berlin and Stuttgart, 319 pp.

Massee, G. (1890) A monograph of the Thelephoreae. – Part II. Journal of the Linnean Society. Botany 27: 95–205.

Montagne, C. (1842) Cryptogamae nilgherienses seu Plantarum cellularium in montibus peninsulae indicae Neel-Gherries dictis à cl. Perrottet collectarum Enumeratio. Annales des sciences naturelles. Botanique II 18, 12–23.

Natarajan, K. and Kolandavelu, K. (1998) *Resupinate Aphyllophorales of Tamil Nadu, India.* C.A.S. in Botany, University of Madras, Madras, 133 pp.

Parmasto, E. (2001a) Hymenochaete cruenta and H. sphaericola, two sibling species of Hymenochaetales (Hymenomycetes, Basidiomycota). Czech Mycology 52 (4): 307–315.

Parmasto, E. (2001b) Hymenochaetoid fungi (Basidiomycota) of North America. Mycotaxon 79: 107–176.

Parmasto, E. (2004). Two new species of Hymenochaete (Hymenomycetes, Basidiomycota) from India. Folia Cryptogamica Estonica 40: 41–42.

Petch, T. (1925) Notes on Ceylon Thelephoraceae, etc. Annals of the Royal Botanic Gardens, Peradeniya 9 (3): 259–298.

Rattan, S.S. (1977) The resupinate Aphyllophorales of the North Western Himalayas. *Biblioth. Mycol.* 60. J. Cramer, Vaduz, 427 pp.

Saccardo, P.A. (1888) *Sylloge fungorum.* 6. Patavii, 929 pp.

Saksena, S.B. and Vyas, K.M. (1964) The wood decaying fungi of Sangar, Madhya Bharati, Journal of University of Saugar 11–31(B): 15–28.

Sharma, J.R. (1995) *Hymenochaetaceae of India.* Botanical Survey of India, Calcutta, 219 pp.

Theissen, F. (1911) Fungi aliquot Bombayensis, a rev. Ed. Blatter collecti. Annales Mycologici 9: 153–159.

Thind, K.S. and Adlakha, K.L. (1956) Thelephoraceae of the Mussoorie Hills-I. Indian Journal of Mycological Research 2 (1 & 2): 57–64.

4. Endolithic Fungi

Stjepko Golubic[1], Gudrun Radtke[2] and Therese Le Campion-Alsumard[3]
[1]Biological Science Center, Boston University, Boston, MA 02215, USA
[2]Hessisches Landesamt fuer Umwelt und Geologie, Rheingaustr. 186, D-65203 Wiesbaden, Germany
[3]Centre d'Oceanologie de Marseille, Universite d'Aix Marseille II, URA CNRS 41, Station Marine d'Endoume, F-13007 Marseille, France

SUMMARY Fungi are an important constituent of microbial endolithic assemblages. Their activity includes penetration of limestone, mollusk shells and other carbonate substrates (euendoliths), where they may exploit mineralized organic matter, engage in symbiotic relations with algae, or attack endolithic algae and organisms that form carbonate skeletons such as corals. Endolithic fungi leave specific boring traces, which can be identified in the fossil record and described as trace fossils (ichnofossils). Their distribution is light independent and extends to abyssal oceanic depths. In lichen-type associations with algae they occupy barren cliffs and are important factors in bioerosion of stone monuments in terrestrial environments. They are also among the first occupants of rocks emerging from under retreating glaciers. Important, but insufficiently studied is the role of aggressive endolithic fungi in skeletons of corals where they are ubiquitous and globally distributed. Their relationship with the coral coelenterate and endolithic algae is in a state of equilibrium in healthy, growing reef corals, but may turn detrimental to coral health when reefs are exposed to environmental stress.

Introduction

The existence of microbial populations that penetrate and dwell inside hard mineral substrates, primarily carbonates and phosphates has been noticed since the mid 19th century, but the early reports were not clear in distinguishing between phototrophic (endolithic cyanobacteria and algae) and heterotrophic (mostly lower and higher fungi) microboring organisms. This distinction became important when the interest for endoliths and associated bioerosion activity extended to the fossil records and opened the possibility of using them as paleoenvironmental, specifically paleobathymetric indicators (Budd and Perkins, 1980; Vogel et al. 1995, 2000). Phototrophic endoliths dominate within the illuminated (euphotic) zone in the oceans, where they settle in distinct zones. Distribution patterns are less prominent in light-independent heterotrophs, which follow the distribution of organic substrates for food and occur in all depths including abyssal (Golubic et al. 1984; Hook and Golubic, 1992; Raghukumar et al. 1992). The literature on endoliths and bioerosion is summarized in Radtke et al. (1997).

Endolithic mode of life has a long history. The oldest microboring organisms identified as cyanobacteria were discovered to penetrate ancient stromatolites in 1500 My old rocks of Dahongyu Formation in China (Zhang and Golubic, 1987). Microbial endoliths participate in many geologically significant processes, including bioerosion of limestone and other calcareous substrates, in production of fine grain sediment (Torunski, 1979) and in modification of sediment grains by micritization (Schneider and Torunski, 1982). The significance of biofilms in sediment stabilization and construction is today widely acknowledged and so are endolithic horizons within calcareous substrates which act destructively as internal biofilms (Golubic and Schneider, 2003). Some ancient microborings were diagenetically obstructed or filled with subsequent precipitates, but a surprisingly large proportion remained vacant, so that they can be cast in polymerizing resins and studied in comparison with modern boring patterns (Golubic et al. 1970). Only the shape and distribution pattern of these borings remained, but they do reveal the boring behavior of their makers which, in turn, followed their environmental and trophic requirements. Although boring patterns often characterize the organisms that made them, they are considered traces, and thus are subject to separate classifications, following the rules of ichnological nomenclature (Ekdale et al. 1984; Tavernier et al. 1992).

Mineralized skeletons of a broad range of marine animals are organized by embedded organic products, which serve as food source to endolithic fungi. Whilst the algae are limited to illuminated parts of the ocean, fungi and other heterotrophs are not depth limited. Studying marine endolithic fungi in aphotic depths is actually preferable for practical reasons, because they cannot be confused with microborings produced by algae. In shallower ranges, which include the coastal waters (down to a depth of 300 m in clear oligotrophic oceans), the distinction between endolithic fungi and algae is often difficult, because of convergent evolution of boring and reproductive behavior among unrelated organisms exploiting similar environments.

Boring Behavior of Fungi

Tunnels of different but fairly constant diameters are produced by fungal hyphae, but similar tunnels are produced by filamentous (and sometimes coccoid) cyanobacteria and eukaryotic algae. Both phototrophs and heterotrophs maintain contact with the substrate surface and often spread parallel to it. The tunnels may be branched in different ways and at different frequencies in both fungi and algae. Fungi reproduce by forming bag-like sporangia and so do endolithic green and red algae. In some cases, sporangial bags are a part of an elaborate system of filaments, in others they occur isolated as separate stages in the life cycle of the organisms. Examples exist both among fungi and algae.

There are, however, marked differences between the boring behavior of phototrophs and heterotrophs. Fungi show a preference to proliferate within organic lamellae of a bivalve shell (see Schneider, 1976; Pl. 8, Fig. 1). They are often observed to criss-cross the shell without an apparent obstacle, whereas endolithic algae within the same shell tend to proliferate within the mineral portion of the shell and are often confined and crowded in particular crystallites, apparently unable to digest and penetrate the surrounding organic lamellae (Golubic et al. 1975, Fig. 12.8). The distinct boring behavior of fungal endoliths suggests the presence of specific enzymes applied to digest organic matter invested in the bivalve shell, which are lacking in endolithic phototrophs.

Endolithic Fungi in Shells

Endolithic fungi in shells were recognized and described as early as 1889 by the French phycologists Eduard Bornet and Charles Flahault (Bornet, 1891). They have provided the descriptions of *Ostracoblabe implexa* (Fig. 1) a fungus characterized by spindle-shaped swellings along hyphae and *Lithopythium gangliiforme* (Fig. 2) with spherical swellings interconnected by hyphae. However, the systematic position of these organisms remains uncertain. The fungal passages through the mineralized interior of shells and shell fragments are visualized by casting the borings in polymerizing resins such as araldite (Golubic et al. 1970), and the obtained tunnel replicas are observed by SEM.

Additional endolithic fungi related to chytrids were discovered in beach sand in Australia by Zebrowski (1936), who described several new genera, including *Dodgella* and *Conchyliastrum*. Their borings were later classified separately as trace fossils and named *Saccomorpha* (Fig. 3) and *Polyactina*, respectively, on the grounds that similar borings may be produced by unrelated organisms (Radtke, 1991). In fact, the original descriptions of these organisms provided hardly more information than the shape and size of their boreholes.

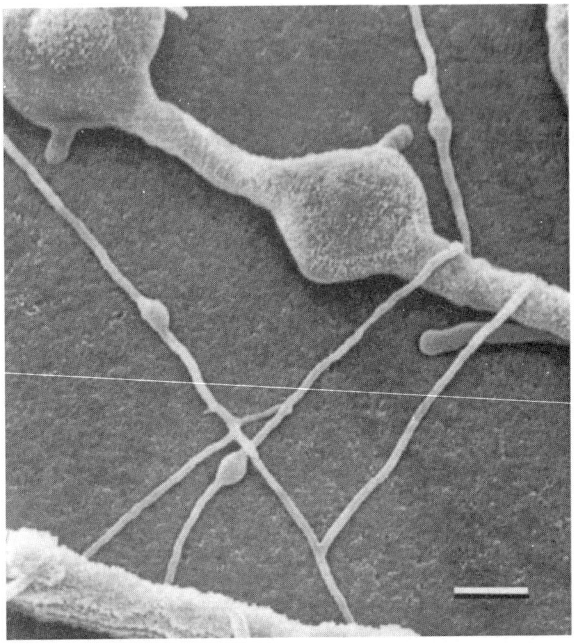

Fig. 1. Resin replica of the boring trace of a lower fungus *Ostracoblabe implexa* inside a shell of *Anomia ephippium*, apparently parasitising on a much larger endolithic green alga *Phaeophila*. Note the characteristic spindle-shaped swelling of the fungus. SEM, Scale bar is 10 μm.

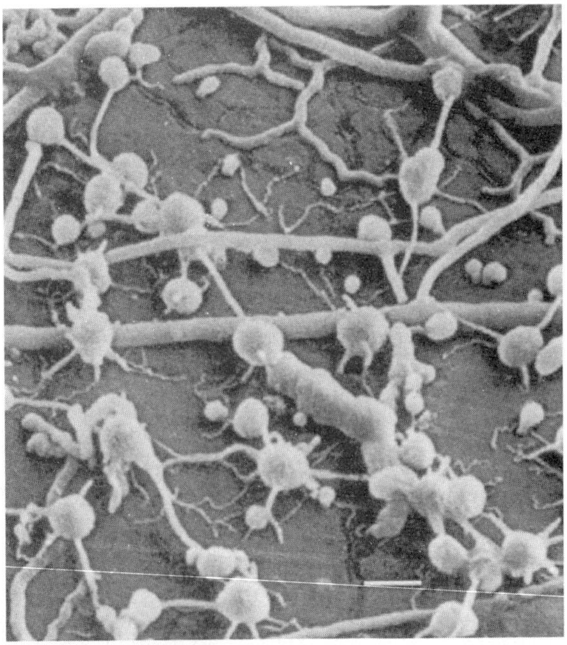

Fig. 2. Boring traces of endolithic fungus *Lithopythium gangliiforme* consist of spherical sporangial swellings connected by fine ca. 1 μm wide hyphae. In addition submicron exploratory hyphae sometimes emanate from the swellings. Wider tubular borings in the picture are produced by other phototrophic and heterotrophic endoliths. SEM of resin casts; Scale bar is 10 μm.

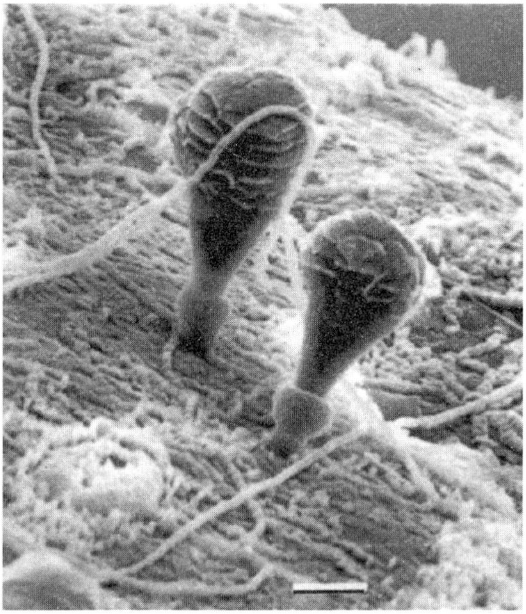

Fig. 3. Boring traces in shell fragments, named *Saccomorpha clava,* with characteristic collar around the neck that connects the sporangial swelling with the substrate surface. The borings were probably produced by the fungus *Dodgella inconstans* Zebrowski. SEM of resin replica; Scale bar is 10 μm.

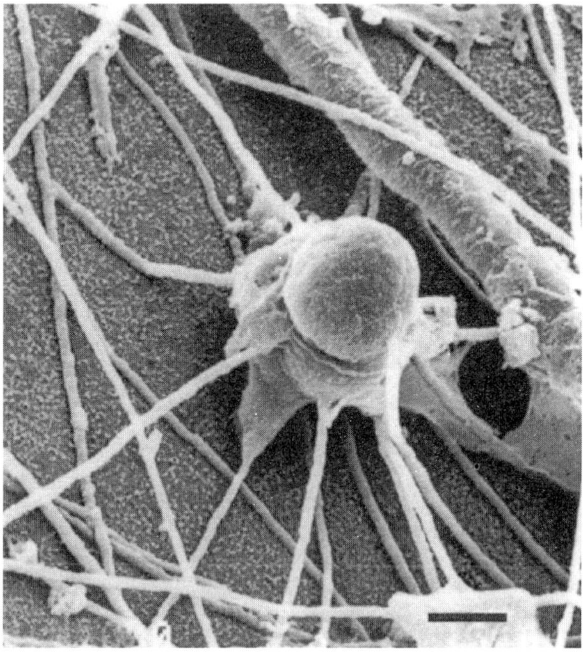

Fig. 4. View of the shell borings consisting of a fungal sporangial swelling with radiating hyphae. A larger tubular boring of an unidentified heterotrophic borer passes in the upper right of the picture. SEM of resin replica; Scale bar is 10 μm.

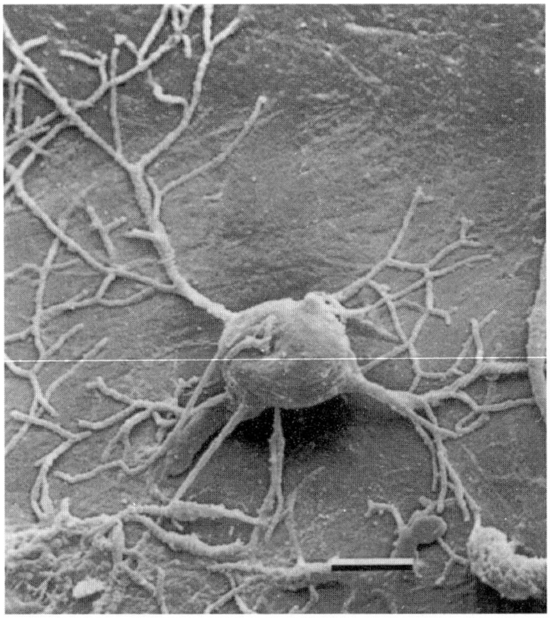

Fig. 5. Sporangial swelling of a lower fungus with a wide tubular connection with the substrate surface, apparently serving for spore release and fine, dichotomously branched and tapering exploratory hyphae. SEM of resin replica; Scale bar is 10 µm.

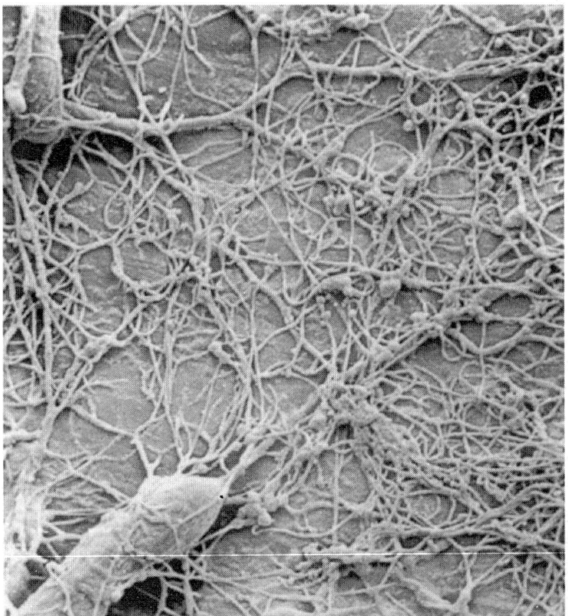

Fig. 6. The same type of endolithic fungus as in Fig. 5 produced enlarged and ramified sporangial swelling and a dense network of branched hyphae permeating the shell's interior. SEM of resin replica; Scale bar is 10 µm.

Fungal hyphae produce fine borings of uniform diameter which are quite similar to those produced by filamentous cyanobacterium *Plectonema terebrans,* but their association with particularly shaped sporangial swellings helps their recognition as fungal (Fig. 4). Dichotomously branched and finely tapered ramifications are common in endolithic fungi but not in endolithic cyanobacteria and algae. Such fine, exploratory hyphae with possible trophic function originate from bag-shaped swellings (Fig. 5) and may, over time, permeate throughout the host shell (Fig. 6). Bag-shaped swellings of some endolithic fungi are connected to the substrate surface by larger diameter tunnels, which probably serve for dispersal of spores.

Endolithic Lichens

Some fungi depend on phototrophs for food. They may be associated with phototrophs in saprophytic, parasitic or symbiotic relationships and such associations may determine their distribution as indirectly light dependent. Lichens are the most prominent symbiotic associations of fungi co-evolved with cyanobacteria and algae, which dominate on rock surfaces on land but also extend into coastal marine environments. Different lichens form and penetrate marine coastal limestone by incorporating different endolithic cyanobacteria: *Hyella caespitosa* and *H. balani* respectively (Le Campion-Alsumard and Golubic, 1985). 'Lichenization' is a gradual process, usually taking place at the margins of algal distribution, where the conditions for independent growth of each partner are suboptimal. A large proportion of fungi which engage in lichen associations are poorly known, because taxonomic determination requires full development of lichen morphology, which may remain incomplete due to environmental conditions. On the other hand, new species of lichens have been described with neither alga nor fungus identified (Jaag, 1945). Boring activity of fungal hyphae in various stages of integration into lichens is quite efficient, so that their exploratory hyphae can be found distant from the algal component and several mm below the rock surface. Geologists studying microfossils in rocks collected from exposed outcrops are advised to remove cm-thick layers of the rock surface to avoid the possibility of confusing fossil structures with recent endolithic fungi. Epilithic and endolithic lichens are often the first pioneers in colonization of newly exposed rock surfaces. Their slow progress in colonization and development has been successfully timed in studies at the front of retreating alpine glaciers (Pohl, 2000; Golubic and Schneider, 2003).

Fungi in Corals

Important but largely ignored is the role of endolithic fungi in reef corals. Coral reefs are ecologically most complex, geologically most important and currently endangered ecosystems. Corals, as one of the essential constructive elements in the reefs are consortia of at least four members. Most research concentrates on the relationships between the coelenterate and endosymbiotic symbiodinia (zooxanthellae), and is concerned with the effects of coral bleaching. Much less is known about the almost universally present endolithic algae and fungi in the coral skeleton. Endolithic algae in corals have been considered neutral or even beneficial to coral hosts (Fine and Loya, 2002), whereas fungi are considered parasitic and associated with coral diseases (e.g. Raghukumar and Raghukumar, 1991). Both of these endolithic members grow from the interior skeleton upward and outward, keeping pace with the rate of coral growth and skeletal accretion (Lukas, 1973, 1974; Le Campion-Alsumard et al. 1995a). These endoliths seem to infect the coral soon after its larva settles and they can be detected as the coral grows to the tips of its structural spines (pali), a few hours after these have been deposited by the polyp (Le Campion-Alsumard et al. 1995b Fig. 2).

The endolithic algae inhabit coral skeletons as a convenient shelter, sufficiently illuminated but protected by the polyps from the competing turf algae and small grazers, whereas the endolithic fungi are there primarily for food, attacking both the coral polyps and endolithic algae. Kendrick et al. (1982) isolated a large number of fungi from corals, of which they could identify 12 genera, including *Aspergillus*. However, they could not determine which of their isolates were the dominant endoliths in the skeletons of growing corals. *Aspergillus*-type conidiophores of endolithic fungi were observed in skeletons of *Porites lobata* from Mayotte island, Mosambique channel, Indian Ocean (Priess et al. 2000, Fig. 4d). The fungal hyphae permeated the skeletal carbonate and entered the pore spaces, where the conidiophores developed. Such conidiophores were also found in the pores of *Porites lobata* from Moorea island near Tahiti, French Polynesia (Fig. 7). Fungal attacks on corals has been interpreted as infections deriving from terrestrial sources, such as windblown Saharan sands. *Aspergillus sydowii* has been identified as a pathogen on sea fan corals (*Gorgonia* spp.) in the Caribbean reefs, however, the identity with terrestrial strains could not be established (Alker et al. 2001). In the view of global, ubiquitous and dominant presence of endolithic fungi in corals, it is more likely that we deal here with a large, previously ignored populations of genuinely marine fungi (consult also Kohlmeyer and Kohlmeyer, 1979).

Concentration of endolithic fungi were found associated with black bands in the interior of the coral skeleton (Bak and Laane, 1987). Applying high resolution SEM imagery of fractured coral skeleton and resin-replicas of boreholes, combined with light microscopy of decalcified skeletons and petrographic thin sections helped to establish the relationship between endolithic fungi and algae (Lukas, 1973; Fig. 13). It was found, that fungal hyphae attack and penetrate algal filaments apparently using them as alternate hosts (Fig. 8). This activity is accompanied by the release of a dark tannin-like substance that stains the hyphae, algal filaments and to some extent the skeletal aragonite. This process turns green bands caused by algae into black bands and serves as an indicator of fungal activity (Priess et al. 2000).

A slowing down in coral growth (e.g. in association with bleaching or other environmental stress) results in concentration of endolithic algae in the skeleton around live polyps, which may be followed by similar concentrations of parasitic fungi. If the coral recovers, concentric green or black bands stain the skeleton and date such crises like tree rings (Le Campion-Alsumard et al. 1995a).

The fungal attack on polyps is evident from the coral's defense response. As the polyps try to patch up the perforations caused by the fungal hyphae, they deposit conical structures made of dense carbonate over the borehole (Le Campion-Alsumard et al. 1995b). This action is similar to formation of pearls in pearl oysters and other bivalves, but here it is always located against the skeletal wall where the intruders are coming from. As the corals grow they move upward while vacating the skeletal chambers. At that time the fungal hyphae exit the skeleton and enter the pore space (Fig. 9). The interior of skeletons of growing corals is riddled by endolithic algae, mainly the siphonal chlorophyte *Ostreobium quekettii,* and much finer endolithic fungi which attack both algae and polyps. Fig. 10 illustrates the internal arrangement of borings including the connection with the conical structure shown in upper left corner.

Recent studies have shown that endolithic fungi are not restricted to tropical reef corals, or to corals with cellular endosymbionts, or to those harboring endolithic algae. They have been found in small *Astrangia* colonies near Woods Hole, Massachusetts. In *Astrangia* endosymbionts are facultative and the skeleton is bored by endolithic cyanobacteria rather than the usual chlorophyte *Ostreobium quekettii* (Bentis et al. 2000 and unpublished). Deep *Lophelia pertusa* reefs which grow in the aphotic depths without phototrophic symbionts or endoliths are bored by endolithic fungi, thus extending their occurrence to higher latitudes (Freiwald, 2003).

There has been recently an increase in concerns regarding the health of coral reefs and diseases of reef corals with an attempt to diagnose and identify their causative agents. In this context, insufficient attention has been paid to endolithic fungi. Based on the information available to date, it is important to realize that endolithic fungi and algae are almost always present in healthy corals. Within the coral consortium, an apparent equilibrium is being maintained, in which all members are able to survive. However, environmental stress of different kinds may cause serious disturbance to this interrelationships which, in turn, may be critical for the health of the coral. Stressed corals may fall victim to their own endolithic fungi, or their activity may weaken the animal and make it more vulnerable to additional, possibly bacterial, pathogens.

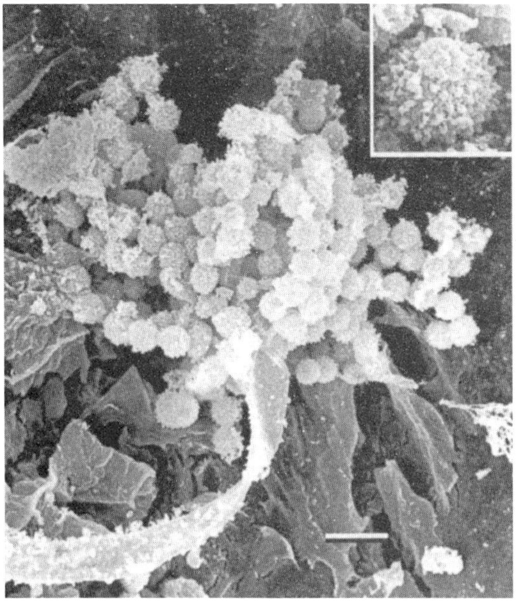

Fig. 7. Conidiophore with conidia (insert) of *Aspergillus* sp., in the pore space within the skeleton of the massive coral *Porites lutea,* Moorea island, F.P. SEM of fractured coral skeleton; Scale bar is 10 μm (3 μm for the insert).

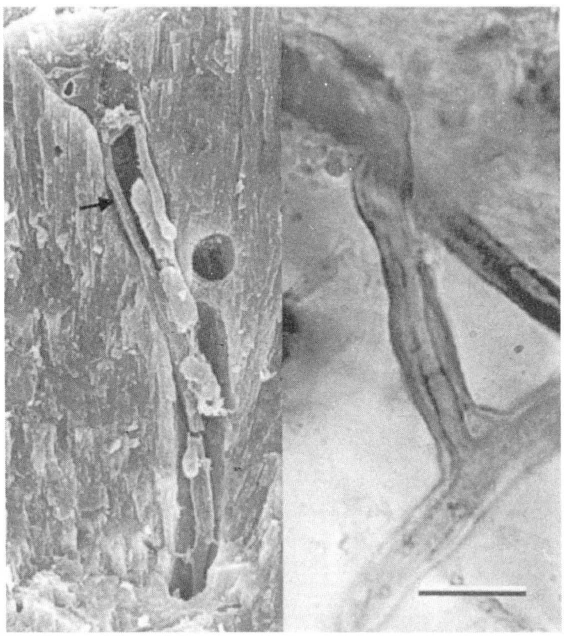

Fig. 8. Endolithic fungi inside the filaments of the endolithic alga *Ostreobium quekettii.* The picture left shows fractured aragonite *Porites* skeleton with one borehole of *Ostreobium* in transversal and one in longitudinal section. The latter contains fungal hyphae and remains of the algal cell wall (arrow; SEM of critical point dried sample). The picture right shows fungal hypha inside algal filament following extraction by HCl (Transmitted light micrograph). Scale bar for both pictures is 10 μm.

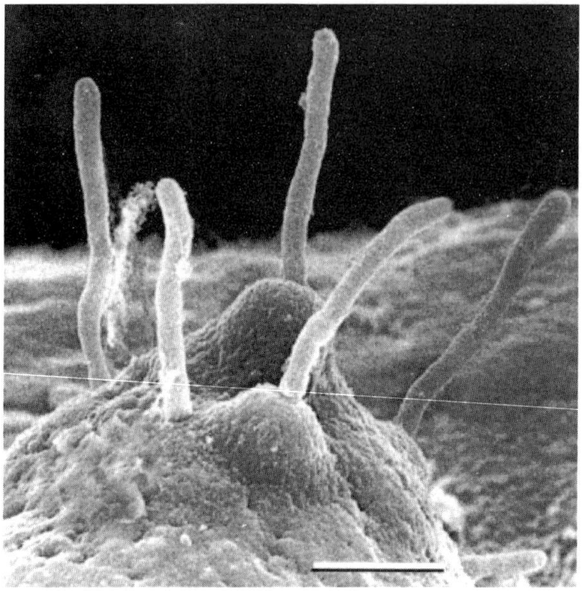

Fig. 9. Hyphae of endolithic fungi protruding into the pores of the skeleton of the massive coral *Porites lutea,* Moorea island, F.P. Note the conical carbonate buildup at the base of the hyphae, deposited by the polyp in defense of fungal attack. SEM of critical-point dried preparation of coral skeletal wall; Scale bar is 10 μm.

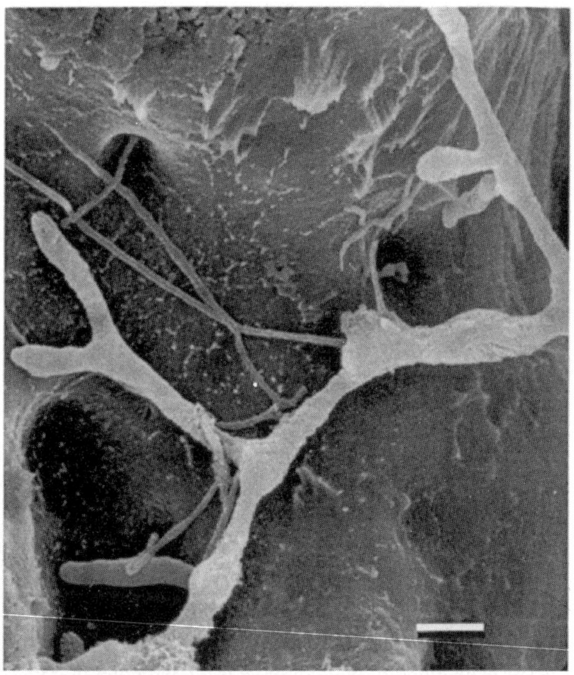

Fig. 10. Resin casts of endolith tunnels within coral skeleton. The large branched tunnel is produced by the chlorophyte *Ostreobium quekettii.* The fine tunnels, less than one micrometer in diameter are made by fungi. In the upper left of the picture is the interior view of a cone with its fungal hypha. SEM of resin replica; Scale bar is 10 μm.

Future Perspectives

Fungi have historically been studied mainly in terrestrial environments, where their activities as plant pathogens and as beneficial partners of plants (mycorhiza associations) attracted most attention. However, fungi play significant ecological roles in the oceans and their impact on marine ecosystems down to abyssal depths and their integrations have only started to be explored. Lichens have traditionally been studied as separate entities. Today they are understood as one of many integrations of fungi into systems of higher complexity, thereby acquiring new properties. There is a need, particularly in ecological research, to bridge traditional disciplinary boundaries, because the most interesting problems lie between them. Endolithic fungi encompass a well defined group of ecological niches that are wide-spread and diversified in the marine environments. There is indication that they have a long history and are firmly integrated in the marine ecosystems. However, there is only a limited knowledge about the distribution of endolithic fungi, their diversity and taxonomic identity - and the mechanisms by which they penetrate carbonate substrates are completely unknown. There is also indication that fungal activities may be critical for the health of coral reefs and marine algal stands (Bentis et al. 2000), inviting research with practical application. Traces of fungal endolithic activity are preserved in fossil records. Their study as trace fossils and the recognition of their identity opens an evolutionary perspective, which should complement phylogenetic reconstructions based on nucleic acid sequences. An assessment of biodiversity of endolithic and other fungi will require intensified field research as well as application of newly developed analytical techniques, which include molecular tools that proved increasingly successful in microbial ecology.

References

Alker, A.P., Smith, G.W. and Kim, K. (2001) Characterization of *Aspergillus sydowii* (Thom et Church): a fungal pathogen of Caribbean sea fan corals. Hydrobiologia 460: 105–111.

Bak, R.P.M. and Laane, R.W.P.M. (1987) Annual black bands in skeletons of reef corals (Scleractinia), Marine Ecology Progress Series 38: 169–175.

Bentis, C.J., Kaufman, L. and Golubic, S. (2000) Endolithic fungi in reef-building corals (Order: Scleractinia) are common, cosmopolitan, and potentially pathogenic. Biological Bulletin 198: 254–260.

Bornet, E. (1891) Sur l'*Ostracoblabe implexa* Born. et Flah. Jardin botanique 5(1): 397–400.

Budd, D.A. and Perkins, R.D. (1980) Bathymetric zonation and paleoecological significance of microborings in Puerto Rican shelf and slope sediments. Journal of Sedimentary Petrology. 50(3): 881–904.

Ekdale, A.A., Bromley, R.G. and Pemberton, S.G. (1984) Ichnology: The use of trace fossils in sedimentology and stratigraphy, SEPM Short Course Notes. 15: 1–317.

Fine, M. and Loya, Y. (2002) Endolithic algae: an alternative source of photoassimilates during coral bleaching. Proceedings of the Royal Society of London. Series B, Biological Sciences. 269: 1205–1210

Freiwald, A. (2003) Reef-forming cold-water corals. In: Ocean Margin Systems, (eds), Wefer, G., Billett, D., Hebbeln, D., Jørgensen, B.B., Schluter, M. and van Weering, T., Springer-Verlag, Heidelberg-Berlin-New York, 365–385.

Golubic, S., Brent, G. and Le Campion, T. (1970) Scanning electron microscopy of endolithic algae and fungi using a multipurpose casting-embedding technique. Lethaia, 3: 203–209.

Golubic, S., Campbell, S.E., Drobne, K., Cameron, B., Balsam, W.L., Cimerman, F. and Dubois, L. (1984) Microbial endoliths: a benthic overprint in the sedimentary record, and paleobathymetric cross-reference with Foraminifera. Journal of Paleontology 58(2): 351–361.

Golubic, S., Perkins, R.D. and Lukas, K.J. (1975) Boring microorganisms and microborings in carbonate substrates. In: The Study of Trace Fossils. (ed), Frey R.W., Springer-Verlag, Heidelberg-Berlin-New York, 229–259.

Golubic, S. and Schneider, J. (2003) Microbial endoliths as internal biofilms. In: Fossil and Recent Biofilms, (eds), Krumbein, W.E., Dornieden, T. and Volkmann, M. Kluwer Publ. Co., 249–263.

Hook, J.E. and Golubuc, S. (1992) Microbial shell destruction in deep-sea mussels, Florida escarpment. P.S.Z.N.I. Marine Ecology 14(1): 81–89.

Jaag, O. (1945) Untersuchungen über die Vegetation und Biologie der Algen des nackten Gesteins in den Alpen, im Jura und im Schweizerischen Mittelland. Beitrage Kryptogamen Flora Schweiz 9: 1–560.

Kendrick, B.W., Risk, M.J., Michaelides, J. and Bergman, K. (1982) Amphibious microborers: bioeroding fungi isolated from live corals. Bulletin of Marine Science 32, 862–867.

Kohlmeyer, J. and Kohlmeyer, E. (1979) Marine mycology: The higher fungi. Academic Press. New York, 690 pp.

Le Campion-Alsumard, T. and Golubic, S. (1985) Hyella caespitosa Bornet et Flahault and Hyella balani Lehman (Pleurocapsales, Cyanophyta) : a comparative study. Archiv für Hydrobiologie- Supplementbände. 71(1/2), Algological Studies 38/39: 119–148.

Le Campion-Alsumard, T., Golubic, S. and Hutchings, P. (1995a) Microbial endoliths in skeletons of live and dead corals: *Porites lobata* (Moorea, French Polynesia). Marine Ecology Progress Series 117: 149–157.

Le Campion-Alsumard, T., Golubic, S and Priess, K., (1995b) Fungi in corals: symbiosis or disease? Interaction between polyps and fungi causes pearl-like skeleton biomineralization. Marine Ecology Progress Series 117: 137–147.

Lukas, K.J. (1973) Taxonomy and ecology of the Recent endolithic microflora of reef corals with the review of the literature on endolithic microphytes, Unpublished Ph.D. dissertation, University of Rhode Island, 154pp.

Lukas, K.J. (1974) Two species of the chlorphyte *Ostreobium* from skeletons of Atlantic and Caribbean reef corals. Journal of Phycology. 10: 331–335.

Pohl, W. (2000) Wechselwirkungen zwischen endolithischen Biofilmen und Karbonatgesteinen in alpinen Gebieten Mitteleuropas. Ph.D. dissertation, University of Göttingen. URL: http://webdoc.gwdg.de/diss/2000/pohl/.

Priess, K., Le Campion-Alsumard, T., Golubic, S., Gadel, F. and Thomassin, B.A., (2000) Fungi in corals: black bands and density-banding of *Porites lutea* and *P. lobata* skeleton. Marine Biology 136: 19–27.

Radtke, G. (1991) Die mikroendolithischen Spurenfossilien im Alt-Tertiär West-Europas und ihre palökologische Bedeutung. Courier Forschungsinstitut Senckenberg 138: 1–185.

Radtke, G., Hofmann, K. and Golubic, S. (1997) A bibliographic overview of micro- and macroscopic bioerosion. Courier Forschungsinstitut Senckenberg 201: 307–340.

Raghukumar, C. and Raghukumar, S. (1991) Fungal invasion of massive corals. P.S.Z.N.I. Marine Ecology. 4(3): 251–260.

Raghukumar, C., Raghukumar, S., Sharma, S. and Chandramohan, D. (1992) Endolithic fungi from deep sea calcareous substrata: Isolation and laboratory studies. In: Oceanography of the Indian Ocean, (ed), Desai, B.N., 3–9.

Schneider, J. and Torunski, H. (1982) Biokarst on limestone coasts, morphogenesis and sediment production. Marine Ecology 4: 45–63.

Schneider, J. (1976) Biological and inorganic factors in the destruction of limestone coasts. Contrib. Sedimentology. 6: 1–112 .

Tavernier, A., Campbell, S.E. and Golubic, S. (1992) A complex marine shallow-water boring trace: Dendrorete balani n. ichnogen. et ichnospec. Lethaia, 25(3): 303–310.

Torunski, H. (1979) Biological erosion and its significance for the morphogenesis of limestone coasts and for near shore sedimentation (Northern Adriatic). Senckenbergiana Maritima 11: 193–266.

Vogel, K., Bundschuh, M., Glaub, I., Hofmann, K., Radtke, G. and Schmidt, H. (1995) Hard substrate ichnocoenoses and their relations to light intensity and marine bathymetry. Neues Jahrbuch für Geologie und Palaeontologie, Abhandlungen 195, (Seilacher Festschrift): 46–91.

Vogel, K., Gektidis, M., Golubic, S., Kiene, W.E. and Radtke, G. (2000) Experimental studies on microbial bioerosion at Lee Stocking Island, Bahamas and One Tree Island, Great Barrier Reef, Australia: implications for paleoecological reconstructions. Lethaia 33(3): 190–204.

Zebrowski, G. (1936) New genera of Cladochytriaceae. Annals Missouri Botanical Garden, 23: 553–564. +12 figs.

Zhang Y. and Golubic, S. (1987) Endolithic microfossils (cyanophyta) from early Proterozoic stromatolites, Hebei, China. Acta Micropaleontologica Sinica 4(1):1–12, pl. 1–3.

5. Rock-Inhabiting Fungi: From Biodeterioration to Human Disease

Evgenia V. Bogomolova
Komarov Botanical Institute of Russian Academy of Sciences
2 Prof. Popov Str., 197376, St. Petersburg, Russia

SUMMARY This article is a general introduction to the rock-inhabiting micromycetes. Small dark-colored compact colonies of unknown microorganisms have been first found on various rock substrates in cold and hot deserts and other stressful habitats. Detailed study revealed the fungal nature of these colonies. Further investigations have shown that in native rock conditions lithobiontic fungi demonstrate a set of specific adaptive morphological features. During floristic studies rock fungi have been found in many regions of the world (European countries, Russia, Ukraine, Israel, Antarctica, etc.).

Besides their fundamental biological and ecological significance, investigations of rock-inhabiting fungi have important practical aspects. Rock fungi belong to the group of major biodeteriorative microorganisms which destroy historical monuments and objects of art made of stone (marble, limestone, sandstone, calcite, granite, etc.). Most studies on these fungi are directed on investigations of biodeteriorative status of museum exhibits and historical buildings. Some methods of studying microbially contaminated stone surfaces are discussed.

The role of micro-organisms in the processes of primary soil formation and the role of these fungi in lithobiontic microbial communities are discussed, including the problem of organic nutrition for heterotrophs. The hypotheses about trophic links between bacteria and fungi on the rock surface and their important role in polymicrobial community formation are presented.

Some data on species compositions of lithobiontic fungi are given. There are several species which may be called "typical rock inhabitants". An example of summarized lithobiontic genera list from Crimea, St. Petersburg and White Sea coast is given. The common morphological features of microcolonial yeast-like fungi as typical rock inhabitants are described. This information is supplemented by data on patterns of spatial distribution of lithobionts on the rock surface.

In conclusion, the less studied aspect of lithobionts is described. It is their possible relation to human opportunistic fungi like species from the genera *Exophiala*, *Aureobasidium*, *Phialophora*, *Alternaria*, *Cladosporium*, etc. Many species from these genera are common for human mycoses and for rock surface, but until now it is unclear whether lithobionts may cause human mycoses in immune-deficient patients, or not and whether new pathogenic fungal species arise from natural habitats. Rock surfaces may, hypothetically serve as natural reservoirs and "generators" for potentially human pathogenic fungal species. Possible natural links between human pathogenic fungi and lithobionts are discussed.

Introduction

The intensive study of rock-inhabiting micromycetes started in early 1980s (Krumbein and Jens, 1981; Staley et al. 1982). Small dark-colored compact colonies of unknown microorganisms were found on various rock substrates in deserts and other stressful habitats. Detailed study revealed the fungal nature of these colonies. Further investigations by different scientists showed that in native rock conditions lithobiontic fungi demonstrate a set of specific adaptive morphological features, i.e., spherical cell shape, very thick and melanized cell walls, compact arrangement of cells into the colony, etc. (Diakumaku et al. 1995; Bogomolova et al. 1998; Sterflinger and Prillinger, 2001). During floristic studies rock fungi have been found in many regions of the world (European countries, Russia, Ukraine, Israel, Antarctica, etc.) (Krumbein and Jens, 1981; Friedmann, 1982; Diakumaku et al. 1995; Bogomolova et al. 1997; Sterflinger and Krumbein, 1997; de Leo et al. 1999; Sterflinger, 2000). Our present knowledge about the geographical distribution of this group is limited only by the lack of investigations. Generally, these fungi are cosmopolites.

Besides their fundamental biological and ecological significance, investigations of rock-inhabiting fungi have important applied aspects. In particular, it is known now that rock-fungi belong to the group of major biodeteriorative microorganisms which destroy historical monuments and objects of art made of stone (marble, limestone, sandstone, calcite, granite, etc.). Most studies on these fungi are directed on investigation of the biodeteriorative status of museum exhibits and historical buildings (Urzi and Krumbein, 1994; Diakumaku et al. 1995; Wollenzien et al. 1995, 1996; Sterflinger and Krumbein, 1997).

The less studied aspect of lithobionts is their possible relation to human opportunistic fungi like species from the genera *Exophiala, Aureobasidium, Phialophora, Alternaria, Cladosporium*, etc. Many species from these genera are common for human mycoses and for rock surface, but until now it is unclear whether lithobionts may cause human mycoses in immune-deficient patients, or not.

Role of Micro-organisms in Stone Degradation Process

Rock weathering is extremely important for the process of primary soil formation, which usually starts with the destruction of rocks and minerals. This process depends on many factors, the main ones being physical weathering and biological degradation. Biological agents of rock degradation are lichens, bacteria, cyanobacteria, green algae, fungi (Webley et al. 1963; Wood and Mac Rae, 1972; Krumbein and Dyer, 1985; Gehrmann et al. 1988; Danin and Ganeva, 1990; May, 1993; Urzi and Krumbein, 1994; Gorbushina et al. 1994; Gomez-Alarcon et al. 1995a, b; Hirsch et al. 1995a; Krumbein, 1995). These organisms usually share habitats and develop complex communities on the rock.

A great impact to the understanding of the role of microorganisms in rock weathering is made by the works of Prof. W.E. Krumbein (Oldenburg University, Germany) and his colleagues (Krumbein and Jens, 1981; Krumbein and Dyer, 1985; Gehrmann et al. 1988; Gehrmann and Krumbein, 1994; Krumbein, 1995; Gorbushina et al. 1994; Diakumaku et al. 1995; Krumbein, 1995; Urzi et al, 1994, 1995; Sterflinger and Krumbein, 1995a, b; Urzi and Krumbein, 1994; Wollenzien et al. 1995). Their investigations deal with conditions in which lithobiontic organisms exist and with the possible role of these organisms in rock destruction and biogenic patina formation. It is shown that rock biodeterioration by microorganisms is affected mainly by temperature, air humidity, pH and the presence of nutrition sources for microbes. Basically, there are two main types of microbial influence on rocks, viz. chemical (through acid and pigment production mainly by fungi and bacteria) and physical (i.e. mechanical action-penetration of hyphae, rhizoids, etc. which destroy the surface layer of rock).

Micromycetes on the Rock Surface

Studies on rock biodegradation caused by fungi started in early sixties and seventies (Henderson and Duff, 1963; Webley et al. 1963; Silverman and Munoz, 1970). It was shown experimentally that some species of imperfect microfungi possess biodegradative activity caused by their extracellular secondary metabolites. It was demonstrated that the main factor for rock biodegradation by microfungi is acid production.

Later studies revealed species compositions of fungal rock communities. Hirsch et al. (1995a) found 9 genera of micromycetes on the surface of stone buildings and monuments in Germany: *Aspergillus, Alternaria, Penicillum, Candida, Aureobasidium, Cladosporium, Phoma, Paecilomyces, Sporobolomyces*. On the monuments in Alcala de Henares (Spain) species from the genera *Alternaria, Penicillium, Phoma, Trichoderma, Mucor, Ulocladium, Dictyodesmium, Phialostelia* have been found (Gomez-Alarcon et al. 1995a). Twenty six species of micromycetes were isolated from deteriorating sandstone of the Salamanca cathedral (Spain) among which the genera *Penicillium, Phoma, Cladosporium, Trichoderma, Fusarium* (de la Torre et al. 1991) dominated.

During the studies of mycobiota of stone buildings and monuments in Germany and the Mediterranean Basin, scientists from the Geomicrobiology group (Oldenburg University, Germany) showed that dark-colored (dematiaceous) fungi are dominant on the rock substrate (Braams, 1992; Gorbushina et al. 1994; Diakumaku et al. 1995; Wollenzien et al. 1995). The total number of micromycetes isolated from different monuments was more than 90 species.

Scientists from the Botany and Biophysics Departments of St. Petersburg State University have been studying the mycobiota and biodeterioration of calcareous rocks from different geographical regions (Greece, Ukraine, Russia) since the nineties. The list of micromycetes species found on rock surfaces number more than 100 (Bogomolova et al. 1996; Gorbushina et al. 1996; Bogomolova et al. 1997; Gorbushina, 1997). Their main attention is directed to the study of stone monuments of the ancient Greek city of Chersonesos (Crimea, Ukraine) and on rock fungi ecology, physiology and morphology (Bogomolova et al. 1996, 1997, 1998, 2002, 2003; Bogomolova, 1999, 2000).

Role of Fungi in Lithobiontic Microbial Communities

Among growth limiting factors that affect lithobiontic fungi the most important is lack of organic nutrition. Heterotrophic fungi may obtain organic substances from other microorganisms, like algae, bacteria and lichens. Some more food may be brought with dust. But, in general, the total amount of organic particles on the rock surface is very limited and therefore the rock surface can be an extremely limiting substrate for fungi. In such conditions, the role of microbial interactions on rocks is very important.

Braams (1992) made an attempt to study interactions between lithobiontic fungi and algae in experimental conditions by the method of joint cultures. It was noted that the dominant type of relations between autotrophs and heterotrophs was neutral. But at the same time some fungal species (*Trichoderma viride*, *Aspergillus fumigatus*) may inhibit algal growth and some algae may stimulate fungal growth. Other authors described the joint growth of fungi and algae on the rock surface in native conditions (Urzi and Krumbein, 1994; Urzi et al. 1995; Grondona et al. 1997) and their possible role in polymicrobial communities.

Some authors hypothese about trophic links between bacteria and fungi on the rock surface and their important role in polymicrobial community formation (Urzi and Krumbein, 1994; Urzi et al, 1995). Fungi excrete a number of metabolites which may serve as nutrients for other microorganisms, in particular, for bacteria (Mirchink, 1984, 1988; Yanagita, 1990). Other investigators (Somova et al. 1998) have found a spectrum of specific enzymes in fungi and bacteria which may serve for lysis of microbial cell walls and hydrolysis of microbial exopolysaccharides. This fact supports the hypothesis of strong trophic relations in lithobiontic microbial communities.

In general, rock surfaces are usually colonized by complex polymicrobial communities (Krumbein and Jens, 1981; Krumbein and Dyer, 1985; Urzi et al. 1995). Rock weathering under the action of such communities goes faster than under the action of single groups of microorganisms.

Microcolonial Yeast-Like Fungi as Typical Rock Inhabitants

The classical work in lithobiontic fungi is by Staley et al. (1982). Microcolony is the specific form of fungal growth on native rock surface inherent to many lithobiontic species despite their taxonomic position. Authors note the wide distribution and abundance of fungal microcolonies on rocks in arid conditions (deserts). Microcolonies are defined as compact clusters of spherical dark-colored cells which are situated in various spaces, holes, cracks on rock surfaces.

Further investigations (Gorbushina et al. 1994; Wollenzien et al. 1995; Gorbushina et al. 1996; Gorbushina, 1997; Bogomolova et al. 1998) revealed presence of fungal microcolonies on different types of rocks and minerals.

Microcolonial rock-inhabiting fungi are characterized by common morphological features and by similar appearance in culture conditions. On agar media they form dark-colored, slimy or dry, elevated, sometimes cerebriform colonies of restricted size (about 15 mm dia after 3 months of growth). Vegetative cells are usually meristematic (i.e. grow by formation of longitudinal or transverse septae), or reproduce by budding. Some species may form endoconidia. Other species demonstrate no conidiation at all. It is important that some species of microcolonial fungi are first described on rock: *Sarcinomyces petricola, Coniosporium apollinis, Coniosporium perforans, Coniosporium uncinatum, Phaeococcomyces chersonesos* (Wollenzien et al. 1996; Sterflinger et al. 1997; de Leo et al. 1999; Bogomolova and Minter, 2003).

There are a number of works devoted to the determination of the taxonomical position of rock-fungi by molecular and physiological methods (de Hoog and Hermanides-Nijhof, 1977; Uijthof, 1996; Sterflinger and Gorbushina, 1997; de Hoog et al. 1999; Haase et al. 1999; Sterflinger et al. 1999; Untereiner et al. 1999; Zalar et al. 1999a, b). Authors place most of these fungi to the following orders: Chaetothyriales, Dothideales, Pleosporales (Sterflinger et al. 1999). Generally, most of these micromycetes have ascomycetous affinity.

Some microcolonial fungi are connected by synanamorphs with non-taxonomic group of black yeasts (dark-colored species with yeast-like stage of growth). Among them there are species from the genera *Exophiala, Aureobasidium, Hormonema, Moniliella, Rhinocladiella* etc. (de Hoog and Hermanides-Nijhof, 1977; Hermanides-Nijhof, 1977; de Hoog 1979; de Hoog and McGinnis, 1987). They belong to the families Dothideaceae and Herpotrichiellaceae (de Hoog, 1993; Yurlova, 1997).

Studies on mycobiota of calcareous rock carried out by our group have shown the dominance of microcolonial fungi on marble and limestone (Figs. 1 and 2) in different geographical regions (Bogomolova et al. 1996; Gorbushina et al. 1996; Gorbushina, 1997; Bogomolova et al. 1998; Bogomolova, 1999, 2000). Morphological, cultural and ecological peculiarities of lithobiontic microcolonial fungi have been studied experimentally (Gorbushina, 1997; Bogomolova, 1999, 2000; Bogomolova et al. 2002, 2003).

Biodeterioration Caused by Rock-inhabiting Fungi

The role of microorganisms in the processes of biodeterioration of stone monuments and sculptures is the subject of interest of many authors (Staley et al. 1982; Gorbushina et al. 1994; Urzi and Krumbein, 1994; Diakumaku et al. 1995; Urzi et al. 1995; Wollenzien et al. 1995). In the main part of the articles, the presence of bacteria, algae, fungi and lichens on the surface of the stone is noted. The activities of different organisms in this community leads to changes of stone color and chemical composition and surface destruction. Environmental conditions also play an important role in the destructive processes. Air pollution has an essential influence on the development of microbial communities (Urzi and Krumbein, 1994; Urzi et al. 1995). Studies over the last ten years, which were carried out at St. Petersburg University, show the important role of fungi in the processes of biodeterioration of marble monuments (Gorbushina et al. 1994; Bogomolova et al. 1996).

Over the years, scientists have paid more attention to the problem of biodeterioration of historical (archaeological) monuments (Wood and Mac Rae, 1972; Bassi and Giacobini, 1973; Bassi and Chiatante, 1976; Sorlini et al. 1982; Danin and Ganeva, 1990; May, 1993; Urzi et al. 1994; Wollenzien et al. 1995;

Fig. 1. Calcite surface showing active colonization by fungi (black colonies) and green algae.

Hirsch et al. 1995a, b; Saiz-Yimenez, 1995; Arroyo et al. 1995; Gomez-Alarcon et al. 1995a, b; Young, 1996; Grondona et al 1997).

Krumbein and colleagues studied biodeterioration of stone monuments situated in the Mediterranean region and in Germany (Braams, 1992; Gorbushina et al. 1994; Urzi and Krumbein, 1994; Diakumaku et al. 1995; Sterflinger and Krumbein, 1995a, b; Urzi et al. 1995). This group developed a set of methods for the study of lithobiontic microorganisms. It was shown that polluted air in big cities contains different chemical substances (organic acids, derivatives of nitrogen and sulfur, dust), that may be utilized by rock-inhabiting heterotrophs. This, in turn, leads to increase of aggressiveness of lithobiontic communities towards the rock surface.

Microbially Contaminated Stone Surfaces

There are a number of special methods for rock fungi investigation. For most of these methods stone

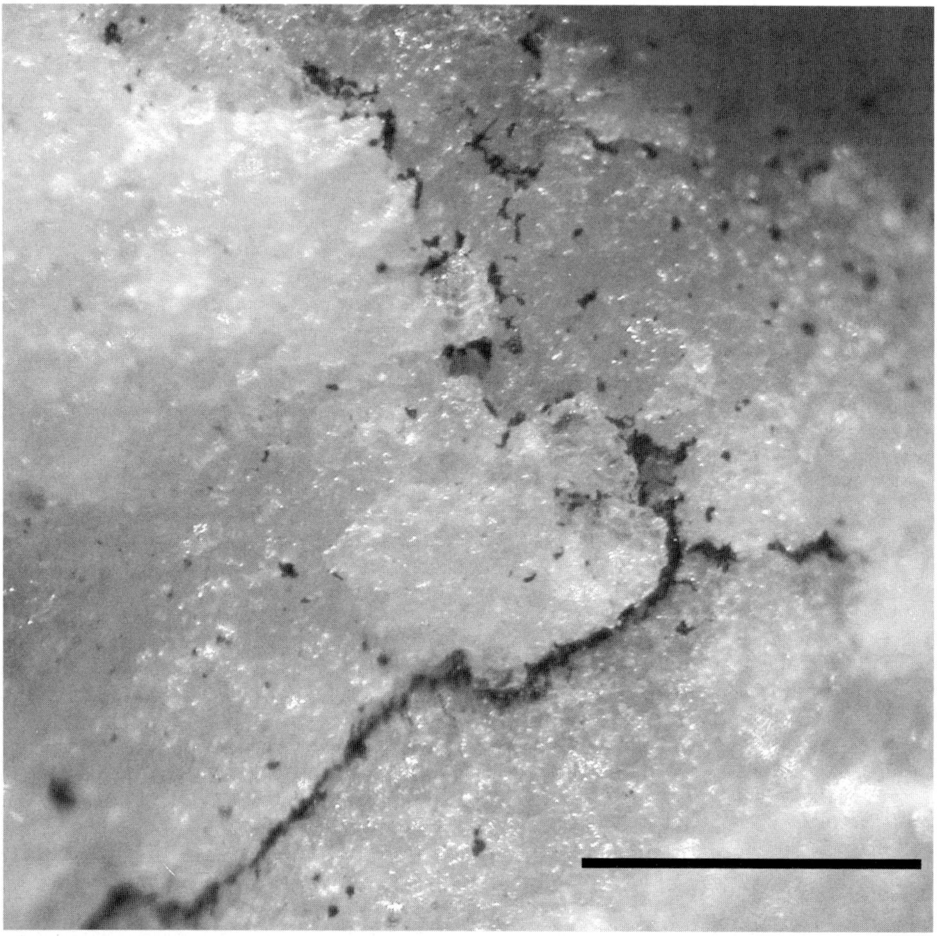

Fig. 2. Marble colonized by black fungi along the cracks and holes. Note the greater extent of calcite destruction by joint fungal-algal community. Bar-1 cm.

material is needed which is not allowed at all from the objects of high cultural value. A new and original non-destructive method was described by Sterflinger and Krumbein (1995a). This method uses replicas of microbially contaminated stone surfaces and makes optical and SEM analyses possible without sampling of the material.

The most applicable experimental methods include: optical analysis, scanning electron microscopy (SEM) (Bassi and Chiatante, 1976; Gorbushina et al. 1994; Diakumaku et al. 1995; Wollenzien et al. 1995) and method of stone thin sections for the study of fungal structures in deeper layers of rock (Gehrmann et al. 1988). A variety of techniques have been applied to isolate fungi from rock samples into pure culture. One of the best methods is the direct isolation of microcolonial fungi with a sterile needle under a dissecting microscope. Acid production by stone-inhabiting fungi may be evaluated by visual control of calcium carbonate dissolution in the vicinity of fungal colonies (Gorbushina et al. 1994; Diakumaku et al. 1995; Wollenzien et al. 1995).

There is a wide spectrum of methods for studying the protective properties of biocides. Mechanical, physical, biological and chemical treatment methods are in use at present (Urzi and Krumbein, 1994).

Chemical methods are the most frequently employed. Biocides must have no coloration, no chemical or physical reaction capable of modifying the characteristics of stone. The influence of biocides on different strains of fungi isolated from stone must be studied *in vitro* before practical application to avoid possible side-effects (bright pigment production, stimulation, etc.). A mini-review of methods for the study of lithobiontic microbiota has been reported (Hirsch et al. 1995b).

Species Compositions of Lithobiontic Fungi

The species compositions of lithobiontic fungi may vary from region to region and depend on the type of rock (Bogomolova et al. 1996; 1997). However, there are several species which may be called "typical rock inhabitants", i.e., those which are present on rock throughout the year with no dependence on season; which can form microcolonies on rock and have specific clump-like morphology and melanized thick walled cells (Diakumaku et al. 1995; Wollenzien et al. 1995, 1996; Bogomolova et al. 1998; Sterflinger and Prillinger, 2001). Other species (mostly soil and epiphytic hyphomycetes) demonstrate abundant growth on rock only in wet seasons or on the rocks placed in favorable conditions (wet, warm, with no extra light, covered with soil particles and/or other organic matter). As all other season they may be present on the rock surface only in the form of spores or other propagules. Below is an example of summarized lithobiontic genera list from Crimea, St. Petersburg and White Sea coast (Bogomolova et al. 1997). Typical lithobionts are shown in bold.

Zygomycota: *Absidia, Chaetocladium, Mucor, Rhizopus*
Ascomycota: *Melanospora, Sordaria*
Deuteromycota: *Acremoniella, Acremonium, Alternaria, Arthrinium, Aspergillus,* ***Aureobasidium,*** *Botrytis, Cephalosporium, Chloridium, Ciliciopodium,* ***Cladosporium, Coniosporium,*** *Curvularia, Epicoccum,* ***Exophiala,*** *Fusarium, Gliomastix, Hormiscium,* ***Hormonema,*** *Humicola, Moniliella,* ***Monodictys,*** *Paecilomyces, Papulospora, Penicillium, Pestalotia,* ***Phaeococcomyces, Phaeosclera, Phaeotheca, Phialophora, Pithomyces,*** *Polyscytalum, Pseudobotrytis, Pycnidiella, Rhinocladiella, Rhinocladium, Rhizoctonia,* ***Sarcinomyces,*** *Scopulariopsis, Scytalidium, Stachybotrys, Staphylotrichum, Stemphylium, Taeniolella,* ***Torula,*** *Trichocladium, Trichoderma, Trichothecium,* ***Trimmatostroma,*** *Ulocladium.*

Other authors (Sterflinger and Prillinger, 2001) add some more species to the list of typical lithobionts (*Epicoccum nigrum, Phoma glomerata, P. macrostoma, Coniothyrium* sp.) and generally this list is growing with every new investigation.

Spatial Patterns of Lithobiontic Fungi on the Rock

Lithobiontic fungi may inhabit different parts of rocks. Depending on ecological conditions and rock surface structure, microorganisms may inhabit various microniches. There are epilithic forms (inhabiting directly on the surface of the rock, endolithic forms) inhabiting empty spaces on the surface layer of rock, chasmoendolithic forms (inhabiting various empty spaces and cracks inside the rock) (Friedmann, 1982; Friedmann and Ocampo-Friedmann, 1984). Some colonies may be found as deep as 2 cm into the micro pores and cracks. It is still under discussion, whether these fungi create small holes by their own acids and ferments, or just inhabit pre-existing ones.

Studies on the growth patterns of lithobiontic fungi on different types of surface micro relief have been done by Bogomolova et al. (2002). Quantitative estimation of the spatial structure of colonies has been done by computer image analysis. Three types of colonization patterns have been observed depending on glass surface micro relief. The experiments have shown that substrate micro topography predetermines colony morphology and pattern of substrate colonization by fungi.

Experimental studies of the influence of rocks and their mineral composition and surface structure on the morphology of lithobiontic fungi colonies has been done by Bogomolova et al. (2003). The 19 samples of natural rocks (marble, granite) and minerals, which are the components of these rocks (calcite, feldspar, quartz, mica), from several quarries, were inoculated with 2 strains of lithobiontic fungi in laboratory conditions and incubated for one year under changing temperature and humidity. It was shown that the intensity of rock colonization by the fungi and the spatial distribution of the fungal colonies depends both on the chemical composition of the rock or minerals and its surface structure (micro topography). It was found that calcite marble is less tolerant to fungal attack than dolomite marble. Granite was shown to be much more tolerant to fungal biodeterioration.

Possible Links Between Human Pathogenic Fungi and Lithobionts

It is well known that human pathogenic fungi may cause various diseases (the most common are diseases of skin and nails). The much less known fact is that degradation of stone monuments and human mycoses (fungal diseases) are caused by the same organisms – microscopic fungi. Moreover, investigations have shown that these fungi are in many cases from the same genera and are even belonging to the same species; take, for example, the group of black yeasts (de Hoog, 1993). Lists of human-pathogenic opportunistic fungi may be found at http://www.doctorfungus.org and in the Atlas of clinical fungi (de Hoog and Guarro, 1995). These are the species from the genera *Exophiala, Aureobasidium, Phialophora, Alternaria, Cladosporium*, etc. The spectrum of mycoses caused by opportunistic fungi includes chromoblastomycosis, phaeohyphomycosis, deep mycoses of central nervous system and practically all organs and tissues. Deep mycoses occur mostly in immuno-deficient patients. Certainly, it does not mean that ancient monuments deteriorated by lithobiontic fungi are directly infectious to humans. However, this information makes one wonder what meaning it has for humans? and what are the conditions that make both humans and stone subject to fungal attack.

What are the similarities between the human being as a "substrate" and a stone monument? On the stone surface the fungus is faced with high temperatures, low nutrition, very little water, high concentration of mineral salts leached from marble (i.e., high osmotic pressure), intensive ultraviolet sunlight irradiation, etc. What are the conditions inside the human body? These are high temperature (only few fungi are able to grow at 37°C), high osmotic pressure, immune response, many nutrients. At a first glance, there are not many common features between the human body and the marble surface except difficult conditions of survival in both places, and certain benefits. But fungi are evolutionary, very old and primitive organisms. Their responses to external challenges are not specific, but more generalized. Whatever influences they undergo, they respond morphologically (by change of cell form, by physical isolation of cell from the external aggressive factors), and physiologically by generation of progeny that are hopefully, able to tolerate adverse conditions. Both the human body and the stone provide some advantages to low-competitive fungal species and at the same time keep a pressure that enforces them to generate adaptive features for survival. The most important common features between rock and medical fungi are tolerance to high temperature and osmotic pressure, dimorphism, black color due to melanin formation and thick cell walls. The environmental pressure on fungal strains that are living on stone surfaces is quite strong, and taking into account fungal variability, it can result in the occurrence of new mutant strains in this ecological niche. Certainly, changes and mutations may arise in human pathogenic fungi, too.

In medical mycology, there is now a situation when every year more and more species of fungi are reported as the agents of human mycoses; especially for developing countries. Sometimes these species have never been referred to as pathogens. Scientists are looking for natural sources of such "newly established" pathogens. The medical problem is caused by the dramatically increasing number of AIDS patients, wide use of antibiotics, bad living conditions, increase of the number of immuno-suppressed individuals, patients with organ transplants, cancer, hormonal therapy, diabetes and other susceptible groups. On the other hand, new pathogenic fungal species can arise in nature. A potential "candidate" to be a pathogen must possess a set of adaptive features allowing it to invade and colonize the human body. Conclusion is that rock surface can serve as a natural reservoir and "generator" for potentially human pathogenic fungal species. The investigation of links between medical fungi and environmental rock-fungi is indeed one of the actual tasks for mycology.

Nowadays, application of medical advances for protection of historical monuments has already started. Common drugs for humans and monuments are currently used, for instance, for protection of cultural art forms from fungal attack. We must keep our historical heritage for future generations. Understanding the general principles of fungal diseases in humans and monuments may serve for the development of measures for monument protection and principles for human anti-fungal treatment. But the process of "reverse" movement from study of environmental rock-fungi to medical application is still slow. Revealing the origins of new pathogenic fungal strains may be useful for preventive actions against species with high virulence.

References

Arroyo, G., Arroyo, I. and Vivar, C. (1995) Microbiological analysis of mortars from church of San Juan del Mercada at Benavente. Spain. Science of the Total Environment 167: 221–229.

Bassi, M. and Chiatante, D. (1976) The role of pigeon excrements in stone biodeterioration. International Biodeterioration Bulletin 12(3): 73–79.

Bassi, M. and Giacobini, C. (1973) Scanning electron microscopy: A new technique in the study of the microbiology of works of art. International Biodeterioration Bulletin 9: 57–68.

Bogomolova, E.V. (1999) Morphological potential of *Phaeococcomyces sp* (strain Ch 49)—a typical representative of microcolonial lithobiontic fungi. Mikologiya i Fitopatologiya 33(2): 95–100. (in Russian).

Bogomolova, E.V. (2000) Polymorphism in lithobiontic micromycetes: inducing factors and adaptive role, PhD thesis, St. Petersburg State University, Russia (in Russian).

Bogomolova, E.V., Belostotskaya, G.B. and Panina, L.K. (2002) A tendency of spatial colony structure formation in lithobiontic fungi depending on topography of substrate surface. Mikologiya i Fitopatologiya 36 (6): 1–8 (in Russian).

Bogomolova, E.V. and Minter, D.W. (2003) A new microcolonial rock-inhabiting fungus from marble in Chersonesos (Crimea, Ukraine), Mycotaxon 86: 195–204.

Bogomolova, E.V., Olkhovaya, E.A., Panina, L.K. and Soukharjevsky, S.M. (2003) Experimental study of influence of rocks and minerals chemical composition and surface structure over the lithobiontic fungi colonies morphology. Mikologiya i Fitopatologiya 37 (3): 1–13 (in Russian).

Bogomolova, E.V., Vlasov, D.Yu. and Panina L.K. (1998) Morphology of microcolonies of epilithic black yeast from the genus *Phaeococcomyces* de Hoog. Doklady Botanical Sciences 361–363: 19–20.

Bogomolova, E.V., Vlasov, D.Yu. and Sagulenko, E.S. (1997) Comparative analysis of marble-specific mycobiotes. Mikologiya i Fitopatologiya 31(5): 9–15 (in Russian).

Bogomolova, E.V., Vlasov, D.Yu., Zelenskaya, M.S. and Panina, L.K. (1996) Species composition of micromycetes inhabiting the antique marble surface in Chersonesos. Vestnik St. Petersburg State University 3(4): 49–54 (in Russian).

Braams, J. (1992) Ecological studies on the fungal microflora inhabiting historical sandstone monuments, Thesis for the awarding of the degree of a Doctor of Science, University of Oldenburg, Germany.

Danin, A. and Ganeva, G. (1990) Deterioration of limestone walls in Jerusalem and marble monuments in Rome caused by cyanobacteria and cyanophilous lichens. International Biodeterioration Bulletin 26: 397–417.

de Hoog, G.S. (1979). The black yeasts, II: *Moniliella* and allied genera. Studies in Mycology 19: 90.

de Hoog, G.S. (1993) Evolution of black yeasts: possible adaptation to the human host, Antonie van Leeuwenhoek 63: 105–109.

de Hoog, G.S. and Guarro, J. (eds). (1995) Atlas of clinical fungi. Baarn: Centraalbureau voor Schimmelcultures.

de Hoog, G.S. and Hermanides-Nijhof, E.J. (1977) Survey of the black yeasts and allied fungi. Studies in Mycology 15: 178–221.

de Hoog, G.S. and McGinnis, M.R. (1987) Ascomycetous black yeasts. In: The expanding realm of yeast-like fungi. Studies in Mycology 3: 187–199.

de Hoog, G.S., Zalar, P., Urzi, C., de Leo, F., Yurlova, N.A. and Sterflinger, K. (1999) Relationships of dothideaceous black yeasts and meristematic fungi based on 5.8S and ITS2 rDNA sequence comparison. Studies in Mycology 43: 31–37.

de la Torre, M.A., Gomez-Alarcon, G., Melgarejo, P. and Saiz-Jimenez, C. (1991). Fungi in weathered sandstone from Salamanca Cathedral, Spain. Science of the Total Environment 107: 159–168.

de Leo, F., Urzi, C. and de Hoog, G.S. (1999) Two *Coniosporium* species from rock surfaces. Studies in Mycology 43: 70–79.

Diakumaku, E., Gorbushina, A.A., Krumbein, W.E., Panina, L. and Soukharjevski, S. (1995) Black fungi in marble and limestone—an aesthetical, chemical and physical problem for the conservation of monuments. Science of the Total Environment 167: 295–304.

Friedmann, E.I. (1982) Endolithic microorganisms in the Antarctic cold desert. Science 215: 1045–1053.

Friedmann, E.I. and Ocampo-Friedmann, R. (1984) Endolithic microorganisms in extreme dry environments: analysis of a lithobiontic microbial habitat. In Current perspectives in microbial ecology, (eds), Klug, J.J. and Reddy, C.A. Washington, DC: American Society for Microbiology 177–185.

Gehrmann, C.K. and Krumbein, W.E. (1994) Interactions between epilithic and endolithic lichens and carbonate rocks. In Proceedings of the III International Symposium on the conservation of monuments in the Mediterranean Basin (eds), Fassina, V., Ott, H., and Zezza, F.: 311–316. Venice: Beni Artistici e Storici di Venezia.

Gehrmann, C.K., Krumbein, W.E. and Petersen, K. (1988) Lichen weathering activities on minerals and rock surfaces. Studia Geobotanica 8: 33–45.

Gomez-Alarcon, G., Cilleros, B., Flores, M. and Lorenzo, J. (1995a) Microbial communities alteration processes in monuments at Alcala de Henares, Spain. Science of the Total Environment 167: 231–236.

Gomez-Alarcon, G., Munoz, M., Arino, X. and Ortega-Calvo, J.J. (1995b) Microbial communities in weathered sandstones: The casa of carracosa del Campo Church, Spain. Science of the Total Environment 167: 249–254.

Gorbushina, A.A. (1997) Biological peculiarities of marble-deteriorating micromycetes, PhD thesis, St. Petersburg State University, Russia (in Russian).

Gorbushina, A.A., Krumbein, W.E., Hamman, C.H., Panina, L., Soukhajevski, S., Wollenzien, U. (1994) Role of black fungi in color change and biodeterioration of antique marbles. Geomicrobiology Journal 11: 205–221.

Gorbushina, A.A., Panina, L.K., Vlasov, D.Yu. and Krumbein, W.E. (1996) Marble-deteriorating fungi in Chersonesos. Mikologiya i Fitopatologiya 30 (4): 23–27 (in Russian).

Grondona, I., Monte, E., Rives, V. and Vicente, M.A. (1997) Lichenized association between Septonema termes sp. nov., a coccoid cyanobacterium, and a green alga with unforeseen biopreservation effect of Villamayor sandstone at "Casa Lis" of Salamanca, Spain. Mycological Research 101(12): 1489–1495.

Haase, G., Sonntag, L., Melzer-Krick, B. and de Hoog, G.S. (1999) Phylogenetic inference by SSU-gene analysis of members of the Herpotrichiellaceae with special reference to human pathogenic species. Studies in Mycology 43: 80–97.

Henderson, M.E.K. and Duff, R.B. (1963) The release of metallic and silicate irons from minerals rocks and soils by fungi activity. Journal of Soil Science 14: 236–246.

Hermanides-Nijhof, E.J. (1977) *Aureobasidium* and allied genera. Studies in Mycology 15: 141–177.

Hirsch, P., Eckhardt, F.E. and Palmer, R.J. (1995a) Fungi active in weathering of rocks and stone monuments. Canadian Journal of Botany 73(1): 1384–1390.

Hirsch, P., Eckhardt, F.E.W. and Palmer, R.J. (1995b) Methods for the study of rock-inhabiting microorganisms—A mini review. Journal of Microbiological Methods 23: 143–167.

Krumbein, W.E. (1995) A neglected carbon sink? Biodegradation of rocks. In Microbial Diversity and Ecosystem Functions. (eds), Allsopp, D., Colwell, R.R. and Hawksworth, D.L. Wallingford: CAB International 113–123.

Krumbein, W.E. and Dyer, B. (1985) This planet is alive—Weathering and biology, a multi-faceted problem. Chemistry of Weathering 149: 143–160.

Krumbein, W.E. and Jens, R. (1981) Biogenic rock varnishes of the Negev Desert (Israel): an ecological study of iron and manganese transformation by cyanobacteria and fungi. Oecologia 50: 25–38.

May, E. (1993) Microbial communities on stone: detection, distribution and activity. Biodeterioration Abstract 7: 154–162.

Mirchink, T.G. (1984) Soil fungi as a component of biogeocoenosis. In Soil microorganisms as a component of biogeocoenosis, (ed), Mishustin E.N. Moscow: Nauka. (in Russian) 114–131.

Mirchink, T.G. (1988) Soil mycology. Moscow: Moscow State University (in Russian)

Saiz-Yimenez, C. (1995) Microbial melanins in stone monuments. Science of the Total Environment. 167: 273–286.

Silverman, M.P. ana Munoz, E.F. (1970) Fungal attack on rock: solubilization and altered infrared spectra. Science 169: 985–987.

Somova, N.G., Dobrovolskaya, T.G., Zenova, G.M. and Ivanovsky, R.N. (1998) Microbial growth on the surfaces of stone buildings: Synecological analysis. Mikrobiologiya 67(5): 687–693.

Sorlini, C., Allievi, L., Sacchi, M. and Ferrari, A. (1982) Microorganisms present in deterioration materials of the "Palazzo della Ragio" in Milan. International Biodeterioration Bulletin 18(4): 105–110.

Staley, J.I., Palmer, F. and Adams, J.B. (1982) Microcolonial fungi: common inhabitants on desert rock. Science 215: 1093–1095.

Sterflinger, K. (2000) Fungi as geologic agents, Geomicrobiology Journal 17: 97–124.

Sterflinger, K., de Baere, R., de Hoog, G.S., de Wachter, R., Krumbein, W.E. and Haase, G. (1997). *Coniosporium perforans* and *C. apollinis*, two new rock-inhabiting fungi isolated from marble in the Sanctuary of Delos (Cyclades, Greece). Antonie van Leeuwenhoek 72: 349–363.

Sterflinger, K., de Hoog, G.S. and Haase, G. (1999) Phylogeny and ecology of meristematic ascomycetes. Studies in Mycology 43: 5–22.

Sterflinger, K. and Gorbushina, A.A. (1997) Morphological and molecular characterization of a rock inhabiting and rock decaying dematiaceous fungus isolated from antique monuments of Delos (Cyclades, Greece) and Chersonesos (Crimea, Ukraine). Systematic Applied Microbiology 20: 329–335.

Sterflinger, K. and Krumbein, W.E. (1995a) Precision replicas of microbially contaminated surfaces for optical and SEM-analyses. Journal of Microbiological Methods 23: 301–308.

Sterflinger, K. and Krumbein, W.E. (1995b) Multiple stress factors affecting growth of rock-inhabiting black fungi. Botanica Acta 108(6): 490–496.

Sterflinger, K. and Krumbein, W.E. (1997) Dematiaceous fungi as a major agent for biopitting on Mediterranean marbles and limestones. Geomicrobiology Journal 14: 219–230.

Sterflinger, K. and Prillinger, H. (2001) Molecular taxonomy and biodiversity of rock fungal communities in an urban environment (Vienna, Austria). Antonie van Leeuwenhoek 80: 275–286.

Uijthof, J.M.J. (1996) Relationships within the black yeast genus *Exophiala* based on ITS1 sequences. Mycological Research 100(10): 1265–1271.

Untereiner, W.A., Gerrits van den Ende, A.H.G., de Hoog, G.S. (1999) Nutritional physiology of species of *Capronia*. Studies in Mycology 43: 98–106.

Urzi, C. and Krumbein, W.E. (1994) Microbiological impacts on the culture heritage. In Report of the Dahlem Workshop on durability and change: the science, responsibility and cost of sustaining cultural heritage (eds), Krumbein, W.E., Brimblecombe, P., Cosgrove, D.E. and Staniforth, S. Chichester: Wiley. 107–135.

Urzi, C., Krumbein, W.E., Pernice, A. and Wollenzien, U. (1994) Repertamento di funghi demaziacei da manufatti in Pietra: distribuzione, ecologiaa e tassonomia. L'Igiene Moderna 102: 329–341.

Urzi, C., Wollenzien, U., Criseo, G. and Krumbein, W.E. (1995) Biodiversity of the rock inhabiting microbiota with special reference to black fungi and black yeasts. In Microbial diversity and ecosystem function. (eds), Allsopp, D., Colwell, R.R. and Hawksworth, D.L. Wallingford: CAB International. 289–301.

Webley, D.M., Henderson, M.E.K. and Taylor, I.F. (1963) The microbiology of rocks and weathered stone. Journal of Soil Science 14: 102–112.

Wollenzien, U., de Hoog, G.S., Krumbein, W.E. and Uijthof, J.M.J. (1996) *Sarcinomyces petricola*, a new microcolonial fungus from marble in the Mediterranean basin. Antonie van Leeuwenhoek 71: 281–288.

Wollenzien, U., de Hoog, G.S., Krumbein, W.E. and Urzi, C. (1995) On the isolation of microcolonial fungi occurring on and in marble and other calcareous rocks. Science of the Total Environment 167: 287–294.

Wood, P.A. and Mac Rae, I.C. (1972) Microbial activity in sandstone deterioration. International Biodeterioration Bulletin 8(1): 25–27.

Yanagita, T. (1990) Natural microbial communities. Tokyo: SPR–Verlag.

Young, P. (1996) Pollution-fueled "biodeterioration" threatens of historic stone. Environ. Science and Technology 30(5): 206–208.

Yurlova, N.A. (1997) Taxonomy of the genus *Aureobasidium* Viala et Boyer. Mikologiya I Fitopatologiya 31(5): 67–76 (in Russian).

Zalar, P., de Hoog, G.S. and Gunde-Cimerman, N. (1999a) *Trimmatostroma salinum*, a new species from hypersaline water. Studies in Mycology 43: 57–62.

Zalar, P., de Hoog, G.S. and Gunde-Cimerman, N. (1999b) Ecology of halotolerant Dothideaceous black yeasts. Studies in Mycology 43: 38–48.

6. Diversity of Aquatic Hyphomycetes in the Tropics

Iracema Helena Schoenlein-Crusius[1] and Elaine Malosso[2]
[1]Instituto de Botânica, Seção de Micologia e Liquenologia, Av. Miguel Stéfano 3687, Água Funda, CEP: 04301-903 São Paulo, Brazil
[2]Escola Superior de Agricultura "Luiz de Queiroz", Departamento de Solos e Nutrição de Plantas, Av. Pádua Dias 11, Agronomia, CEP 13418-900 Piracicaba, SP, Brazil

SUMMARY Aquatic Hyphomycetes are anamorphic typical aquatic fungi with tetraradiate, star-like or sigmoid conidia, being common in running freshwater. Their importance in the decay of allochtonous organic matter has been pointed out in many studies and the geographic distribution of the species has been considered worldwide. Tropical regions are expected to be rich sources of new species, through, little data are effectively published. Although many surveys have been conducted in several regions of the world, the knowledge about the diversity of this relevant fungal group seems far from completion, especially in the tropics. In view of the high rates of forest destruction and climatic changes observed in recent years there is an urgent necessity to evaluate their impact on the diversity and metabolic features of the species, in order to find possible new bioindicators or species with physiological capabilities to improve the recovery of the ecosystems. Training of a larger numbers of specialists and the improvements in isolation and identification techniques and the combination of traditional techniques with molecular procedures may allow relevant advances in the conservation policies of tropical areas. This article presents taxonomical concepts, some historical data and considerations of methods and techniques to study aquatic Hyphomycetes.

Biodiversity and Fungal Diversity: Reality of Tropical Regions

Biodiversity, as a concept, has been defined as the extent of biological variation on earth and can be expressed in terms of ecosystems, populations or species (Hawksworth, 1991).

The estimates of fungal biodiversity are usually linked with evolution patterns, ecosystem functions and human progress. Thus, the knowledge of the general diversity of the native mycota is of utmost importance when considering that each fungal species has its own niche in the habitat. They can be found, for example, participating in the food webs as organic matter decomposers and contributors to nutrient cycling or as symbionts with plants, to cite some of the many important roles played by fungi that help to maintain the energy balance of the ecosystems. In addition, due to their high adaptability and biochemical potential, the discovery of new fungal species, which may be used for industrial and pharmaceutical purposes, has been intensively investigated (Young, 1999).

As Hyde and Hawksworth (1997) pointed out very well, one of the serious problems concerned with the knowledge of the fungal diversity in the tropics is that most of the areas are situated in developing countries where, in general, there are few active mycologists. This leads to the studies of very diverse tasks like the survey of fungal diversity, isolation, culturing, identification of known species, description of new species and search for new bio-products by, perhaps, a handful of competent professionals.

The lack of trained personnel is a serious problem causing delays in attending to the different demands and compromising the discovery of new fungal species and the unravelling of their metabolic potentials.

In tropical countries, like Brazil for example, several urgent needs, such as finding new agricultural areas, stopping soil and water pollution, protecting pristine areas and treating already polluted ones, and decreasing the high rates of forest destruction, are hampered by the absence of basic knowledge about the diversity and/or potential of the fungi and insufficient governmental financial support for environmental matters. In addition, there is a pressing need to follow the directions of international environmental policies about biodiversity conservation.

In the tropics, the destruction of the ecosystems by burning, deforestation and pollution would affect the diversity of fungal species that have not been described yet. About 100,000 km^2 of forests have been lost each year.

At this rate, many new species are probably going to be extinct before scientists have the opportunity to isolate and describe them. Tropical regions are expected to be rich sources of new species though little data are effectively published (Hawksworth, 1991). According to Rossman et al. (1998), only a few mycologists work in the tropics, although knowledge and interest in tropical microfungi is increasing. Therefore, it is extremely necessary to encourage studies of the Taxonomy and Ecology of fungi in the tropical ecosystems.

Aquatic Mycota in Freshwater Habitats

In the aquatic environment, two kinds of fungi may be present: the zoosporic and non-zoosporic fungi. The former have specialized structures for mobility that improve fungal dispersion. The non-zoosporic fungi produce resting spores as a means to resist the changes in their environments such as soils getting saturated or waterlogged during the rainy season. The Ascomycota and Basidiomycota (higher fungi), yeasts, aquatic Hyphomycetes and terrestrial anamorphic fungi may compose the non-zoosporic mycota of aquatic environments.

The fungi that have been considered to be truly aquatic are those whose reproduction is only possible in the water, such as the zoosporic fungi and aquatic Hyphomycetes (also called "Ingoldian fungi") and some species of Ascomycota and Basidiomycota.

The aquatic mycota have important functions in the aquatic environments. These fungi are responsible for the decomposition of organic matter (cellulose, lignin, chitin, keratin, protein, humic and fulvic acids, different kinds of sugar, etc.), for the accumulation and gradual release of minerals, for the increase of the palatability of submerged substrates utilised by the invertebrates (Bärlocher and Kendrick, 1974), and, in some cases, for the removal of pollutants from water.

The fungi in freshwater habitats may assume several functions, according to their degree of tolerance and adaptability to the water. The resident fungi live in the water during their whole life, whereas amphibious fungi may live in the water or in the air. The transient fungi are usually carried to the water by rain, wind or fallen leaves. Sometimes the latter may disappear after submergence or may remain dormant. Their activity in the decomposition of submerged substrates is as yet open to question. For more details on the activities of transient fungi in submerged conditions (see Park 1972 a, b).

According to their form and life-cycle, a classification of the freshwater fungi into the following groups was proposed: Ingoldian Hyphomycetes (fungi that present conidia with hydrodynamic shape and are exclusively dependent on aquatic environment for reproduction); aero-aquatic Hyphomycetes (fungi that may endure submerged conditions but reproduce out of aquatic environments), and terrestrial-aquatic Hyphomycetes and submerged aquatic Hyphomycetes, which are observed in aquatic and terrestrial environments as facultative organisms (Goh and Hyde, 1996a). Later on, three groups were better distinguished among the former: Ingoldian fungi, aero-aquatic fungi and lignicolous aquatic fungi (Goh, 1997).

Aquatic Hyphomycetes

The group of aquatic Hyphomycetes comprises those fungi that produce conidia exclusively in the aquatic environment or in the interstitial water among soil particles. Their main habitats are streams with clear, clean, well-aerated waters with moderate turbulence; however, they can also be found in reservoirs and lakes with different kinds or levels of pollution. Their typical substrates are leaves, twigs and branches of deciduous trees, which have fallen into the water, submerged macrophytes and eventually healthy roots of riparian trees (Dix and Webster, 1995).

The conidia of the aquatic Hyphomycetes may be found trapped in foam, dispersed in the water, floating on the water surface, or associated with the decomposing organic matter (Ingold, 1975).

From the taxonomical point of view, the aquatic Hyphomycetes constitute an artificial phylogenetically heterogeneous group. They are anamorphs of the Ascomycota and Basidiomycota (Alexopoulos et al. 1996; Webster, 1992) and the identification of the species has been mainly based on the morphological features of the conidia (Webster, 1992; Marvanová, 1997), which is probably the result of convergent evolution (Webster, 1992).

It is estimated that there are about 300 named species of Hyphomycetes reported from freshwater habitats. They are distributed over approximately 143 genera (280 species) known from the tropical regions (Goh, 1997). Shearer (1993) presented an excellent work about freshwater Ascomycetes, including techniques for their study, systematics, distribution, ecology and evolution.

The term "tetraradiate fungi" has also been frequently used to name this fungal group as many species produce conidia with a radiate or star-like shape. These conidia consist of a central part from which three or four arms are projected in divergent positions (Subramanian, 1983; Dix and Webster, 1995), however, there are also species with sigmoid, fusiform, coiled and spherical conidia, which are usually large, with sizes ranging between 50 and 100 μm.

The hydrodynamic shape of these conidia is one of the most striking features of aquatic Hyphomycetes and confers to them a higher ability to remain suspended in the water for extended periods of time. This increases the chances of the propagules to become attached to organic substrates available for colonisation.

The aquatic Hyphomycetes has also been named "Ingoldian fungi" in honour to Prof. C.T. Ingold, who was one of the most important pioneers in the study of this fungal group. He started observing aquatic Hyphomycetes in the 1940's and isolated and described many species in several aquatic environments in the United Kingdom and other countries. Many of the species described by Prof. Ingold are now considered cosmopolitan (Ingold, 1975; Marvanová, 1997).

The richness of Hyphomycetes in the aquatic environments is probably influenced by the type of substrates, the chemical and physical changes in the streams (Hasija and Singhal, 1991), and some complex interactions among several climatic and limnological factors (Bärlocher, 1992; Marvanová, 1997). During the autumn in temperate climates, streams receive a great supply of tree leaves, which may be densely colonised by a high diversity of aquatic Hyphomycetes (Bärlocher, 1992). However, in the tropics the analyses of the seasonal behaviour of these fungi become more complex due to the constant input of leaf litter and the heavy rainfall throughout the year instead of the four marked seasons found in temperate regions (Schoenlein-Crusius and Milanez, 1996).

What is Really Aquatic?

In tropical and equatorial climates the leaf litter and the soil are constantly wetted by heavy, daily rainfalls, which may easily pass the mark of 3000 mm/year in certain regions. In such conditions, it may be difficult to distinguish aquatic from terrestrial fungi only on basis of their habitat characteristics.

It has been shown that some typical geofungi isolated from the tropical regions with high levels of rainfall, like *Epicoccum nigrum* Link and *Trichoderma viride* Rifai, are able to produce spores in submerged cultures (Schoenlein-Crusius and Milanez, 1998). Furthermore, species of *Beltrania, Camposporium, Chaetendophragmia, Cladosporium, Cryptophiale, Dactylella, Dictyochaeta, Kionochaeta, Phaeoisaria, Subulispora, Tetraploa* and other usually terrestrial genera of Hyphomycetes are usually found on submerged litter as "facultative aquatic fungi" (Goh, 1997).

Conversely, some typical aquatic species such as *Sporidesmium* and *Articulispora* have been isolated from non-aquatic environments. The reason for this behaviour is as yet uncertain, however, it could be related to adaptation or tolerance of the fungi to drastic changes in their habitat and/or in the climatic conditions of the ecosystem.

Despite some difficulty with the ecological definition of what organisms are truly aquatic, the recognition of their importance has continuously increased over the years. They are considered predominant in the leaf decomposition process in aquatic environments and able to degrade several plant cell polymers such as cellulose, hemicellulose, pectin (Chamier, 1985) and probably lignin (Hasija and Singhal, 1991; Marvanová, 1997).

Aquatic Hyphomycetes may also be found as endophytes of submerged roots of riparian trees (Iqbal et al. 1995), on root surfaces (Waid, 1954), and in roots of diseased plants (Nemec, 1969). Due to their ability to degrade a broad range of organic, complex substrates, these fungi are known to produce enzymes of great economic interest like amylase, cellulase, pectinase, protease, pyrocatechol oxidase, triacyl glycerol lipase and xylanase (Chandrashekar and Kaveriappa, 1991; Suberkropp, 1992; Marvanová, 1997).

In addition to the breakdown of leaf tissues by many different enzymes, the aquatic Hyphomycetes are well known for enriching the nitrogen and protein in the substrates they colonise, therefore increasing the palatability of the leaves to the leaf-eating invertebrates. Thus, the biomass of the aquatic Hyphomycetes constitutes an important protein source in the diet of aquatic invertebrates (Bärlocher, 1992; Suberkropp, 1992; Dix and Webster, 1995; Sridhar and Bärlocher, 2000).

Excellent reports on the function of aquatic Hyphomycetes in natural environments and the description of the methods used in the ecological studies concerning this group may be found in Bärlocher (1992), Suberkropp (1992), Dix and Webster (1995) and Marvanová (1997).

Some Historical Data and Geographical Distribution

The aquatic Hyphomycetes have been characterized as one of the most active groups of fungi in the mineralization of leaf litter. Consequently, they play an important part in the dynamics of nutrient cycling in the aquatic environments (Gessner and Chauvet, 1994). Therefore, our knowledge of the diversity of aquatic Hyphomycetes species in terms of their geographic distribution has been considered very relevant.

In the tropical regions such knowledge is yet fragmented and characterized by a great need of complementary information.

The recognition of the stauroid and scolecoid conidial forms in water occurred for the first time in the 1890's (De Wildeman, 1893, 1894, 1895), when *Clavariopsis, Lemnonniera* and *Tetracladium* were described. Some 50 years later, in the decade of 1940, many pioneering works on aquatic Hyphomycetes were reported by Ingold, who first recognized 16 species in decaying alder leaves submerged in a little brook in Leicestershire, England (Ingold, 1942). Afterwards, other 19 species were described (Ingold, 1943a, b; 1944, 1952, 1958b, 1959a, 1964, 1974, 1975) followed by a large number of studies involving the aquatic Hyphomycetes in Britain (Ingold, 1964, 1974; Iqbal and Webster, 1973a, b; Shearer and Webster, 1985).

In 1975, a milestone in the studies of the aquatic Hyphomycetes in the UK was reached with the publication of the "Guide to aquatic Hyphomycetes" (Ingold, 1975), which is a very useful book containing a compilation of several species descriptions with excellent drawings. In Germany, Kegel (1906) studied the aquatic Hyphomycetes and described *Varicosporium elodeae*. Nilsson (1958, 1962b) described a great number of species in Sweden and has also published a volume (Nilsson, 1964) containing important ecological information of the Scandinavian Hyphomycetes. In Austria, Regelsberger et al. (1987) studied species diversity of aquatic Hyphomycetes in four streams and Eggenschwiller and Bärlocher (1983) conducted a survey in 16 streams in France, Germany and Switzerland.

In Spain (Pais Vasco) Descals et al. (1978) described many aquatic Hyphomycetes species and Chauvet et al. (1997) verified the influence of eucalyptus on the sporulation activity and the diversity of the aquatic Hyphomycetes during the decomposition process of submerged leaf litter in Spanish streams. A survey of the aquatic Hyphomycetes was also carried out in the Gredos and Bejar mountains (Descals et al. 1995a, b) and the Catalan Pyrenees (Descals, 1997).

Bärlocher et al. (1995) isolated aquatic Hyphomycetes from alder and eucalyptus leaves in stream in Portugal. Later, Bärlocher and Graça (2002) characterized the fungal communities of eight Portuguese streams, comparing native deciduous forest with stands of *Eucalyptus*.

Chauvet (1991) and Fabre (1998a, b) studied the distribution of aquatic Hyphomycetes in rivers of southwestern France (River Tech, Adour and Nive), with emphasis on spatial and temporal distributions, conidial concentration, species richness and diversity.

The spatial and temporal differences among species (Fabre, 1998b) and the relationships between spatial and temporal dynamics of the communities (Fabre, 1998c) were also investigated. In the same country, Mothe-Jean-Louis (1997) reported 65 species of aquatic Hyphomycetes during a 5 year study of River Osse.

In the Eastern Europe, Dudka (1974) listed 89 species of aquatic Hyphomycetes for Ukraine and reported 196 species in the book "Aquatic Fungi Imperfecti of USSR" (Dudka, 1985). Several studies of the mycoflora of running waters were conducted in Poland in different types of springs (Czeczuga et al. 1989), in the lowlands of the River Biebrza (Czeczuga et al. 1990a), in the River Czarna Hancza and River Marycha (Czeczuga et al. 1990b), and also in lotic and lentic systems (Czeczuga, 1991a, b, c), including the lake Sniardwy and its neighbouring lakes (Czeczuga, 1991d).

In North America, the studies of the aquatic Hyphomycetes started in the 1950's, with Ranzoni's work on Californian streams (Ranzoni, 1953). Petersen (1962, 1963a, b) carried out studies in the eastern United States and described several aquatic Hyphomycetes. Scott and Umphlett (1963) described fungi isolated from Virginia and Dyko (1976) found aquatic Hyphomycetes in humicolous ecosystems, collecting leaf litter samples in several sites of the north American states of Georgia, Kentucky, North Carolina and Tennessee. Anastasiou (1964), Goos (1970, 1978) and Ranzoni (1979) also described aquatic Hyphomycetes from the Hawaiian islands.

Several new species of Hyphomycetes from Canadian streams (Bärlocher, 1987), especially from the Catamaran Brook in New Brunswick, were taxonomically described by Marvanová and Bärlocher (1988, 1989). Colonization dynamics, seasonal patterns and logging effects on aquatic Hyphomycetes were also investigated, verifying the importance of the immersion time of substrates and season on the colonization dynamics (Garnett and Giberson, 2000).

Bärlocher (2000) conducted a 5 years study, verifying that significant correlations between the 1st and the 4th years proved weak or non-significant in subsequent periods. During this study 87 taxa of aquatic Hyphomycetes were found making Catamaran Brook one of the richest streams in number of species currently known.

Aquatic Hyphomycetes were also observed in Nigeria (Ingold, 1956, 1959b; Alasoadura, 1968a, b, c), southern Rhodesia (Ingold, 1960), Uganda (Ingold, 1958a), Sierra Leoa (Singh, 1976; Singh and Musa, 1977), Ethiopia (Bhat and Chien, 1990) and South Africa (Greathead, 1961, Webster et al. 1994).

Asian scientists also showed interest in the study of aquatic Hyphomycetes. Several surveys of aquatic Hyphomycetes were conducted in Japan (Tubaki, 1957; Ando and Tubaki, 1984a, b, c; Ando and Kawamoto, 1986a, b; Ando, 1992). Young-Nian and Liang (1988) reported several new records for the Sichuan Province of China. Tubaki et al. (1983) has also investigated the aquatic Hyphomycetes in Thailand. In India, representative surveys of aquatic Hyphomycetes were carried out by Subramanian and Bhat (1981), Sridhar and Kaveriappa (1988, 1989) and Rajashekar and Kaveriappa (1993). Sridhar et al. (1992) published an excellent compilation of the Indian aquatic Hyphomycetes species. Freshwater Hyphomycetes were studied on decaying debris in Pakistan (Iqbal et al. 1979), where the influence of periodicity was later verified (Iqbal, 1992) and isolation methods reported (Iqbal, 1993; 1994).

In human modified environments, seventeen species of Aquatic Hyphomycetes were found as endophytes in the roots of riparian trees near a brick wall of a canal in Lahore. These results contributed to extend the concepts of aquatic Hyphomycetes distribution to environments subjected to anthropic influence (Iqbal et al. 1995).

From the Pacific Islands, Malaysia has the greatest number of publications dealing with the aquatic Hyphomycetes (Nawawi, 1975, 1985a, b, c, d, e, 1987a, b; Nawawi and Kuthubutheen, 1987a, b, 1988a, b, c, d, 1989a, b, 1990). There are also several reports from Australia (Cowling and Waid, 1963; Shaw, 1972; Shaw and Sutton, 1985; Goh and Hyde, 1996b, c, d), Papua-New Guinea and Solomon Island (Tubaki, 1965; Matsushima, 1971a, b; Shaw and Sutton, 1985) and New Zealand (Aimer and Segedin, 1985).

One of the first studies on aquatic Hyphomycetes in the tropics was carried out in Central America (Jamaica) by Hudson and Ingold (1960), followed by Hudson (1961). In the same continent the studies were carried out in Cuba (Marvanová and Marvan, 1969), in Puerto Rico (Padgett, 1976; Betancourt and Caballero, 1983; Betancourt and Garcia, 1987; Betancourt and Justiniano, 1989) and in the Dominican Republic (Betancourt et al. 1986). The records of aquatic Hyphomycetes in South America are scarce and come mainly from Brazil, Argentina, Ecuador, Peru, Chile and Venezuela (Schoenlein-Crusius and Grandi, 2003).

In Argentina, most of the surveys were performed in the subtropical region of the River Santiago in Buenos Aires (Arrambarri et al. 1987a, b, c) and others carried out at different localities involving the sampling of leaf litter or woody debris, from which many uncharacteristic aquatic fungi were isolated (Arambarri and Spinedi, 1984; Godeas, 1985; Cabello et al. 1990; Cazau et al. 1990, 1993; Arrambarri et al. 1997; Cabello et al. 1998; Arambarri et al. 2001). In Chile, samples of wood, leaf litter and insects were collected from several streams and lakes in the temperate region of Osorno (Burgos and Riffart, 1982). Fourteen taxa of Ingoldian fungi were observed, nine were identified to species level and had their taxonomic features documented through spore drawings. In Venezuela, aquatic Hyphomycetes were isolated from submerged leaf litter, branches and petioles in streams, pools and waterfalls in the surroundings of Caracas (Nilsson, 1962a).

Matsushima (1993, 1995) greatly contributed to the knowledge of Hyphomycetes in South America. In his studies of decaying palm tree material, he collected petioles, leaf litter and fruits, and also soil samples, in the vicinity of some rivers in the equatorial Amazon region of Ecuador and Peru. Many fungi with star-like conidia, considered aquatic, were isolated in these studies but it is not clear if they were obtained from really submerged substrates. In Peru, 30 taxa were described. Among them, eight known species considered aquatic according to Ingold (1975) or Marvanová (1997) were found. These Hyphomycetes were fully described and documented through excellent photographs and drawings.

To date, almost all sites where aquatic Hyphomycetes were studied in Brazil are situated in the state of São Paulo. These studies involved collections in streams, medium size rivers, waterfalls, lakes and reservoirs under the subtropical climatic conditions of the Southeast Brazil. The Brazilian studies of Hyphomycetes in freshwater started at the end of the 1980's with the investigation of fungal succession on submerged leaves (Schoenlein-Crusius and Milanez, 1989; Schoenlein-Crusius et al. 1990).

These initial observations of tetraradiate conidia were followed by species descriptions and ecological remarks on their occurrence (Schoenlein-Crusius et al. 1992, 1999). Although the number of studies carried out and taxa found in Brazil are greater than in any other South American country, novel species are yet to be found. A comparison of the results reported from the different South American countries showed that some species such as *Anguillospora longissima* (Sacc. and Sydow) Ingold, *Articulispora tetracladia* Ingold, *Clavariopsis aquatica* de Wildeman, *Lunulospora curvula* Ingold and *Triscelophorus monosporus* Ingold are cosmopolitan in this continent and have also been frequently reported from most subtropical and tropical regions thus far studied.

Studies of the aquatic Hyphomycetes and zoosporic fungi associated with submerged leaf litter in the Brazilian Cerrado were intensified through surveys carried out in the River Monjolinho, in the municipality of São Carlos in the State of São Paulo (Malosso, 1995). Afterwards, the fungal diversity, biomass production and ergosterol content of submerged leaves were compared between the rivers Monjolinho and Jacaré-Guaçú (lotic systems) and those with the reservoir "Represa do Guarapiranga" (lentic system) in the city of São Paulo. The results of these comparisons did not show a clear correlation between ergosterol/biomass content and fungal diversity. However, the aquatic Hyphomycetes were more frequent in the rivers than in the reservoir regardless of the eutrophication level of the environment (Malosso, 1999).

The Amazonian region (in Brazil) is where most of the rivers, streams, and ponds are located. However, only one record of *Helicosporium* sp. has been reported (Silva, 2002) and the aquatic Hyphomycetes communities of that ecosystem is practically unknown. In Brazil, as probably in other countries of South America, the interest of researchers and students in the taxonomy of aquatic Hyphomycetes arose from the observation of these species in ecological studies of leaf litter decomposition and fungal succession. However, the scarcity of taxonomical keys that include the tropical species and the difficulty in obtaining cultures on solid media have discouraged further initiatives. The quality of the current taxonomical descriptions and drawings must also be improved.

Marvanová (1997) published an identification key to tropical and subtropical species, based on Indian and Malaysian material. The publication of identification keys and the improvement of the isolation techniques applied to the aquatic Hyphomycetes in tropical waters may stimulate the interest of the taxonomists to the systematics and ecology of this fungal group elsewhere.

In conclusion, the use of a combination of techniques is recommended that allows for the isolation of increasing numbers of fungal taxa from the aquatic environment in future studies. The adaptation of field collection and isolation procedures to obtain representatives of all taxonomic groups is imperative. It is also evident that there is a need to integrate the results of diversity surveys and ecology studies carried out by taxonomists and ecologists from different research groups as this kind of association may give faster and better founded answers to the questions raised by the need to protect water sources, problems of pollution, loss of species and conservation policies.

Methods and Techniques to Study Aquatic Hyphomycetes

Though aquatic Hyphomycetes have been known to scientists for over a 100 years, our knowledge of this group is as yet incomplete. To fill in the gaps it is necessary to learn more about their biology, ecological functions and mainly, the phylogenetic relationships of the group. The need of agreement over the taxonomic position of the aquatic Hyphomycetes is imperative. To achieve such goals, the isolation of species in pure cultures is a key step.

The hydrodynamic spores of the aquatic Hyphomycetes offer an advantage in the studies aiming at species isolation, as they are considerably lighter and can be easily retained in traps. In general, the foam formed by the movement of the water against natural barriers like stones, twigs and logs, especially in lotic systems, constitutes a natural trap for the specially shaped conidia of the Hyphomycetes. Therefore, the direct observation of microscope slides containing foam drops allows a rapid overview of the diversity of the aquatic Hyphomycetes. These fungi can also be obtained by incubating in sterile distilled water, for a few days, samples of freshly collected submerged leaf litter. During incubation, one should search for the conidial development at the veins and margins of the leaf substrates.

Marvanová (1997) found that the spreading of foam or conidial suspensions onto the surface of object glasses covered with a thin layer of culture media like malt extract agar or soil extract agar with Bengal Rose was effective. This setting was then incubated for 24h at 15-18°C and followed by new isolations of the germinating spores onto Petri dishes containing sterile solid media. To follow the changes of the fungal community during the decomposition processes in aquatic environments, the use of leaf and twig packs (litter bags) may be effective. In these cases the sites and time since submersion can be closely monitored.

To isolate the aquatic Hyphomycetes, a number of leaf disks or fragments are usually cut off, incubated in well-aerated water and directly observed under the microscope to find the fungal structures. Leaf mapping is also an interesting technique, which consists of cutting the collected leaves into squares of pre-established sizes (e.g. 2 × 2 or 6 × 6 mm) and incubating a number of standardised squares in diluted salt solution or sterile distilled water. This procedure enhances sporulation, which can then be used to estimate the occurrence of different species of fungi on the leaves and to draw leaf colonisation maps. This technique has been discussed in detail by Shearer and Lane (1983) and Bärlocher (1992).

Although the isolation methods seem simple, obtaining pure cultures of aquatic Hyphomycetes has been a hard and time-consuming work. The isolation process requires practice in handling fine instruments, a steady hand and a good microscope.

Following the recommendations of Ingold (1975) to isolate a great diversity of aquatic Hyphomycetes, several types of leaves have been directly collected from small, well aerated streams (Schoenlein-Crusius and Milanez, 1990) or placed in nylon-net litter bags and submerged in several points of lakes (Schoenlein-Crusius and Milanez, 1989; Schoenlein-Crusius et al. 1990) and streams in the Brazilian Atlantic rainforest (Schoenlein-Crusius et al. 1992). The leaves or litter bags were collected monthly, brought to the laboratory, washed under tap water and incubated in Petri dishes containing sterile distilled water for 3 or 5 days at room temperature (22-25°C). These procedures were effective for the observation of conidia from different species and to follow the development of conidia of some species.

In studies of fungal succession (Schoenlein-Crusius and Milanez, 1989; Schoenlein-Crusius et al. 1990, Schoenlein-Crusius and Milanez, 1998), where terrestrial fungi were also isolated by the method of leaf disk washing (Pugh et al. 1972), the growth of aquatic Hyphomycetes was observed on washed leaf disks incubated in the same way as the leaf fragments described above. During these studies, obtaining pure cultures was extremely different, not only because of the massive presence of bacteria, yeasts and protozoa but also due to the fast growth of several zoosporic fungi on the same culture media used for the germination of the conidia of the aquatic Hyphomycetes.

It has been very common to detect good growth of *Achlya*, *Saprolegnia* and *Gonapodyia* on the surface of submerged leaf litter samples from the rainforest of Southeast Brazil. In addition the growth of some aquatic Hyphomycetes like *Tripospermum* and *Triscelophorus* has also been observed on baits like cellophane, corn straw, etc., originally used to isolate chytrids. In these cases the use of antibiotics or selective media has proved of little use. The incubation of the spores for germination at low temperatures, generally ranging from 5 to 10°C, resulted in better isolations of aquatic Hyphomycetes due to slower growth rates of contaminants at lower temperatures.

The description of several methods and interesting tips to isolate aquatic Hyphomycetes are available from the works of Marvanová (1997), Webster and Descals (1981), Nawawi (1985b), Dix and Webster (1995) and Ingold (1975).

Mycology is a relatively young science compared to chemistry or mathematics and has, until recently, relied on the recovery and analysis of pure cultures although the limitations of culture methods are widely known. In parallel to the studies involving isolation and taxonomical description of species of aquatic Hyphomycetes, several studies (Iversen, 1973; Bärlocher and Kendrick, 1974; Suberkropp et al. 1976; Findlay and Arsuffi, 1989) attempted to quantify the fungal biomass associated with decomposition of substrates.

It is clear that the culture-based techniques will continue to be used in fungal ecology; however, culture-free techniques involving the analysis of specific substances of microbial origin has become increasingly popular (Newell, 1994; Bååth, 2001; Malosso et al. 2004).

Starting with the work of Seitz et al. (1977) when ergosterol was used to quantify fungal biomass in stored grains, this substance has since then been frequently used as a specific fungal biomarker (Seitz et al. 1979; Lee et al. 1980; Matcham et al. 1985; Gessner et al. 1991; Gardner et al. 1993). This method is used to estimate live fungal biomass because it is based on the ergosterol content of labile cell and vacuole membranes and not of cell walls (Newell, 1992; Wallander et al. 1997). It has been said that a comparison of ergosterol with cell wall components such as glucosamine and mannose can be used to quantify the relative amounts of live and dead fungal mass (Newell, 1992, 1994; Newell et al. 1995).

Ergosterol assays have been proposed as the most promising method for quantifying mycelial mass (Newell, 1992; Gessner and Schmitt, 1996). This is primarily because ergosterol is highly specific to eumycotic fungi, is quickly oxidised in dead fungal mass, and the ergosterol-to-biomass conversion factor remains relatively constant between plant litter samples (Gessner and Newell, 1997).

Liquid phase ergosterol extraction followed by high performance liquid chromatography (HPLC) analysis has been used to study fungal dynamics in natural environments (Seitz et al. 1979; Newell and Fallon, 1991; Bååth, 2001) and is based on the extraction of ergosterol into an organic solvent followed by injection of the extracts into an HPLC equipped with a detector set at wavelength 282 nm. This wavelength maxima is due to the conjugated double bond found on the second ring of the ergosterol backbone allowing its specific detection (Seitz et al. 1979; Newell, 1992).

The quantification of ergosterol as a biomass measure of the aquatic Hyphomycetes has been extensively employed in the last decade (Gessner and Schwoerbel, 1991) and there have been many studies of fungal biomass on submerged decomposing leaves in temperate regions (Suberkropp et al. 1993; Gessner and Chauvet, 1994; Baldy et al. 1995; Maamri et al. 1998; Robinson, 2001; Pascoal and Cassio, 2004).

The estimates of fungal growth rates and dynamics through the incorporation of an isotope into the ergosterol molecule have also become common in recent years (Newell, 1993; Newell et al. 1995; Suberkropp, 1995; Suberkropp and Weyers, 1996; Bååth, 2001; Malosso et al. 2004).

Gessner et al. (2003) have written a comprehensive paper on the methods used to detect and quantify aquatic Hyphomycetes in streams, including the analysis of ergosterol, ATP and chitin for biomass measurements. However, there is a lack of such studies in tropical regions with the work by Mathuriau and Chauvet (2002) being the only internationally published paper dealing with ergosterol analysis in tropical aquatic Hyphomycetes to date. In Brazil, the first study to use ergosterol as an index of aquatic Hyphomycetes biomass was carried out by Malosso (1999), followed by other, presently unpublished, M.Sc. dissertations.

According to Gessner et al. (2003), there have been three main approaches commonly used to study the structure of aquatic Hyphomycetes communities, all three based on the observation of conidia. However, there are a number of community structure analyses techniques commonly used in soils that could be applied to aquatic systems, for example, the analysis of the membrane components phospholipid fatty acids (PLFA).

It is known that phospholipids are found in the membranes of all living cells and are actively synthesized during microbial growth and not found as storage products (Tunlid and White, 1992). Phospholipids can be extracted from samples in a one phase chloroform:methanol:buffer (Bligh and Dyer, 1959) and analysed, after transesterification and purification, by gas chromatography. The most general method for PLFA interpretation is to use the entire profile of the fatty acids as fingerprint as this permits a detailed assessment of changes at the community level (Bossio and Scow, 1998). With the appropriate statistics, similarities among communities or the impact of environmental conditions on a community can be quantified and tested (Bossio and Scow, 1998).

In taxonomy, the PLFA profile has been regarded as a useful tool in the characterization of fungal species (Sumner and Morgan, 1969; Viljoen et al. 1986; Hammonds and Smith, 1986; Sancholle and Dalpé, 1993). Bentivenga and Morton (1994) and Graham et al. (1995) used the fatty acid approach to overcome the difficulty of studying the systematics of obligatory symbiotic glomalean endomycorrhizal fungi. They suggest PLFA as a reliable taxonomic marker for fungi and is largely independent of culture conditions, the plant host and prolonged storage.

Stahl and Klug (1996) presented a broader study of fungal fatty acid composition including members of the three main groups (Zygo-, Asco- and Basidiomycota) and the oomycetes and mitosporic fungi.

Multivariate analysis of their data showed that all fungal species analysed presented distinct profiles either by producing different fatty acids or changing the relative concentration of them (or both). From the type of information obtained with the PLFA analysis, the diversity of different systems and distribution of certain groups of organisms may be addressed by taking into account the signature fatty acids present in the profiles (Zelles et al. 1995; White et al. 1996).

Despite the fact that PLFA have been largely (and successfully) employed in the fungal studies in soils, there is no record in the literature of this technique being used to analyse aquatic Hyphomycetes communities in aquatic systems or submerged leaf litter.

Considerable attention has also been focused on the genetic aspects of fungal populations in an attempt to improve our knowledge of fungal taxa (Bennett, 1985; Bruns et al. 1991) and their phylogenetic relationships (Suh and Nakase, 1995; Bills et al. 1999; Pryor and Gilbertson, 2000).

Nucleic acid sequencing methods have developed so quickly that comparative sequencing of homologous genes has rapidly become a standard technique in molecular systematics and phylogeny and is now considered one of the most powerful methods applied in microbial systematics (Ludwig, 1991).

It has been suggested that differences in ribosomal RNA (rRNA) sequences can be used to define, more precisely, fungal species (Valente et al. 1999). The ribosomal DNA/RNA approach has also been used many times to explore the diversity and to analyse the structure of microbial communities (Muyzer and Smalla, 1998; Pennanen et al. 2001) and has been shown to be faster and more efficient in the detection and quantification of fungi both in medical (Kappe et al. 1998; Posteraro et al. 2000; Turin et al. 2000) and ecological (Kowalchuk et al. 1997; Ma et al. 2000) research.

In ecological studies, fungal communities can be investigated by rDNA sequence analysis (White et al. 1990, Kowalchuk et al. 1997; Kowalchuk, 1999), where the possibility of selective amplification of fungal 18S rDNA from total DNA (Smit et al. 1999), thus avoiding cultivation, is seen as an advantage. However, the PCR approach is not without problems because of the possibility of incomplete cell lysis, incomplete extraction of DNA from samples, and PCR biases (Muyzer and Smalla, 1998).

Where the objective of the community analysis is to identify community members or the presence of particular genes, cloning techniques allow separation of each copy of the target gene from a mixed population into the cloning vector. The cloning technique can detect both known and unknown DNA sequences in a sample, but can be very time consuming and in diverse microbial communities it may be necessary to analyse several thousand clones from a single sample to obtain a characteristic inventory of the community (Clement et al. 1998). However, it is not always necessary or desirable to identify individual community members; patterns derived from the diversity of specific genes can be used to rapidly characterize environmental samples.

One of these pattern analysis methods is the denaturing gradient gel electrophoresis (DGGE), introduced by Muyzer et al. (1993). In DGGE, PCR-amplified ribosomal DNA fragments of the same length but with different sequences are separated on polyacrylamide gels. This separation is based on the decreased electrophoretic mobility of a partially melted double-stranded DNA molecule over a linear gradient of DNA denaturant, consisting of a mixture of urea and formamide (Muyzer and Smalla, 1998).

A number of studies have used DGGE to analyse fungal communities. Kowalchuk et al. (1997) used DGGE to study fungal infection in a sand-stabilizing species of grass in coastal sand dune areas in the Netherlands and Smit et al. (1999) to identify fungal community members by comparing the relative mobility of bands of community DNA fingerprints with DNA sequences of separate clones obtained from wheat rhizospheres.

Later Vainio and Hantula (2000) have shown good separation using DGGE of DNA fragments of 390 bp from wood-colonizing fungi and reported that the DGGE analysis of environmental samples collected from Norway spruce stumps was generally in accordance with the investigation of mycelial cultures isolated from the same decomposing substrate. Based on these results, Pennanen et al. (2001) have shown that DGGE can be used to differentiate samples from environments varying in fungal diversity. The method was shown to be reproducible and can be extended to samples from extremely heterogeneous environments. Fungal communities causing biodeterioration of art objects that may contain high salt concentrations, heavy metal compounds and mineral pigments have also been successfully assessed using DGGE (Möhlenhoff et al. 2001).

However, there is no technique without a drawback. In the case of DGGE it must be remembered, when interpreting the fingerprints, that bands at the same position in the denaturant gel have the same melting behaviour, but not necessarily the same sequence and only sequencing of the bands can prove sequence identity (Muyzer and Smalla, 1998).

DGGE also presents two other important limitations: (i) it is generally considered possible to separate only relatively small fragments, upto 500 bp (Myers et al. 1985) which limits the amount of sequence information available for subsequent phylogenetic inferences when bands are excised and sequenced and (ii) DGGE technique tends to display only the rDNA fragments of the predominant taxa in the community due to PCR bias. Despite these concerns DGGE is considered a reliable method for investigating the temporal and spatial distribution of fungal populations.

Molecular methods have already been used in studies of aquatic Hyphomycetes (Baschien et al. 2001; McArthur et al. 2001; Marvanová et al. 2002; Bauer et al. 2003; Laitung et al. 2004), including a DGGE separation of fragments of the 18S rDNA (Nikolcheva et al. 2003). These studies show some positive results but also give evidence of the necessity of optimisation of the techniques and primers for the aquatic Hyphomycetes. In addition, as for the PLFA, there is no study reported in the literature of the aquatic Hyphomycetes using molecular methods in tropical regions.

Although many studies on the aquatic Hyphomycetes have been conducted in several regions of the world, the knowledge about the diversity of this relevant fungal group seems far from complete, especially in the tropics.

Beside the interest in enhancing the knowledge about the geographical distribution and diversity of aquatic Hyphomycetes in conserved, pristine or well preserved areas, there is an urgent necessity to evaluate the effects of pollution and climate changes (among many other environmental impacts) on the diversity and metabolic features of the species, in order to find may be new bioindicators or species with physiological properties to improve the recovery of impacted or destroyed ecosystems.

To achieve these aims much large numbers of trained specialists are needed who can combine the use of traditional isolation and identification techniques with molecular procedures and ecological approaches.

References

Aimer, R. D. and Segedin, B. P. (1985) Some aquatic hyphomycetes from New Zealand streams. New Zealand Journal of Botany 23: 273–299.

Alasoadura, S. O. (1968a) *Flabellospora crassa* n. gen. n. sp. an aquatic hyphomycete from Nigéria. Nova Hedwigia 15: 415–418.

Alasoadura, S. O. (1968b) *Flabellospora verticillata*, a new species of aquatic hyphomycetes from Nigéria. Transactions of the Mycological Society of Japan 15: 419–421.

Alasoadura, S. O. (1968 c) Some aquatic hyphomycetes from Nigéria. Transactions of the British Mycological Society 51: 535–540.

Alexopoulos, C. J., Mims, C. W. and Blackwell, M. (1996) Introductory Mycology, 4th Edition John Wiley & Sons Inc., New York.

Anastasiou, C. J. (1964) Some aquatic Fungi Imperfecti from Hawaii. Pacific Science 18: 202–206.

Ando, K. (1992) A study of terrestrial aquatic hyphomycetes. Transactions of the Mycological Society of Japan 33: 415–425.

Ando, K. and Kawamoto, I. (1968a) *Arborispora*, a new genus of Hyphomycetes. Transactions of the Mycological Society of Japan 27: 119–128.

Ando, K. and Kawamoto, I. (1986b) Material for the fungus flora of Japan. Transactions of the Mycological Society of Japan 27: 321–326.

Ando, K. and Tubaki, K. (1984a) Some undescribed hyphomycetes in the rain drops from intact leaf surface. Transactions of the Mycological Society of Japan 25: 21–37.

Ando, K. and Tubaki, K. (1984b) Some underscribed hyphomycetes in rainwater draining from intact trees. Transactions of the Mycological Society of Japan 25: 39–47.

Ando, K. and Tubaki, K. (1984 c) Materials for the fungus flora of Japan. Transactions of the Mycological Society of Japan 25: 395–398.

Arambarri, A. M. and Spinedi, H. A. (1984) Micoflora de la hojarasca de *Nothofagus pumilio*. III. Darwiniana 25(1–4): 321–330.

Arambarri, A. M., Cabello, M. N. and Cazau, M. C. (1997) *Gyrothrix flagelliramosa* sp. nov., a new hyphomycete from Argentina. Mycological Research 101 (12): 1529–1530.

Arambarri, A. M., Cabello, M. N. and Cazau, M. C. (2001) *Dictyosporium triramosum*, a new Hyphomycete from Argentina. Mycotaxon 78: 185–189.

Arambarri, A. M., Cabello, M. N. and Mengascini, A. (1987a) Estudio sistematico de los Hyphomycetes del Rio Santiago (prov. Buenos Aires, Argentina). Darwiniana 28 (1–4): 293–301.

Arambarri, A. M., Cabello, M. N. and Mengascini, A. (1987b) New Hyphomycetes from Santiago River (Buenos Aires Province, Argentina). Mycotaxon 29: 29–35.

Arambarri, A. M., Cabello, M. N. and Mengascini, A. (1987 c) New Hyphomycetes from Santiago River II. (Buenos Aires Province, Argentina). Mycotaxon 30: 263–267.

Bååth, E. (2001) Estimation of fungal growth rates in soil using ^{14}C-acetate incorporation into ergosterol. Soil Biology and Biochemistry 33: 2011–2018.

Baldy, V., Gessner, M. O. and Chauvet, E. (1995). Bacteria, fungi and breakdown of leaf litter in a large river. Oikos 74: 93–102.

Bärlocher, F. (1987). Aquatic hyphomycete spora in 10 streams of New Brunswick and Nova Scotia. Canadian Journal of Botany 65: 76–79.

Bärlocher, F. (1992) Research on aquatic Hyphomycetes: historical background and overview. In: Bärlocher, F. (ed), The ecology of aquatic Hyphomycetes. Springer Verlag, Berlin, pp. 1–15.

Bärlocher, F. (2000) Water-borne conidia of aquatic hyphomycetes: seasonal and yearly patterns in Catamaran Brook, New Brunswick, Canada. Canadian Journal of Botany 78: 157–167.

Bärlocher, F. and Graça, M. A. S. (2002) Exotic riparian vegetation lowers fungal diversity but not decomposition in Protuguese streams. Freshwater Biology 47: 1123–1135.

Bärlocher, F. and Kendrick, B. (1974) Dynamics of the fungal population on leaves in a stream. Journal of Ecology 62: 761–791.

Bärlocher, F., Canhoto, C. and Graça, M. A. S. (1995) Fungal colonization of alder and eucalyptus leaves in a Portuguese stream. Archiv für Hydrobiologie 133: 457–470.

Baschien, C., Manz, W., Neu, T. R. and Szwezyki, U. (2001) Fluorescence *in situ* hybridization of freshwater fungi. International Review of Hydrobiology 86: 371–381.

Bauer, R., Begerow, D., Oberwinkler, F. and Marvanova, L. (2003) *Classicula*: the teleomorph of *Naiadella fluitans*. Mycologia 95: 756–764.

Bennett, J. W. (1985) Prospects for a molecular mycology. In: Bennett, J. W. and Lasure, L. L. (eds.) Gene Manipulations in Fungi. Orlando, Academic Press, pp. 515–527.

Bentivenga, S. P. and Morton, J. B. (1994) Stability and heritability of fatty acid methyl ester profiles of glomalean endomycorrhizal fungi. Mycological Research 98: 1419–1426.

Betancourt, C. and Caballero, M. (1983) Aquatic hyphomycetes (Deuteromycotina) from Los Chorros, Utuado, Puerto Rioco. Caribbean Journal of Science 19: 41–42.

Betancourt, C. and Garcia, J. (1987). Aquatic hyphomycetes of Dona Juana stream in the Toro Negro state forest, Villalba, Puerto Rico. Caribbean Journal of Science 23 (2): 278–285.

Betancourt, C. and Justiniano, J. (1989) Hongos ingoldianos presentes en el arroyo de Oro, Mayaguez, Puerto Rico. International Journal of Mycology and Lichenology 4: 1–8.

Betancourt, C., Cruz, J. and Garcia, J. (1987) Los hifomicetos acuaticos de la quebrada Dona Juana em el bosque estatal de Toro Negro, Villalba, Puerto Rico. Caribbean Journal of Science 23: 278–284.

Betancourt, C., Cruz, J.; Garcia, J. and Galarza, L. (1986) Estudio preliminar de los hifomicetos acuaticos (deuteromicotina) de la Republica Dominicana. Caribbean Journal of Science 22: 49–51.

Bhat, D. J. and Chien, C-Y. (1990) Water-borne hyphomycetes found in Ethiopia. Transactions of the Mycological Society of Japan, *Tokyo* 31: 147–157.

Bills, G. F.; Platas, G.; Pelaez, F. and Masurekar, P. (1999) Reclassification of a pneumocandin-producing anamorph, *Glarea lozoyensis* gen. et sp. nov., previously identified as *Zalerion arboricola*. Mycological Research 103: 179–192.

Bligh, E. G. and Dyer, W. J. (1959) A rapid method of total lipid extraction and purification. Canadian Journal of Biochemistry and Physiology 37: 911–917.

Bossio, D. A. and Scow, K. M. (1998) Impacts of carbon and flooding on soil microbial communities: Phospholipid fatty acid profiles and substrate utilization patterns. Microbial Ecology 35: 265–278.

Burgos, J. E. and Riffart, R. G. (1982) Hongos saprobios en Chile: ambiente limnico. Noticiario Mensual del Museo Nacional de Historia Natural de Chile (306–307): 5–8.

Bruns, T. D., White, T. J. and Taylor, J. W. (1991) Fungal molecular systematics. Annual Review of Ecology and Systematics 22: 525–564.

Cabello, M. N., Arambarri, A. M. and Cazau, M. C. (1998) *Minimidochium parvum*, a new species of hyphomycete from Argentina. Mycological Research 102 (3): 383–384.

Cabello, M. N., Cazau, M. C. and Arambarri, A. M. (1990) New Hyphomycetes from Santiago River. III. (Buenos Aires Province, Argentina). Mycotaxon 38: 15–19.

Cazau, M. C., Arambarri, A. M. and Cabello, M. N. (1990) New Hyphomycetes from Santiago River. IV (Buenos Aires Province, Argentina). Mycotaxon 38: 21–25.

Cazau, M. C., Arambarri, A. M. and Cabello, M. N. (1993) New Hyphomycetes from Santiago River. VI. (Buenos Aires Province, Argentina). Mycotaxon 46: 235–240.

Chamier, A. C. (1985) Cell-wall degrading enzymes of aquatic hyphomycetes: a review. Botanical Journal of the Linnean Society 91: 67–81.

Chandrashekar, K. R. and Kaveriappa, K. M. (1991) Production of extra-cellular cellulase by *Lunulospora curvula* and *Flagellospora penicillioides*. Folia Microbiologica 33: 55–58.

Chandrashekar, K. R., Sridhar, K. R. and Kaveriappa, K. M. (1990) Periodicity of water-borne hyphomycetes in two streams of western Ghat forests (India). Acta for Hydrochemistry and Hydrobiology 18 (2): 187–204,

Chauvet, E. (1991). Aquatic hyphomycete distribution in South Western France. Journal of Biogeography 18: 699–706.

Chauvet, E., Fabre, E., Elósegui, A. and Pozo, J. (1997) The impact of eucalyptus on the leaf-associated aquatic hyphomycetes in Spanish streams. Canadian Journal of Botany 75: 880–887.

Clement, B. G., Kehl, L. E., De Bord, K. L. and Kitts, C. L. (1998) Terminal restriction fragment patterns (TRFPs), a rapid, PCR-based method for the comparison of complex bacterial communities. Journal of Microbiological Methods 31: 135–142.

Cowling, S. W. and Waid, J. S. (1963) Aquatic hyphomycetes in Australia. Australian Journal of Science 26: 122–123.

Czeczuga, B. (1991a) Studies of aquatic fungi. XXIII. The mycoflora of lake Wigry and seven adjacent lakes. Archiv fur Hydrobiologie 120 (4): 495–510.

Czeczuga, B. (1991b) Studies of aquatic fungi. part 16: Aquatic fungi of the river Pisa and its tributary, river Skroda. Acta for Hydrochemistry and Hydrobiology 19 (1): 57–65.

Czeczuga, B. (1991c) Studies of aquatic fungi. XXII Mycoflora of the river Wegorapa and its tributary, Goldapa-Jarka. Acta for Hydrochemistry and Hydrobiology 19 (5): 517–528.
Czeczuga, B. (1991d) Studies of aquatic fungi. XVIII Aquatic fungi in lake Sniardwy and eighteen neighbouring lakes. Internationale Internationale Revue der Gesamten Hydrobiologie und Hydrographie 76 (1): 121–135.
Czeczuga, B., Brozozowska, K. and Woronowicz, L. (1990a) Studies of aquatic fungi. XIII. Mycoflora of the river Czarna Hancza and its tributaries, river Marycha. Internationale Revue der Gesamten Hydrobiologie und Hydrographie 75 (2): 245–255.
Czeczuga, B., Woronowicz, L. and Brozozowska, K. (1989) Studies on aquatic fungi. IX. Mycoflora of different types of springs. Acta Hydrobiologica 31 (3/4): 273–283.
Czeczuga, B., Woronowicz, L. and Brzosowska, K. (1990b) Studies of aquatic fungi. XII. Aquatic fungi of the low land river Biebrza. Acta Mycologica 26 (1): 77–83.
De Wildeman, E. (1893) Notes mycologiques. Fasc. II. Annales de la Société Belge de Microscopie 17: 35–68,
De Wildeman, E. (1894) Notes mycologiques. Fasc. III. Annales de la Société Belge de Microscopie 18: 135–161.
De Wildeman, E. (1895) Notes mycologiques. Fasc. II. Annales de la Société Belge de Microscopie 19: 193–206.
Descals, E.; Sanders, P. F. and Ugalde, U. O. (1978) Hifomicetos ingoldianos del Pais Vasco. Munibe 29: 237–260.
Descals, E., Peláez, F. and Llorca, L. V. L. (1995a) Fungal spora of streams foam from central Spain. I. Conidia identifiable to species. Nova Hedwigia 60 (3–4): 533–550.
Descals, E. , Peláez, F. and Llorca, L. V. L. (1995b) Fungal spora of streams foam from central Spain. II. Chronology, spore frequency and unknown forms. Nova Hedwigia 60 (3–4): 551–569.
Descals, E. (1997) Ingoldian fungi from the Catalan Pyrenees: pure culture studies. Mycotaxon 63: 431–466.
Dix, N. J. and Webster, J. (1995) Fungal Ecology. Chapman and Hall, London, pp, 549.
Dudka, I. O. (1974) Ukranian aquatic Hyphomycetes. Naukova Dumka, Kiev, USSR, pp, 240.
Dudka, I. O. (1985) Aquatic Fungi Imperfecti of USSR. Mycologia 79: 821–831.
Dyko, B. J. (1976) A preliminary study of aquatic hyphomycetes on leaves in forest and leaf stream leaf litter. Journal of the Tennessee Academy of Science 51: 7–8.
Eggenschwiller, S. and Bärlocher, F. (1983) Aquatic Hyphomycetes in 16 streams in France, Germany and Switzerland. Transactions of the British Mycological Society 81: 371–379.
Fabre, E. (1998a) Aquatic Hyphomycetes in three rivers of southwestern France. I. Spatial and temporal changes in conidial concentration, species richness, and community diversity. Canadian Journal of Botany 76 (1): 99–106.
Fabre, E. (1998b) Aquatic Hyphomycetes in three rivers of southwestern France. II. Spatial and temporal differences between species. Canadian Journal of Botany 76 (1): 107–114.
Fabre, E. (1998c) Aquatic Hyphomycetes in three rivers of southwestern France. III. Relationships between communities spatial and temporal dynamics. Canadian Journal of Botany 76 (1): 115–121.
Findlay, S. E. G. and Arsuffi, T. L. (1989) Microbial growth and detritus transformations during decomposition of leaf litter in a stream. Freshwater Biology 21: 261–270.
Gardner, R. M., Tindall, W. G., Cline, S. M. and Brown, K. L. (1993) Ergosterol determination in activated sludge and its application as a biochemical marker for monitoring fungal biomass. Journal of Microbiological Methods 17: 49–60.
Garnett, H., Bärlocher, F. and Giberson, D. (2000) Aquatic hyphomycetes in Catamaran Brook: colonization dynamics, seasonal patterns, and logging effects. Mycologia 92: 29–41.
Gessner, M. O. and Chauvet, E. (1994) Importance of stream microfungi in controlling breakdown rates of leaf litter. Ecology 75: 1807–1817.
Gessner, M. O. and Newell, S. Y. (1997) Bulk quantitative methods for the examination of eukaryotic organoosmotrophs in plant litter. In: (ed), Hurst, C. J. Manual of Environmental Microbiology. Washington DC, American Society for Microbiology Press, 295–308.
Gessner, M. O. and Schmitt, A. L. (1996) Use of solid–phase extraction to determine ergosterol concentrations in plant tissue colonized by fungi. Applied and Environmental Microbiology 62: 415–419.
Gessner, M. O. and Schwoerbel, J. (1991) Fungal biomass associated with decaying leaf litter in a stream. Oecologia 87: 602–603.
Gessner, M. O., Bärlocher, F. and Chauvet, E., (2003) Qualitative and quantitative analyses of aquatic hyphomycetes in streams. Fungal Diversity Research Series 10: 127–157.

Gessner, M. O., Bauchrowitz, M. A. and Escautier, M. (1991) Extraction and quantification of ergosterol as a measure of fungal biomass in leaf litter. Microbial Ecology 22: 285–291

Godeas, A. M. (1985) Hifomicetes (Deuteromycotina) acuaticos de Tierra del Fuego. Physis, Secc. B, 43 (104): 7–9.

Goh, T. K. (1997) Tropical Freshwater Hyphomycetes. *In* Biodiversity of Tropical Microfungi. (ed), Hyde, K. D. Hong Kong University Press, 189–227.

Goh, T. K., Hyde, K. D. (1996a) Biodiversity of freshwater fungi. Journal of Industrial Microbiology 17: 328–345.

Goh, T. K. and Hyde, K. D. (1996b) *Brachydesmiella anthostomelloidea*, a new species of dematicaeous hyphomycete from Australia. Mycological Research 100: 1364–1366.

Goh, T. K. and Hyde, K. D. (1996c) *Helicoon gigantisporum* sp. Nov. From Australia and an amended key to the genus. Mycological Research 100: 1485–1488.

Goh, T. K. and Hyde, K. D. (1996d) *Cryptophiale multiseptata* sp. nov. On submerged wood in Australia and keys to the genus. Mycological Research 100: 999–1004.

Goos, R. D. (1970) "*In vitro*" sporulation in *Actinospora megalospora*. Transactions of the British Mycological Society 55: 335–337.

Goos, R. D. (1978) Occurrence of *Triscelophorus monosporus* in upland site on Oahu, Hawaii. Mycologia 70: 188–189.

Graham, J. H., Hodge, N. C. and Morton, J. B. (1995) Fatty acid methyl ester profiles for characterization of glomalean fungi and their endomycorrhizae. Applied and Environmental Microbiology 61: 58–64.

Greathead, S. K. (1961) Some aquatic Hyphomycetes in South Africa. Journal of South African Botany 27: 195–228.

Hammonds, P. and Smith, S.N. (1986) Lipid composition of a psychophilic, a mesophilic and a thermophilic *Mucor* species. Transactions of the British Mycological Society 86: 551–560.

Hasija, S. K. and Singhal, P. K. (1991) Degradation of Plant Litter by Aquatic Hyphomycetes. In: Handbook of Applied Mycology: soils and plants. (eds), Arora, D. K., Rai, B., Mukerji, K. G. and Knudsen, G. Marcel Dekker New York, 481–505.

Hawksworth, D. L. (1991) The fungal dimension of biodiversity: magnitude, significance and conservation. Mycological Research 95: 641–655.

Hudson, H. J. (1961). *Heliscus submersus* sp. nov. An aquatic hyphomycete from Jamaica. Transactions of the British Mycological Society 44: 91–94.

Hudson, H. J. and Ingold, C. T. (1960). Aquatic hyphomycetes from Jamaica. Transactions of the British Mycological Society 43: 469–478.

Hyde, K. D. and Hawksworth, D. L. (1997) Measuring and monitoring the biodiversity of fungi. In: Biodiversity of Tropical Microfungi (ed), Hyde, K. D. Hong Kong University Press, Hong Kong, 11–28.

Ingold, C. T. (1942) Aquatic hyphomycetes of decaying alder leaves. Transactions of the British Mycological Society 25: 339–417.

Ingold, C. T. (1943a) Further observations on aquatic hyphomycetes. Transactions of the British Mycological Society 26: 104–115.

Ingold, C. T. (1943b) Triscelophorus monosporus n. gen., n. sp. an aquatic hyphomycete. Transactions of the British Mycological Society 26: 148–152.

Ingold, C. T. (1944) Some new aquatic hyphomycetes. Transactions of the British Mycological Society 27: 35–47.

Ingold, C. T. (1952) *Actinospora megalospora* n. sp. an aquatic hyphomycete. Transactions of the British Mycological Society 35: 66–71.

Ingold, C. T. (1956) Stream spora in Nigeria. Transctions of the British Mycological Society 39: 108–110.

Ingold, C. T. (1958a) Aquatic hyphomycetes from Uganda and Rhodesia. Transactions of the British Mycological Society 41: 109–114.

Ingold, C. T. (1958b) New aquatic hyphomycetes: *Lemonniera brachycladia*, *Anguillospora crassa* and *Fluminispora ovalis*. Transactions of the British Mycological Society, Cambridge 41 (3): 365–372.

Ingold, C. T. (1959a) *Polycladium equisetii* gen. nov.; sp. nov.: an aquatic hyphomycete on *Equisetum fluviatile*. Transactions of the British Mycological Society 42 (1): 112–114.

Ingold, C. T. (1959 b) Aquatic spora of Ono forest, Nigeria. Transactions of the British Mycological Society 42 (4): 479–485.

Ingold, C. T. (1960) Aquatic hyphomycetes in Southern Rhodesia. Proceedings and Transcations of the Rhodesia Scientific Association 48: 49053.

Ingold, C. T. (1964) A new species of Ingoldia from Britain. Transactions of the British Mycological Society 47: 103–107.

Ingold, C. T. (1974) *Tricladium chaetocladium* sp. Nov. An aquatic hyphomycete from Britain. Transactions of the British Mycological Society 63: 624–626.

Ingold, C. T. (1975) An Illustrated Guide to Aquatic and Water–borne Hyphomycetes (Fungi Imperfecti) with notes on their Biology. Freshwater Biological Association, Ambleside, pp. 96.

Ingold, C. T. (1979) Advances in the study of the so-called aquatic hyphomycetes. American Journal of Botany 66: 218–226.

Iqbal, S. H. (1992) Periodicity of freshwater hyphomycete conidia in some streams of Pakistan. Transactions of the Mycological Society of Japan 33: 293–304.

Iqbal, S. H. (1993) Effiency of artificial foam in trapping conidia of Ingoldian fungi. Annales Botanici Fennici. 30: 153–160.

Iqbal, S. H. (1994) Species diversity of freshwater hyphomycetes in some streams of Pakistan. Comparison of sampling techniques. Mycoscience 35: 331–343.

Iqbal, S. H. and Webster, J. (1973a). Aquatic hyphomycete spora of the River Exe and its tributaries. Transactions of the British Mycological Society 61: 331–346.

Iqbal, S. H. and Webster, J. (1973b) The trapping of aquatic hyphomycete spora by air bubbles. Transactions of the British Mycological Society 60: 37–48.

Iqbal, S. H., Bhatty, S. F. and Malik, K. S. (1979) Freshwater hyphomycetes on decaying plant debris submerged in some streams in Pakistan. Transactions of the Mycological Society of Japan 20: 51–61.

Iqbal, S. H., Firdaus-e-Bareen and Noureen Yousaf (1995) Freshwater Hyphomycete communities in a canal. 1. Endophytic hyphomycetes of submerged roots of trees sheltering a canal bank. Canadian Journal of Botany 73: 538–543.

Iversen, T. M. (1973) Decomposition of autumn-shed beech leaves in a spring-brook and its significance for the fauna. Archiv fur Hydrobiologie 73: 305–312.

Kappe, R., Okeke, C. N., Fauser, C., Maiwald, M. and Sonntag, H. G. (1998) Molecular probes for the detection of pathogenic fungi in the presence of human tissue. Journal of Medical Microbiology 47: 811–820.

Kowalchuk, G. A. (1999) New perspectives towards analysing fungal communities in terrestrial environments. Current Opinion in Biotechnology 10: 247–251.

Kowalchuk, G. A., Gerards, S. and Woldendorp, J. W. (1997) Detection and characterization of fungal infections of *Ammophila arenaria* (Marram Grass) roots by denaturing gradient gel electrophoresis of specifically amplified 18S r RNA. Applied and Environmental Microbiology 63: 3858–3865.

Kegel, W. (1906) *Varicosporium elodeae*, ein Wasserpilz mit auffallender Konidienbildung. Berliner Deutsche Botanische Gesellschaft 68: 465–515.

Laitung, B., Chauvet, E., Feau, N., Feve, K., Chikhi, L. and Gardes, M. (2004) Genetic diversity in *Tetrachaetum elegans*, a mitosporic aquatic fungus. Molecular Ecology 13: 1679–1692.

Le-John, H. B. (1965) Sierra Leone freshwater hyphomycetes. Transactions of the British Mycological Society 48: 261–265.

Lee, C., Howarth, R. W. and Howes, B. L. (1980) Sterols in decomposing *Spartina alterniflora* and the use of ergosterol in estimating the contribution of fungi to detrital nitrogen. Limnology and Oceanography 25: 290–303.

Ludwig, W. (1991) DNA sequencing in bacterial systematics. In: Nucleic Acid Techniques in Bacterial Systematics (eds), Stackebrandt, E., Goodfellow, M. Chichester, John Wiley & Sons, 69–94.

Ma, L.-J., Rogers, S. O. and Catranis, C. M. (2000) Detection and characterization of ancient fungi entrapped in glacial ice. Mycologia 92: 286–295.

Maamri, A., Chauvet, E., Chergui, H., Gourbiere, F. and Pattee, E. (1998) Microbial dynamics on decaying leaves in a temporary Moroccan river. I – Fungi. Archiv fur Hydrobiologie 144: 41–59.

Malosso, E. (1995) Ocorrência de Hyphomycetes (Fungi Imperfecti) e fungos zoospóricos em ambientes aquáticos (Rio do Monjolinho, São Carlos, SP). Universidade Federal de São Carlos, Centro de Ciências Biológicas e da Saúde. São Carlos, SP, Brazil. pp. 54, B.Sc. Hons. Dissertation.

Malosso, E. (1999) Hyphomycetes em ambientes aquáticos lótico e lêntico – Ocorrência e biomassa. Universidade Federal de São Carlos, Programa de Pós-Graduação em Ecologia e Recursos Naturais. São Carlos, SP, Brazil. pp. 98, M.Sc. Dissertation.

Malosso, E., English, L., Hopkins, D. W. and O'Donnell, A. G. (2004) Use of ^{13}C-labelled plant materials and ergosterol, PLFA and NLFA analyses to investigate organic matter decomposition in Antarctic soil. Soil Biology and Biochemistry 36: 165–175.
Marvanová, L. (1984) Conidia in water of the protected area Slovensky. Biologia, Bratislava, 39: 821–832.
Marvanová, L. (1997) Freshwater Hyphomycetes: a survey with remarks on tropical taxa. In: Tropical Mycology, (eds), Janardhanan, K. K., Rajendran, C., Natarajan, K., Hawksworth, D. L. Science Publishers, Inc., Enfield USA, 169–226.
Marvanová, L. and Bärlocher, F. (1988) Hyphomycetes from Canadian streams. I. Basidiomycetous anamorphs. Mycotaxon 32: 339–351.
Marvanová, L. and Bärlocher, F. (1989) Hyphomycetes from Canadian streams. II. Three new taxa. Mycotaxon 35: 85–99.
Marvanová, L. and Marvan, P. (1969) Aquatic hyphomycetes in Cuba. Céska Mykologica 23: 135–140.
Marvanová, L., Marvan, P. and Ruzicka, J. (1967) *Gyoerffyella* Kol., a genus of the hyphomycetes. Persoonia 5: 29–44.
Marvanova, L., Landvik, S., Fisher, P. J., Moss, S. T. and Ainsworth, A. M. (2002) A new fungus with arthroconidia from foam. Nova Hedwigia 75: 255–269.
Matcham, S. E., Jordan, B. R. and Wood, D. A. (1985) Methods for assessment of fungal growth on solid substrate. Society for Applied Bacteriology Technical Series 19: 5–18.
Mathuriau, C. and Chauvet, E. (2002) Breakdown of leaf litter in a neotropical stream. Journal of the North American Benthological Society 21: 384–396.
Matsushima, T. (1971a) Mycological reports from New Guinea and the Solomon Islands. 7. Some interesting Fungi imperfecti. Bulletin of the National Science Museum, Tokyo. 14: 460–480.
Matsushima, T. (1971b) Microfungi of the Solomon Islands and Papua New Guinea. Shionogi Research Laboratory, Shiomogi and Co., Ltd. Kobe 1–78.
Matsushima, T. (1993) Matsushima Mycological Memoirs n°. 7, 75p+131 plates, September.
Matsushima, T. (1995) Matsushima Mycological Memoirs n°. 8, 54p+120 plates, August.
McArthur, F. A., Bärlocher, M. O., MacLean, N. A. B., Hiltz, M. D. and Bärlocher, F. (2001) Asking probing questions: Can fluorescent in situ hybridization identify and localise aquatic hyphomycetes on leaf litter? International Review of Hydrobiology 86: 429–438.
Möhlenhoff, P., Müller, L., Gorbushina, A. A. and Petersen, K. (2001) Molecular approach to the characterisation of fungal communities: methods for DNA extraction, PCR amplification and DGGE analysis of painted art objects. FEMS Microbiology Letters 195: 169–173.
Mothe-Jean-Louis, A. M. (1997) Dynamique pluriannuelle des hyphomycetes aquatiques de L'Osse, riviere realimentee du sudouest de la France. Canadian Journal of Botany 75: 155–169.
Muyzer, G., De Waal, E. C. and Uitterlinden, A. G. (1993) Profiling of complex microbial populations by denaturating gradient gel electrophoresis analysis of polymerase chain reaction-amplified genes coding for 16S r RNA. Applied and Environmental Microbiology 59: 695–700.
Muyzer, G. and Smalla, K. (1998) Application of denaturing gradient gel electrophoresis (DGGE) and temperature gradient gel electrophoresis (TGGE) in microbial ecology. Antonie van Leeuwenhoek 73: 127–141.
Myers, R. M., Fischer, S. G., Lerman, L. S. and Maniatis, T. (1985) Nearly all single base substitutions in DNA fragments joined to a GC–clamp can be detected by denaturing gradient gel electrophoresis. Nucleic Acids Research 13: 3131–3145.
Nawawi, A. (1975) Another hyphomycete with branched conidia. Transactions of the British Mycological Society 64: 243–246.
Nawawi, A. (1985a) Aquatic hyphomycetes and other water-borne fungi from Malaysia. Malayan Nature Journal 39: 75–134.
Nawawi, A. (1985b) Basidiomycetes with branched, water-borne conidia. Botanical Journal of the Linnean Society 91: 51–60.
Nawawi, A. (1985c) Another aquatic hyphomycete genus from foam. Transactions of the British Mycological Society 85: 174–177.
Nawawi, A. (1985d) More *Tricladium* species from Malaysia. Transactions of the British Mycological Society 85: 177–182.

Nawawi, A. (1985e) Some interesting hyphomycetes from water. Mycotaxon 24: 217–226,

Nawawi, A. (1987a) *Clavariopsis azianii*, sp. Nov., a new aquatic hyphomycete from Malaysia. Transactions of the British Mycological Society 88: 428–432.

Nawawi, A. (1987b) Diplocladiella appendiculata sp. Nov. A new aero-aquatic hyphomycete. Mycotaxon 28: 297–302.

Nawawi, A. and Kuthubutheen, A. J. (1987a) *Triscelosporium verrucosum* gen. et sp. nov. A dematiaceous aeroaquatic hyphomycete with tetraradiate conidia. Mycotaxon 29: 285–290.

Nawawi, A. and Kuthubutheen, A. J. (1987b) *Tetrabrachium elegans* gen. et sp. nov., a hyphomycete with tetraradiate conidia. Mycotaxon 29: 291–297.

Nawawi, A. and Kuthubutheen, A (1988a) *Tricladiospora,* a new genus of dematicaeous hyphomycetes with staurosporous conidia from submerged decaying leaves. Transactions of the British Mycological Society 90: 482–487.

Nawawi, A. and Kuthubutheen, A. J. (1988b) Another new hyphomycete from leaf litter. Mycotaxon 31: 339–343.

Nawawi, A. and Kuthubutheen, A. J. (1988c) *Tricladiomyces geniculatus* sp. nov., a conidial basidiomycete. Transactions of the British Mycological Society 90: 670–673.

Nawawi, A. and Kuthubutheen, A. J. (1988d) Additions to *Condylospora* (Hyphomycetes) from Malaysia. Mycotaxon 33: 329–338.

Nawawi, A. and Kuthubutheen, A. J. (1989a) *Quadricladium aquaticum* gen et sp nov., an aquatic hyphomycete with tetraradiate conidia. Mycotaxon 34: 489–495.

Nawawi, A. and Kuthubutheen, A. J. (1989b) A new taxon in *Colispora* (Hyphomycetes) from Malaysia. Mycotaxon 34: 497–501.

Nawawi, A. and Kuthubutheen, A. J. (1990) *Obeliospora,* a new genus of phialosporus hyphomycetes with appendaged conidia. Mycotaxon 37: 395–400.

Nemec, S. (1969) Sporulation and identification of fungi isolated from root-rot diseased strawberry plants. Canadian Journal of Botany 47: 945–947.

Newell, S. Y. (1992) Estimating fungal biomass and productivity in decomposing litter. In: The Fungal Community. Its Organization and Role in the Ecosystem. (eds), Carroll, G. C. and Wicklow, D. T. 2nd edition. New York, Marcel Dekker, pp. 521–562.

Newell, S. Y. (1993) Membrane-containing fungal mass and fungal specifc growth rate in natural samples. In: Handbook of Methods in Aquatic Microbial Ecology. (eds), Kemp, P. F., Sherr, B. F., Sherr, E. B. and Cole, J. J. Boca Raton, Lewis Publishers, pp. 579–586.

Newell, S. Y. (1994) Total and free ergosterol in mycelia of saltmarsh ascomycetes with access to whole leaves or aqueous extracts of leaves. Applied and Environmental Microbiology 60: 3479–3482.

Newell, S. Y. and Fallon, R. D. (1991) Toward a method for measuring instantaneous fungal growth rates in field samples. Ecology 72: 1547–1559.

Newell, S. Y., Moran, M. A., Wicks, R. and Hodson, R. E. (1995) Productivities of microbial decomposers during early stages of decomposition of leaves of a freshwater sedge. Freshwater Biology 34: 135–148.

Nikolcheva, L. G., Cockshutt, A. M. and Bärlocher, F. (2003) Determining diversity of freshwater fungi on decaying leaves: Comparison of traditional and molecular approaches. Applied and Environmental Microbiology 69: 2548–2554.

Nilsson, S. (1958) On some swedish freshwater Hyphomycetes. Svensk. Botanisk Tidskrift 56: 291–318,

Nilsson, S. (1962a) Some Aquatic Hyphomycetes from South America. Svensk Botanisk Tidskrift 56 (2): 351–361.

Nilsson, S. (1962b) Second note on Swedish freshwater hyphomycetes. Botaniska Notiser 115: 73–86.

Nilsson, S. (1964) Freshwater Hyphomycetes: taxonomy, morphology and ecology. Symbolae Botanice Upsalense 18: 1–130.

Padgett, D. E. (1976) Leaf decomposition by fungi in a tropical rainforest stream. Biotropica 8 (3): 166–178.

Park, D. (1972a) Methods of detecting fungi in organic detritus in water. Transactions of the British Mycological Society 58 (2): 281–290.

Park, D. (1972b) On the ecology of heterotrophic micro-organisms in freshwater. Transactions of the British Mycological Society 58 (2): 291–299.

Pascoal, C. and Cassio, F. (2004) Contribution of fungi and bacteria to leaf litter decomposition in a polluted river. Applied and Environmental Microbiology 70: 5266–5273.

Pennanen, T., Paavolainen, L. and Hantula, J. (2001) Rapid PCR-based method for the direct analysis of fungal communities in complex environmental samples. Soil Biology and Biochemistry 33: 697–699.

Petersen, R. H. (1962) Aquatic hyphomycetes from North America. I. Aleurosporae (Part 1) and key to the genera. Mycologia 54: 117–151.

Petersen, R. H. (1963a) Aquatic hyphomycetes from North America. II. Aleuroposrae (Part 2), and Blastosporae. Mycologia 55: 18–29.

Petersen, R. H. (1963b) Aquatic hyphomycetes from North America III. Phialosporae and miscellaneous species. Mycologia 55: 570–581.

Posteraro, B., Sanguinetti, M., Masucci, L., Romano, L., Morace, G. and Fadda, G. (2000) Reverse cross blot hybridization assay for rapid detection of PCR-amplified DNA from *Candida* species, *Cryptococcus neoformans,* and *Saccharomyces cerevisiae* in clinical samples. Journal of Clinical Microbiology 38: 1609–1614.

Pugh, G. J. F., Buckey, N. G. and Mulder, J. (1972) The role of phylloplane fungi in the early colonization of leaves. Symposia Biologica Hungarica 11: 329–333.

Pryor, B. M. and Gilbertson, R. L. (2000) Molecular phylogenetic relationships amongst *Alternaria* species and related fungi based upon analysis of nuclear ITS and mt SSU rDNA sequences. Mycological Research 104, 1312–1321.

Rajashekar, M. and Kaveriappa, K. M. (1993) Ecological observations on water-borne hyphomycetes of Cauvery River and its tributaries in India. Archiv für Hydrobiologie 12: 487–497.

Ranzoni, F. V. (1953) The aquatic hyphomycetes of California. Farlowia 4: 353–398.

Ranzoni, F. V. (1979) Aquatic hyphomycetes from Hawaii. *Mycologia* 71: 786–795.

Regelsberger, B., Messner, K. and Descals, E. (1987) Species diversity in aquatic Hyphomycetes in four Austrian streams. Mycotaxon 30: 439–454.

Robinson, C. H. (2001) Cold adaptation in Arctic and Antarctic fungi. New Phytologist 151: 341–353.

Rossman, A. Y., Tulloss, R. E., O'Dell, T. E. and Thorn, R. G. (1998) Protocols for an all taxa biodiversity inventory of fungi in a Costa Rican Conservation Area. Borne, N. C., Parkway Publishing Inc. 194p.

Sancholle, M. and Dalpé, Y. (1993) Taxonomic relevance of fatty acids of arbuscular mycorrhizal fungi and related species. Mycotaxon 49: 187–193.

Schoenlein-Crusius, I. H. and Grandi, R. A. P. (2003) The diversity of aquatic Hyphomycetes in South America. Brazilian Journal of Microbiology 34: 183–193.

Schoenlein-Crusius, I. H. and Milanez, A. I. (1989) Sucessão fúngica em folhas de *Ficus microcarpa* L. F, submersas no Lago Frontal situado no Parque Estadual das Fontes do Ipiranga, São Paulo. Revista de Microbiologia 20 (1): 95–101.

Schoenlein-Crusius, I. H. and Milanez, A. I. (1990) Aquatic Hyphomycetes in São Paulo State, Brazil. I First observations. Hoehnea 17: 111–115.

Schoenlein-Crusius, I. H. and Milanez, A. I. (1996) Diversity of aquatic fungi in Brazilian Ecosystems. In: Biodiversity in Brazil, a first approach, (eds), Bicudo, C. and Menezes, N. A. pp. 31–48.

Schoenlein-Crusius, I. H. and Milanez, A. I. (1998) Fungal succession on leaves of *Alchornea triplinervia* (Spreng.) M. Arg. Submerged in a stream of an Atlantic Rainforest in the state of São Paulo, Brazil. Revista Brasileira de Botânica 21 (3): 253–259.

Schoenlein-Crusius, I. H., Pires-Zottarelli, C. L. A. and Milanez, A. I. (1990) Sucessão fúngica em folhas de *Quercus robur* L. (carvalho) submersas em um lago situado no município de Itapecerica da Serra, SP. Revista de Microbiologia 21 (1): 61–67.

Schoenlein-Crusius, I. H., Pires-Zottarelli, C. L. A. and Milanez, A. I. (1992) Aquatic fungi in leaves submerged in a stream in the Atlantic Rainforest. Revista de Microbiologia 23 (3): 167–171.

Schoenlein-Crusius, I. H., Pires-Zottarelli, C. L. A., Milanez, A. I. and Humphreys, R. D. (1999) Interaction between the mineral content and the occurrence number of aquatic fungi in leaves submerged in a stream in the Atlantic rainforest, São Paulo, Brazil. Revista Brasileira de Botânica 22 (2): 133–139.

Scott, W. W. and Umphlett, C. J. (1963) Some new and unusual fungi from Virginia. II. Aquatic Hyphomycetes. Virginia Journal of Science 14: 47–64.

Seitz, L. M., Mohr, H. E., Burroughs, R. and Sauer, D. B. (1977) Ergosterol as an indicator of fungal invasion in grains. Cereal Chemistry 54: 1207–1217.

Seitz, L. M., Sauer, D. B., Burroughs, R., Mohr, H. E. and Hubbard, J. D. (1979) Ergosterol as a measure of fungal growth. Phytopathology 69: 1202–1203.

Shaw, D. (1972) *Ingoldiella hamata* gen. et sp. nov. A fungus with clamp connections from a stream in North Queensland. Transactions of the British Mycological Society 59: 255–259.

Shaw, D. and Sutton, B. C. (1985) A new aero-aquatic hyphomycete from Papua New Guinea and Australia. Botanical Journal of the Linnean Society 91: 25–36.

Shearer, C. A. (1993) The freshwater Ascomycetes. Nova Hedwigia 56 (1–2): 1–33.

Shearer, C. A. and Lane, L. C. (1983) Comparison of three techniques for the study of aquatic hyphomycete communities. Mycologia 75: 498–506.

Shearer, C. A. and Webster, J. (1985) Aquatic hyphomycete communities in the River Teign. I. Longitudinal distribution patterns. Transactions of the British Mycological Society 84: 489–501.

Silva, M. I. L. (2002) Micobiota de água e do solo das margens de igarapés situados na área de mata do campus da Universidade do Amazonas, AM. São Paulo, 175p. (Ph.D. Thesis, Instituto de Ciências Biológicas, USP).

Singh, N. and Musa, T. M. (1977) Terrestrial occurrence and the effect of temperature on growth, sporulation and spore germination of some tropical aquatic hyphomycetes. Transactions of the British Mycological Society 68: 103–106.

Singh, N. (1976) *Pyramidospora herculiformis* sp. nov., a new aquatic hyphomycete from Sierra Leone. Transactions of the British Mycological Society 66: 347–350.

Smit, E., Leeflang, P., Glandorf, B., van Elsas, J. D. and Wernars, K. (1999) Analysis of fungal diversity in the wheat rhizosphere by sequencing of cloned PCR-amplified genes encoding 18S rRNA and temperature gradient gel electrophoresis. Applied and Environmental Microbiology 65: 2614–2621.

Sridhar, K. R. and Bärlocher, F. (2000) Initial colonization, nutrient supply, and fungal activity on leaves decaying in streams. Applied and Environmental Microbiology 66 (3): 1114–1119.

Sridhar, K. R. and Kaveriappa, K. M. (1988) Colonization of leaf litter by aquatic hyphomycetes in a tropical stream. Archiv für Hydrobiologie 112: 627–630.

Sridhar, K. R. and Kaveriappa, K. M. (1989) Observations on aquatic hyphomycetes of the western Ghat streams, India. Nova Hedwigia 49 (3–4): 455–467.

Sridhar, K. R., Chandrashekar, K. R. and Kaveriappa, K. M. (1992) Research on the Indian subcontinent. In: The Ecology of Aquatic Hyphomycetes (ed), Bärlocher, F., Springer Verlag, 182–211.

Suberkropp, K. (1992) Aquatic Hyphomycete Communities. In: The Fungal Community—Its organization and role in the ecosystem. (ed), Carroll, G. C. and Wicklow, D. T. Marcel Dekker, New York, Mycology Series No. 9, 729–747.

Suberkropp, K. (1995) The influence of nutrients on fungal growth, production, and sporulation during leaf breakdown in streams. Canadian Journal of Botany 73 (Suppl. 1): 1361–1369.

Suberkropp, K. and Weyers, H. (1996) Application of fungal and bacterial production methodologies to decomposing leaves in streams. Applied and Environmental Microbiology 62: 1610–1615.

Suberkropp, K., Gessner, M. O. and Chauvet, E. (1993) Comparison of ATP and ergosterol as indicators of fungal biomass associated with decomposing leaves in streams. Applied and Environmental Microbiology 59: 3367–3372.

Suberkropp, K., Godshalk, G. L. and Klug, M. J. (1976) Changes in chemical composition of leaves during processing in a woodland stream. Ecology 57: 720–727.

Suh, S. O. and Nakase, T. (1995) Phylogenetic analysis of the ballistosporus anamorphic genera *Udeniomyces* and *Bullera*, and related basidiomycetous yeasts, based on 18S rDNA sequence. Microbiology 141: 901–906.

Sumner, J. L. and Morgan, E. D. (1969) The fatty acid composition of sporangiospores and vegetative mycelium of temperature-adapted fungi in the Order Mucorales. Journal of General Microbiology 59: 215–221.

Subramanian, C. V. (1983) Hyphomycetes – Taxonomy and Biology. Academic Press, London, 502 p.

Subramanian, C. V. and Bhat, D. J. (1981) Conidia from freshwater foam samples from the Western Ghats, Southern India. Kavaka 9: 45–62.

Stahl, P. D. and Klug, M. J. (1996) Characterization and differentiation of filamentous fungi based on fatty acid composition. Applied and Environmental Microbiology 62: 4136–4146.

Tubaki, K. (1957) Studies on the Japanese hyphomycetes. (III). Aquatic group. Bulletin of the National Science Museum 3: 249–268.

Tubaki, K. (1965) Short note on aquatic spora in East New Guinea. Transactions of the Mycological Society of Japan 16: 357–360.

Tubaki, K., Watanabe, K. and Manoch, L. (1983) Aquatic hyphomycetes from Thailand. Transactions of the Mycological Society of Japan 6: 11–16.

Tunlid, A. and White, D. C. (1992) Biochemical analysis of biomass, community structure, nutritional status, and metabolic activity of microbial communities in soil. In: Stotzky, G. and Bollag, J. M. (eds.) Soil Biochemistry Vol. 7, New York, Marcel Dekker, p. 229–262.

Turin, L., Riva, F., Galbiati, G. and Cainelli, T. (2000) Fast, simple and highly sensitive double-rounded polymerase chain reaction assay to detect medically relevant fungi in dermatological specimens. European Journal of Clinical Investigation 30: 511–518.

Vainio, E. J. and Hantula, J. (2000) Direct analysis of wood-inhabiting fungi using denaturing gradient gel electrophoresis of amplified ribosomal DNA. Mycological Research 104: 927–936.

Valente, P., Ramos, J. P. and Leoncini, O. (1999) Sequencing as a tool in yeast molecular taxonomy. Canadian Journal of Microbiology 45: 949–958.

Viljoen, B. C., Kock, J. L. F. and Lategan, P. M (1986) Fatty acid composition as a guide to the classification of selected genera of yeasts belonging to the Endomycetales. Journal of General Microbiology 132: 2397–2400.

Waid, J. S. (1954) Occurrence of aquatic hyphomycetes upon the root surfaces of beech grown in woodland soils. Transactions of the British Mycological Society 37: 420–421.

Wallander, H., Massicotte, H. B. and Nylund, J. E. (1997) Seasonal variation in protein, ergosterol and chitin in five morphotypes of *Pinus sylvestris* L. ectomycorrhizae in a mature Swedish forest. Soil Biology and Biochemistry 29: 44–53.

Webster, J. (1992) Anamorph-teleomorph relationships. In: Bärlocher, F. (ed), The ecology of aquatic Hyphomycetes. Springer-Verlag, Berlin, pp. 99–117.

Webster, J. and Descals, E. (1981) Morphology distribution and ecology of conidial fungi in freshwater habitat. In: Biology of conidial fungi. Vol. I. Ed. Cole, G. C. and Kendrick, B. Academic Press London, p. 295–355.

Webster, J., Marvanová, L. and Eicker, A. (1994) Spores from foam from South African rivers. Nova Hedwigia 59: 379–398.

White, D. C., Stair, J. O. and Ringelberg, D. B. (1996) Quantitative comparisons of *in situ* microbial biodiversity by signature biomarker analysis. Journal of Industrial Microbiology 17: 185–196.

White, T. J., Bruns, T., Lee, S. and Taylor, J. (1990) Amplification and direct sequencing of fungal ribosomal RNA genes for phylogenetics. In: PCR Protocols. (eds.). Innins, M. A., Gelfand, D. H., Sninsky, J. J. and White, T. J. San Diego, Academic Press, pp. 315–322.

Yong-Nian, Y., Jin-Liang, L. and Jin-Qin, Y. (1988) Notes on some aquatic hyphomycetes from Sichuan province of China. Acta Mycologica Sinica 7 (3): 138–148.

Young, R. N. (1999)Importance of biodiversity to the modern pharmaceutical industry. Pure and Applied Chemistry 71: 1655–1661.

Zelles, L., Rackwitz, R., Bai, Q. Y., Beck, T. and Beese, F. (1995) Discrimination of microbial diversity by fatty acid profiles of phospholipids and lipopolysaccharides in differently cultivated soils. Plant and Soil 170: 115–122.

7. Molecular Identification and Detection of Plant Pathogenic and Toxigenic *Fusarium* Fungi

Tapani Yli-Mattila and Sari Paavanen-Huhtala
Laboratory of Plant Physiology and Molecular Biology, Department of Biology, University of Turku, 20014 Turku

SUMMARY *Fusarium* species are probably the most economically important phytopathogen and mycotoxigenic genus of filamentous fungi. *Fusarium* taxonomy has been constantly revised due to problems in species identification. Due to the growing interest in the mycotoxins and the routine identification of plant pathogenic and toxigenic *Fusarium* strains, a reliable and rapid system for identification of toxigenic strains is important. Molecular and biochemical methods are most promising for this purpose. Molecular analyses with e.g. RAPD-PCR, UP-PCR or RAMS primers can be used for screening differences and similarities between several isolates of different species. When species-specific markers are found, the PCR products can be sequenced and more specific SCAR primers can be designed. Alternatively, known DNA variable sequences, such as β-tubulin or ribosomal ITS and IGS sequences of *Fusarium* isolates can be used for designing species-specific primers. Several other molecular methods, such as AFLP, DGGE, ARMS-PCR and different hybridisation analyses can also be used for molecular identification and detection of species. Quantitative PCR methods make it possible to quantify the DNA and RNA amounts of different *Fusarium* species directly from plants or soil. The molecular results obtained can also be used for taxonomical studies together with traditional morphological characters, crossing experiments and metabolite analyses.

Introduction

Fusarium Taxonomy and Species Concepts: History

Fusarium species are widely distributed in soil and on organic substrates and have been isolated from all over the world (Booth, 1971). Since there have been problems in species identification, *Fusarium* taxonomy has been constantly revised and updated. Morphological, biological and phylogenetic species concepts have been applied within *Fusarium* (O'Donnell, 1996; O'Donnell et al. 2000). The basis for *Fusarium* systematics is the monograph of Wollenweber and Reinking (1935). The number of morphological species, which have been recognised within the genus *Fusarium,* has varied from the nine species of Snyder and Hansen (1945) to approximately 100 taxa (Gerlach and Nirenberg, 1982). The division of species into sections and their number, has also changed, especially after the new DNA-based techniques and phylogenetic species concepts (Taylor et al. 2000) have developed. Current classifications (Booth, 1971; Gerlach and Nirenberg, 1982; Nelson et al. 1983) are still based mainly on morphological species concepts.

Certain species, such as *F. oxysporum,* are divided according to the host plants attacked, into special forms or races of the fungus, while most of the other *Fusarium* species are less specialised. The *formae speciales* are physiologic strains, which are indistinguishable from saprophytic strains of the same species but which show different physiological properties in their ability to parasitise specific hosts (Booth, 1971).

In several cases a correlation between *formae speciales* and vegetative compatibility groups (Leslie, 1993) or molecular markers (Bulat and Mironenko, 1996; Baayen et al. 2000) has been found.

Morphological species are defined on the basis of differences in morphological features, which are often difficult to find between closely related species. The subjective nature of this concept and the lack of sexual stage (Fig. 1) in most species may also cause problems in identification. Biological species concept (Matuo and Snyder, 1973; Leslie et al. 2001) also has limitations as a tool for identification of unknown *Fusarium* isolates, since this concept has been developed only for a few species and it is a very slow method to use.

Fusarium Diseases and Toxins

Fusaria may cause vascular wilt, seedling blight, root rot, crown rot and/or leaf blight (Fig. 2). *Fusarium* usually infects plants through the roots or in cereals through heads during flowering (Mesterhazy, 1997; Perkowsky et al. 1997). In northern Europe fusaria are so common in agricultural environments that they are often present in 100% of grain samples examined (Uoti and Ylimäki, 1974; Chelkowski, 1989). These fungi cause considerable damage to cereal crops, berries and horticultural plants in many countries. Several species produce mycotoxins, that contaminate cereals and other agricultural commodities and are harmful for human and animal health. Various species produce zearalenone (*F. graminearum, F. culmorum*), trichothecenes [nivalenol (NIV): *F. poae, F. graminearum, F. culmorum*; deoxynivalenol (DON): *F. graminearum, F. culmorum;* T-2/HT-2: *F. sporotrichioides, F. langsethiae*], fumonisins (*F. verticillioides*); the last of these are economically the most important *Fusarium* mycotoxins for grain farmers, existing primarily in moulded corn (Moss, 1996, 1998; Alexander et al. 1997; Logrieco et al. 2002). Newer mycotoxins (Logrieco et al. 2002; Thrane et al. 2004; Jestoi et al. 2004) include moniliformin, enniatins (*F. avenaceum, F. arthrosporioides, F. tricinctum)* and beauvericin (*F. poae* and *F. sporotrichioides*). Toxin production has been shown to be temperature dependent; the age, origin and substrate of a particular isolate can also influence toxin production (Joffe, 1986; Llorens et al. 2004). In addition, there may be interactions between different *Fusarium* and other fungal species in grains (Kosiak et al. 2004).

Most *Fusarium* toxins have been shown to be highly heat stable. This is why they can persist in a process and end up in the final product in food industry. Because of global warming and new agricultural techniques new harmful and more toxigenic *Fusarium* species, for example *Fusarium graminearum*, may cause problems in new areas of Northern Europe and North America. On the other hand, the increase of organic crop production and the social and economic pressure to protect environment will decrease the possibilities for plant protection using chemicals.

Present Detection Methods

Nowadays detection of fungi in cereals and agricultural commodites is carried out using conventional morphological, microbiological and antibody-based techniques (Figs. 3 and 4). The use of such methods is restricted by the fact, that their resolution is not good enough, they can detect only living micro-organisms or they are semi-quantitative. Problems may arise from dead fungal biomass that cannot be detected, but the toxins or gushing factors may still be present. In addition, it is often difficult to grow cultures from the fungal isolates found in nature. Biochemical analyses (Thrane, 1989) and isoenzymes (Bonde et al. 1993) can in some cases be used for strain identification, but they are less sensitive and more laborius than DNA-based methods. Metabolite profiling methods, which are formed by enzymes coded by the genome (Nielsen et al. 2004), often give similar results as DNA-based methods (Thrane et al. 2001). Pathogenicity tests on test plants can also be used for identification of plant pathogenic fungal strains, but this method is not specific and it is so slow that the results often cannot be utilised during the same growth season.

During the last years the molecular biological studies of bacterial and viral plant diseases and direct (cultivation-independent) detection methods in studies of soil bacteria (Van Elsas et al. 2000) have progressed

Fig. 1. *Gibberella zeae* (anamorph *Fusarium graminearum*) mycelia and perithecia in wheat in Russia (Gagkaeva).

Fig. 2. Naturally by *Fusarium* spp. infected oats (left) (Gagkaeva); artificially by *F. culmorum* infected Mahti spring wheat (middle) and artificially by *F. sporotrichioides* infected Scarlett spring barley (right).

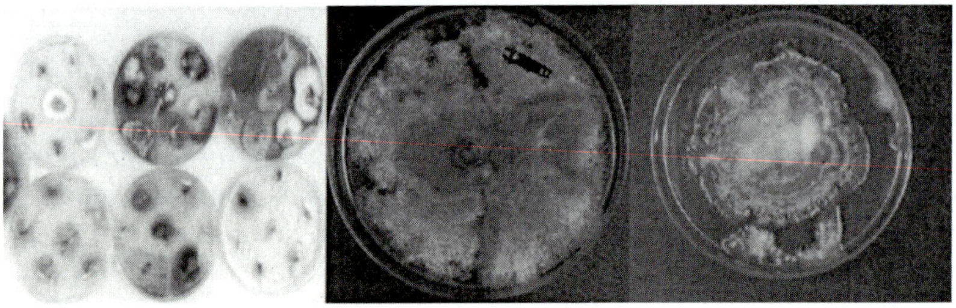

Fig. 3. Plating the grains on different media (left) (Konstantinova); pure culture of *F. culmorum* on PDA medium (middle) and pure culture of *F. avenaceum* on PDA medium (right).

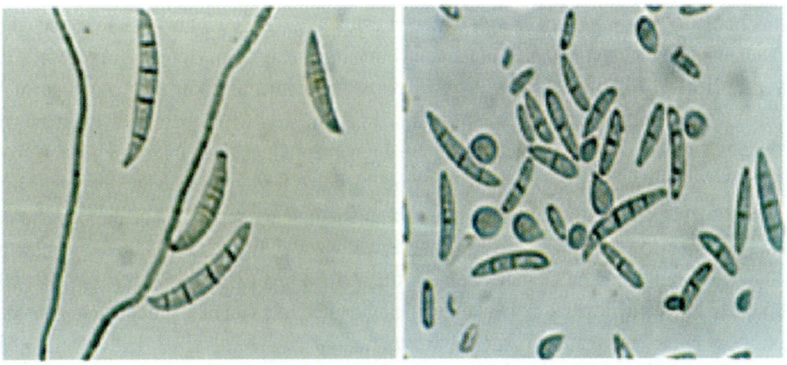

Fig. 4. Macroconidia of *F. culmorum* (left) and macro- and microconidia of *F. sporotrichioides* (right) (Konstantinova).

more quickly than those of fungal plant diseases and soil fungi. This is due to the fact, that it is often more difficult to extract DNA from fungi, fungal genomes are larger and more complex to study and there have been fewer molecular biologists among plant mycologists than among plant virologists and bacteriologists. At the moment several plant mycologists finally have adapted and developed a lot of molecular methods and are ready to use them for direct detection of fungi.

It has been found that in many cases the virulence of plant pathogenic fungi is connected to the enzymes and toxins they produce (Annis and Goodwin, 1997; Desjardins and Hohn, 1997; Nicholson et al. 2003). So, it is important to investigate toxins and enzymes, which participate in pathogenesis and/or control the synthesis of toxins, in plant pathogenic fungal strains. Some of these factors are probably connected to (a)virulence genes of plant pathogenic fungal strains. Toxins and other secondary metabolites can also be used as markers in epidemiological, population and evolution biology studies. The molecular detection of moulds directly in food matrixes using molecular techniques also needs further development, because moulds can cause health problems and enormous economical losses when they contaminate food and feed.

Methods

Protein Extraction and Isozyme Analysis

Inocula are taken from the mycelia obtained and placed on potato dextrose agar (PDA) plates. Often single-spore isolates are used to be sure about the homogeneity of the cultures. Each isolate is grown for 4-6 days on plates either containing cellophane membrane (Yli-Mattila et al. 1996) or without the membrane (Paavanen-Huhtala et al. 2000) at 25°C in constant light.

The mycelia are scraped off the cellophane membrane and ground to a fine powder under liquid nitrogen with a mortar and pestle. Proteins can be extracted with STEB buffer (Rosendahl and Sen, 1992). Proteins are electrophoresed in polyacrylamide gels and gel preparation and enzyme staining of gels is performed according to Loxdale et al. (1983), Sen and Hepper (1986) and Shields et al. (1983). The positions of isozyme bands are determined and the gels are photographed and dried under vacuum and heat (Yli-Mattila et al. 1996).

Metabolite Profiling Analyses

In metabolite analyses the mycelia are usually grown on Yeast Extract Sucrose agar (YES), which is a very effective general secondary metabolite production medium (Smedsgaard, 1997). Culture extracts are prepared by cutting plugs of ca. 6 mm diameter from different parts of the colonies. The plugs are transferred to 1.5 ml vials and a solvent mixture of methanol-dichloromethane-ethyl acetate (1:2:3) containing 1% formic acid is added. The plugs are extracted ultrasonically. The extracts are transferred to clean vials and the organic phase is evaporated to dryness. The residues are re-dissolved ultrasonically in methanol containing formic acid, hydrochloric acid and water. Evaporation is necessary to remove the ethyl acetate. The samples are filtered before HPLC analysis (Smedsgaard, 1997). The identification of fungal species takes place by image analysis of HPLC chromatograms (Thrane et al. 2001).

DNA Extraction

DNA is often extracted from fresh mycelia with SDS lysis buffer preceded by one or two phenol and/or phenol/chloroform extraction and RNAase treatment (Lee and Taylor, 1990, Karjalainen and Kammiovirta 1994, Yli-Mattila et al. 1996). Alternatively, to eliminate the use of toxic phenol the extraction buffer contains Proteinase K and denaturation of proteins is done using chloroform/octanol without RNAase according to Yli-Mattila et al. (1998). DNA can also be extracted using SDS lysis buffer together with one phenol and one chloroform/octanol extraction (Paavanen-Huhtala, 2000). The DNA yield of one chloroform

/octanol extraction varies from 1 to 20 ng/µl in a total volume of 25 µl. DNA isolation from young mycelia of *Fusarium* is easy, in the sense that they do not produce disturbing amounts of polysaccharides or other inhibitory substances which would require extra purification steps.

A better yield can be obtained, if the extraction is performed with commercial kits, e.g. with PUREGENE® DNA Isolation Kit D-6000A (Gentra Systems, Minneapolis, USA) using a modification of the manufacturers' protocol (Van Elsas et al. 2000; Yli-Mattila et al. 2004a). DNA can be extracted from the plant material with the chloroform/octanol method, with the DNeasy Plant Mini Kit (Qiagen, Hilden, Germany) according to the instructions of the manufacturer, or with the CTAB methods (Weising et al. 1995 or Doohan et al. 1998).

Good results for PCR from ground grains have been obtained by the method of Kamalay et al. (1990) modified by Konstantinova et al. (2002). In this method 1 g of ground grains is extracted three times with 5 ml of cold (4°C) chloroform:methanol (2:1,v/v) and filtered. When the grain powder is dried, 5 ml of extraction buffer (100 mM Tris-HCl, pH 8.0; 50 mM EDTA, pH 8.0; 500 mM NaCl; 10 mM β-mercaptoethanol) is added and after short vortexing 650 µl of 10% SDS is added after which is the 30 minutes incubation in 65°C. KOAc is added to a final concentration of 1.5 M and incubation is continued on ice for 30 min. Precipitated proteins are removed by centrifuging (12000 rpm, 30 min). Supernatant is filtered and equal amount of cold (–20°C) isopropanol is added. Samples are incubated at –20°C for 30 min and centrifuged with a maximum speed at 0°C for 20 min. The DNA pellet is dried and dissolved in 500 µl of TE buffer by incubating at 65°C for an hour. Purification is continued by extracting once with phenol and twice with chloroform:isoamyl alcohol. NH_4OAc is added to a final concentration of 2 M and 2 volumes of 70% ethanol are added. Samples are centrifuged at maximum speed for 15 min after which DNA is dried and dissolved in TE buffer.

Another extraction method for fungal material situated on the surface of the grains is modified from Taylor et al. (2001). In this method (Yli-Mattila et al. 2004b), 10 g of grains are first soaked for 5 min and then vortexed for one min in 20 ml of TE buffer (10 mM Tris, 1 mM EDTA, pH 7.5). The mixture is then transferred to Bioreba bags, in which the grains are treated with a Bioreba hand homogeniser. Filtered seed soak (15 ml) is centrifuged (10000 rpm at 4°C) and DNA is extracted with the Dneasy® Plant Mini Kit of Qiagen, GenElute™ Plant Genomic DNA Kit of Sigma or with the chloroform/octanol method. Suitable methods for direct extraction of fungal DNA from soil have been described for example by Van Elsas et al. (2000).

RAPD-PCR

RAPD primers can be either synthesized by, e.g. Applied Biosystems DNA synthesizer or purchased from different companies. Amplification conditions can be modified from those of Williams et al. (1990). The overall changes include decreasing the reaction volume from 50 µl to 25 or 20 µl and using different reaction buffers (Karjalainen and Kammiovirta, 1994; Yli-Mattila et al. 1996; Paavanen-Huhtala et al. 1999; Paavanen-Huhtala, 2000). The annealing temperatures and times are also optimised according to the species, primers and thermal cyclers used. Amplification products are usually electrophoresed in agarose gels. Bands are detected visually from photographs or by image analysis from the negatives and recorded in a 0/1 (absent/present) matrix.

In normal RAPD amplifications, 0.1-2 ng of DNA is usually used. In the sensitivity testing of strain-specific primers, an amplification product was obtained when the amount of DNA was 5 pg or more (Paavanen-Huhtala et al. 2000). This result is based on evaluation of the DNA amount by comparison to standards after electrophoresis and dilution assay. Using these methods, the exact concentrations of small amounts of DNA are difficult to determine. The result obtained, however, is in accordance with the results of the study of Schilling et al. (1996), in which 50 pg of *F. culmorum* and 5 pg of *F. graminearum* DNA were sufficient to yield PCR products with species-specific primers. In the primer sensitivity testing of Tooley et al. (1997), 10 pg of *Phytophthora infestans* DNA with ethidium bromide staining produced a detectable PCR product.

Ribosomal DNA RFLP

The ITS1-5.8S rDNA-ITS2 region can be amplified with primers ITS1 and ITS4 (White et al. 1990), while the IGS region can be amplified with primers CNS1 and CNL12 (Appel and Gordon, 1996; Yli-Mattila et al. 2004a; Fig. 5). There are suitable primers available also for amplification of conservative regions of ribosomal DNA (White et al. 1990). The digestion of amplified rDNA (e.g. Konstantinova and Yli-Mattila, 2004) is performed according to the instructions of the manufacturer for restriction endonucleases.

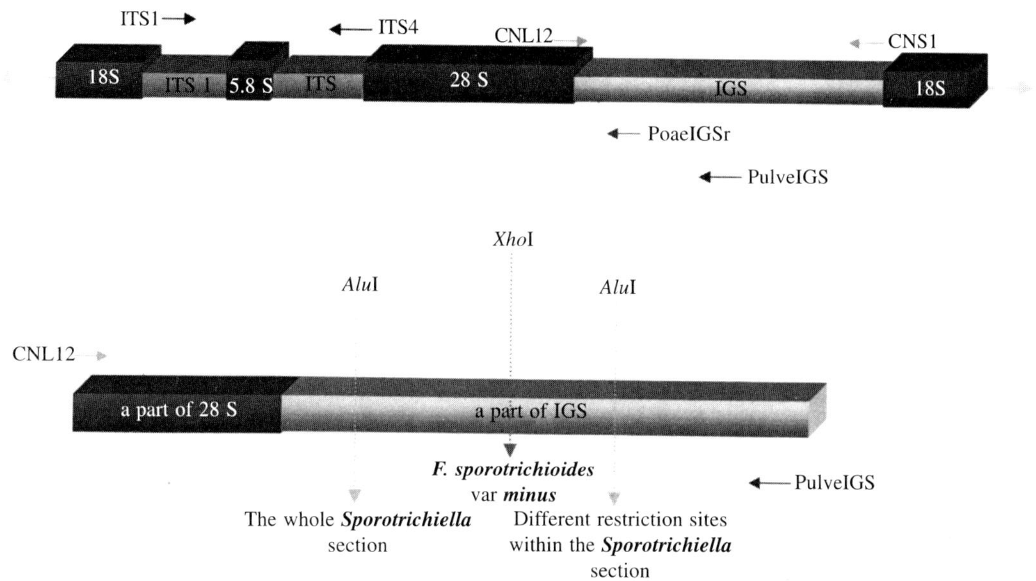

Fig. 5. Species- and strain-specific primers and restriction enzymes for ribosomal DNA region (Konstantinova).

Sequencing and Design of Specific Primers

PCR products can be purified and sequenced by automatic sequencing (Yli-Mattila et al. 1997b, 2004a). The RAPD-PCR products have to be cloned before sequencing, if they have identical ends, while the PCR products from Sequencing with Arbitrary Primer pairs (SWAPP) method (Burt *et al.* 1994) can be sequenced directly like other PCR products. The DNA fragments to be cloned can be cut from special agarose gel after electrophoresis and the DNA can be extracted using β-agarase enzyme and choloroform/isoamyl alcohol or by simply melting the agarose (Paavanen-Huhtala et al. 2000). Clonings can be performed using different cloning kits following the manufacturer's instructions. Plasmid DNA can be isolated using alkali lysis method or the slightly modified boiling method (Sambrook et al. 1989), or with the Quantum Prep kit (Bio-Rad Laboratories, Richmond, CA, USA). After isolation of plasmid DNA with the boiling method, DNA is purified (Paavanen-Huhtala, 2000). DNA fragments ligated into plasmids are sequenced using primers designed for the plasmids. DNA sequences can be aligned using the Clustal W program (Thompson et al. 1994). Similar DNA sequences are searched from DNA Data Banks.

Specific primers can be designed as sequence-characterised amplified region (SCAR) primers according to Paran and Michelmore (1993). It is better that these primers contain as much as possible of the original RAPD primers and not the inner part of the PCR product (Paavanen-Huhtala et al. 2000). Species-specific primers have also been designed based on known differences in ribosomal DNA sequences (Konstantinova and Yli-Mattila, 2004, Fig. 5).

RAMS, UP-PCR and Dot-blot Hybridization

Amplification with random amplified micro satellite (RAMS) primer and electrophoresis of amplification products is done according to Hantula et al. (1996) and Paavanen-Huhtala et al. (2000). In the amplification of universally primed (UP) primers, the optimised reaction mixture of 25 μl contains 0.4 mM dNTPs, 40 ng of primer, 2.5 μl 10x Dynazyme reaction buffer and 0.5 U Dynazyme polymerase (Finnzymes, Espoo, Finland). The program for UP-PCR is described in Paavanen-Huhtala et al. (2000). The dot-blot hybridisation with UP-PCR amplification products is done according to Yli-Mattila et al. (1998), Bulat et al. (1998) and Yli-Mattila et al. (2004a).

Data Treatment and Phenetic and Phylogenetic Analyses

The data produced by the isozyme, RAPD-PCR and rDNA RFLP can be analysed by the unweighted pair group method with arithmetic mean (UPGMA) of NTSYS-pc 1.70 (Rolf, 1989) and Wagner (PENNY program) parsimony and neighbor-joining (NJ, NEIGHBOR program) analysis of the software package PHYLIP 3.5 (Felsenstein, 1993). The fragment sizes can be estimated and the banding patterns within a gel and between the gels compared by using different image analyses (Konstantinova and Yli-Mattila, 2004). The consensus trees are constructed using CONSENSE program and the reliability of the phylogenetic trees constructed by the NJ-method is explored by a bootstrap re-sampling using the SEQBOOT program of PHYLIP 3.5 (Yli-Mattila et al. 1997a; Paavanen-Huhtala et al. 1999; Konstantinova and Yli-Mattila, 2004). In addition to the most parsimony trees, it is also advisable to examine trees that are longer. The program RNA (Farris, 1994) can be used for testing the degree of support for individual branches by constructing the strict consensus tree of all trees one step longer than the most parsimonious one(s), then two steps longer, and so forth (Bremer, 1988). The number of extra steps, at which the support for the clade disappears, is its Bremer or branch support (Farris et al., 1994). The congruence of the data matrices and tree topologies obtained from different data can be compared with programs KON (Farris et al. 1994) and Component 2.0 (Page, 1993a, b) as was done by Paavanen-Huhtala et al. (1999).

Due to the large size of the data matrix, it is not always possible to use algorithms that would ensure the detection of all of the most parsimonious trees. Therefore, the heuristic search mode is used. To find the most parsimonious trees, parsimony analysis with the heuristic search mode of PAUP 3.1.1 can be used (Swofford, 1993) together with programs Hennig 86 (Farris, 1988) and joyride of the Random Cladistics package of Siddall (1995). For phylogenetic analyses of the aligned sequence data, programs DNADIST and DNAPENNY of PHYLIP 3.5 can also be used. POY 2.7 program (Wheeler, 1996; Gladstein and Wheeler, 2001) can be used for analysing DNA sequences without previous alignment and excluding of gaps (Yli-Mattila et al. 2002, 2004a).

Discussion

Fingerprinting Methods

Since the morphological identification of species or strains in fungi is sometimes difficult, new methods, based on protein or DNA, have been developed. The advantages of these new methods are their rapidity, their sensitivity and usually their cost-effectiveness. They may also be suitable for integration into large-scale certification, and standardised test kits can be produced (Janse, 1995). The new methods, however, still contain some pitfalls. For example, specificity and reproducibility may be unknown or tested only to a limited extent. The results may also be affected by changes in experimental conditions; in addition, in PCR-based methods the PCR products obtained are not necessarily derived from the target DNA (Janse, 1995). Modern methods should thus be seen as a welcome addition to already existing ones.

Of the methods based on proteins, isozyme analysis has been a successful tool in studying evolution or genetic variability. Its application to fungal taxonomy and genetics was realised on a larger scale at the beginning of the 80s (Micales et al. 1986). The first type of DNA-based fingerprinting method used was restriction fragment length polymorphism (RFLP), based on the length differences of DNA fragments produced by cutting the DNA with restriction enzymes. With the introduction of the polymerase chain reaction (PCR) (Saiki et al. 1985) a faster approach to genome characterisation was found. Three groups described a polymorphism assay independently and simultaneously. Williams et al. (1990) called their assay "random amplified polymorphic DNA" (RAPD), while Welsh and McClelland (1990) called theirs "arbitrarily primed PCR" (AP-PCR). Both of these assays are based on the observation that a single oligonucleotide of a randomly chosen DNA sequence, subjected to conditions resembling those of the polymerase chain reaction, will prime the amplification of several DNA fragments (Rafalski et al. 1991). The third PCR fingerprinting method was developed simultaneously and was called "universally primed PCR" (UP-PCR) (Bulat and Mironenko, 1990). The random amplified micro satellites (RAMS) technique is a newer genetic fingerprinting method for studies of intraspecific variation (Zietkiewicz et al. 1994; Hantula et al. 1996). Sequencing too has become an increasingly popular method during the last years, mainly due to the developed automation of sequencing. There are also other PCR fingerprinting methods, resembling the above methods but differing mainly in the length of the primers and annealing temperatures used in amplification. More information about other fingerprinting methods can be found else where, e.g. in Caetano-Anolles (1994), Vos et al. (1995), Pejic et al. (1998), Goldstein and Schlötterer (1999), Manninen (2000) and Mach et al. (2004).

Isozyme Analysis

Fingerprints produced by this method are based on the presence of multiple molecular forms (isozymes) of certain enzymes. These forms have similar enzymatic properties, but differ slightly in amino-acid sequence and therefore in electrophoretic mobility (Bonde et al. 1993). Detectable isozymes arise from different alleles at a single locus, from multiple loci coding for a single enzyme, or from post-translational processing and formation of secondary isozymes (Micales et al. 1986). Some enzymes may also have several stable configurations varying in tertiary and quaternary structure, and these will frequently have different electrophoretic mobilities.

Sufficient quantities of material are needed in order to extract enough enzymes for detection (Micales et al. 1986). This is usually not a problem with fungi, except with obligate pathogens, which cannot live without a living host in artificial growth media (Bonde et al. 1993). The protein extraction method of Rosendahl and Sen (1992) has proved to be a viable method. The protein concentration from ca. 0.3 g of fresh mycelia, according to spectrophotometric determination (Bradford, 1976), is approximately 4-8 μg/μl (the final volume is ca. 50-100 μl; Paavanen-Huhtala, 2000). If the sample material is grown in culture, conditions of light, temperature and nutrition should be uniform in order to minimise the differential expression of induced enzymes (Micales et al. 1986). Different electrophoretic techniques, such as starch gel, polyacrylamide gel electrophoresis and isoelectric focusing, may be used to separate isozymes in their native form (Fig. 6). Detection is performed with an appropriate colour-forming substrate for a particular

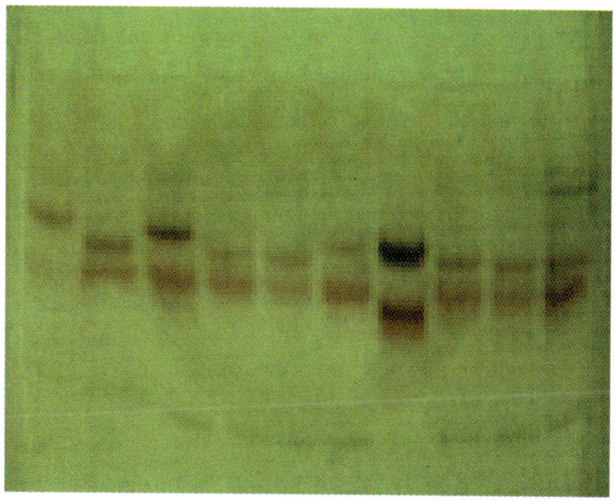

Fig. 6. Esterase patterns of *Fusarium avenaceum* isolates.

enzyme. The technique is relatively simple and cheap, depending on the electrophoretic technique used, and numerous genetic loci may be compared. The isozyme analysis only detects significant differences in enzyme structure, which are usually at the species level (Bonde et al. 1993). The method is relatively time-consuming and standardisation is not easy. Isozyme patterns may or may not be congruent to RFLP patterns (Bonde et al. 1993).

Metabolite Prolifing

Nielsen et al. (1999) have identified fungal strains by applying chemical image analysis techniques to chromatographic matrices of crude fungal culture extracts separated by high performance liquid chromatography with diode array detection (HPLC-DAD). By a multidiciplinary approach it has been possible to characterise *Trichoderma* isolates at species level (Lubeck et al. 2000, Thrane et al. 2001). This method also seems useful for analysing isolates of *F. avenaceum/F. arthrosporioides/F. tricinctum* species complex (Yli-Mattila et al. 2005).

RFLP

Restriction patterns are generated by cleavage of DNA with restriction enzymes and by size separation of the resulting fragments via gel electrophoresis. Patterns obtained by RFLP are based on alterations in the DNA sequence that changes the recognition sequence of a restriction enzyme (Janse, 1995). Total DNA can be analysed in an agarose gel and fragments visualised with ethidium bromide. However, the fragments are usually too numerous and overlap with each other. Because of this problem, probes and Southern blot hybridisation are used, which render visible only the fragments, which hybridise to the probe (Janse, 1995). The RFLP assay can be used to identify co-dominant traits, since more than one allele can be detected at a locus (Rafalski et al. 1991). The fragment patterns may be compared in two ways: the pattern of fragments is analysed or a mapping approach is employed, using the pattern of fragments to deduce a map of the enzyme sites (Bruns et al. 1991). By means of mapping, the physical relationships of restriction fragments can be determined. However, incomplete digestion of DNA can easily result in errors in the map. Large quantities of DNA are required for RFLP analysis and the technique is difficult to automate. Moreover, in order for this to be a fast and useable method, sizeable laboratories and specialised equipment are needed (Pejic et al. 1998).

One variation of the RFLP method is ribosomal DNA RFLP analysis (rDNA RFLP, ribotyping) in which PCR amplification of rDNA is followed by digestion with restriction enzymes. Ribosomal DNA has proved to be a useful region to study, because sequence data are available and it can be used for any organism. Ribosomal RNA genes are tandemly repeated multigene families. In fungi each repeat contains a copy of the coding 18S (small subunit, SSU), 5.8S and 28S (large subunit, LSU) genes, which have relatively few differences between species. Much more variable regions are the non-coding internal transcribed spacer (ITS) regions flanking the 5.8S gene and intergenic spacer regions (IGS) between 28S and 18S genes (White et al. 1990). The IGS regions are even more variable than ITS regions. IGS regions may contain 5S gene or tandem arrays of this gene (Duchesne and Anderson, 1990; Liew et al. 1998). Of the non-coding regions the externally transcribed spacer (ETS) and ITS are transcribed, while the non-transcribed spacer (NTS) is not. Depending on which rDNA region is the target for investigation, rDNA can be used to differentiate genera, species or subspecies (Bruns et al. 1991). ITS-RFLP results showed identical restriction profiles for all investigated *F. oxysporum* and *F. avenaceum* isolates and did not separate *F. tricinctum* isolate t43 from *F. avenaceum* isolates (Paavanen-Huhtala et al. 1999; Paavanen-Huhtala, 2000), while IGS patterns have showed variation within *F. oxysporum* (Appel and Gordon, 1995), *F. avenaceum* and *F. tricinctum* (Yli-Mattila et al. 2002) and *F. poae*, *F. langsethiae* and *F. sporotrichioides* (Yli-Mattila et al. 2004a, Konstantinova and Yli-Mattila, 2004). The RFLP patterns of IGS sequences have been used for identification of closely related *F. poae, F. sporotrichioides* and *F. langsethiae* isolates (Konstantinova and Yli-Mattila, 2004; Fig. 7), while RFLP patterns of ITS sequences are useful for identification of *F. avenaceum, F. oxysporum, F. redolens* and *F. commune* isolates (Paavanen-Huhtala et al, 1999).

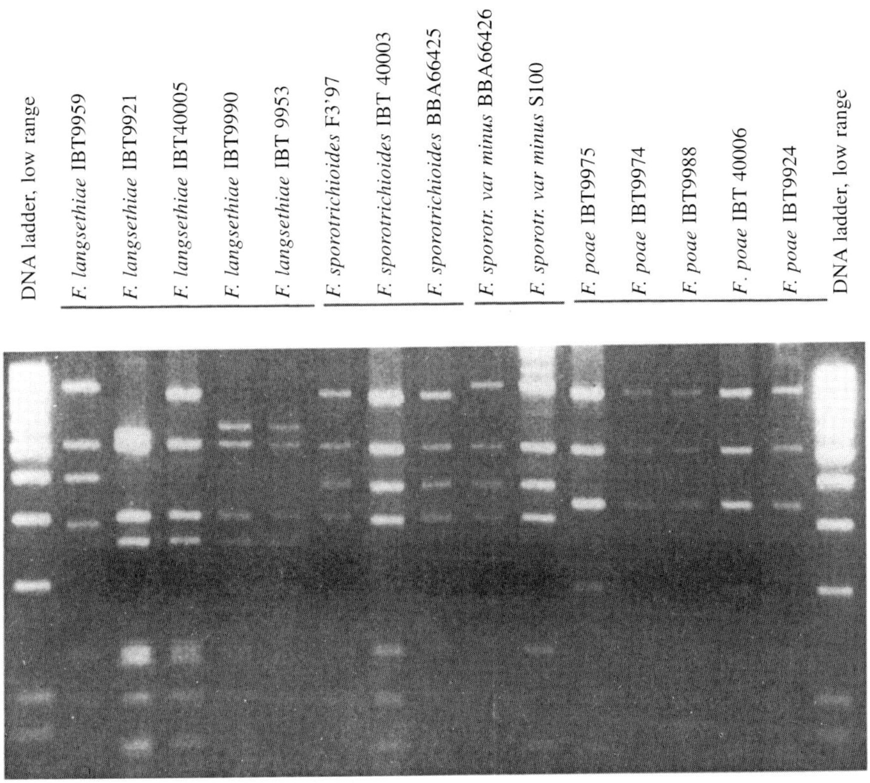

Fig. 7. IGS-RFLP patterns after digestion of CNL 12/CNS1 amplification products of *F. langsethiae, F. sporotrichioides, F. sporotrichioides var minus* and *F. poae* with *Msp*I (Konstantinova and Yli-Mattila, 2004).

Thus, while IGS-RFLP analysis does not give as much information as IGS sequencing, it is faster and cheaper and gives useful phylogenetic information about *Fusarium* species. When PCR and RFLP techniques are used together, it is possible to find even a species-specific mitochondrial rDNA marker, based on single nucleotide polymorphism (SNP), as for example between *Tuber* species (Mabru et al. 2004).

Random Amplified Polymorphic DNA (RAPD) and Arbitrarily Primed PCR (AP-PCR)

RAPD primers are short, typically 10 bp in length, and are used in low stringent amplification conditions (annealing 35-40 °C). A single primer will support DNA amplification from a genomic template if binding sites of the correct orientation on opposite strands of the template exist within an amplifiable distance (up to several thousand nucleotides) of each other (Rafalski et al. 1991). The nature of the fragments that are amplified is highly dependent on the primer sequence and on the DNA sequence of the genome being assayed (Rafalski et al. 1991). Sources of polymorphisms may include single nucleotide changes, deletions of a priming site, or insertions that prevent amplification or change the size of amplified fragment (Rafalski et al. 1991).

RAPD markers are well suited for gene mapping in plant and animal breeding applications and for DNA fingerprinting, with particular utility for studies of population genetics (Williams et al. 1990, Fig. 8). Like most molecular markers, the information content of an individual RAPD marker is very low, but when a large number of these anonymous markers are used to define the genome they begin to have utility (Williams et al. 1990). Nearly all RAPD markers are dominant. This means that a RAPD assay detects the presence of only one allele at a locus and it cannot be used to distinguish heterozygous loci (Rafalski et al. 1991). The absence of an amplified fragment represents all other alleles at that locus that fail to prime amplification. Co-dominant RAPD markers, observed as different-sized DNA segments amplified from the same locus, have been observed only rarely (Williams et al. 1990). In the study of Devos and Gale (1992) it was observed that the majority of amplification products using random decamer primers arose from repetitive DNA sequences. However, in the study of Crowhurst et al. (1991) most of the amplification products arose from single-copy sequences. These different results are due to the differences in genome size and in the level of repetitive DNA in the genomes of the species investigated.

653

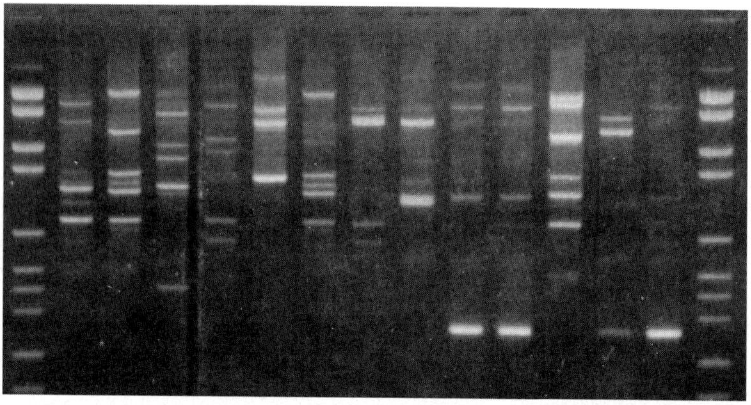

Fig. 8. RAPD-PCR patterns of *Beauveria* strains with primer OPA-1. The stains are: 1, 2, 8, 12, 13, 33, 135, 139, 252, 317, 384, 2624, 2990 (Mitina and Yli-Mattila, 2002). Molecular markers (2176, 1766, 1230, 1033, 653, 517, 473, 394, 298, 234, 220 and 154 bp) are on the left and on the right.

The advantages of the RAPD assay are that it does not require preliminary knowledge of the target DNA sequence or preparation of species-specific probes and filters (Williams et al. 1990). Also the amount of DNA required is small. Technically the RAPD assay is a simple and rapid method, in which radioactivity is not needed (Foster et al. 1993). Provided that a thermo cycling instrument is available, the start-up costs for the RAPD assay are minimal. A universal set of primers can be used for genomic analysis in a wide variety of species (Williams et al. 1990).

Concerns over the use of the RAPD assay are mainly due to problems related to reproducibility (Devos and Gale, 1992; Majer et al. 1996; Skov, 1998). RAPD reactions have been shown to be sensitive to changes in reaction components and to the temperature conditions under which amplifications are performed (Tommerup et al. 1995). This may be due to the short primers, which require a low annealing temperature; this makes annealing to the template less stable and sustainable, and can lead to mismatches (Lübeck, 1997). With highly controlled PCR conditions, the results can be made more reliable. Bowditch et al. (1993) have suggested that only strong amplification products should be chosen over faint products when scoring individuals or populations, since the strongest ones are more reproducible. Contamination of the DNA sample with foreign DNA can result in false amplification products, since the decamer primers amplify DNA from virtually all sources (Bachmann, 1994). Impurities in the DNA or amplification mixture which inhibit PCR amplification are also a significant source of false-negative results and nonreproducibility

(Bowditch et al. 1993). These risks can be avoided by repeating the RAPD assays, preferably with DNA from different isolations or with diluted DNA. One disadvantage is that fragments of equal size in different individuals may be the result of amplification from a different region of the genome. This is more probable when distant relatives are examined (Bowditch et al. 1993). If fragment size is used as an indicator of homology, then gel resolution becomes critical for the correct classification of fragments (Rieseberg, 1996). Homology can be ascertained by hybridisation, by restriction enzyme cutting or by sequencing the amplification product.

The primers used in AP-PCR are 20-34 bp in length, and the primers mainly used are gene-specific and have originally been designed for other purposes, e.g. primers M13, T7, T3 (Welsh and McClelland, 1990). Thus, AP-PCR requires no prior knowledge of the molecular biology of the organism under study. The amplification initiated at low annealing temperatures similar to the RAPD assay, but the following amplification cycles have more stringent conditions (annealing 60°C), including radioactive incorporation (Welsh and McClelland, 1990). Each primer yields a different pattern of AP-PCR products, allowing the differentiation of even closely related strains of the same species (Welsh and McClelland, 1990). The banding patterns resulting after polyacrylamide gel separation and autoradiography detection are more complex than the fingerprints revealed by RAPD. AP-PCR is also more expensive and labour-intensive compared to RAPD, because of the use of radionucleotide incorporation.

To increase the reproducibility of RAPD markers, sequence-characterised amplified region (SCAR) DNA analysis has been developed (Paran and Michelmore, 1993). SCAR primers are extended RAPD primers designed after sequencing the RAPD fragment and are thus gene-specific like any common PCR primer. SCAR markers are therefore less prone to experimental variation than RAPD markers (Parent and Pagé, 1998). SCAR DNA analysis is technically similar to standard PCR analysis, with more stringent annealing temperatures than in RAPD analysis. By using species-specific SCAR primers it is possible to detect *Fusarium* species directly from, e.g. grain samples (Fig. 8) without the need of isolating *Fusarium* pure cultures from the grains.

From the RAPD technique, Burt et al. (1994) have developed a method called sequencing with arbitrary primer pairs (SWAPP). In this method two different arbitrary primers are used, and products, which are not present when either primer is used alone are sought out because they have a different primer at each end and can be directly sequenced (Burt et al. 1994). According to Williams et al. (1993), fewer additional bands are obtained with primer pair combinations compared to two separate reactions using each primer individually. The SWAPP method allows distinguishing between heterozygotes and homozygotes by generating co-dominant markers, which allow searching for length polymorphisms (Burt et al. 1994; Paavanen-Huhtala, 2000).

Universally-Primed PCR (UP-PCR)

The UP-primers consist of two parts: the core (5′-end, 6-8 nt) and the variable 3′-end (8-10 nt). The 5′-end is capable of generating numerous PCR products with any kind of template DNA and it stabilises the primer hybridisation at high annealing temperatures (Bulat and Mironenko, 1992; Bulat et al. 1998). In the core, mismatches are possible. The 3′-end of the primer is the most important part of the primer. It has to match the target sequence with 100% reliability. The 3′-end of a primer being random by nature is designed to avoid the amplification of phylogenetically conserved regions of the genome (Bulat and Mironenko, 1990). When the UP-PCR technique was elaborated (Bulat and Mironenko, 1990, 1992), the random sequences at the 3′-end of UP primers were tested experimentally against the panel of representative organisms, to ensure that every primer will work with any DNA source. They were also tested against known databases to avoid their strong matching to any known encoding sequences. Due to the design of primers, priming is primarily targeted to intergenic, more variable areas of the genomes (Bulat et al. 1998). The primers are used with annealing temperatures between 52 and 60°C, with fast ramping PCR machines, and amplification products are separated on agarose or polyacrylamide gels. UP-PCR requires pure cultures of the strains, similar to RAPD-PCR and AP-PCR. On the whole, UP-PCR patterns are less variable compared with RAPD profiles (Bruns et al. 1991). It can be more difficult to find strain-specific fragments for closely related strains among UP-PCR patterns compared to RAPD patterns; in the latter case the variation may also be due to the non-reproducibility of RAPDs. UP-PCR products can also be used directly as hybridisation probes (Yli-Mattila et al. 1997a) and are useful in direct sequencing in cases of pair-wise combination of primers (Bulat et al. 1998).

The UP-PCR technique is more reproducible than the RAPD method. In the study of Lübeck et al. (1999) exactly the same banding profiles were obtained in two different laboratories using different thermal cyclers and PCR protocols with up to 4°C difference in the annealing temperatures. UP-PCR provides more information per primer applied, so that large amounts of primers need not be tested (Lübeck, 1997). According to the results of Bulat, Alekhina, Lebedev and Yli-Mattila (unpublished results, personal communication) the UP-PCR protocol (high primer annealing temperature with low Mg/dNTPs ratio buffer and high ramping speed) also gives more fragments per primer and the fragments are more reproducible than in common RAPD-PCR (low annealing temperature with higher Mg/dNTPs ratio buffer and low ramping rate), when used with RAPD-PCR primers. UP-PCR can be used for the identification and genotyping of strains and for the delineation of species, subspecies and races (Bulat et al. 1995). The selective amplification of evolutionarily less conserved areas allows discrimination of individuals belonging to closely related strains, depending on the primers chosen (Lübeck, 1997).

Random Amplified Micro Satellites (RAMS)

Zietkiewicz et al. (1994) have demonstrated a method for measuring genetic diversity in plants and animals by means of primers based on micro satellites. The technique combines most of the benefits of micro satellite analysis with the universality of RAPD analysis. RAMS was proposed as a short name for the technique by Hantula et al. (1996). The primers are usually about 18 bases long, including a 5′-end degenerate anchor to ensure annealing at the beginning of the micro satellite (Hantula et al. 1996). The DNA fragments amplified in RAMS reactions consist of two micro satellites located close enough to be amplified in PCR and the area between them (Zietkiewicz et al. 1994). RAMS should detect polymorphisms with greater likelihood than RAPD (Hantula et al. 1996). Sources of variability in RAMS fragments can be mutations at the priming site, the outcome of an insertion/deletion event within the amplified fragment, or variability in the number of micro satellite repeats (Hantula et al. 1996).

Denaturing gradient gel electrophoresis (DGGE) and amplification refractory mutation system (ARMS)-PCR: Denaturing gradient gel electrophoresis (DGGE) is based on the restriction-enzyme independent detection of DNA sequence variations like single-base substitutions (Eng and Vijg, 1997). In DGGE analysis the sequence alterations cause changes in the melting temperature of double stranded DNA, which can be analysed by a linearly increasing gradient of DNA denaturants established in polyacryalamide gel (Van Elsas et al. 2000; Mach et al. 2004). The separation power of denaturing gels is good and DGGE has been shown to reliably separate strains of different *Fusarium* species even from mixtures and in the presence of their natural hosts corn, wheat and oats (Mach et al. 2004). DGGE also allow the resolution of mixtures of PCR products of several fungal species in soil (Van Elsas et al. 2000).

Amplification refractory mutation system (ARMS)-PCR enables the design of multiple primers against specific nucleotide areas in a given gene fragments bearing single-base changes or small deletions by multiplex PCR (Ferrie et al. 1992). It is possible to differentiate between closely related *F. langsethiae, F. sporotrichioides* and *F. poae* strains according to the number of amplicons produced by two primer pairs of ARMS-PCR designed for known ß-tubulin sequences (Mach et al. 2004). ARMS-PCR is suitable only for identification of pure cultures, but it is relatively simple and quick to use.

Amplified fragment length polymorphism (AFLP) analysis: Amplified fragment length polymorphism (AFLP) analysis is a combination of RFLP analysis and PCR, resulting in a highly informative fingerprints (Vos et al. 1995, Fig. 9). No previous information on the genome is required and AFLP protocol needs only minor adaptation to the group of organisms studied. In the first step, genomic DNA is digested with a restriction enzyme. Adaptors of known sequences are then ligated to both ends of all restriction fragment and PCR is performed using specifically designed primers allowing only a subset of restriction fragments to be amplified. The 5′-ends of the primers is made complementary to the adaptors, while the 3′-ends extend for a few arbitrarily chosen nucleotides into the restriction fragment. Exact matching of the 3′-end of a primer is necessary for amplifications. Therefore, only those restriction fragments are amplified, the ends of which basepair with the 3′ primer extensions. For example, 1/16 of the total set of fragments in the case of one-nucleotide extension and 1/256 in the case of two-nucleotide extension are amplified. The amplification products are separated on highly resolving sequencing gels. Typically 50-100 restriction fragments are detected in each AFLP reaction. The fragments are labelled by radioactive labels or they can be detected by silver staining method (Creste et al. 2001; Abd-Elsalam et al. 2002, 2003a, b). Fluorogenic labels of primers can be used for automated fluorescence analysis (Grundmann et al. 1995; Schmidt et al. 2004b). The bands of interest can be cut out, sequenced and transformed into SCARs by designing specific primers like RAPD bands (Schmidt et al. 2003).

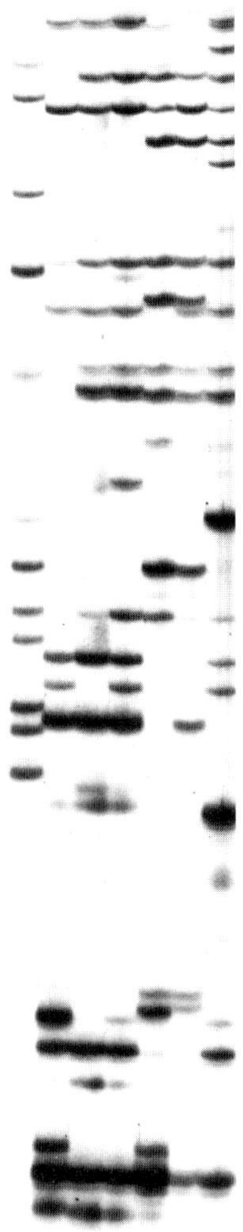

Fig. 9. AFLP patterns obtained in *F. avenaceum* and *F. arthrosporioides* isolates (Yli-Mattila, Paavanen-Huhtala, Klemdal, unpublished results).

In addition to molecular fingerprinting and studying relationships among fungal isolates at the population and species level AFLP analysis is useful for high-resolution mapping by bulked segregant analysis (Thomas et al. 1995; Cnops et al. 1996) and marker-assisted selection (Dahleen et al. 2003). It can also be adapted for use with cDNA synthesized from mRNA (Bachem et al. 1996). AFLP has been found suitable for analysing genetic diversity between closely related *Fusarium* isolates (Kiprop et al. 2002; Sivaramakrishnan et al. 2002; Arroyo-Garcia et al. 2003; Abd-Elsalam et al. 2002, 2003a, b; Zeller et al. 2003; Schmidt et al. 2004a). AFLP generally resolves taxa with high resolution, but like RAPD-PCR and UP-PCR it is not as well suited for distinguishing fungi at the level above the species. AFLP and UP-PCR methods provide similar results, but AFLP markers give better resolution, especially if AFLP is run on polyacrylamide gel and UP-PCR on ordinary agarose gel (Abd-Elsalam et al. 2002).

Quantitative PCR: The most suitable present techniques for accurate DNA quantification are the PCR coamplification of target and competitor DNA (Gilliland et al. 1990) and the real-time PCR product quantification initially developed for monitoring fluorescence emission of ethidium bromide at each cycle (Higuchi et al. 1992). Later variations of the real-time quantitative PCR include new DNA-binding dyes such as the SYBR Green (Morrison et al, 1998) fluorescent oligonucleotides (Whitcombe et al. 1999), molecular beacons (Tyagi and Kramer, 1996) fluorescence resonance energy transfer (FRET) probes (Wittwer et al. 1997) and TaqMan probes (Heid et al. 1996). The quantification of the template is more accurate with fluorescence labels than with SYBR Green. Two primers allow amplification of the product to which a third dual-labelled fluorogenic oligonucleotide, the TaqMan probe, will anneal. During amplification the Taq polymerase releases the 5′ fluorescent tag from the annealed TaqMan probe, yielding a real-time fluorescence directly proportional to the concentration of the target sequence. In TaqMan technology, the primers and probes are designed to fit the fixed reaction conditions, while in the end point PCR the primers are often designed first and the reaction conditions are optimised for the primers.

By now competitor DNA (Nicholson et al. 1998, 2003), solid phase PCR assay (DIAPOPS, Niessen and Vogel, 1998) and SYBR Green real-time PCR product amplification (Schnerr et al. 2001; Fig. 10) have

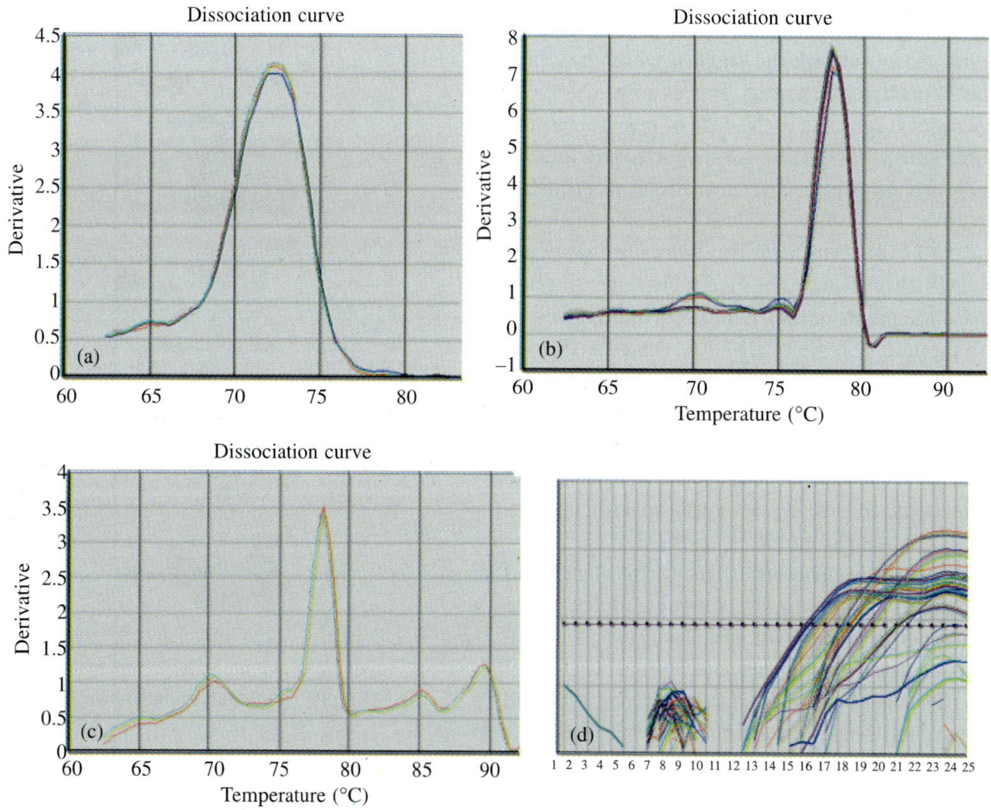

Fig. 10. Quantitative PCR with SYBR Green (T. Yli-Mattila and S. Paavanen-Huhtala, unpublished results). (a) Negative controls with JIA primers specific for *F. avenaceum*. Only primer dimers are formed. (b) Positive controls with *F. avenaceum* specific JIA primers. Only specific product is formed. (c) Grain sample with *F. avenaceum* specific JIA primers. Specific product and a little bit of primer dimers and two non-specific PCR products are formed. (d) The cycle number required for the threshold value of *F. avenaceum*-specific PCR product in real-time quantitative PCR depends on the amount *F. avenaceum* DNA in grains (T. Yli-Mattila and S. Paavanen-Huhtala, unpublished results).

been used for quantification of *Fusarium* fungi, but several research groups are now also developing TaqMan real-time PCR quantification methods for *Fusarium* fungi. Competitive PCR requires the construction of an internal standard which has the same primer sites as the target DNA but the resulting PCR product is a different size. So, the target and internal standard PCR products are distinguishable on a agarose gel after electrophoresis (Gilliland et al, 1990). Real time PCR does not require post-PCR handling, which reduces the risk of cross-contamination and handling errors. It is also faster and less labour intensive (Heid et al. 1996). The disadvantage of the real time system is the cost of the specialised equipment that is capable of real time fluorescent measurement of PCR reactions and the specialised reaction kits compared with cheaper gel image analysis system and standard reaction kits in competitive PCR.

Methods have also been developed for reverse transcription (RT)-PCR detection and quantification of the *Tri5* gene encoding trichodiene synthase, which catalyzes the first reaction in the trichothecene biosynthetic pathway (Doohan et al. 1999). The advantage of RT-PCR is that it will only detect mRNA, which is short-lived. To detect toxigenic *Fusarium* fungi at harvest a PCR assay is more appropriate, as these fungi usually produce the mycotoxins prior to harvest and during harvesting only target DNA but not mRNA will be present. A RT-PCR test would be more appropriate to detect the presence of growing mycotoxigenic *Fusarium* fungi on foodstuffs during storage and processing.

DNA Sequencing

There are several factors to consider in selecting a region for sequence analysis. For example, the region should be evolving at an appropriate rate for the comparison in question, and the region should be present ideally as a single copy (Bruns et al. 1991). Most sequence studies in fungi have focused on rRNA genes (Figs. 11 and 12). Their popularity is due primarily to the presence of universally conserved regions that serve as ideal primer sites (Bruns et al. 1991). According to the sequences obtained, it is possible to design specific primers directed to areas with more variable sequences and facilitate the identification of closely related organisms (Fig. 5). Sequencing results from different laboratories can be directly compared. The publication of sequences and their deposition in electronic databases (e.g. GenBank, EMBL and DNA Databank of Japan) facilitate the confirmation of results and their application to other taxa without the need to obtain strains or to repeat experiments (Bruns et al. 1991).

```
Fa17/Fa38/Fa49

    Far1/Far2       CAACCCTCAAGCCCCCGGGTTTGGTGTTGGGGATCGGCTCTGCC-TTCTGG-CGG

    Fa28            ..............T.............................-......-...

    Fa37            ............................................-..AC..-...

    Ft43            ............................................-.CAC..-...

    Ft92            ............................................C-T....G...

    Fr10/Fr11       ..............T.......................GAGC.T-......-...

    Fo1/Fo2         ..............T.......................GAGC..-..GC..-...

    Fo5/Fo6/Fo7/    ............A.---A.C...........AC...CG.TAATT-CG---.-...
```

Fig. 11. Alignment of the most divergent sequences of ITS2 region in *F. arthrosporioides* (Far1, Far2), *F. avenaceum* (Fa17, Fa28, Fa37, Fa38, Fa49), *F. oxysporum* (Fo5, Fo6, Fo7, X94173), *F. commune* (Fo1, Fo2), *F. redolens* (FR10, FR11) and *F. tricinctum* (Ft43, Ft92) isolates.

```
IBT9931/ CCCCAACTTCGA---CTCTC--CGAGGAGAGCTTGGTAGGGGG-------------------------

IBT9933  --------------------------GAC------AGACACGCAAGGCA-G--CTCCGGCTGCCTG-

BBA69074 ...........---.....--....................-------------------------

         --------------------------...------..............-.--..T..........-

BBA70812 ...........---.....--....................A-C----------------------

         -------------------------A.GC------....CACA......-.--..T........C.-

BBA70813 ...........---.....--....................A-C.---------------------

         -------------------------A.GC------....CACA......-.--..T........C.-

IBT9959  .......GA...GAA...G.AATAGC...G..G.A........G-TCTGAGTG--TGTGTGTGTGTG

         TGTGTGTGTGTGTGTGTGTGTGTGT-G.G.TC--AA.NG.G......-..-CTCA.AT.......C.-

IBT9951  .......GA...GAA...G.AA.AGC.A.G..G.A........G-TCTGAGTG--TGTGTGT-------

         -------------------------G.G.TT--AA.AG.GT........-.TCG.AT.......GA-

IBT9961  .......GA...GAA...G.AA.AG.GA.G..G.A........G-TCTGAGTG--TGTGTGT-------

         -------------------------G.G.TTTTAA..G.GT........-.TCG.AT.......GA-

IBT40011 .......GA...GAA...G.AA.AG.GA.G..G.A........G-TCTGAGTG--TGTGTGT-------

         -------------------------G.G.TT--AA..G.GT........-.TCG.AT.......GA-

IBT9956  .......GA...GAA...G.AA.AGC...G..G.A........GGTCTGAGTG--TGTGTGT-------

         -------------------------G.G.TT--AA..G.GT........-.TCG.AT.......GA-

IBT9946  .......NA...GAA...G.AATAG....G..G.AN..C....G-TCTGAGTG--TGTGTGT-------

         -------------------------G.G.TC--AA..G.GT........-.TCN.AT.T..N..GA-

IBT8804  .......GA...GAA...G.AATAG....G..G.A........G-TCTGAGTG--TGTGTGT-------

         -------------------------G.G.TC--AA..G.GT........-.TCG.A........GA-

IBT9965  .......GA...GAA...G.AA.AG....G..G.A........G-TCTGAGTG--TGTGTGT-------

         -------------------------G.G.TC--AA..G.GT........A.TCG.AT.......GA-

IBT9971  .......GA...GAA...G.AA.AG....G..G.A........G-TCTGAGTG--TGTGTGTGT-----

         -------------------------G.G.TC--AAA.G.GT........-.TCG.AT....G..GA-

BBA69072 .......GA...GAA...G.AA.AG....G..G.A........G-TCTG.GTG--TGTGTGTGTGTGT-

         -------------------------G.G.TC--A...G.GT........-.TCG.AT....G..GA-

IBT1906  .......GA...GAA...G.AA.AG....G..G.A........G-TCT.AGTG--TGTGTGTGT-----

         -------------------------G.G.TC----AAG.GT........-.TCG.AT....G..CGA

IBT40003 .......GA...GAA...G.AA.AG.G..G..G.A........G-TCTG.GTG--TGTGTGTGT-----

         -------------------------G.G.TC--AA.GG.GT...G....-.TCG.AT....G..GA--

IBT1929  .......GA...GAA...G.AA.AG.G..G..G.A........G-TCTG.NTGAGTGTGTGTGTGTGT

         -------------------------GTG.TC--AAANG.GT........-.TCT.AT....G..GA-

BBA62425 .......GA.G.GAA...G.AA.AG.G..G..G.A........G-TCCG.GTG--TGTGTGT-------

         -------------------------GGG.TCAAAA.GG.GT........-.TCG.AT....G..GA-

BBA62426 .......GA.G..AA...G.AA.AG.G..G..G.A........G-TCCG.GTG--TGTGTGT-------

         -------------------------GGG.TCAAAA.GG.GT........-.TCG.AT....G..GA-
```

Fig. 12. Alignment of IGS sequences of *F. poae* (IBT9931, IBT9933, BBA69074), *F. kyushuense* (BBA70812, BBA70813), *F. langsethiae* (IBT9959, IBT9951, IBT9961, IBT40011) and *F. sporotrichioides* (IBT 9946, IBT8804, IBT9965, IBT9971, BBA69072, IBT1906, IBT4993, IBT1929, BBA62426, BBA62426). Isolates BBA62425 and BBA62426 belong to *F. sporotrichioides* var. *minus* (Yli-Mattila et al. 2004a).

Molecular Taxonomy as Compared to Traditional Morphological Taxonomy

Fusarium genus is divided into several sections on the basis of asexual (anamorph) characters. The sexual states of *F. avenaceum* (sections *Arthrosporiella*; *Roseum*), *F. graminearum* (sections *Discolor*; *Fusarium*) and *F. equiseti* (section *Gibbosum*) belong to the teleomorph genus *Gibberella* (Booth, 1971; Gerlach and Nirenberg, 1982). Sexual states have not been found in *F. culmorum* (sections *Discolor*; *Fusarium*), *F. oxysporum* and *F. redolens* (section *Elegans*). According to Guadet et al. (1989) *F. culmorum*, *F. oxysporum* and *F. redolens* are so closely related to *F. graminearum* on the basis of rRNA sequence analysis that they all belong to the teleomorph genus *Gibberella*.

According to the UPGMA dendrogram of RAPD-PCR results (Yli-Mattila et al. 1996), the *F. avenaceum, F. redolens, F. equiseti* and *F. oxysporum* isolates studied, except for two *F. avenaceum* isolates, form their own clusters at a similarity level of 40%. At the same similarity level, *F. culmorum* and *F. graminearum* isolates group together, as was expected according to the previous studies (Booth, 1971; Koopmann et al. 1994; Schilling et al. 1994). This is also in accordance with the latest phylogenetic studies of O'Donnell et al. (2000) based on gene genealogies. Isolates of *F. equiseti* group together with *F. graminearum/F. culmorum* group, though not very closely, while isolates of *F. oxysporum* and *F. redolens* are loosely grouped together in the RFLP analysis of 5.8S-ITS2 sequences (Paavanen-Huhtala et al. 1999), which is in accordance with latest molecular data (Skovgaard et al. 2003).

The nuclear ribosomal repeat includes both highly conserved genes and more variable regions (Taylor et al. 2000). The intergenic spacer region (IGS) appears to be the most variable region of ribosomal DNA (Hillis and Dixon, 1991). Thus, it has been possible to perform phylogenetic analyses (Yli-Mattila et al. 2002) and design species-specific primers for *F. poae* and *F. sporotrichioides/F. langsethiae* based on IGS sequences (Konstantinova and Yli-Mattila, 2004).

The most divergent part of the ITS2 sequences are seen in Fig. 11, which includes the sequences presented in the paper of Yli-Mattila et al. (1997b) complete with the sequencing results of Yli-Mattila et al. (2002). The Finnish *F. oxysporum* isolate Fo5 has an identical ITS1 sequence with the accession number X94173, while Fo6 differs in one nucleotide (Paavanen-Huhtala et al. 1999). The ITS1 and ITS2 sequences (Paavanen-Huhtala et al. 1999) of the Finnish *F. redolens* isolates Fr10 and Fr11 are identical with accession number X94169 (strain CBS 360.87, Waalwijk et al. 1996b) of *F. redolens.* On the basis of these results, isolates Fo5 and Fo6 can be considered as typical *F. oxysporum* isolates and Fr10 and Fr11 as typical *F. redolens* isolates. On the other hand, only the ITS1 sequence of Fo1 and Fo2 isolates are identical with that of *F. oxysporum* accession number X94173, while their ITS2 region is different. The separation between typical *F. oxysporum* isolates and isolate Fo1 was already supported by the results of RAPD-PCR analysis of Yli-Mattila and Hyvönen (1996), in which Fo1 did not group with other *F. oxysporum* isolates but formed a group of its own. When the ITS1-5.8S-ITS2 sequence of Fo1 is compared with *Fusarium* NRRL 22903 isolate used in the studies of O'Donnell and Cigelnik (1997) and O'Donnell et al. (1998), there are differences only in two nucleotides and the mtSSU and β-tubulin sequences of these two isolates are identical (Yli-Mattila et al. 2002; Skovgaard et al. 2003). *Fusarium* sp. NRRL 22903 was first identified as *F. oxysporum*-like species, but in the later analyses of Skovgaard et al. (2003), it, together with the Finnish isolates Fo1 and Fo2 and several other isolates have been reidentified as a new species *Fusarium commune* (Skovgaard et al. 2003).

In the morphological studies of Gerlach and Nirenberg (1982), *F. redolens* was considered to be distinct from *F. oxysporum.* Other researchers, such as Nelson et al. (1983), considered that these two species are conspecific. Ribosomal DNA RFLP analysis (Waalwijk and Baayen, 1995) and the DNA sequence analyses (Waalwijk et al. 1996b; Skovgaard et al. 2003), however, have demonstrated convincingly that these two species are not conspecific. Waalwijk et al. (1996a, b) and O'Donnell and Cigelnik (1997) have found differences between the morphological and molecular classification of several *Fusarium* species based on

ITS2 sequences. According to these researchers the incongruity may be due to the hypothetical coexistence of several ITS types in the ancestral *Fusarium* species, which would mean that the ITS2 groupings do not accurately reflect the phylogeny of *Fusarium* species. Different versions of ribosomal genes within single genomes or among strains have also been found in several other groups of fungi (Martin, 1990).

The partial 28S rDNA region (ca. 400 nucleotides) is very similar in the sequenced *F. oxysporum* (Fo5), *F. redolens* (Fr10), *F. commune* (Fo1, Fo2), *F. avenaceum* (Fa28, Fa37) and *F. tricinctum* (Ft43) isolates, the maximum difference being only three nucleotides between *F. redolens* and *F. avenaceum* isolates. This result may indicate that the 28S region has also insufficient variation to separate different *Fusarium* species from each other, since *F. oxysporum* and *F. avenaceum* species differ clearly from each other at least in morphology.

Most *F. avenaceum* and *F. tricinctum* isolates have identical ITS1 region, while ITS2 region is more variable (Yli-Mattila et al. 2002). At the time of the publication of Yli-Mattila et al. (1996) and Paavanen-Huhtala et al. (1999) isolate Ft43 was still identified as *F. avenaceum.* According to morphological studies (H. Nirenberg, personal communication), isolates Fa17, Fa28, Fa37 and Fa49 belong to *F. avenaceum* (section *Arthrosporiella/Roseum*), although isolate Fa49 has a slightly different morphology, while Ft43 was identified as *F. tricinctum* (Corda) Sacc. (section *Sporotrichiella*). In the isozyme and RAPD analyses isolates Ft43 and Fa49 differed from other *F. avenaceum* isolates (Yli-Mattila et al. 1996). According to the β-tubulin sequence data (Yli-Mattila et al. 2002) Ft43 differs clearly from *F. avenaceum* isolates in 23 nucleotides, of which 10 are in common with isolate Fa49. The differences between isolate Ft43 and Fa17 or Fa28 are also supported by the dot blot hybridisation results of UP-PCR (Yli-Mattila et al. 1997a, 2002), and according to the *F. avenaceum* specific primer pair JIAf/r, isolate Ft43 does not belong to *F. avenaceum* species (Yli-Mattila et al. 2002)). The recent sequencing results of the IGS region and β-tubulin gene indicate that isolate Fa49 may be an intermediate between *F. avenaceum/F. arthrosporioides* and *F. tricinctum* clusters (Yli-Mattila et al. 2002).

According to the recent UP-PCR results based on cross dot blot hybridisation and rDNA sequence (ITS and IGS) analyses, *F. tricinctum* should be moved from section *Sporotrichiella* to section *Roseum/Arthrosporiella*, which includes for instance *F. avenaceum* and *F. arthrosporioides* species (Yli-Mattila et al. 1997a, 2002; S.A Bulat, T. Yli-Mattila and I.A. Alekhina, unpublished results). Also Bateman et al. (1996) have previously obtained equivalent results, and found section *Sporotrichiella* to be a genetically diverse group. Although the β-tubulin sequence data (Yli-Mattila et al. 2002) differentiated Ft43 clearly from other *F. avenaceum* and *F. arthrosporioides* isolates, it also showed that *F. avenaceum* and *F. tricinctum* isolates are closely related.

In the study concerning the development of strain-specific primers for pathogenic *F. avenaceum* isolates (Paavanen-Huhtala, 2000; Yli-Mattila et al. 2004b), one *F. arthrosporioides* (Far1) isolate proved to be closely related to *F. avenaceum* isolates. This was verified with the *F. avenaceum* specific JIAf/r primer pair (Yli-Mattila et al. 2004b), and is also in agreement with previous rDNA sequencing studies (Yli-Mattila et al. 2002). According to morphological studies this isolate is intermediate between *F. avenaceum* and *F. arthrosporioides* (H. Nirenberg, personal communication).

Development of Strain- and Species-Specific Primers

Specific primers have been developed for fungal species (Schilling et al. 1996; Doohan et al. 1998; Ward and Adams, 1998; Möller et al. 1999; Bulat et al. 2000; Schmidt et al. 2003; Konstantinova and Yli-Mattila, 2004) especially to identify pathogenic species in infected plant and animal material, and soil in addition to pure cultures. The development of strain-specific primers, in contrast, has proved to be difficult, because more variation is needed than is found for instance in the ITS regions (Paavanen-Huhtala, 2000; Yli-Mattila et al. 2004a). RAPD profiling has proved to be capable of revealing a great many more species- and strain-specific polymorphisms than ITS sequencing. It is also clear that the use of SCARs derived from RAPD bands does not automatically result in species- or strain-specific primer pairs (Turner et al. 1998). In the study of Möller et al. (1999) the development of species-specific primers for *F. subglutinans* turned out to be difficult, in spite of a careful selection of RAPD fragments. According to the authors, one possible reason, why specific primers amplified the DNA of other anamorphs of the '*Gibberella fujikuroi* species complex', was that the genome of *F. subglutinans* is the largest among the anamorphs and may share many repeats with the other species.

The *F. avenaceum*-specific primer pair JIAf/r (Turner et al. 1998) amplifies a fragment from all *F. avenaceum* and *F. arthrosporioides* isolates tested, but not from *F. tricinctum* isolates (Yli-Mattila et al. 2004b). All *F. avenaceum, F. arthrosporioides* and *F. tricinctum* isolates give a product with primer pair FA-ITSf/r (Turner et al. 1998; Yli-Mattila et al. 2004b). Thus, by using the combined results of these primer pairs, it is possible to identify both *F. tricinctum* and *F. avenaceum/F. arthrosporioides* isolates (Fig. 13).

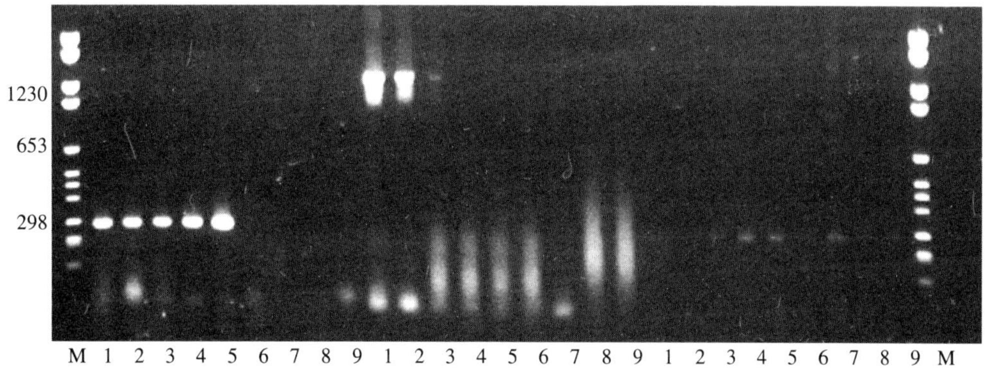

Fig. 13. PCR products obtained with primer pairs JIAf/r (ca. 300 bp), Fa17f/r (ca. 1550 bp) and FaITSf/r (ca. 270 bp) in *F. arthrosporioides* strains 1-3, *F. avenaceum* strains 4-5 and *F. tricinctum* strains 6-9. M is the molecular marker.

In the development of specific primers for the two most pathogenic *F. avenaceum* isolates (Fa17 and Fa38), there were more difficulties (Paavanen-Huhtala, 2000; Yli-Mattila et al. 2004b) than in the corresponding study of *Gliocladium* (Paavanen-Huhtala et al. 2000). The methods used in both studies were similar and based on same principles. The biggest difficulties were with the specificity of the primers, since none of the developed primers alone were specific; using a combination of primers, however, the desired identification results were obtained (Paavanen-Huhtala, 2000). Seven specific primer pairs were designed. In addition to isolates Fa17 and Fa38, the primer pairs Fa5f/r and Fa8f/r amplified fragments from 19 and 15 other isolates respectively. When these primers were used together in the same amplification tube, two amplification fragments were obtained from six isolates (Paavanen-Huhtala, 2000). On the basis of this result, isolates Fa17 and Fa38 can be identified from the group of 22 isolates which were used at the beginning of the study and which have been used in pathogeny testing, except for two isolates (Yli-Mattila et al. 1997a). Primer pair Fa17f/r was designed to be specific for isolate Fa38, but it also amplified a fragment from four other *F. avenaceum* isolates: Fa17, Fa19, Fa50 and Fa77. According to this result (Yli-Mattila et al. 2002, 2004b), the primer pair Fa17f/r can be used for the identification of the isolate group 17, 19, 38 and 50, which in previous phylogenetic analyses based on UP-PCR (Yli-Mattila et al. 1997a), were closely related and was named as main group II. When the amplification result of primer pair Fa17f/r was included in the amplification results of primer pairs Fa5f/r and Fa8f/r, isolates Fa17 and Fa38 could be identified from 60 *F. avenaceum* isolates (Paavanen-Huhtala, 2000). However, one *F. arthrosporioides* isolate (Far1), which has been found to be closely related to *F. avenaceum*, could not be separated from isolates Fa17 and Fa38. More isolates of main group II were later found among *F. arthrosporioides* isolates. So, the specific primers of main group II are also specific for most *F. arthrosporioides* isolates (Yli-Mattila et al. 2004b; Fig. 13).

One reason why it was more difficult to develop strain-specific primers in *F. avenaceum* compared to the study of *Gliocladium* was that there were two isolates instead of one, for which the specific primers had to be designed. The close relation to the isolates Fa19 and Fa50 made also the separation from other isolates more difficult: one primer pair was enough to identify the isolate group, but three primer pairs were needed to identify isolates Fa17 and Fa38. These two isolates were isolated from different geographic origins and even from different host plants. The main connection between these isolates was their high pathogenicity compared to other isolates in previous studies (Yli-Mattila et al. 1997a). It might thus have been supposed that the common RAPD fragments of these isolates would be related to pathogenicity, but according to the GenBank search none of the fragments sequenced were related to pathogenicity. The study of Woo et al. (1996) indicated that there was no clear relationship between the results of the molecular analyses and the pathogenicity tests. This means that pathogenically similar isolates may be genetically different and that common RAPD fragments may thus be difficult to find.

The specificity of the RAPD fragments could have been increased by raising the annealing temperature and thus improve the possibility to get specific primers. By using longer gels or gels with better resolution, the presence/absence of RAPD fragments could have been determined more accurately. Likewise, the specificity of the RAPD fragment can be checked by hybridization or multiple clones can be chosen and screened by restriction enzyme digestion before choosing the clone according to which SCAR primers are designed (Gu et al. 1999). Though, in the study of Möller et al. (1999), no reliable conclusions could be drawn from Southern hybridizations about the specificity of primers derived from the sequence of the hybridization probe.

By using the species-specific primers it is possible to detect toxigenic *Fusarium* species directly from the DNA extracted from grains (Fig. 13). A good correlation has often been observed between the percentage of infection by the routine morphological testing assays and the intensity of the PCR amplified product (Konstantinova et al. 2002; Yli-Mattila et al. 2004c). According to the preliminary results there is also a clear correlation between the amount of *F. graminearum* DNA and DON in Finnish oats and wheat (T. Yli-Mattila, S. Paavanen-Huhtala, A. Haikara, T. Sarlin, M. Jestoi and A. Rizzo, unpublished results) and *F. poae* DNA and NIV in oats (Yli-Mattila et al. 2004d).

The *Tri5* gene codes for trichodiene synthase, involved in the first specific step of the biosynthesis of all trichothecene mycotoxins (Hohn and Beremand, 1989). Highly conserved regions in the *tri5* sequences of *Fusarium* have been utilised to set up PCR-based assays for the detection of trichothecene-producing *Fusarium* spp. (Doohan et al. 1998; Niessen and Vogel, 1998; Edwards et al. 2001, 2002). These assays have been successfully applied to detect contamination in cereals and cereal products. In addition, quantitative SYBR Green real-time PCR based on the gene-specific primer pair Tox5-1/Tox5-2 was demonstrated to correlate with DON concentrations in wheat samples (Schnerr et al. 2001). Intraspecific variation has also been found in the intron sequence of *tri5* sequences and it has been possible to design species-specific primers based on *tri5* sequences (Niessen et al. 2004). It has also been possible to design specific primers for *F. culmorum* and *F. graminearum* isolates producing either nivalenol (NIV) or DON (Bakan et al. 2002; Kim et al. 2003).

Future Perspectives

In future, the existing molecular methods will be improved and automated and used more in routine identification and detection of *Fusarium* species. While this requires costly equipments, it also allows quicker and more reliable identification and less researchers than the traditional morphological identification of fungal isolates. The computer software and analysing of the molecular data is also expected to improve considerably and the total amount of sequence data in data bases will increase. For example, >90% of the genome of *F. graminearum* (*Gibberella zeae*) has already been sequenced (http://www.broad.mit.edu/annotation/fungi/fusarium/index.htm).

One of the problems is how mycologists can effectively combine morphological, chromatographic and molecular data to produce reliable classification and identification of *Fusarium* isolates. One of the conclusions is multidisciplinary or polyphasic taxonomic studies, in which all available data will be utilised (Seifert et al. 2000; Yli-Mattila et al. 2002; Schmidt et al. 2004a).

The work dealing with the molecular studies of biosynthetic genes of different toxins of *Fusarium* species is in progress and the sequences of these genes will be used for designing new primers, the PCR products of which will correlate e.g. with zearalenone, enniatin and moniliformin concentrations in grains.

Microarray technology is the most promising new technique for large scale identification and detection of different microorganisms (Wilson et al. 2002). Species-specific sequences can be derived from pathogenicity and toxin genes or by shotgun sequencing of cloned fragments from uncharacterized DNA regions throughout the genome. Primers are designed to amplify these regions in multiplex reactions. Primer pairs are screened against the GenBank database for potential similar nucleotide sequences, which is also a good practice for primer pairs used in end point or realtime PCR for species identification. The best diagnostic PCR products are used for identification of each species. The internal diagnostic sequence of the PCR products are used to create probes for the microarray. The diagnostic amplicons for each species are pooled, fragmented and labeled and the labeled mixture is hybridized to an array. This allows for high throughput screening of many microbes simultaneously, while in realtime quantitative PCR the detection of multiple PCR products is limited to the spectral overlap of available probes. Thus, microarray technology is already at the moment specific, but it is not always sensitive and quantitative enough for different purposes as compared to realtime PCR.

Acknowledgements

The investigation was supported by the Academy of Finland (Grant Nos. 34275 and 52104), National Technology Agency of Finland (No. 40168/03) and the Nordic network grant from Nordic Council. We are also grateful to Dr. Sonja Klemsdal for critically reading the manuscript and Drs. Tatjana Gagkaeva, Galina Mitina and Pavlina Konstantinova for the photographs.

References

Abd-Elsalam, K., Khalil, M., Aly, A-H. and Asran-Amal, A-M. (2002) Genetic diversity among *Fusarium oxysporum* f. sp. *vasinfectum* isolates revealed by UP-PCR and AFLP markers. Phytopathologia Mediterranea 41: 252–258.

Abd-Elsalam, K.A., Schnieder, F., Asran-Amal, A., Khalil, M.S. and Verreet, J-A. (2003a) Intra-species genomic groups in *Fusarium semitectum* and their correlation with origin and cultural characteristics. Zeitscfrift für Pflanzenkrankheiten und Pflanzenschutz 110: 409–418.

Abd-Elsalam, K.A., Schnieder, F., Khalil, M.S., Asran-Amal, A. and Verreet, J.-A. (2003b) Use of AFLP fingerprinting to analyze genetic variation within and between populations of *Fusarium* spp. derived from Egyptian cotton cultivars. Journal of Plant Pathology 85: 1–5.

Alexander, N.J., Proctor, R.H., McCormick, S.P. and Plattner, R.D. (1997) Genetic and molecular aspects of the biosynthesis of trichothecenes by *Fusarium*. Cereal Research Communication 25: 315–320.

Annis, S.L. and Goodwin, P.H. (1997) Recent advances in the molecular genetics of plant cell wall-degrading enzymes produce by plant pathogenic fungi. European Journal of Plant Pathology 103: 1–14.

Appel, D.J. and Gordon, T.R. (1995) Intraspecific variation within populations of *Fusarium oxysporum* based on RFLP analysis of the intergenic spacer region of the rDNA. Experimental Mycology 19: 120–128.

Appel, D.J. and Gordon, T.R. (1996) Local and regional variation in populations of *Fusarium oxysporum* based on RFLP analysis of the intergenic spacer region of the ribosomal DNA. Molecular Plant-Microbe Interactions 9: 125–138.

Arroyo-Garcia, R., Cenis, J.L., Tello, J., Martinez-Zapater, J.M. and Cifuentes, D. (2003) Genetic relationships among seven specialized forms of *Fusarium oxysporum* determined by DNA sequencing of the ITS region and AFLPs. Spanish Journal of Agricultural Research 1: 55–63.

Baayen, R.P., O'Donnell, K., Bonants, P.J.M., Cigelnik, E., Kroon, L.P.N.M., Roebroeck, E.J.A. and Waalwijk, C. (2000) Gene genealogies and AFLP analyses in the *Fusarium oxysporum* complex identify monophyletic and nonmonophyletic formae speciales causing wilt and rot disease. Phytopathology 90: 891–900.

Bachem, C.W.B, Van der Hoeven, R.S., de Bruijn, S.M., Vrengdenhil, D., Zabeau, M. and Visser, R.G.F. (1996) Visualisation of gene expression using a novel method of RNA fingerprinting based on AFLP: analysis of gene expression during potato tuber development. The Plant Journal 9: 745–753.

Bachmann, K. (1994) Tansley Review No. 63: Molecular markers in plant ecology. New Phytologist 126: 403–418.

Bakan, B., Giraud-Delville, C., Pinson, L., Morlar-Richard, D., Fournier, E. and Brygoo, Y. (2002) Identification by PCR of *Fusarium culmorum* strains producing large and small amounts of deoxynivalenol. Applied and Environmental Microbiology 68: 5472–5479.

Bateman, G.L., Kwasna, H. and Ward, E. (1996) Relationships among *Fusarium* spp. estimated by comparing restriction fragment length polymorphisms in polymerase chain reaction-amplified nuclear rDNA. Canadian Journal of Microbiology 42: 1232–1240.

Bonde, M.R., Micales, J.A. and Peterson, G.L. (1993) The use of isozyme analysis for identification of plant-pathogenic fungi. Plant Disease 77: 961–968.

Booth, C. (1971) *The genus* Fusarium. Commonwealth Mycological Institute, Kew, Surrey, pp. 237.

Bowditch, B.M., Albright, D.G., Williams, J.G.K. and Braun, M.J. (1993) Use of randomly amplified polymorphic DNA markers in comparative genome studies. Methods in Enzymology 224: 294–309.

Bradford, M.M. (1976) A rapid and sensitive method for the quantitation of microgram quantities of protein utilizing the principle of protein-dye binding. Analytical Biochemistry 72: 248–254.

Bremer, K. (1988) The limits of amino acid sequence data in angiosperm phylogenetic reconstruction. Evolution 42: 795–803.

Bruns, T.D., White, T.J. and Taylor, J.W. (1991) Fungal molecular systematics. Annual Review of Ecological System 22: 525–564.

Bulat, S.A., Lubeck, M.., Alekhina, I.A., Jensen, D.F., Knudsen, I.M.B. and Lubeck, P.S. (2000) Identification of a universally primed-PCR-derived sequence-characterized amplified region marker for an antagonistic strain of *Clonostachys rosea* and development of a strain-specific PCR detection assay. Applied and Environmental Microbiology 66: 4758–4763.

Bulat, S.A. and Mironenko, N. (1990) Species identity of the phytopathogenic fungi *Pyrenophora teres* Dreschler and *P. graminea* Ito and Kuribayashi. Mikologiya i Fitopatologia 24: 435–441(in Russian).

Bulat, S.A. and Mironenko, N. (1992) Polymorphism of yeast-like fungus *Aureobasidium pullulans* (De Bary) revealed by universally primed polymerase chain reaction: species divergence. Genetika 28: 19–30 (in Russian).

Bulat, S.A. and Mironenko, N.V. (1996) Identification of fungi and analysis of their genetic diversity by polymerase chain reaction (PCR) with gene-specific and non-specific primers. Russian Journal of Genetics (formerly Soviet Genetics) 32: 143–159.

Bulat, S.A., Mironenko, N.V. and Zholkevich, Yu.G. (1995) Genetic structure of soil population fungus *Fusarium oxysporum* Schlechtend. Fr. Molecular reidentification of the species and genetic differentiation of isolates using polymerase chain reaction technique with universal primers (UP-PCR). Russian Journal of Genetics 31: 271–278.

Bulat, S.A., Lübeck, M., Mironenko, N., Jensen, D.F. and Lübeck, P.S. (1998) UP-PCR analysis and ITS1 ribotyping of strains of *Trichoderma* and *Gliocladium*. Mycological Research 102: 933–943.

Burt, A., Carter, D.A., White, T.J. and Taylor, J.W. (1994) DNA sequencing with arbitrary primer pairs. Molecular Ecology 3: 523–525.

Caetano-Anolles, G. (1994) MAAP: a versatile and universal tool for genome analysis. Plant Molecular Biology 25: 1011–1026.

Chelkowski, J. (1989) Formation of mycotoxins produced by fusaria in heads of wheat, triticale and rye. In: *Fusarium*: Mycotoxins, Taxonomy and Pathogenicity, (ed), Chelkowski J. Elsevier, Amsterdam, 63–81.

Cnops, C., den Boer, B., Gerats, A., Van Montagu, M. and Van Lijsebettens, M. (1996) Chromosome landing at the *Arabidopsis TORNADO1* locus using an AFLP-based strategy. Molecular and General Genetics 253: 32–41.

Creste, S., Tulmann-Neto, A. and Figueira, A. (2001) Detection of single sequence repeat polymorphisms in denaturing polyacrylamide sequencing gels by silver staining. Plant Molecular Biology Reporter 19: 1–8.

Crowhurst, R.N., Hawthorne, B.T., Rikkerink, E.H.A. and Templeton, M.D. (1991) Differentiation of *Fusarium solani* f. sp. *cucubitae* races 1 and 2 by random amplification of polymorphic DNA. Current Genetics 20: 391–396.

Dahleen, L.S., Agrama, H.A., Horsley, R.D., Steffenson, B.J., Schwarz, P.B., Mesfin, A. and Franckowiak, J.D. (2003) Identification of QTLs associated with *Fusarium* head blight resistance in Zhedar 2 barley. Theoretical Applied Genetics 108: 95–104.

Desjardins A.E. and Hohn, T.M. (1997) Mycotoxins in plant pathogenesis. Molecular Plan-Microbe Interactions 10: 147–152.

Devos, K.M. and Gale, M.D. (1992) The use of random amplified polymorphic DNA markers in wheat. Theoretical Applied Genetics 84: 567–572.

Doohan, F.M., Weston, G., Rezanoor, H.N., Parry, D.W. and Nicholson, P. (1999) Development and use of reverse transcription-PCR assay to study expression of Tri5 by *Fusarium* species in vitro and in planta. Applied and Environmental Microbiology 65: 3850–3854.

Doohan, F.M., Parry, D.W., Jenkinson, P. and Nicholson, P. (1998) The use of species-specific PCR-based assays to analyse *Fusarium* ear blight of wheat. Plant Pathology 47: 197–205.

Duchesne, L.C. and Anderson, J.B. (1990) Location and direction of transcription of the 5S rRNA gene in *Armillaria*. Mycological Research 94: 266–269.

Edwards, S.G., O'Callaghan, J. and Dobson, D.W. (2002) PCR-based detection and quantification of mycotoxigenic fungi. Myological Research 106: 1005–1025.

Edwards, S.G., Pirgozliev, S.R., Hare, M.C. and Jenkinson, P. (2001) Quantification of trichothecene-producing *Fusarium* species in harvested grain by competitive PCR to determine efficacies of fungicides against *Fusarium* head blight of winter wheat. Applied and Environmental Microbiology 67: 1575–1580.

Eng, C. and Vijg, J. (1997) Genetic testing: the problems and the promise. Nature Biotechnology 15 : 422–426.

Farris, J.S. (1988) Hennig 86, Version 1.5. Port Jefferson Station, New York.

Farris, J.S., Källersjö, M., Kluge, A.G. and Bult, C. (1994) Testing significance of incongruence. Cladistics 10: 315–320.

Farris, J.S. (1994) RNA, Version 940821. Molekylärsystematiska laboratoriet. Naturhistoriska riksmuseet, S-10405 Stockholm, Sweden.

Felsenstein, J. (1993) PHYLIP (Phylogeny Inference Package), version 3.5. Department of Genetics: University of Washington, Seattle.

Ferrie, R.M., Schwarz, M.J., Robertson, N.H., Vaudin, S., Super, M., Malone, G. and Little, S. (1992) Development, multiplexing, and application of ARMS tests for common mutations in the CFTR gene. American Journal of Human Genetics 18: 71–76.

Foster, L.M., Kozak, K.R., Loftus, M.G., Stevens, J.J. and Ross, I.K. (1993) The polymerase chain reaction and its application to filamentous fungi. Mycological Research 97: 769–781.

Gerlach, W. and Nirenberg, H. (1982) The genus *Fusarium*—a pictorial atlas. P. Parey, Berlin, pp. 406.

Gilliland, G., Perrin, S., Blanchard, K. and Bunn, H.F. (1990) Analysis of cytokine mRNA and DNA: detection and quantification by competitive polymerase chain reaction. Proceedings of the National Academy of Sciences USA 87: 2725–2729.

Gladstein, D. and Wheeler, W. (2001) POY (Phylogeny Reconstruction via Direct Optimization of DNA data), Version 2.7. Department of Invertebrates, American Museum of Natural History, Central Park West, 79th St. New York, NY 10024–5192, USA

Goldstein, D.B. and Schlötterer, C. (1999) Microsatellites: Evolution and Applications. Oxford University Press.

Grundmann, H., Schneider, C.H., Tichy, H.V., Simon, R., Klare, I., Hartung, D. and Daschner, F.D. (1995) Automated laser fluorescence analysis of randomly amplified polymorphic DNA: a rapid method for investigating nosocomial transmission of *Acinetobacter baumanii*. Journal of Medical Microbiology 43: 446–451.

Gu, W., Post, C.M. Aguirre, G.D. and Ray, K. (1999) Individual DNA bands obtained by RAPD analysis of canine genomic DNA often contain multiple DNA sequences. Journal of Heredity 90: 96–98.

Guadet, J., Julien, J., Lafey, J.F. and Brygoo, Y. (1989) Phylogeny of some *Fusarium* species as determined by large-subunit rRNA sequence comparison. Molecular Biology of Evolution 6: 227–242.

Hantula, J., Dusabenyagasani, M. and Hamelin, R.C. (1996) Random amplified microsatellites (RAMS)—a novel method for characterizing genetic variation within fungi. European Journal of Forest Pathology 26: 159–166.

Heid, C.A., Stevens, J., Livak, K.J. and Williams, P.M. (1996) Real time quantitative PCR. Genome Research 6: 986–994.

Higuchi, R., Dollinger, G, Walsh, P.S. and Griffith, R. (1992). Simultaneous amplification and detection of specific DNA sequences. Biotechnology 10: 413–417.

Hillis, D.M. and Dixon, M.T. (1991) Ribosomal DNA: Molecular evolution and phylogenetic inference. Quarterly Reviewed Biology 66: 411–453.

Hohn, T.M. and Beremand, P.D. (1989) Isolation and nucleotide sequence of sesquiterpene cyclase gene from the tricothecene-producing fungus *Fusarium sporotrichioides.* Gene 79: 131–138.

Janse, J.D. (1995) New methods of diagnosis in plant pathology—perspectives and pitfalls. EPPO Bulletin 25: 5–17.

Jestoi, M., Rokka, M., Yli-Mattila, T., Parikka, P., Rizzo, A. and Peltonen, K. (2004) Beauvericin, enniatins and moniliformin in Finnish grains. Food Additives and Contaminants (manuscript submitted).

Joffe, A.Z. (1986) *Fusarium* species: their biology and toxicology, John Wiley & Sons, New York.

Kamalay, J.C., Tejwani, R. and Rufener, G.K. (1990). Isolation and analysis of genomic DNA from single seeds. Crop Science 30: 1079–1084.

Karjalainen, R. and Kammiovirta, K. (1994) Identification of *Heterobasidion annosum* (Fr.) Bref., a root and butt rot disease of trees by PCR fingerprinting. In: Modern assays for plant pathogenic fungi: Identification, detection and quantification, (eds) Schots, A., Dewey, F.M. and Oliver, R. University Press, Cambridge, 111–116.

Kim, H.-S., Lee, T., Dawlatana, M., Yun, S.-H. and Lee, Y.-W. (2003) Polymorphism of trichothecene biosynthesis genes in deoxynivalenol- and nivalenol-producing *Fusarium graminearum* isolates. Mycological Research 107: 190–197.

Kiprop, E.K., Baudoin, J.P., Mwangómbe, A.W., Kimani, P.M., Mergeai, G. and Maquet, A. (2002) Characterization of Kenyan isolates of *Fusarium udum* from Pigeonpea (*Cajanus cajan* (L.) Millsp.) by cultural characteristics, aggressiveness and AFLP analysis. *Journal of* Phytopathology 150: 517–527.

Konstantinova, P., Karadzhova, J., Yli-Mattila, T. and van den Bulk, R. (2002) Molecular detection of *Alternaria alternata* and *Fusarium* spp. In barley seeds and comparison with routine testing assays. Petria 12: 239–247

Konstantinova, P. and Yli-Mattila, T. (2004) IGS-RFLP analysis and development of molecular markers for identification of *Fusarium poae, Fusarium langsethiae, Fusarium sporotrichioides* and *Fusarium kyushuense.* International Journal of Food Microbiology 95: 321–331.

Koopmann, B., Karlovsky, P. and Wolf, G. (1994) Differentiation between *Fusarium culmorum* and *Fusarium graminearum* by RFLP and with species specific DNA probes. In: Modern assays for plant pathogenic fungi: Identification, detection and quantification, (eds), Schots, A., Dewey, F.M. and Oliver, R. CAB International, Oxford, 37–46.

Kosiak, B., Torp, M., Skjerve, E. and Andersen, B. (2004) *Alternaria* and *Fusarium* in Norwegian grains of reduced quality—a matched pair sample study. International Journal of Food Microbiology (in press).

Lee, S.B. and Taylor, J.W. (1990) Isolation of DNA from fungal mycelia and single spores. In: PCR protocols: A Guide to Methods and Applications, (eds), Innis, M.A., Gelfand, D.H., Sninsky, J.J. and White, T.J. Academic Press, San Diego, 282–287.

Leslie, J.F. (1993) Fungal vegetative compatibility. Annual Review of Phytopathology 31: 127– 150.

Leslie, J.F., Zeller, K.A. and Summerell, B.A. (2001) Icebergs and species in populations of *Fusarium.* Physiological and Molecular Plant Pathology 59: 107–117.

Liew, E.C.Y., Maclean, D.J. and Irwin, J.A.G. (1998) Specific PCR based detection of *Phytophthora medicaginis* using the intergenic spacer region of the ribosomal DNA. Mycological Research 102: 73–80.

Llorens, A., Mateo, R., Hinojo, M.J., Valle-Algarra, F.M. and Jiménez, M. (2004) Influence of environmental factors on the biosynthesis of type B tricothecenes by isolates of *Fusarium* spp. from Spanish crops. International Journal of Food Microbiology (94:43–54).

Logrieco, A., Rizzo, A., Ferracane, R. and Ritieni, A. (2002) Occurrence of beauvericin and enniatins in wheat affected by *Fusarium avenaceum* head blight. Applied Environmental Microbiology 68: 82–85.

Loxdale, H.D., Castañera, P. and Brookes, C.P. (1983) Electrophoretic study of enzymes from cereal aphid populations.

I. Electrophoretic techniques and staining systems for characterising isoenzymes from six species of cereal aphids (*Hemiptera: Aphididae*). Bulletin of Entomological Research 73: 645–657.

Lübeck, M. (1997) Marker genes and PCR based approaches as tools for monitoring fungi in soil with emphasis on UP-PCR and fungi from the genera *Trichoderma* and *Gliocladium*, PhD Thesis, The Royal Veterinary and Agricultural University, Copenhagen, Denmark.

Lübeck, M., Alekhina, I.A., Lübeck, P.S., Jensen, D.F. and Bulat, S.A. (1999) Delineation of *Trichoderma harzianum* into two different genotypic groups by a highly robust fingerprinting method, UP-PCR and UP-PCR product cross-hybridization. Mycological Research 103: 289–298.

Lübeck, M., Poulsen, S.K., Lübeck, P.S., Jensen, D.F. and Thrane, U. (2000) Identification of *Trichoderma* strains from building materials by ITS1 ribotyping, UP-PCR fingerprinting an UP-PCR cross hybridization. FEMS Microbiology Letter 185: 129–134.

Mabru, D., Douet, J.P., Mouton, A., Dupré, C., Ricard, J.M., Médina, B., Castroviejo, M. and Chevalier, G. (2004) PCR-RFLP using a SNP on the mitochondrial Lsu-rDNA as an easy method to differentiate *Tuber malanosporum* (Perigord truffle) and other truffle species in cans. International Journal of Food Microbiology 94: 33–42.

Mach, R.L., Kullnig-Gradinger, C.M., Farnleitner, A.H., Reischer, G., Adler, A. and Kubicek, C.P. (2004) Specific detection of *Fusarium langsethiae* and related species by DGGE and ARMS-PCR of a β-tubulin (tub1) gene fragment. International Journal of Food Microbiology (in press).

Majer, D. Mithen, R., Lewis, B., Vos, P. and Oliver, R.P. (1996) The use of AFLP fingerprinting for the detection of genetic variation in fungi. Mycological Research 100: 1107–1111.

Manninen, O. (2000) Genetic mapping of traits important in barley breeding, PhD thesis, University of Helsinki, Finland.

Martin, F.N. (1990) Variation in the ribosomal DNA repeat unit within single-oospore isolates of the genus *Pythium.* Genome 33: 585–591.

Matuo, T. and Snyder, W.C. (1973) Use of morphology and mating populations in the identification of formae speciales in *Fusarium solani*. Phytopathology 63: 562–565.

Mesterhazy, A. (1997) Methodology of resistance testing and breeding against *Fusarium* head blight in wheat and results of the selections. Cereal Research Communications 25: 631–636.

Micales, J.A., Bonde, M.R. and Peterson, G.L. (1986) The use of isozyme analysis in fungal taxonomy and genetics. Mycotaxon 27: 405–449.

Mitina, G. and Yli-Mattila, T. (2004) RAPD-PCR, UP-PCR and rDNA sequence analyses of entomopathogenic *Verticillium lecanii* and *Beauveria bassiana* strains. Journal of Russian Phytopathological Society 3: 7–17.

Morrison, T.M., Weiss, J.J. and Wittwer, C.T. (1998) Quantification of low-copy transcripts by continuous SYBR green I monitoring during amplification. *Biotechniques* 24: 954–962.

Moss, M.O. (1996) Mycotoxins. Mycological Research 100: 513–523.

Moss, M.O. (1998) Recent studies of mycotoxins. Journal of Applied Microbiology Symposium Supplement 84: 62S–76S.

Möller, E.M., Chelkowski, J. and Geiger, H.H. (1999) Species-specific PCR assays for the fungal pathogens *Fusarium moniliforme* and *Fusarium subglutinans* and their application to maize ear rot. Journal of Phytopathology 147: 497–508.

Nelson, P.E., Toussoun, T.A. and Marasas, W.F.O. (1983) *Fusarium* species. An Illustrated Manual for Identification, Pennsylvania State University Press, University Park, Pennsylvania, USA.

Nicholson, P., Simpson, D.R., Weston, G., Rezanoor, H.N., Lees, A.K., Parry, D.W. and Joyce, D. (1998) Detection and Quantification of *Fusarium culmorum* and *Fusarium graminearum* in cereals using PCR assays. Physiological and Molecular Plant Pathology 53: 17–37.

Nicholson, P., Chandler, E., Draeger, R.C., Gosman, N.E., Simpson, D.R., Thomsett, M. and Wilson, A.H. (2003) Molecular tools to study epidemiology and toxicology of fusarium head blight of cereals. European Journal of Plant Pathology 109: 691–703.

Nielsen, N-P.V., Smedsgaard, J. and Frisvad, J.C. (1999) Full second-order chromatographic/spectrometric data matrices for automated sample identification and component analysis by non-data-reducing image analysis. Analytical Chemistry 71: 727–735.

Nielsen, K.F., Smedsgaard, J., Larsen, T.O., Lund F., Thrane U. and Frisvad J.C. (2004) Chemical identification of fungi: Metabolite profiling and metabolomics. In: Fungal Biotechnology in Agricultural, Food and Environmental Applications, (ed), Arora, D.K. Marcel Dekker, Inc., New York, 19–35.

Niessen M.L. and Vogel, R.F. (1998) Group specific PCR-detection of potential tricothecene-producing Fusarium species in pure culture and cereal samples. Systematic and Applied Microbiology 21: 618–631.

Niessen, L., Schmidt, H. and Vogel, R.F. (2004) Use of tri5 gene sequences for PCR detection and taxonomy of trichothecene producing species in the *Fusarium* section *Sporotrichiella.* International Journal of Food Microbiology (in press).

O'Donnell, K. (1996) Progress towards a phylogenetic classification of *Fusarium.* Sydowia 48: 57–70.

O'Donnell, K. and Cigelnik, E. (1997) Two divergent intragenomic rDNA ITS2 types within a monophyletic lineage of the fungus *Fusarium* are nonorthologous. Molecular Phylogenetics and Evolution 7: 103–116.

O'Donnell, K., Cigelnik, E. and Nirenberg, H.I. (1998) Molecular systematics and phylogeography of the *Gibberella fujikuroi* species complex. Mycologia 90: 465–493.

O'Donnell, K., Kistler, H.C., Tacke, B.K. and Casper, H.H. (2000) Gene genealogies reveal global phylogeographic structure and reproductive isolation among lineages of *Fusarium graminearum*, the fungus causing wheat scab. Proceedings of the National Academy of Sciences of the United States of America 97: 7905–7910.

Paavanen-Huhtala, S. (2000) Molecular-based assays for the determination of diversity and identification of *Fusarium* and *Gliocladium* fungi. Ph.D. Thesis, Annales Universitatis Turkuensis AII 139, University of Turku, Finland, Painosalama OY, Turku, pp. 198,

Paavanen-Huhtala, S., Avikainen, H. and Yli-Mattila, T. (2000) Development of strain-specific primers for a strain *Gliocladium catenulatum* used in biological control. European Journal of Plant Pathology 106: 187–198.

Paavanen-Huhtala, S., Hyvönen, J., Bulat, S.A. and Yli-Mattila, T. (1999) RAPD-PCR, isozyme, rDNA RFLP and rDNA sequence analyses in identification of Finnish *Fusarium oxysporum* isolates. Mycological Research 103: 625–634.

Page, R.D.M. (1993a) COMPONENT, Version 2.0, Cladistics 9: 351–353.

Page, R.D.M. (1993b) User's manual for COMPONENT, Version 2.0, The Natural History Museum, London.

Paran, I. and Michelmore, R.W. (1993) Development of reliable PCR-based markers linked to downy mildew resistance genes in lettuce. Theoretical and Applied Genetics 85: 985–993.

Parent, J-G. and Pagé, D. (1998) Identification of raspberry cultivars by sequence characterized amplified region DNA analysis. HortScience 33: 140–142.

Pejic, I., Ajmone-Marsan, P., Morgante, M., Kozumplick, V., Castiglioni, P., Taramino, G. and Motto, M. (1998) Comparative analysis of genetic similarity among maize inbred lines detected by RFLPs, RAPDs, SSRs, and AFLPs. Theoretical Applied Genetics 97: 1248–1255.

Perkowski, J., Kiecana, I., Schumacher, U., Muller, H.M., Chelkowski, J. and Golinski, P. (1997) Head infection and accumulation of *Fusarium* toxins in kernels of 12 barley genotypes inoculated with *Fusarium graminearum* isolates of two chemotypes, European Journal of Plant Pathology 103: 85–90.

Rafalski, J.A., Tingey, S.V. and Williams, J.G.K. (1991) RAPD markers—a new technology for genetic mapping and plant breeding. AgBiotech News Information 3: 645–648.

Rieseberg, L.H. (1996) Homology among RAPD fragments in interspecific comparisons. Molecular Ecology 5: 99–105.

Rolf, J.F. (1989) NTSYS-pc. Numerical taxonomy and multivariate system, Version 1.70, Applied Biostatistics Inc., New York.

Rosendahl, S. and Sen, R. (1992) Isozyme analysis of mycorrhizal fungi and their mycorrhiza, Methods in Microbiology 24: 169–194.

Saiki, R., Scharf, S., Faloona, F., Mullis, K.B., Horn, G.T., Erlich, H.A. and Arnheim, N. (1985) Enzymatic amplification of β-globin genomic sequences and restriction site analysis for diagnosis of sickle cell anaemia. Science 230: 1350–1354.

Sambrook, J., Fritch, E.F. and Maniatis, T. (1989) Molecular Cloning: A laboratory Manual. 2nd ed. Cold Spring Harbour Laboratory Press, New York

Schilling, A.G., Möller, E.M. and Geiger, H.H. (1994) RAPDs of *Fusarium culmorum* and *F. graminearum*: Application for genotyping and species identification. In: Modern assays for plant pathogenic fungi: Identification, detection and quantification, (eds), Schots. A., Dewey. F.M. and Oliver, R. CAB International, Oxford, 47–56.

Schilling, A.G., Möller, E.M. and Geiger, H.H. (1996) Polymerase chain reaction-based assays for species-specific detection of *Fusarium culmorum, F. graminearum* and *F. avenaceum*. Phytopathology 86: 515–522.

Schmidt H., Adler A., Holst-Jensen A., Klemsdal S.S., Logrieco A., Mach R.L., Nirenberg H.I., Thrane U., Torp M., Vogel, R.F., Yli-Mattila T. and Niessen, L. (2004a) An integrated taxonomic study of *Fusarium langsethiae, Fusarium poae* and *Fusarium sporotrichioides* based on the use of composite datasets. International Journal of Food Microbiology 95: 341–349.

Schmidt, H. Niessen, L. and Vogel, R.F. (2004b) AFLP fingerprinting analysis of species in the *Fusarium* section *Sporotrichiella*. International Journal of Food Microbiology 95: 297–304.

Schmidt, H., Ehrmann, M., Vogel, R.F., Taniawski, M.H. and Niessen, L. (2003) Molecular typing of *Aspergillus ochraceus* and construction of specific SCAR-primers based on AFLP. Systematic and Applied Microbiology 26: 138–146.

Schnerr, H., Niessen, L. and Vogel, R.F. (2001) Real time detection of the *tri5* gene in *Fusarium* species by LightCycler™-PCR using SYBR® Green I for continuous fluorescence monitoring. International Journal of Food Microbiology 71: 53–61.

Seifert, K., Gams, W., Crous, P.W. and Samuels, G.J. Eds. (2000) Molecules, Morphology and Classification: Towards Monophyletic Genera in the Ascomycetes. Studies in Mycology, No. 45, CBS, Baarn/Delft, The Netherlands.

Sen, R. and Hepper, C.M. (1986) Characterization of vesicular-arbuscular mycorrhizal fungi (*Glomus* spp.) by selective enzyme staining following polyacrylamide gel electrophoresis. Soil Biology and Biochemistry 18: 29–34.

Shields, C.R., Orton, T.J. and Stuber, C.W. (1983) An outline of general resource needs and procedures for the electrophoretic separation of active enzymes from plant tissue. In: Isozyme in plant genetics and breeding, Part, A, (eds), Tanksley, S.D. and Orton, T.J. Elsevier Science Publishers B.V., Amsterdam, 443–458.

Siddall, M. (1995) *Random cladistics*. MS-DOS program package distributed by the author.

Sivaramakrishnan, S., Kannan, S. and Singh, S.D. (2002) Genetic variability of *Fusarium* wilt pathogen isolates of chickpea (*Cicer arietinum* L.) assessed by molecular markers. Mycopathologia 155: 171–178.

Skov, E. (1998) Are RAPD-markers reproducible between different laboratories? A case study of *Picea abies* (L.) KARST. Silvae Genetica 47: 283–287.

Skovgaard, K., Rosendahl, S., O'Donnell, K. and Nirenberg, H.I. (2003) *Fusarium commune* is a new species identified by morphology and molecular genetic phylogenetic data. Mycologia 95: 630–636.

Smedsgaard J. (1997) Micro-scale extraction procedure for standardized screening of fungal metabolite production in cultures. Journal of Chromatografia A 760: 264–270.

Snyder, W.C. and Hansen, H.N. (1945) The species concept in *Fusarium* with reference to *Discolor* and other sections. American Journal Botany 32: 657-666.

Swofford, D.L. (1993) PAUP: *Phylogenetic Analysis Using Parsimony*, version 3.1.1. Illinois Natural History Survey: Champaign, Illinois.

Taylor, E.J.A., Stevens, E.A., Bates, J.A., Morreale, D., Lee, D., Kenyon, D.M. and Thomas, J.E. (2001) Rapid-cycle PCR detection of *Pyrenophora graminea* from barley seed. Plant Pathology 50: 347–355.

Taylor, J.W., Jacobson, D.J., Kroken, S., Kasuga, T., Geiser, D.M., Hibbett, D.S. and Fisher, M.C. (2000) Phylogenetic species recognition and species concepts in fungi. Fungal Genetics and Biology. 31: 21–32.

Thomas, C.M., Vos, P., Zabeau, M., Jones, D.A., Norcott, K.A., Chadwick, B.P. and Jones, J.D.G. (1995) Identification of amplified restriction fragment polymorphism (AFLP) markers tightly linked to the tomato *Cf-p* gene for resistance to *Cladosporium fulvum*. The Plant Journal 8: 785–794.

Thompson, J.D., Higgins, D.G. and Gibson, T.J. (1994) Clustal W: Improving the sensitivity of progressive multiple sequence alignment through sequence weighting, position-specific gap penalties and weight matrix choice. Nucleic Acids Research 22: 4673–4680.

Thrane, U. (1989) *Fusarium* species and their specific profiles of secondary metabolites. In: *Fusarium*: Mycotoxins, Taxonomy and Pathogenicity, (ed), Chelkowski, J. Elsevier Science Publishers B.V., Amsterdam.

Thrane, U., Poulsen, S.B., Nirenberg, H.I. and Lieckfeldt, E. (2001) Identification of *Trichoderma* strains by image analysis of HPLC chromatograms. FEMS Microbiology Letters 203: 249–255.

Thrane, U., Adler, A., Clasen, P., Langseth, W., Nielsen, K.F. (2004) Chemical diversity of *Fusarium poae, F. sporotrichioides* and *F. langsethiae* sp. Nov. International Journal of Food Microbiology 95: 257–266.

Tommerup, I.C., Barton, J.E. and O'Brien, P.A. (1995) Reliability of RAPD fingerprinting of three basidiomycete fungi, *Laccaria, Hydnangium* and *Rhizoctonia*. Mycological Research 99: 179–186.

Tooley, P.W., Bunyard, B.A., Carras, M.M. and Hatziloukas, E. (1997) Development of PCR primers from internal transcribed spacer region 2 for detection of *Phytophthora* species infecting potatoes. Applied and Environmental Microbiology 63: 1467–1475.

Turner, A.S., Lees, A.K., Rezanoor, H.N. and Nicholson, P. (1998) Refinement of PCR-detection of *Fusarium avenaceum* and evidence from DNA marker studies for phenetic relatedness to *Fusarium tricinctu.* Plant Pathology 47: 278–288.

Tyagi, S. and Kramer, F.R. (1996) Molecular Beacons: probes that fluoresce upon hybridization, Nature Biotechnology 14: 303–308.

Uoti, J. and Ylimäki, A. (1974). The occurence of *Fusarium* species in cereal grain in Finland. Annales Agriculturae Fenniae 13: 5–17.

Van Elsas, J.D., Duarte, G.F., Keijzer-Wolters, A. and Smit, E. (2000) Analysis of the dynamics of fungal communities in soil via fungal-specific PCR of soil DNA followed by denaturing gradient gel electrophoresis. Journal of Microbiological Methods 15: 133–151.

Vos, P., Hogers, R., Bleeker, M., Reijans, M., van de Lee, T., Hornes, M., Frijters, A., Pot, J., Peleman, J., Kuiper, M. and Zabeau, M. (1995) AFLP: a new technique for DNA fingerprinting. Nucleic Acids Research 23: 4407–4414.

Waalwijk, C. and Baayen, R.P. (1995) Identification of races of *Fusarium oxysporum* f. sp. *dianthi* by DNA fingerprinting and vegetative compatibility. EPPO Bulletin 25: 117–124.

Waalwijk, C., Baayen, R.P., de Koning, J.R.A. and Gams, W. (1996a) Ribosomal DNA analyses challenge the status of *Fusarium* sections *Liseola* and *Elegans.* Sydowia 48: 90–104.

Waalwijk, C., de Koning, J.R.A., Baayen, R.P. and Gams, W. (1996b) Discordant groupings of *Fusarium* spp. from sections *Elegans, Liseola* and *Dlaminia* based on ribosomal ITS1 and ITS2 sequences. Mycologia 88: 361–368.

Ward, E. and Adams, M.J. (1998) Analysis of ribosomal DNA sequences of *Polymyxa* species and related fungi and the development of genus- and species-specific PCR primers. Mycological Research 102: 965–974.

Weising, K., Nybom, H., Wolff, K. and Meyer, W. (1995) DNA fingerprinting in plants and fungi, CRC Press, Boca Raton.

Welsh, J. and McClelland, M. (1990) Fingerprinting genomes using PCR with arbitrary primers, Nucleic Acids Research 18: 7213–7218.

Wheeler, W.C. (1996) Optimization alignment. The end of multiple sequence alignment in phylogenetics? Cladistics 12: 1–9.

Whitcombe, D., Theaker, J., Guy, S.P., Brown, T. and Little, S. (1999) Detection of PCR products using self-probing amplicons and fluorescence. Nature Biotechnology 17: 804–807.

White, T.J., Bruns, T., Lee, S. and Taylor, J. (1990) Amplification and direct sequencing of fungal ribosomal RNA genes for phylogenetics. In: PCR protocols: A Guide to Methods and Applications, (eds), Innis, M.A., Gelfand, D.H., Sninsky, J.J. and White, T.J. Academic Press, San Diego, 315–322.

Williams, J.G.K., Kubelik, A.R., Livak, K.J., Rafalski, J.A. and Tingey, S.V. (1990) DNA polymorphisms amplified by arbitrary primers are useful as genetic markers. Nucleic Acids Research 18: 6531–6535.

Williams, J.G.K., Hanafey, M.K., Rafalski, J.A. and Tingey, S.V. (1993) Genetic analysis using random amplified polymorphic DNA markers. Methods in Enzymology 218: 704–740.

Wilson, W.J., Strout, C.L., DeSantis, T.Z., Stilwell, J.L., Carrano, A.V. and Andersen, G.L. (2002) Sequence-specific identification of 18 pathogenic microorganisms using microarray technology. Molecular and Cellular Probes 16: 119–127.

Wittwer, C.T., Herrmann, M.G., Moss, A.A. and Rasmussen, R.P. (1997) Continuous fluorescence monitoring of rapid cycle DNA amplification. *Biotechniques* 22: 130–138.

Wollenweber, H.W. and Reinking, O.A. (1935) *Die Fusarien, ihre Beschreibung, Schadwirkung und Bekämpfung.* P. Parey, Berlin, pp. 355.

Woo, S.L., Zoina, A., Del Sorbo, G., Lorito, M., Nanni, B., Scala, F. and Noviello, C. (1996) Characterization of *Fusarium oxysporum* f. sp. *phaseoli* by pathogenic races, VCGs, RFLPs and RAPD. Phytopathology 86: 966–973.

Yli-Mattila T. and Hyvönen, J. (1996) RAPD-PCR analysis of *Fusarium* strains—cladistic evaluation of the results. Sydowia 48: 184–195.

Yli-Mattila, T., Mach, R., Alekhina, I.A., Bulat, S.A., Koskinen, S., Kullnig-Gradinger, C.M., Kubicek, C. and Klemsdal, S.S. (2004a) Phylogenetic relationsip of *Fusarium langsethiae* to *Fusarium poae* and *F. sporotrichioides* as inferred by IGS, ITS, β-tubulin sequence and UP-PCR hybridization analysis. International Journal of Food Microbiology 95: 267–285.

Yli-Mattila, T., Mironenko, N.V., Alekhina, I. Hannukkala, A. and Bulat, S.A. (1997a) Universally primed polymerase chain reaction analysis of *Fusarium avenaceum* isolated from wheat and barley in Finland. Agricultural Food Science in Finland 6: 25–36.

Yli-Mattila, T., Mironenko, N.V., Alekhina, I., Hannukkala, A., Hyvönen, J. and Bulat, S.A. (1998) Identification of *Fusarium avenaceum* isolates by RAPD-PCR and UP-PCR. In: Advances in the detection of plant pathogens by polymerase chain reaction, (eds) Manceau, C. and Spak, J. Proceedings of COST 823 Workshop Ceske Budejovice, 20-21 June, 1996, 114–123.

Yli-Mattila, T., Paavanen-Huhtala, S., Hannukkala, A., Alekhina, I., Hyvönen, J. and Bulat, S.A. (1997b) RAPD-PCR, UP-PCR, rDNA RFLP, rDNA sequence and isozyme analyses of *Fusarium* isolates in Finland. Cereal Research Communications 25: 629–630.

Yli-Mattila, T., Paavanen S., Hannukkala A., Parikka P., Tahvonen R. and Karjalainen R., (1996) Isozyme and RAPD-PCR analyses of *Fusarium avenaceum* strains from Finland. Plant Pathology 45: 126–134.

Yli-Mattila T., Paavanen-Huhtala S., Bulat S.A., Alekhina I.A. and Nirenberg H.I. (2002) Molecular, morphological and phylogenetic analysis of the *Fusarium avenaceum/F. arthrosporioides/F. tricinctum* species complex—a polyphasic approach. Mycological Research 106: 655–669.

Yli-Mattila, T., Paavanen-Huhtala, S., Parikka, P., Jestoi, M., Klemsdal, S. and Rizzo, A. (2005) Genetic variation metabolites and mycotoxins of *Fusarium avenaceum* and related species. Mycotoxin Research (submitted).

Yli-Mattila, T., Paavanen-Huhtala, S., Parikka, P., Konstantinova, P. and Gagkaeva, T. (2004b) Molecular and morphological diversity of *Fusarium* species in Finland and northwestern Russia. European Journal of Plant Pathology 110: 573–585.

Yli-Mattila, T., Paavanen-Huhtala, S., Parikka, P., Hietaniemi, V., Jestoi, M. and Rizzo, A. (2004c) Toxigenic fungi and mycotoxins in Finnish cereals. In: An overview on Toxigenic Fungi and Mycotoxins in Europe, (eds), Logrieco, A and Visconti, A. Kluwer Academic Publishers, The Netherlands, 83–100.

Yli-Mattila, T., Paavanen-Huhtala, S., Parikka, P., Hietaniemi, V., Jestoi, M. and Rizzo, A. (2004d). Real-time PCR detection and quantification of *Fusarium poae* as compared to mycotoxin production in grains in Finland. In: Canty, S.M., Boring, T., Wardwell, J. and Ward, R.W. (Eds), Proceedings of the 2nd International Symposium on *Fusarium* Head Blight, incorporating the 8th European *Fusarium* Seminar; 2004, 11–15; Orlando, Fl, USA. Michigan State University, East Lansing, MI, pp. 422–425.

Zeller, K.A., Summerell, B.A., Bullock, S. and Leslie, J.F. (2003) *Gibberella konza* (*Fusarium konzum*) sp. nov. from prairie grasses, a new species in the *Gibberella fujikuroi* species complex. Mycologia 95: 943–954.

Zietkiewicz, E., Rafalski, A. and Labuda, D. (1994) Genome fingerprinting by simple sequence repeat (SSR)-anchored polymerase chain reaction amplification. Genomics 20: 176–183.

8. Evolutionary Relationships Among Economically Important *Aspergillus* Species

János Varga
Department of Microbiology, Faculty of Sciences, University of Szeged,
P.O. Box 533, H-6701 Szeged, Hungary

SUMMARY Members of the filamentous fungal genus *Aspergillus* have long been used by mankind for the production of various fermented foods, secondary metabolites, organic acids and enzymes. At the same time, *Aspergillus* species are also opportunistic pathogens of emerging importance, and produce a range of mycotoxins including aflatoxins, ochratoxins and patulin which are harmful to humans and animals. *Aspergillus* taxonomy is in process of change due to advancement in molecular taxonomic methods. In this review, we wished to give a general overview of the molecular studies carried out on species belonging to sections involving economically important mycotoxin producing biseriate species (*Aspergillus* sections *Nigri*, *Circumdati*, *Flavi* and *Terrei*). Molecular tools enabled us to revise the taxonomy of *Aspergillus* sections *Nigri*, *Circumdati* and *Flavi*. A similar approach is underway for clarifying the taxonomy of *Aspergillus* section *Terrei* and other *Aspergillus* sections not covered in this review. Phylogenetically unrelated species were found to produce the same mycotoxins. For example, aflatoxins have been produced under laboratory conditions by species belonging to three different sections, while ochratoxin A has been produced by a variety of unrelated species. Further studies are needed to clarify the taxonomy of Aspergilli and the biosynthesis of these mycotoxins in order to develop suitable methods for their detection.

Introduction

The name *Aspergillus* was first applied by Micheli in 1729 to molds bearing conidial heads and stalks (Raper and Fennell, 1965). Today, *Aspergillus* is among the most economically important of the fungal genera. Isolates of *Aspergillus* are used for the production of soy sauce, several organic acids and enzymes, and biologically active metabolites such as lovastatin (Campbell-Platt and Cook, 1989; Bennett and Klich, 1992; Pariza and Johnson, 2001; Manzoni and Rollini, 2002). Aspergilli are frequently encountered in plant products. The most important consequence of their presence is mycotoxin contamination. The main mycotoxins produced by Aspergilli are the aflatoxins, ochratoxin A and patulin, which are produced by a variety of *Aspergillus* species.

Partly because of its economic importance, the genus *Aspergillus* has one of the better described taxonomies among filamentous fungi. In this review, we wish to give an overview of the taxonomic relationships among economically important *Aspergillus* species. The molecular techniques applied include PCR-based methods, RFLP techniques, and phylogenetic analysis of ribosomal RNA gene sequences together with morphological and physiological features. The fungal groups treated involve sections including biotechnologically important (mostly) biseriate species of subgenus *Aspergillus* (sections *Nigri*, *Flavi*, *Circumdati* and *Terrei*; Fig. 1).

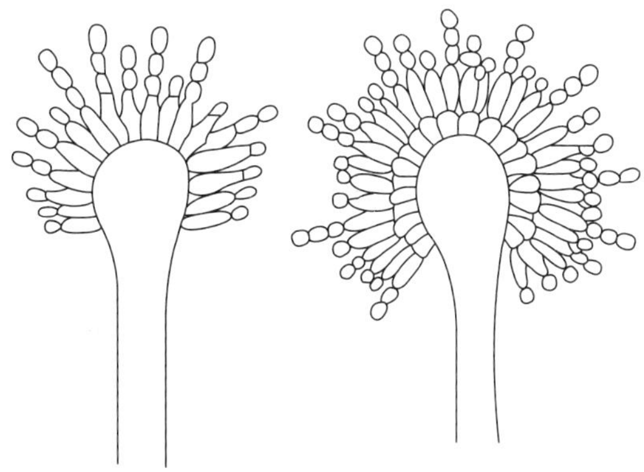

Fig. 1. Typical uniseriate and biseriate conidial heads of Aspergilli.

Economical Importance: The Janus Faces of Aspergilli

Aspergillus species can be both harmful and beneficial for mankind. They are responsible for a number of plant and human diseases and produce several toxins. At the same time, Aspergilli are widely used in the food and pharmaceutical industry for the production of various acids, enzymes and other compounds useful for humans.

Aspergillus Species as Pathogens of Plants and Animals

Although not considered to be major causes of plant disease, *Aspergillus* species are responsible for several disorders in various plants. For example, chlorosis of almond, albinism of citrus, black rot of onions, crown rot of peanut, vine canker of grapes are caused by Aspergilli (Varga et al. 2004; Michailides et al. 2002). The most common plant pathogens are *Aspergillus niger* and *A. flavus* (Varga et al. 2004). In contrast with specialized plant pathogens such as powdery mildews, rusts and some *Fusarium* species, *Aspergillus* species are opportunistic pathogens without host specialization (St. Leger et al. 2000). While only a limited number of *Aspergillus* species can invade living plant tissues, several are encountered as storage moulds on plant products (Raper and Fennell, 1965; Kozakiewicz, 1989).

Several *Aspergillus* species are considered as opportunistic human and animal pathogens. Diseases caused by *Aspergillus* species are increasing in importance, especially among immunocompromised hosts. Clinical manifestations are variable, ranging from allergic to invasive disease, largely depending on the status of the host's immune system. *Aspergillus fumigatus* is of prevailing importance. This species may cause several diseases, such as allergic bronchopulmonary aspergillosis, aspergilloma, and invasive aspergillosis, a usually fatal disease particularly in immunocompromised patients (Brookman and Denning, 2000). Nowadays, *A. fumigatus* is responsible for more than 90% of invasive aspergillosis in humans (Latgé, 2001). Invasive aspergillosis is a major cause of death at leukemia treatment centers and bone marrow transplantation (BMT) and solid-organ transplantation units. Other species like *A. niger, A. terreus, A. flavus* can also be the causative agents of different diseases like sinusitis, osteomyelitis, keratitis, bronchiectasis, otomycosis and ophthalmitis (Marr et al. 2002).

Mycotoxins Produced by *Aspergilli*

Aspergillus species can contaminate foods and feeds at different stages including pre- and postharvest stages, processing and handling. The most important aspect of food and feed spoilage is the formation of mycotoxins, which may have harmful effects on human and animal health. Several *Aspergillus* mycotoxins have been identified as contaminants in foods and feeds, the economically most important of which are the

Table 1. Some economically important mycotoxins produced by *Aspergillus* species in various agricultural products (Larsen et al. 2001; Bayman et al. 2002; Varga et al. 2003, 2004)

Mycotoxins	Agricultural product	Species
Aflatoxins	Peanut, corn, cotton	*A. flavus, A. parasiticus, A. nomius*
Ochratoxins	Cereals	*P. verrucosum*
	Meat, cheese	*P. nordicum*
	Grape, wine	*A. niger, A. carbonarius*
	Coffee, spices	*A. ochraceus, A. niger, A. carbonarius*
	Figs	*A. alliaceus*
Patulin	Cereals, malt	*P. expansum, A. clavatus*
	Apple, pear	*P. expansum*

aflatoxins, ochratoxins and patulin (Table 1, Fig. 2). Recent studies indicate that these compounds can be produced by a number of *Aspergillus* species (Varga et al. 2003, 2004). However, only a few of these mycotoxin producers can be regarded as potential health hazards either because they produce only traces of the given mycotoxin (e.g. small amounts of aflatoxins by *A. ruber*, or ochratoxins by *A. wentii* and *A. terreus*), or are encountered rarely if at all in food products (e.g. aflatoxin producing *A. ochraceoroseus, Emericella venezuelensis* and *E. acristata* and ochratoxin-producing *A. auricomus* isolates) (Pitt, 2000).

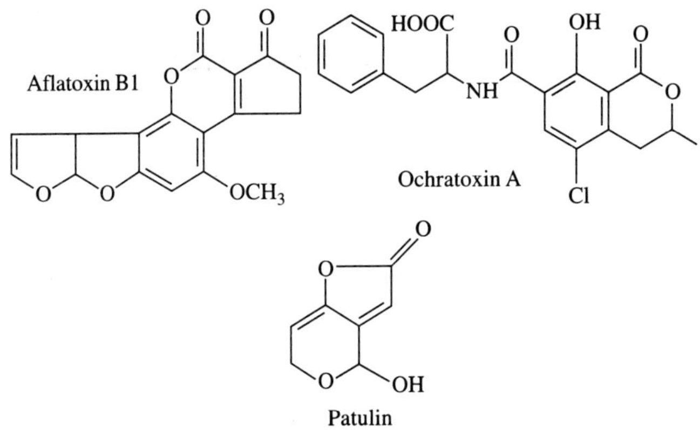

Fig. 2. Structures of the most common mycotoxins produced by Aspergilli.

However, new data indicate that some species recently reported to be mycotoxin producers can be regarded as main sources of mycotoxin contamination in various food products (Abarca et al. 2001; Pitt, 2000; Bayman et al. 2002). For example, although ochratoxin-producing abilities of black Aspergilli have only recently been discovered, these fungi are now considered as major sources of ochratoxin contamination in wine, raisins and coffee (Pitt, 2000).

Aspergillus Species in the Food and Pharmaceutical Industry

Aspergillus species are used in oriental fermentation processes for the production of soy sauce, miso, sake, mirin, shochu and katsuobushi (Hara et al. 1992; Wolf, 1997; Ng, 2003). Besides microorganisms like *Rhizomucor pusillus, R. miehei* or *Endothia parasitica, Aspergillus oryzae* is also extensively used for rennet production in cheese manufacture (Wolf, 1997).

Aspergillus species have the capacity to secrete large amounts of proteins, metabolites and organic acids into their culture medium. This property has been widely exploited by the food and beverage industries where compounds secreted by Aspergilli have been used for decades. This has led to the GRAS (generally regarded as safe) status for some of their products. Filamentous fungi like *A. awamori*, *A. niger* and *A. oryzae* are therefore suitable organisms for the production of commercially interesting homologous and heterologous proteins. Consequently, Aspergilli are frequently used in the fermentation industry for the production of various organic acids (citric acid, gluconic acid, itaconic acid, malic acid, tartaric acid) (Bigelis and Arora, 1992) and enzymes (α- and γ-amylases, catalase, cellulase, α- and β-galactosidases, β-glucanase, glucose oxidase, hemicellulase, insulinase, invertase, lipase, peroxidase, pectinase, acid and alkaline proteases, tannase, urate oxidase and xylanase) (Lowe, 1992). *A. oryzae* and black Aspergilli are used as hosts for heterologous expression of a number of economically important proteins like antibody fragments, interleukin 6, manganese peroxidase, prochymosin, glucoamylase and others (Gouka et al. 1997, Punt et al. 2002, Joosten et al. 2003). Besides, genetically modified black *Aspergillus* isolates are used in the fermentation industry for the production of various enzymes including arabinofuranosidase, chymosin, glucose oxidase, pectin esterase, phospholipase, phytase and xylanase (Pariza and Johnson, 2001).

Various *Aspergillus* species are used in the pharmaceutical industry in biotransformations (Kieslich, 1997). The species most frequently involved include *A. alliaceus, A. flavus, A. niger, A. ochraceus, A. oryzae* and *A. terreus*. Besides, *A. terreus* and some other *Aspergillus* species produce lovastatin, a cholesterol-lowering compound (Kennedy et al. 1999).

Taxonomic Outline of the *Aspergillus* Genus

Teleomorphs of Aspergilli belong to different genera all of which are the members of family Trichocomaceae of the Eurotiales order (Table 3; Alexopoulos et al. 1996). This order is monophyletic as proved by several molecular studies (Geiser and LoBuglio, 2001). Raper and Fennell (1965) described 18 species groups within the *Aspergillus* (form) genus mainly based on cultural and morphological features, which were treated as sections belonging to six subgenera by Gams et al. (1985). Phylogenetic studies of ribosomal RNA gene sequences led to the acceptance of three subgenera with a total of 15 sections and the so-called '*Warcupiella* group', a treatment currently accepted by most *Aspergillus* researchers (Table 2; Peterson, 2000). Our studies focused on the phylogenetic analysis of sequences of the intergenic spacer region and the 5.8S rRNA gene (ITS region) and the D1-D2 region of the 28S rRNA gene in various Aspergilli (Fig. 3). This article deals with those sections which involve economically important biseriate species (Table 2).

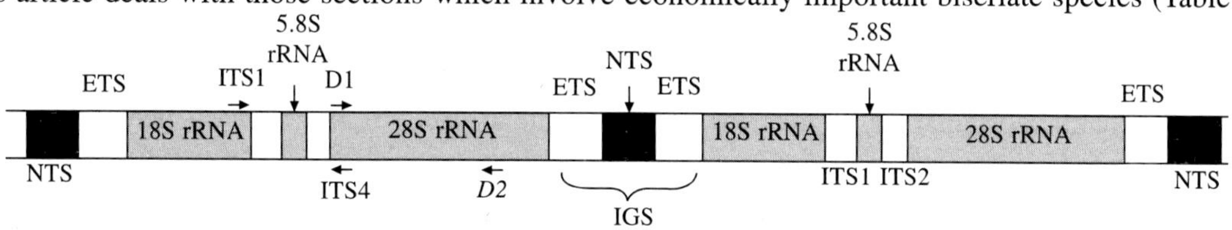

Fig. 3. Organization of the rRNA gene cluster in Aspergilli, and location of the primers used to amplify the ITS region and the D1-D2 region of the 28 S rDNA.

Aspergillus Section *Nigri*

Black Aspergilli (*Aspergillus niger* species group, Raper and Fennell, 1965; *Aspergillus* section *Nigri*, Gams et al. 1985) have a significant impact on modern society. Many species cause food spoilage, and several are used in the fermentation industry to produce hydrolytic enzymes, such as amylases or lipases, and organic acids, such as citric acid and gluconic acid (Raper and Fennell, 1965; Bennett and Klich, 1992; Pariza and Johnson, 2001; Kozakiewicz, 1989). They are also candidates for genetic manipulation in the biotechnology industries since *A. niger* has been granted the GRAS (generally regarded as safe) status by

Table 2. Classification of *Aspergillus* species based on rDNA sequences (the sections examined are in bold type)

Sections	Assignment to subgenus		Teleomorph
	Gams et al. (1985)	Peterson (2000)	
Aspergillus	*Aspergillus*	*Aspergillus*	Eurotium
Restricti	*Aspergillus*	*Aspergillus*	–
Cervini	*Fumigati*	*Aspergillus*	–
Terrei	***Nidulantes***	***Aspergillus***	***Fennellia***
Flavipedes	*Nidulantes*	*Aspergillus*	*Fennellia*
Nigri	***Circumdati***	***Aspergillus***	–
Circumdati	***Circumdati***	***Aspergillus***	***Neopetromyces***
Flavi	***Circumdati***	***Aspergillus***	***Petromyces***
Cremei	*Circumdati*	*Aspergillus*	*Chaetosartorya*
Candidi	*Circumdati*	*Aspergillus*	–
Wentii	*Circumdati*	–	–
Fumigati	*Fumigati*	*Fumigati*	*Neosartorya*
Clavati	**Clavati**	***Fumigati***	**Neocarpenteles**
Nidulantes	*Nidulantes*	*Nidulantes*	*Emericella*
Versicolores	*Nidulantes*	–	–
Usti	*Nidulantes*	–	–
Sparsi	*Circumdati*	*Nidulantes*	–
Ornati	*Ornati*	*Nidulantes*	*Sclerocleista*
Warcupiella group	*Ornati*	*Warcupiella* gr.	*Warcupiella*

the Food and Drug Administration of the US government. Although the main source of black Aspergilli is soil, members of this section have been isolated from various other sources (Varga et al. 2004).

Black Aspergilli are one of the more problematic groups for identification. Raper and Fennell (1965) described 12 species of the black Aspergilli. Al-Musallam (1980) revised the taxonomy of the *A. niger* group by taking mainly morphological features into account. She recognized seven species within this group (*A. japonicus*, *A. carbonarius*, *A. ellipticus*, *A. helicothrix*, *A. heteromorphus*, *A. foetidus* and *A. niger*), and described *A. niger* itself as an aggregate consisting of seven varieties and two formae. Kozakiewicz (1989) distinguished *A. ellipticus*, *A. heteromorphus*, *A. japonicus*, *A. helicothrix*, *A. atroviolaceus* (treated as *A. aculeatus* or *A. japonicus* var. *aculeatus* in other classifications) and *A. carbonarius*, a species exhibiting echinulate conidial ornamentations, which distinguished it from the rest of black *Aspergillus* strains, which displayed verrucose conidia. Within the verrucose category, *A. fonsecaeus*, *A. acidus* (*A. foetidus* var. *acidus*), *A. niger* var. *niger*, *A. niger* var. *phoenicis*, *A. niger* var. *ficuum*, *A. niger* var. *tubingensis*, *A. niger* var. *pulverulentus*, *A. niger* var. *awamori*, *A. citricus* (*A. foetidus*) and *A. citricus* var. *pallidus* (*A. foetidus* var. *pallidus*) were recognised. In recent years, several publications have dealt with the application of different phenotypic and genotypic markers for clarifying the taxonomy of black Aspergilli. Among the genotypic approaches, nuclear and mitochondrial DNA (mtDNA) polymorphisms and PCR-based techniques led to the recognition of four species within the *A. niger* species complex (*A. niger, A. tubingensis, A. brasiliensis, A. foetidus*) (Kusters-van Someren et al. 1991; Varga et al. 1993, 1994; Megnegneau et al. 1993; Accensi et al. 2001; Varga et al. 2000c). Sequence comparisons of nuclear genes encoding various extracellular enzymes supported these results (Bussink et al. 1991; de Graaf et al. 1994; Gielkens et al. 1997). Yokoyama et al. (2001) also distinguished two clusters within the *A. niger* species complex based on phylogenetic analyses of sequences of the mitochondrial cytochrome b gene (Fig. 4). Several well-known species names such as *A. awamori*, *A. usamii*, *A. phoenicis* and *A. ficuum* have been reduced to synonymy. Regarding other black *Aspergillus* species, phylogenetic analyses of sequences of the intergenic spacer region and the 5.8S rRNA gene (ITS region) and the D1-D2 region of the 28S rRNA gene indicate that, apart from those mentioned earlier, at least five other species belong to this section (*A. heteromorphus, A. ellipticus, A. carbonarius, A. japonicus* and *A. aculeatus* (Fig. 4) (Hamari et al. 1997; Kevei et al. 1996; Parenicova et al. 2001; Varga et al. 2000c). The uniseriate species *A. japonicus, A. aculeatus* and isolate CBS 114.80, which is considered to represent a new species, form one well-defined clade, while the biseriate species are on a separate branch (Parenicova et al. 2001; Fig. 4). Although some toxins and other toxic agents, e.g. oxalic acid, nigragillin, malformins and naphtho-γ-pyrones (Schuster et al. 2002), have been reported to be produced by black Aspergilli, the production of ochratoxins is of real economic importance. Until recently, ochratoxin production has been observed only in *A. niger* and *A. carbonarius* (Ueno et al. 1991; Abarca et al. 1994; Ono et al. 1995; Horie, 1995; Téren et al. 1996; Wicklow et al. 1996; Accensi et al. 2001). These species are now considered as major sources of ochratoxin contamination in tropical and subtropical foods including dried vine fruits, wines and coffee (Pitt, 2000; Cabanes et al. 2002; Abarca et al. 2003). Recent data indicate that other black Aspergilli including *A. tubingensis* and *A. japonicus* isolates also have the capacity to produce ochratoxins (Battilani et al. 2003; Dalcero et al. 2002; Abarca et al. 2004; Varga J., Téren J., Pál K., unpublished observations).

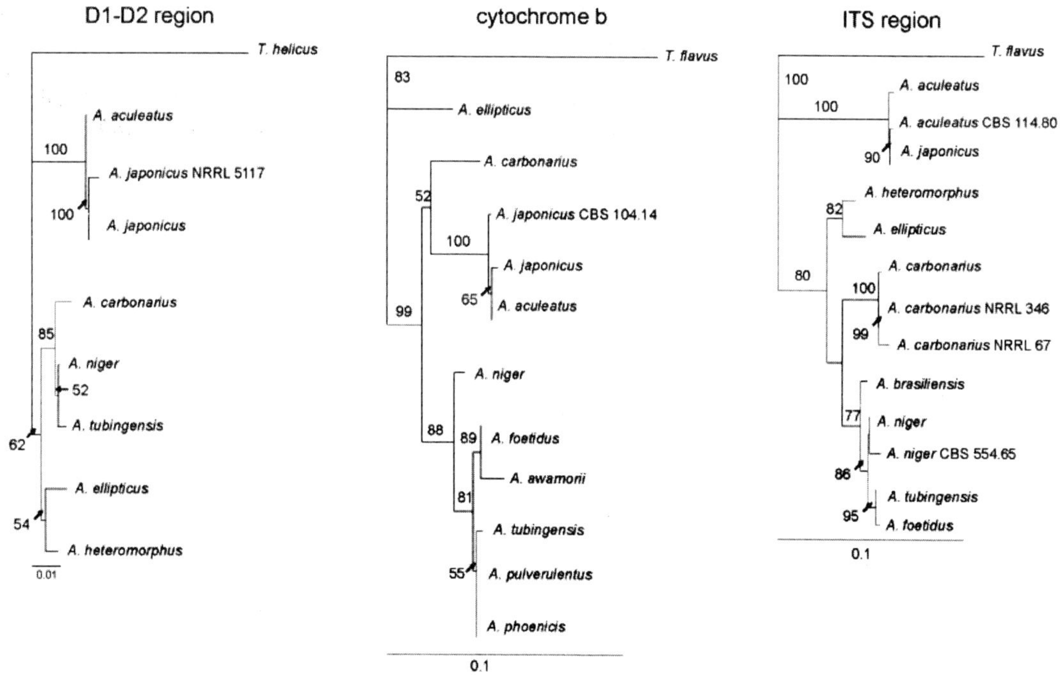

Fig. 4. Neighbor-joining trees of *Aspergillus* section *Nigri* based on ribosomal RNA gene and cytochrome b sequences (adopted from Peterson, 2000; Yokoyama et al. 2001; Parenicova et al. 2001). Numbers above branches indicate bootstrap values; only values >50% are shown.

Aspergillus Section *Flavi*

Aspergillus section *Flavi* historically includes species with conidial heads in shades of yellow-green to brown and dark sclerotia. Isolates of the so-called domesticated species, such as *A. oryzae, A. sojae* and

A. tamarii are used in oriental food fermentation processes and as hosts for heterologous gene expression (Campbell-Platt and Cook, 1989). Genetically modified *A. oryzae* strains are used for the production of enzymes including lactase, pectin esterase, lipase, protease and xylanase (Pariza and Johnson, 2001). Several species of section *Flavi* produce aflatoxins, among which aflatoxin B_1 is the most toxic of the many naturally occurring secondary metabolites produced by fungi. Aflatoxins are mainly produced by *A. flavus* and *A. parasiticus*, which coexist with and grow on almost any crop or food.

Regarding the intra- and interspecific variability of *Aspergillus* section *Flavi*, ITS and 28S rDNA sequences of type strains or representative isolates of the species and subspecies currently assigned to this section have recently been analysed (Peterson, 2000; Rigó et al. 2002). Phylogenetic analysis of sequence data indicated that species of *Aspergillus* section *Flavi* form distinct clades (Fig. 5). The main clades identified, based on sequence data could also be distinguished based on colony colour and their ubiquinone systems. The '*A. flavus*' clade includes species characterised with Q-10(H_2) as their main ubiquinone, and conidial colours in shades of green, bearing dark sclerotia. Studies on the genetic variability of *A. flavus* indicate that the name is currently applied to a paraphyletic group of isolates that may produce aflatoxins B or G, and have large or small sclerotia (Geiser et al. 2000). It was suggested that isolates with small sclerotia, able to produce both aflatoxins B and G (group II), deserve recognition as a new species (Geiser et al. 2000; Ehrlich et al. 2003). The other group (group I) includes isolates producing only aflatoxin B, and having large or small sclerotia. This group also includes isolates of *A. oryzae*, and has previously been described as having a recombining population structure (Geiser et al. 1998). Although several lines of evidence suggest that *A. oryzae* and *A. sojae* are morphological variants of *A. flavus* and *A. parasiticus*, respectively, it was suggested that these taxa should be retained as separate species because of the regulatory confusion that conspecificity might generate in the food industry (Geiser et al. 1998). The '*A. tamarii*' clade

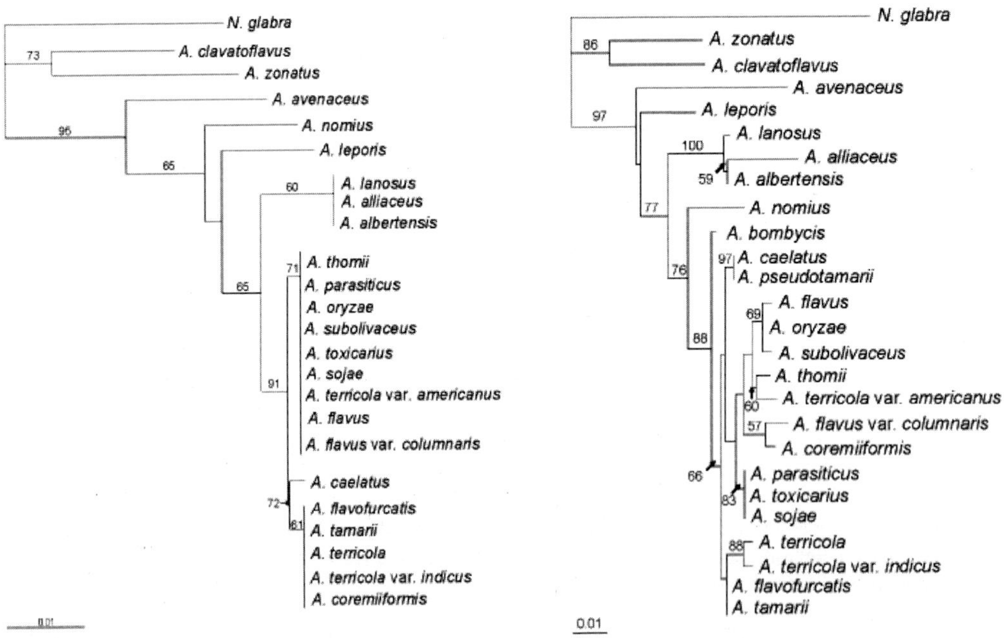

Fig. 5. Neighbor-joining trees of species of *Aspergillus* section *Flavi* based on ITS (right) and D1-D2 sequences (left) (Rigó et al. 2000; Peterson, 2000). Numbers above branches indicate bootstrap values; only values >50% are shown.

contains species with ubiquinone system Q-10(H_2), and conidia in shades of olive to brown, while the '*A. alliaceus*' clade consists of species with the Q-10 ubiquinone system, and conidia in shades of ochre (Rigó et al. 2002; Kuraishi et al. 1990). Two species of this clade, *Petromyces alliaceus* and *P. albertensis,* produce high amounts of ochratoxin (50-300 $\mu g\ ml^{-1}$), and are considered to be responsible for ochratoxin contamination of figs (Bayman et al. 2002; Varga et al. 1996). Isolates of *P. alliaceus* and *P. albertensis* were described as producers of ochratoxins A and B (Ciegler 1972; Varga et al. 1996), and *P. alliaceus* may cause chronic otorrhoea (Koenig et al. 1985). Besides their harmful properties, *Petromyces* species also have beneficial effects on mankind. *P. alliaceus* strains produce asperlicins, potent cyclic peptide antagonists of cholecystokinin, which is a neurotransmitter involved in the hormonal regulation of pancreatic and gastric secretion, gallbladder contraction, and gut motility (Liesch et al. 1988). *P. alliaceus* strains are also used for steroid and alkaloid transformations (Burkhead et al. 1994), and for the production of pectin degrading enzyme preparations (Mikhailova et al. 1995).

The recently described aflatoxin-producing species *A. pseudotamarii* and *A. bombycis* are closely related to *A. caelatus* and *A. nomius*, respectively (Ito et al. 2001; Peterson et al. 2001). Physiological properties and mycotoxin-producing abilities of these taxa justify their treatment as separate species (Table 3; Ito et al. 2001; Peterson et al. 2001). Recent data indicate that *A. nomius* is a paraphyletic group likely to contain several other species (Egel et al. 1994; Feibelman et al. 1998; Cotty and Cardwell, 1999; Ehrlich et al. 2003). Two other species, *A. avenaceus* and *A. leporis,* were found to form separate lineages not closely related to any of the main clades identified. It is proposed that *A. clavatoflavus* and *A. zonatus* be excluded from *Aspergillus* section *Flavi*, in accordance with previous suggestions (Kozakiewicz, 1989). Phylogenetic analysis of partial 28S rRNA gene sequences supported these findings (Fig. 5).

Table 3. Mycotoxin producing abilities of the economically important species of *Aspergillus* section *Flavi* (Varga et al. 2003; Scheidegger and Payne, 2003)

Species	Aflatoxins		CPA*	Kojic acid	Others
	B	G			
A. avenaceus	–	–	–	–	Avenaciolide
A. bombycis	+	+	–	+	–
A. caelatus	–	–	–	+	–
A. flavus group I	+	–	±	+	Aspergillic acid, paspaline, nominine
A. flavus group II	+	±	+	+	Aspergillic acid, paspaline, nominine
A. lanosus	–	–	–	+	Griseofulvin
A. leporis	–	–	–	+	Leporine, pseurotin, antibiotic Y
A. nomius	+	+	–	+	Aspergillic acid, tenuazoic acid, pseurotin, nominine
A. oryzae	–	–	±	+	–
A. parasiticus	+	+	–	+	Aspergillic acid, parasiticol, parasiticolide A
A. pseudotamarii	+	–	+	+	–
A. sojae	–	–	–	+	–
A. tamarii	–	–	+	+	Fumigaclavine A
"*A. argentinicus*"	+	+	+	–	ND
P. alliaceus	–	–	–	+	Ochratoxins, asperlicine, nominine, kotanins, paspaline

*CPA: cyclopiazonic acid; ND: not determined

In the past few years, isolates with unusual mycotoxin profiles have been encountered in some food products. For example, isolates resembling *A. flavus*, but producing aflatoxins B, G and cyclopiazonic acid were found in peanut fields in Argentina (Vaamonde et al. 2003), while *A. tamarii*-like isolates producing the same pattern were found in black pepper and black tea samples in the Czech Republic (Ostry, 2000). Molecular analysis of these isolates is in progress.

Aflatoxin-producing species (*A. flavus, A. parasiticus, A. nomius, A. bombycis* and *A. pseudotamarii*) were scattered throughout the dendrogram indicating that aflatoxin-producing ability was lost (or gained) several times during evolution. Another aflatoxin-producing species, *A. ochraceoroseus* was found to be not related to any of the species belonging to either sections *Flavi* or *Circumdati* (Klich et al. 2000, 2003). This is the only species known to accumulate aflatoxin B_1 and sterigmatocystin simultaneously. This species has previously been assigned either to *Aspergillus* sections *Wentii, Cremei* or *Circumdati* (Christensen, 1981; Samson, 1979; Kozakiewicz, 1989). However, ITS and 28S rDNA data indicate that this species is closely related to section *Nidulantes* (Klich et al. 2003). Its distant relationship to *Aspergillus* section *Flavi* is supported by the observation that *A. ochraceoroseus* DNA hybridised only weakly if at all to the *A. flavus* and *A. parasiticus* aflatoxin biosynthetic gene probes (Klich et al. 2000). Additionally, the order of genes of the aflatoxin biosynthetic gene cluster of *A. ochraceoroseus* was more similar to that of the sterigmatocystin gene cluster of *A. nidulans* than to that of *A. parasiticus* (Klich et al. 2000). These data indicate that the aflatoxin and sterigmatocystin biosynthetic pathway genes in *A. ochraceoroseus* are different from known pathway genes. Additionally, aflatoxin production has also been observed recently in *Emericella venezuelense* and *E. acristata* isolates, which are also closely related to *A. nidulans* (Klich et al. 2001; J.C. Frisvad, personal communication).

Aspergillus Section *Circumdati*

Aspergillus section *Circumdati* historically includes species with biseriate conidial heads in shades of yellow to ochre. Species of *Aspergillus* section *Circumdati* are economically important as ochratoxin-producing spoilage organisms. Species of *Aspergillus* section *Circumdati* are economically important on account of a number of aspects. Several species produce diverse mycotoxins harmful for animals and humans including ochratoxin A, xanthomegnin and viomellein (Varga et al. 1996; Durley et al. 1975). Isolates of *A. ochraceus* or *A. sclerotiorum* are useful for biochemical transformation of steroids, alkaloids or phenazines (Chen et al. 1994). Some species, e.g. *A. sclerotiorum* and *A. melleus* are important sources of proteolytic enzymes (Luisetti et al. 1991), and other metabolites (Matsukuma et al. 1992; Lin et al. 1995), while sclerotia of several species contain antiinsectant compounds (Whyte et al. 1996; Ooike et al. 1997). The most frequently encountered species of the section, *A. ochraceus* is predominately isolated from desert and cultivated soil, but is also a post-harvest pathogen of several agricultural products including cereals, coffee beans, corn and pecans (Kozakiewicz, 1989). *A. ochraceus* is known to cause allergic bronchopulmonary aspergillosis (Novey and Wells, 1978).

Interspecific variability of species assigned to this section was examined using phenotypic features, sequences of the ITS region and the D1-D2 region of the 28S rRNA gene (Fig. 6). Phylogenetic analysis of sequence data indicated that *Aspergillus campestris, A. lanosus, A. dimorphicus* and *A. sepultus* belong to *Aspergillus* sections *Candidi, Flavi* and *Cremei*, respectively (Fig. 6) (Peterson, 1995; Varga et al. 2000d). Two teleomorphic species previously assigned to this section, *Petromyces alliaceus* and *P. albertensis*, together with the asexual *A. lanosus* were found to belong to *Aspergillus* section *Flavi* (Varga et al. 2000a). These results were also supported by phenotypic data, and by the main ubiquinones observed in these species (Kuraishi et al. 1990). Species of the revised *Aspergillus* section *Circumdati* formed two main clades, which could also be distinguished based on phenotypic methods. A sexually reproducing ochratoxin producing species, *Neopetromyces muricatus* was also found to belong to this section (Varga et al. 2000a; Frisvad and Samson, 2000). All these species are characterised by the Q-10(H_2) ubiquinone system. *Aspergillus auricomus* and *A. elegans* did not belong to any of these clades. Ochratoxin-producing abilities of the isolates examined did not correlate with their taxonomic relationships based on ITS sequence data. However, a clade found not to produce ochratoxins during an earlier study of genetic variability of *A. ochraceus* was also identified in this study (Varga et al. 2000a, 2000b, 2000d). Recently a new species, *A. persii* has been assigned to this section (Zotti and Corte, 2002), and several other isolates from tropical and subtropical areas were found to differ significantly from described species of this section (Frisvad, personal communication). Molecular analysis of these isolates should be carried out to determine their taxonomic positions.

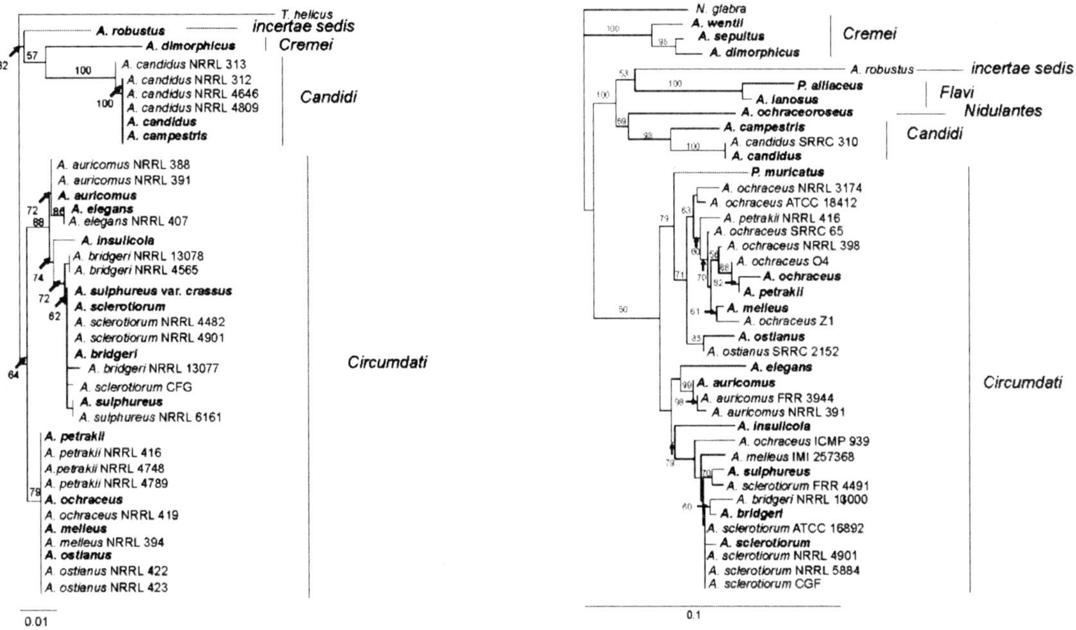

Fig. 6. Neighbor-joining trees of species of *Aspergillus* section *Circumdati* based on D1-D2 (left) and ITS sequences (right; Varga et al. 2000d; Peterson, 2000). Type strains are set in bold type, and the sections are indicated on the right. Numbers above branches indicate bootstrap values; only values >50% are shown.

Aspergillus Section *Terrei*

Aspergillus terreus Thom is an ubiquitous fungus in our environment. Strains of this cosmopolitan species are frequently isolated from desert and grassland soils and compost heaps, and as contaminants of plant products like stored corn, barley and peanuts (Kozakiewicz, 1989). *A. terreus* is an economically important species from a number of aspects. *A. terreus* isolates are used in the fermentation industry for the production of itaconic acid and itatartaric acid and for enzyme production (Bigelis and Arora, 1992; Lowe, 1992). *A. terreus* isolates produce a range of secondary metabolites some of which have properties valuable for mankind, including lovastatin, a cholesterol lowering drug, the antitumour metabolites quadrone and asterriquinone, acetylcholinesterase inhibitors like territrem B and terreulactone, and cyclosporin A (Kaji et al. 1998; Schimmel et al. 1998; Chen et al. 1999; Sallam et al. 2003). Other secondary metabolites produced by *A. terreus* isolates are considered as mycotoxins, including patulin, ochratoxins, citrinin, emodin and sulochrine (Turner and Aldridge, 1983). This species is also an important human pathogen (Lass-Florl et al. 2000).

Two regions of the ribosomal RNA gene cluster, namely the ITS region and the D1-D2 region of the

26S rRNA gene were analysed (Fig. 7; Peterson, 2000; Varga et al. unpublished data). Based on analysis of the ITS region, three main clusters were identified, viz. *A. terreus* and its subspecies and *A. allahabadii* form one cluster; *A. flavipes* clusters together with *A. iizukae* and an *A. niveus* isolate; and *A. carneus* isolates form a third cluster together with *A. obscurus*. *Aspergillus anthodesmis*, an *A. niveus* isolate and *A. janus* var. *brevis* form distinct branches on the tree. The high bootstrap values supported these findings. As seen, while the D1-D2 sequences of most *A. carneus* and *A. niveus* isolates were identical to those of *A. terreus* isolates, their ITS sequences are more informative, i.e. they allowed the separation of *A. terreus*, *A. carneus* and *A. niveus* isolates (Fig. 7). Our data indicate that *Aspergillus* sections *Terrei* and *Flavipes* should be further analysed using molecular methods. Although Peterson (2000) suggested that a revised *Aspergillus* section *Terrei* should include *A. carneus* and *A. niveus* isolates based on D1-D2 sequence data, phylogenetic analysis of ITS sequences indicates that at least *A. carneus* isolates form a well-defined cluster not closely related to *A. terreus*. Lovastatin production of the isolates did not correlate with their positions on any trees based on sequence data, indicating that lovastatin production was lost (or gained) several times during evolution. Additionally, more *A. niveus* isolates should be examined to clarify their taxonomic position.

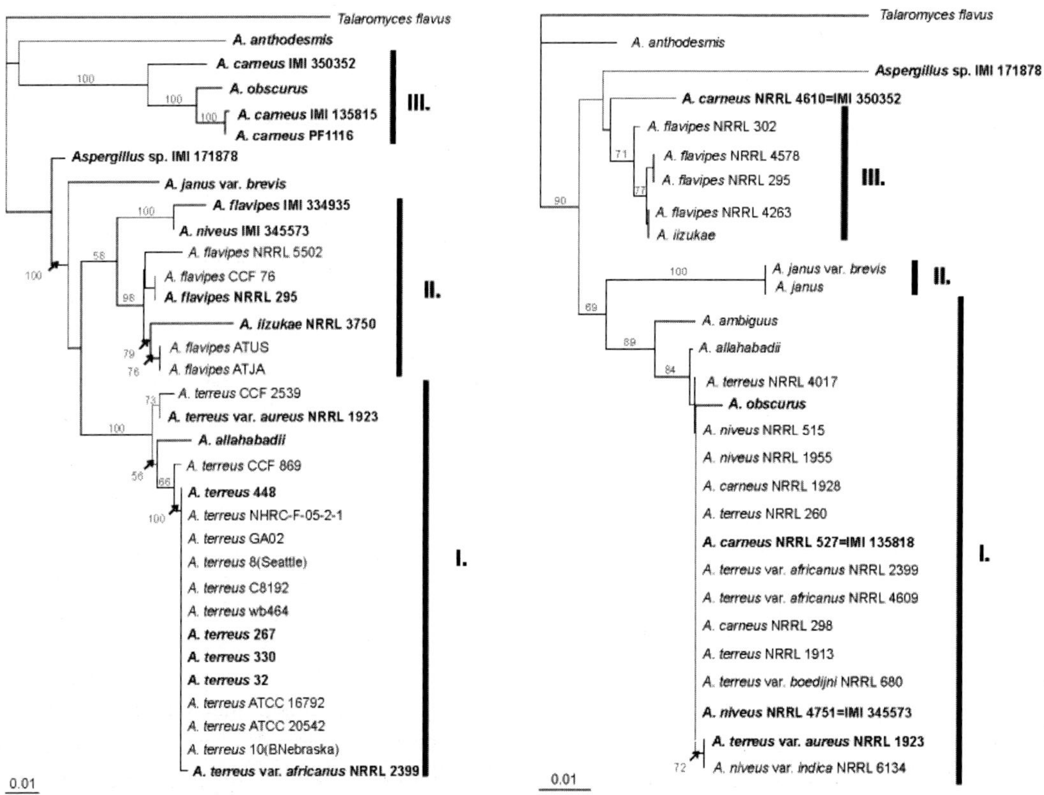

Fig. 7. Neighbor-joining trees of species of *Aspergillus* section *Terrei* based on ITS (left) and D1-D2 sequences (right; Peterson, 2000; Varga et al. in preparation). Numbers above branches indicate bootstrap values; only values >50% are shown.

Conclusions and Future Prospects

Phylogenetic analysis of sequences of the ribosomal RNA gene cluster and the β-tubulin gene clarified the taxonomic relationships of toxigenic Aspergilli causing pre- and postharvest contamination of agricultural products. Molecular data enabled us to clarify the taxonomy of black Aspergilli, *A. flavus* and its relatives, and sections *Circumdati* and Terrei including ochratoxin and patulin-producing species, respectively. Phylogenetically unrelated species can produce the same mycotoxins. Aflatoxin and its precursors or intermediates (e.g. sterigmatocystins) are produced by species belonging to at least three different sections, while ochratoxin is produced by species assigned to seven different sections. In addition, ochratoxins are also produced by some penicillia (Ciegler, 1972; Varga et al. 2001). Based on these observations, mycotoxin-producing abilities of the isolates appear to have been lost (or gained) several times during the evolution of the genus. The organisation of the biosynthetic genes of patulin and ochratoxins is not known, although experiments are in progress in several laboratories to clarify the genetic background of biosynthesis of these mycotoxins. Identification of mycotoxin biosynthetic genes is essential for clarifying the evolution of mycotoxin biosynthesis in Aspergilli, and to develop specific gene probes for the detection of mycotoxin-producing Aspergilli in agricultural products.

Acknowledgements

This research was supported by OTKA grants (F023062, T037217). J. Varga was holder of a Széchenyi fellowship grant.

References

Abarca, M.L., Accensi, F., Bragulat, M.R. and Cabanes, F.J. (2001) Current importance of ochratoxin A-producing *Aspergillus* spp. Journal of Food Protection 64: 903–906.

Abarca, M.L., Accensi, F., Bragulat, M.R., Castella, G. and Cabanes, F.J. (2003) *Aspergillus carbonarius* as the main source of ochratoxin A contamination in dried vine fruits from the Spanish market. Journal of Food Protection 66: 504–506.

Abarca, M.L., Accensi, F., Cano, J. and Cabanes, F.J. (2004) Taxonomy and significance of black aspergilli. Antonie van Leeuwenhoek 86: 33–49.

Abarca, M.L., Bragulat, M.R., Castella, G. and Cabanes, F.J. (1994) Ochratoxin A production by strains of *Aspergillus* niger var. *niger*. Applied and Environmental *Microbiology* 60: 2650–2652.

Accensi, F., Abarca, M.L., Cano, J., Figuera, L. and Cabanes, F.J. (2001) Distribution of ochratoxin A producing strains in the *Aspergillus niger* aggregate. Antonie van Leeuwenhoek 79: 365–370.

Alexopoulos, C.J., Mims, C.W. and Blackwell, M. (1996) Introductory mycology. John Wiley & Sons Inc., New York.

Al-Musallam, A. (1980) Revision of the black *Aspergillus* species. Ph.D. Thesis, Utrecht, The Netherlands.

Battilani, P., Pietri, A., Bertuzzi, A.T., Languasco, L., Giorni, P. and Kozakiewicz, Z. (2003) Occurrence of ochratoxin A-producing fungi in grapes grown in Italy. Journal of Food Protection 66: 633–636.

Bayman, P., Baker, J.L., Doster, M.A., Michailides, T.J. and Mahoney, N.E. (2002) Ochratoxin production by the *Aspergillus ochraceus* group and *Aspergillus alliaceus*. Applied and Environmental Microbiology 68: 2326–2329.

Bennett, J.W. and Klich, M.A. (1992) *Aspergillus*: Biology and industrial applications. Butterworth-Heinemann, Boston, USA.

Bigelis, R. and Arora, D.K. (1992) Organic acids of fungi. In: Handbook of applied mycology Vol. 4. Fungal biotechnology, (eds), Arora D.K., Elander R.P. and Murekji K.G. Marcel Dekker Inc., New York, 357–376.

Brookman, J.L. and Denning, D.W. (2000) Molecular genetics in *Aspergillus fumigatus*. Current Opinion in Microbiology 3: 468–474.

Burkhead, K.D., Peterson, S.W. and Bolen, P.L. (1994) Microbial hyrdoxylation I. Hydroxylation of aniline by *Aspergillus alliaceus*, *A. albertensis* and *A. terreus*. Journal of Industrial Microbiology 13: 233–237.

Bussink, H.J.D., Buxton, F.P. and Visser, J. (1991) Expression and sequence comparison of the *Aspergillus niger* and *Aspergillus tubingensis* genes encoding polygalacturonase II. Current Genetics 19: 467–474.

Cabanes, F.J., Accensi, F., Bragulat, M.R., Abarca, M.L., Castella, G., Minguez, S. and Pons, A. (2002) What is the source of ochratoxin A in wine? International Journal of Food Microbiology 79: 213–215.

Campbell-Platt, G. and Cook, P.E. (1989) Fungi in the production of foods and their ingredients. Journal of Applied Bacteriology 67: S117–S131.

Chen, J.W., Luo, Y.L., Hwang, M.J., Peng, F.C. and Ling, K.H. (1999) Territrem B, a tremorgenic mycotoxin that inhibits acetylcholinesterase with a noncovalent yet irreversible binding mechanism. Journal of Biological Chemistry 274: 34916–3423.

Chen, K.C., Yin, W.S., Tiu, C. and Houng, J.Y. (1994) 11 alpha-hydroxylation of progesterone using modified alginate-immobilized cells. Enzyme and Microbial Technology 16: 551–555.

Christensen, M. (1981) A synoptic key and evaluation of species in the *Aspergillus flavus* group. Mycologia 73: 1056–1084.

Ciegler, A. (1972) Bioproduction of ochratoxin A and penicillic acid by members of the *Aspergillus ochraceus* group. Canadian Journal of Microbiology 18: 631–636.

Cotty, P.J. and Cardwell, K.F. (1999) Divergence of West African and North American communities of *Aspergillus* section *Flavi*. Applied and Environmental Microbiology 65: 2264–2266.

Dalcero, A., Magnoli, C., Hallak, C., Chiachiera, S.M., Palacio, G. and Rosa, C.A.R. (2002) Detection of ochratoxin A in animal feeds and capacity to produce this mycotoxin by *Aspergillus* section *Nigri* in Argentina. Food Additives and Contaminants 19: 1065–1072.

De Graaff, L.H., van den Broeck, H.C., Ooijen, A.J.J. and Visser, J. (1994) Regulation of the xylanase-encoding *xlnA* gene of *Aspergillus tubingensis*. Molecular Microbiology 12: 479–490.

Durley, R.C., Macmillan, J., Simpson, T.J., Glen, A.T. and Turner, W.B. (1975) Fungal products. Part XIII. Xanthomegnin, viomellein, rubrosulphin, and viopurpurin, pigments from *Aspergillus sulphureus* and *Aspergillus melleus*. Journal of the Chemical Society Perkin. 1975: 163–169.

Egel, D.S., Cotty, P.J. and Elias, K.S. (1994) Relationships among isolates of *Aspergillus* sect. *Flavi* that vary in aflatoxin production. Phytopathology 84: 906–912.

Ehrlich, K.C., Montalbano, B.G. and Cotty, P.J. (2003) Sequence comparison of *aflR* from different *Aspergillus* species provides evidence for variability in regulation of aflatoxin production. Fungal Genetics and Biology 38: 63–74.

Feibelman, T.P., Cotty, P.J., Doster, M.A. and Michailides, T.J. (1998) A morphologically distinct strain of *Aspergillus nomius*. Mycologia 90: 618–623.

Frisvad, J.C. and Samson, R.A. (2000) *Neopetromyces* gen. nov. and an overview of teleomorphs of *Aspergillus* subgenus *Circumdati*. Studies in Mycology 45: 201–207.

Gams, W., Christensen, M., Onions, A.H.S., Pitt, J.I. and Samson, R.A. (1985) Infrageneric taxa of *Aspergillus*. In: Advances in *Aspergillus* and *Penicillium* systematics, (eds), Samson R.A. and Pitt J.I. Plenum Press, New York, 55–62.

Geiser, D. M. and LoBuglio, K. F. (2001) The monophyletic Plectomycetes: Ascosphaerales, Onygenales and Eurotiales. In: The Mycota. Vol. VII, Systematics and evolution, (eds), Lemke P., McLaughlin D.J. and McLaughlin E. Springer-Verlag, Berlin, 201–219.

Geiser, D.M., Dorner, J.W., Horn, B.W. and Taylor, J.W. (2000) The phylogenetics of mycotoxin and sclerotium production in *Aspergillus flavus* and *Aspergillus oryzae*. Fungal Genetics and Biology 31: 169–179.

Geiser, D.M., Pitt, J.I. and Taylor, J.W. (1998) Cryptic speciation and recombination in the aflatoxin-producing fungus *Aspergillus flavus*. Proceedings of the National Academy of Sciences, USA 95: 388–393.

Gielkens, M.M.C., Visser, J. and de Graaff, L.H. (1997) Arabinoxylan degradation by fungi: characterization of the arabinoxylan-arabinofuranohydrolase encoding genes from *Aspergillus niger* and *Aspergillus tubingensis*. Current Genetics 31: 22–29.

Gouka, R.J., Punt, P.J. and van den Hondel, C.A. (1997) Efficient production of secreted proteins by *Aspergillus*: progress, limitations and prospects. Applied Microbiology and Biotechnology 47: 1–11.

Hamari, Z., Kevei, F., Kovács, É., Varga, J., Kozakiewicz, Z. and Croft, J.H. (1997) Molecular and phenotypic characterization of *Aspergillus japonicus* strains with special regard to their mitochondrial DNA polymorphisms. Antonie van Leeuwenhoek 72: 337–347.

Hara, S., Kitamoto, K. and Gomi, K. (1992) New developments in fermented beverages and foods with *Aspergillus*. In: *Aspergillus*: biology and industrial applications, (eds), Bennett J.W. and Klich M.A. Butterworth-Heinemann, Stoneham, 133–153.

Horie, Y. (1995) Productivity of ochratoxin A of *Aspergillus carbonarius* in *Aspergillus* section *Nigri*. Nippon Kingakkai Kaiho 36: 73–76.

Ito, Y., Peterson, S.W., Wicklow, D.T. and Goto, T. (2001) *Aspergillus pseudotamarii*, a new aflatoxin producing species in *Aspergillus* section *Flavi*. Mycological Research 105: 233–239.

Joosten, V., Lokman, C., van den Hondel, C.A.M.J.J. and Punt, P.J. (2003) The production of antibody fragments and antibody fusion proteins by yeasts and filamentous fungi. Microbial Cell Factories 2: 1–15.

Kaji, A., Saito, R., Nomura, M., Miyamoto, K. and Kiriyama, N. (1998) Relationship between the structure and cytotoxic activity of asterriquinone, an antitumor metabolite of *Aspergillus terreus*, and its alkyl ether derivatives. Biological Pharmacology Bulletin 21: 945–949.

Kennedy, J., Auclair, K., Kendrew, S.G., Park, C., Vederas, J.C., Hutchinson, C.R. (1999). Modulation of polyketide synthase activity by accessory proteins during lovastatin biosynthesis. Science 284: 1368–1372.

Kevei, F., Hamari, Z., Varga, J., Kozakiewicz, Z. and Croft, J.H. (1996) Molecular polymorphism and phenotypic variation in *Aspergillus carbonarius*. Antonie van Leeuwenhoek 70: 59–66.

Kieslich, K. (1997) Biotransformations. In: Fungal biotechnology, (ed), Anke T. Chapman & Hall, Weinheim, 297–399.

Klich, M.A., Cary, J.W., Beltz, S.B. and Bennett, C.A. (2003) Phylogenetic and morphological analysis of *Aspergillus ochraceoroseus*. Mycologia 95: 1252–1260.

Klich, M.A., Mendoza, C., Mullaney, E., Keller, N. and Bennett, J.W. (2001) A new sterigmatocystin-producing *Emericella* variant from agricultural desert soils. Systematic and Applied Microbiology 24: 131–138.

Klich, M.A., Mullaney, E.J., Daly, C.B. and Cary, J.W. (2000) Molecular and physiological aspects of aflatoxin and sterigmatocystin biosynthesis by *Aspergillus tamarii* and *A. ochraceoroseus*. Applied Microbiology and Biotechnology 53: 605–609.

Koenig, H., Bievre, C., de Waller, J. and Conraux, C. (1985) *Aspergillus alliaceus*, agent of chronic otorrhoea. Bulletin of the Society of French Medical Mycologists (Bull. Soc. Fr. Mycol. Med.) 14: 85–88.

Kozakiewicz, Z. (1989) *Aspergillus* species on stored products. Mycological Papers 161: 1–188.

Kuraishi, H., Itoh, M., Tsuzaki, N., Katayama, Y., Yokoyama, T. and Sugiyama, J. (1990) The ubiquinone system as a taxonomic aid in *Aspergillus* and its teleomorphs. In: Modern concepts in *Penicillium* and *Aspergillus* classification, (eds), Samson, R.A. and Pitt, J.I., Plenum Press, New York, 407–421.

Kusters-van Someren, M.A., Samson, R.A. and Visser, J. (1991) The use of RFLP analysis in classification of the black Aspergilli: reinterpretation of *Aspergillus niger* aggregate. Current Genetics 19: 21–26.

Larsen, T.O., Svendsen, A. and Smedsgaard, J. (2001) Biochemical characterization of ochratoxin A-producing strains of the genus *Penicillium*. Applied and Environmental Microbiology 67: 3630–3635.

Lass-Florl, C., Rath, P., Niederwieser, D., Kofler, G., Wurzner, R., Krezy, A. and Dierich M.P. (2000) *Aspergillus terreus* infections in haematological malignancies: molecular epidemiology suggests association with in-hospital plants. Journal of Hospital Infections 46: 31–35.

Latgé, J.-P. (2001) The pathobiology of *Aspergillus fumigatus*. Trends in Microbiology 9: 382–389.

Liesch, J.M., Hensens, O.D., Zink, D.L. and Goetz, M.A. (1988) Novel cholecystokinin antagonists from *Aspergillus alliaceus*. II. Structure determination of asperlicins B, C, D, and E. Journal of Antibiotics 41: 878–881.

Lin, A., Huang, K.C., Hwu, L. and Tzean, S.S. (1995) Production of type II. ribotoxins by *Aspergillus* species and related fungi in Taiwan. Toxicon 33: 105–110.

Lowe, D.A. (1992) Fungal enzymes. In: Handbook of applied mycology Vol. 4. Fungal biotechnology, (eds), Arora, D.K., Elander, R.P. and Murekji, K.G. Marcel Dekker Inc., New York, 681–706.

Luisetti, M., Piccioni, P.D., Dyne, K., Donnini, M., Bulgheroni, A., Pasturenzi, L., Donnetta, A.M. and Peona, V. (1991) Some properties of the alkaline proteinase from *Aspergillus melleus*. International Journal of Tissue Reactions 13: 187–192.

Manzoni, M. and Rollini, M. (2002) Biosynthesis and biotechnological production of statins by filamentous fungi and application of these cholesterol-lowering drugs. Applied Microbiology and Biotechnology 58: 555–564.

Marr K.A., Patterson T. and Denning D. (2002) Aspergillosis. Pathogenesis, clinical manifestations, and therapy. Infectious disease clinics of North America 16: 875–894.

Matsukuma, S., Ohtsuka, T., Kotaki, H., Shirai, H., Sano, T., Watanabe, K., Nakayama, N., Itezono, Y., Fujiu, M. and Shimma, N. (1992) A new series of natural antifungals that inhibit P450 lanosterol C-14 demethylase. I. Taxonomy, fermentation, isolation and structural elucidation. Journal of Antibiotics 45: 151–159.

Megnegneau, B., Debets, F. and Hoekstra, R.F. (1993) Genetic variability and relatedness in the complex group of black Aspergilli based on random amplification of polymorphic DNA. Current Genetics 23: 323–329.

Michailides, T.J., Peacock, W., Christensen, P., Morgan, D.P. and Felts, D. (2002). First report of *Aspergillus* vine canker of table grapes caused by *Aspergillus niger*. Plant Disease 86: 75.

Mikhailova, R.V., Sapunova, L.I. and Lobanok, A.G. (1995) Three polygalacturonases constitutively synthesized by *Aspergillus alliaceus*. World Journal of Microbiology and Biotechnology 11: 330–332.

Ng, T.B. (2003) Fungi and fermented food. In: Fungal biotechnology in agricultural, food, and environmental applications, (ed), Arora, D.K. Marcel Dekker, Inc. New York, 223–232.

Novey, H.S. and Wells, I.D. (1978). Allergic bronchopulmonary aspergillosis caused by *Aspergillus ochraceus*. American Journal of Clinical Pathology 70: 840–843.

Ono, H., Kataoka, A., Koakutsu, M., Tanaka, K., Kawasugi, S., Wakazawa, M., Ueno, Y. and Manabe, M. (1995) Ochratoxin A producibility by strains of *Aspergillus niger* group stored in IFO culture collection. Mycotoxins 41: 47–51.

Ooike, M., Nozawa, K., Udagawa, S.-I. and Kawai, K.-I. (1997) Bisindolylbenzenoids from ascostromata of *Petromyces muricatus*. Canadian Journal of Chemistry 75: 625–628.

Ostry, V. (2000) Mycological analysis of foodstuffs. In: Health consequences of the human exposure to foreign substances from food chains, (ed) Ruprich J. National Institute of Public Health, Prague, 55–62 (in Czech).

Parenicova, L., Skouboe, P., Frisvad, J., Samson, R.A., Rossen, L., Hoor-Suykerbuyk, M. and Visser, J. (2001) Combined molecular and biochemical approach identifies *Aspergillus japonicus* and *Aspergillus aculeatus* as two species. Applied and Environmental Microbiology 67: 521–527.

Pariza, M.W. and Johnson, E.A. (2001) Evaluating the safety of microbial enzyme preparations used in food processing: update for a new century. Regulatory Toxicology and Pharmacology 33: 173–186.

Peterson, S.W. (1995) Phylogenetic analysis of *Aspergillus* sections *Cremei* and *Wentii*, based on ribosomal DNA sequences. Mycological Research 99: 1349–1355.

Peterson S.W. (2000) Phylogenetic relationships in *Aspergillus* based upon rDNA sequence analysis. In: Integration of modern taxonomic methods for *Penicillium* and *Aspergillus* classification, (eds) Samson, R.A. and Pitt, J.I., Harwood Academic Publishers, Amsterdam, The Netherlands, 323–355.

Peterson, S.W., Ito Y., Horn, B.W. and Goto, T. (2001) *Aspergillus bombycis*, a new aflatoxigenic species and genetic variation in its sibling species, *A. nomius*. Mycologia 93: 689–703.

Pitt, J.I. (2000) Toxigenic fungi: which are important? Medical Mycology 38 (Suppl): 17–22.

Punt, P.J., van Biezen, N., Conesa, A., Albers, A., Mangnus, J. and van den Hondel, C. (2002) Filamentous fungi as cell factories for heterologous protein production. Trends in Biotechnology 20: 200–206.

Raper, K.B. and Fennell, D.I. (1965) The genus *Aspergillus,* Williams and Wilkins, Baltimore.

Rigó, K., Varga, J., Tóth, B., Mesterházy, Á. and Kozakiewicz, Z. (2002) Phylogenetic analysis of *Aspergillus* section *Flavi* based on sequences of the internal transcribed spacer regions and the 5.8 S rRNA gene. Journal of General and Applied Microbiology 48: 9–16.

Sallam, L.A.R., El-Refai, A.M.H., Hamdy, A.H.A., El-Minofi, H.A. and Ibrahim, S. Abdel-Salam, I.S. (2003) Role of some fermentation parameters on cyclosporin A production by a new isolate of *Aspergillus terreus*. Journal of General and Applied Microbiology 49: 321–328.

Samson, R.A. (1979) A compilation of the Aspergilli described since 1965. Studies in Mycology 18: 1–38.

Scheidegger, K.A. and Payne, G.A. (2003) Unlocking the secrets behind secondary metabolism: a review of *Aspergillus flavus* from pathogenicity to functional genomics. Journal of Toxicology, Toxin reviews, 22: 423–459.

Schimmel, T.G., Coffman, A.D. and Parsons, S.J. (1998) Effect of butyrolactone I on the producing fungus, *Aspergillus terreus*. Applied and Environmental Microbiology 64: 3707–3712.

Schuster, E., Dunn-Coleman, N., Frisvad, J.C. and van Dijck, P.W.M. (2002) On the safety of *Aspergillus niger* – a review. Applied Microbiology and Biotechnology 59: 426–435.

St. Leger, R.J., Steven, E., Screen, S.E. and Pirzadeh, B.S. (2000) Lack of host specialization in *Aspergillus flavus*. Applied and Environmental Microbiology 66: 320–324.

Téren, J., Varga, J., Hamari, Z., Rinyu, E. and Kevei, F. (1996) Immunochemical detection of ochratoxin A in black *Aspergillus* strains. Mycopathologia 134: 171–176.

Turner, W.B. and Aldridge, D.C. (1983) Fungal metabolites II. Academic Press, New York.

Ueno, Y., Kawamura, O., Sugiura, Y., Horiguchi, K., Nakajima, M., Yamamoto, K. and Sato, S. (1991) Use of monoclonal antibodies, enzyme-linked immunosorbent assay and immunoaffinity column chromatography to determine ochratoxin A in porcine sera, coffee products and toxin-producing fungi. In: Mycotoxins, endemic nephropathy and urinary tract tumors, (eds), Castegnaro, M., Plestina, R., Dirheimer, G., Chernozemsky, I.N. and Bartsch, H. IARC Scientific Publishers, Lyon, 71–75.

Vaamonde, G., Patriarca, A., Pinto, V.F., Comerio, R. and Degrossi, C. (2003) Variability of aflatoxin and cyclopiazonic acid production by *Aspergillus* section *Flavi* from different substrates in Argentina. International Journal of Food Microbiology 88: 79–84.

Varga, J., Kevei, F., Fekete, C., Coenen, A., Kozakiewicz, Z. and Croft, J.H. (1993) Restriction fragment length polymorphisms in the mitochondrial DNAs of the *Aspergillus niger* aggregate. Mycological Research 97: 1207–1212.

Varga, J., Kevei, F., Debets, F., Kozakiewicz, Z. and Croft, J.H. (1994) Mitochondrial DNA restriction fragment length polymorphisms in field isolates of the *Aspergillus niger* aggregate. Canadian Journal of Microbiology 40: 612–621.

Varga, J., Kevei, É., Rinyu, E., Téren, J. and Kozakiewicz, Z. (1996) Ochratoxin production by *Aspergillus* species. Applied and Environmental Microbiology 62: 4461–4464.

Varga, J., Kevei, É., Palágyi, A., Tóth, B. and Kozakiewicz, Z. (2000a). Analysis of genetic variability within the *Petromyces* genus. Antonie van Leeuwenhoek 77: 83–89.

Varga, J., Kevei, É., Tóth, B., Kozakiewicz, Z. and Hoekstra, R.F. (2000b) Molecular analysis of variability within the toxigenic *Aspergillus ochraceus* species. Canadian Journal of Microbiology 46: 593–599.

Varga, J., Kevei, F., Hamari, Z., Tóth, B., Téren, J., Croft, J.H. and Kozakiewicz, Z. (2000c) Genotypic and phenotypic variability among black Aspergilli. In: Integration of modern taxonomic methods for *Penicillium* and *Aspergillus* classification, (eds), Samson R.A. and Pitt J.I., Harwood Academic Publishers, Amsterdam, 397–411.

Varga, J., Tóth, B., Rigó, K., Téren, J., Hoekstra, R.F. and Kozakiewicz, Z. (2000d) Phylogenetic analysis of *Aspergillus* section *Circumdati* based on sequences of the internal transcribed spacer regions and the 5.8 S rRNA gene. Fungal Genetics and Biology 30: 71–80.

Varga, J., Rigó, K., Téren, J. and Mesterházy, Á. (2001) Recent advances in ochratoxin research I. Production, detection and occurrence of ochratoxins. Cereal Research Communications 29: 85–92.

Varga, J., Rigó, K., Tóth, B., Téren, J. and Kozakiewicz, Z. (2003) Evolutionary relationships among *Aspergillus* species producing economically important mycotoxins. Food Technology and Biotechnology 41: 29–36.

Varga, J., Juhász, Á., Kevei, F. and Kozakiewicz, Z. (2004) Molecular diversity of agriculturally important *Aspergillus* species. European Journal of Plant Pathology (in press)

Whyte, A.C., Gloer, J.B., Wicklow, D.T. and Dowd, P.F. (1996) Sclerotiamide: a new member of the paraherquamide class with potent antiinsectan activity from the sclerotia of *Aspergillus sclerotiorum.* Journal of Natural Products 59: 1093–1095.

Wicklow, D.T., Dowd, P.F., Alfatafta, A.A. and Gloer, J.B. (1996) Ochratoxin A: an antiinsectan metabolite from the sclerotia of *Aspergillus carbonarius* NRRL 369. Canadian Journal of Microbiology 42: 1100–1103.

Wolf, G. (1997). Traditional fermented food. In: Fungal biotechnology, (ed), Anke T. Chapman & Hall, Weinheim, 3–13.

Yokoyama, K., Wang, L., Miyaji, M. and Nishimura, K. (2001) Identification, classification and phylogeny of the *Aspergillus* section *Nigri* inferred from mitochondrial cytochrome b gene. FEMS Microbiology Letters 200: 241–246.

Zotti, M. and Corte, A.M. (2002) *Aspergillus persii:* a new species in section *Circumdati.* Mycotaxon 83: 269–278.

9. Aquatic Hyphomycetes and their Meiosporic Relatives: Slow and Laborious Solving of a Jig-Saw Puzzle

Ludmila Marvanová
Masaryk University Brno, Faculty of Science, Czech Collection of Microorganisms,
Tvrdého 14, 602 00 Brno, Czech Republic

SUMMARY Aquatic hyphomycetes are anamorphic fungi occurring mainly in freshwater streams. They may be only a part of the life cycle of Asco- and Basidiomycetes. Teleomorphs are known in roughly 10% of described aquatic hyphomycetes. It is presumed, that the known teleomorph/anamorph links are only a small percentage of those, which probably exist. Many of anamorphs have probably lost their teleomorphs and became ana-holomorphs.

Molecular methods involving DNA nucleotide sequencing may contribute to better understanding of the relationships of asexual fungi and to their incorporation into the higher taxonomic categories of Asco- and Basidiomycota. At present, the aquatic hyphomycete holomorphs and several ana-holomorphs are classified in two phylla of fungi. In the phyllum Ascomycota, they can be attributed to 13 families spread over 8 orders which belong to 5 classes, all housed in the subphyllum Pezizomycotina. In Basidiomycota, aquatic hyphomycete holomorphs are encountered in the class Urediniomycetes and in the subclass Hymenomycetidae of Hymenomycetes.

The aquatic hyphomycete holomorphs are not being currently incorporated into molecular phylogenetic studies of ascomycetes and basidiomycetes. Their relationships to other members of the same terrestrial genera remain obscure or unknown. This is partly due to the paucity of data on related taxa in the GenBank. For many aquatic hyphomycetes there is lack of living ex-type cultures derived from anamorphs or teleomorphs. For many anamorphic genera neither isolates of all described species, nor a reasonable number of strains of the same species from geographically remote localities are available.

It should be stressed, that a very good documentation be kept about cultures used for molecular studies, so that in case of work with misidentified isolate it can be traced back.

It will be ideal to sequence more isolates of one species in order to reveal possible cryptic lineages, which can be then eventually defined morphologically. More collaboration between dicaryomycete and aquatic hyphomycete specialists will be advantageous for both groups.

Introduction

One of the striking characters of higher fungi is the pleomorphy, i.e. the existence of morphologically and functionally different phenotypic structures for one genotype. In non- lichenised Asco- and Basidiomycota, differently named teleomorphs for sexual (meiotic) and anamorphs for asexual (mitotic) reproduction are currently recognised (Greuter et al. 2000). A holomorph is then the fungus with all its morphs. A special phenomenon is the pleoanamorphy, involving more asexual states (synanamorphs) connected to one sexual state. (Hennebert, 1987). Another step in fungal life cycle may be the loss of sexual state which results in an anamorphic holomorph, briefly ana-holomorph (Hawksworth et al. 1983). Traditionally, more anamorphs are known in ascomycetes. A special group is the microconidial synanamorphs, i.e. morphologically different conidial states which also serve for asexual reproduction of the fungus, e.g. in *Fusarium* (Gams and

Nirenberg, 1989). The spermatisation process involved in sexual reproduction of some ascomycetes is in many cases ensured by the spermatial states known also as andromorphs (Parbery, 1996). However, the role in fertilisation may not always be restricted to special morphological structures and so the functional differences distinguishing synanamorphs and andromorphs may not be clearly evident in individual cases.

At present, no hierarchic system is employed for grouping of anamorphic fungi and practically the highest taxonomic unit is the genus (Sutton and Hennebert, 1994). The categories Hyphomycetes or Coelomycetes traditionally used in classification of anamorphs are also considered artificial (Kendrick, 1989) and according to some authors now they can serve only as descriptive terms for anamorphs with similar arrangement of fruiting structures (Seifert and Gams, 2001).

Along with the weakening of the formal suprageneric hierarchisation of anamorphic fungi, attempts to amalgamate them in the system of Ascomycota and Basidiomycota are continuing with increasing strength. The molecular techniques exploiting nucleotide sequencing of various gene partitions of anamorphs and teleomorphs has become a very powerful tool.

Incorporation of mitotic fungi into the system of meiotic fungi will broaden our insights into the diversity of biota and its probable developmental pathways.

Aquatic Hyphomycetes

Aquatic hyphomycetes (Ingold, 1942) are anamorphic fungi occurring mainly in freshwater streams. They may be only a part of the life cycle of holomorphic asco- and basidiomycetes, but many of them have probably lost their teleomorphs and became ana-holomorphs.

Among other fungi, aquatic hyphomycetes play an important part in aquatic ecosystems. They colonise submerged, predominantly allochthonous plant matter, mainly tree leaves and to lesser extent also wood and other substrata. Their mycelia invade decaying plant tissues, decompose them and make them more nutritious for aquatic detritus-feeding invertebrates. So, these fungi are the main energy suppliers in streams with low or negligible primary production.

The life in lotic ecosystems exhibited selection pressure on their morphology and led to a whole range of adaptations, manifested most distinctly in the shapes of their propagules. Two shapes prevail among their conidia: scolecoform (sigmoid, worm-like) and stauroform (variously branched, predominantly tetraradiate). It is believed that these configurations of conidia facilitate colonisation of substrata in running waters by making contact with their surface on more that one point (Webster, 1987). Excretion of mucous substances and eventual formation of appressoria helps mycelia to penetrate into the tissues (Read et al.1992 a, b).

Anamorphs

The traditional taxonomy of aquatic hyphomycetes is based on the ontogenesis of the reproductive structures and on the final shape of their conidia.

Basic criteria for classification were laid down in the pioneer publications by Ingold (1942 and others) and later summarised in a brief overview (Ingold, 1975a). The main phases of conidial development concern (1) the conidial initiation (how the conidia develop from a conidiogenous cell), (2) proliferation of the conidiogenous cells during the proceeding production of conidia, (3) the morphogenesis of conidia (branching pattern) and (4) the separation of conidia from the conidiogenous cell. This system of characters, in a more or less refined version, is still in use (Marvanová 1997, Gulis et al. 2005).

Teleomorphs

Teleomorphs are known in roughly 10% of described aquatic hyphomycetes (Marvanová, 1997, 2002, Sivichai and Jones, 2003). Unlike the anamorphs, they are mostly terrestrial or semi-aquatic. Except of

Vibrissea truncorum which has long filiform ascospores, no morphological adaptation of the meiospores to aquatic dispersal is encountered. De Wildeman (1893, 1894, 1895) who first recognized the variously branched hyaline entities in water as spores of fungi, did not know any link to a sexual state. Roughly 50 years later, Ingold (1942) described 10 new species from a small stream, all as anamorphic hyphomycetes. The first teleomorph, *Nectria penicillioides*, was described for *Flagellospora penicillioides* by Ranzoni (1956). In the late 1950s J. Webster started his series of discovering anamorph-teleomorph links (Webster, 1959 etc, see Table 1).

Now we know ca. 35 teleomorph/anamorph connections and they confirm the hypothesis about the morphological convergence, put forward by Ingold (1975b) on the basis of the fact that similar conidial shapes are produced by different patterns of conidiogenesis and by Webster (1987), based on the morphological similarity of conidia formed by unrelated teleomorphic states, e.g. one of the larger genera, *Anguillospora*, has eight teleomorphs in four ascomycete orders. Some genera, comprising an assemblage of species with similar conidia are heterogeneous from the point of view of the character of their colonies as well as of the conidiogenesis: the two large genera *Varicosporium* and *Tricladium* are difficult to differentiate clearly from each other on the basis of morphology. Some other genera, judging from the similarity of their colonies and of their uniform conidiogenesis, seem to be morphologically homogeneous, as if they have diversified from common ancestors after they entered the freshwater environment (Marvanová, 1997). Good examples are *Alatospora, Lemonniera, Tetracladium*. They are well recognisable at the genus level, but the species show many intermediates.

Morphological convergence is not only a speciality of aquatic hyphomycetes, although it is very apparent in this group which attracts attention through its extremely rich diversity of spectacular conidial shapes. The terrestrial anamorphic genus *Phialophora* with phialidic conidiogenesis and simple small globose to ellipsoid conidia is an example 'par excellence'; it comprises teleomorphs in discomycetes, pyrenomycetes and loculoascomycetes (Gams, 2000). In contrast, the plant pathogenic teleomorphic *Mycosphaerella* may be linked to 20-30 hyphomycetous or coelomycetous anamorph genera (Sutton and Hennebert, 1994; Crous et al. 2000).

The teleomorph/anamorph connections in aquatic hyphomycetes are derived from pure culture studies. Two approaches employed in ascomycetes were: (1) inoculation of ascospores from ascomata collected in nature (e.g. Marvanová and Descals, 1996; Kohlmeyer et al. 1998) and (2) inducing a teleomorph in culture started from conidia (e.g. Webster, 1959, 1961).

In freshwater basidiomycetes, most known teleomorphs were obtained only in pure culture initiated from conidia (Nawawi et al. 1977; Marvanová and Stalpers, 1987; Marvanová and Suberkropp, 1990; Bauer et al. 2003). There are two exceptions: *Sistotrema hamatum* (Nawawi and Webster 1982) and *Infundibura adhaerens* (Kirschner, 2004) which were isolated from basidiospores.

Molecular Methods in Taxonomy of Aquatic Hyphomycetes

It is presumed, that the known teleomorph/anamorph links are only a small percentage of those, which probably exist. Molecular methods involving DNA nucleotide sequencing may contribute to better understanding of the relationships at the vertical as well as horizontal level. The ratio of publications on these themes increased considerably in all journals with long tradition in the classical taxonomy.

Application of these methods to aquatic hyphomycetes is relatively recent, with only a few studies on this theme published. In some of them only single members of the aquatic hyphomycete group were incorporated (O'Donnell, 1993; Hagedorn and Scholler, 1999; Marvanová et al. 2002; Bauer et al. 2003; Kirschner, 2004). Studies focused on several species of one or more genera are also emerging (Nikolcheva and Bärlocher, 2002; Campbell et al. 2002a, b; Baschien, 2003; Belliveau and Bärlocher, 2005).

Studies in molecular taxonomy of aquatic hyphomycetes proceed rather slowly. Though the role of this group of fungi in freshwater ecosystems is recognised as essential, they are not studied to the extent comparable to human, animal or plant pathogens, whose impact on human health, industry and agriculture receives more attention and is better supported by funding.

The objective of this article is to put together the published information on teleomorph/anamorph relationships in aquatic hyphomycetes as well as the results of phylogenetic analyses of the affinities of the ana-holomorphic taxa to ascomycetes and basidiomycetes. Though the involvement of morphological convergence in the classification of these fungi is now widely agreed upon, this conclusion has been mainly arrived at by comparing the shapes of conidia. The phases of conidiogenesis may be more conservative, therefore I incorporated them into the scheme. The synanamorphs are often abandoned in evaluations of affinities and often are not distinguished from andromorphs (spermatial states). However, andromorphs can be more informative as regards phylogeny because of their role in the sexual process.

Survey of Teleomorph/Anamorph Connections and of Phylogenetic Affinities in Aquatic Hyphomycetes

Table 1 gives a brief survey of actual knowledge about incorporation of meiosporic as well as mitosporic states of freshwater hyphomycetes into the recent system of Ascomycota and Basidiomycota. For Ascomycota the arrangement of Eriksson et al. (2004) is followed, but the traditional categories: loculoascomycetes, discomycetes, pyrenomycete, are also mentioned, even though delimitation and significance of these groups is now doubted (Kirk et al. 2001).

For Basidiomycota there is no very recent agreement about the taxa above orders, especially in Hymenomycetes. The arrangement in Table 1 is adopted from Sampaio and Bauer (2003) and Weiss et al. (2004); for Hymenomycetes subclass Hymenomycetidae (Swann and Taylor 1995), the traditional orders Stereales and families Sistotremataceae and Atheliaceae are given according to Kirk et al. (2001). The terms 'heterobasidiomycetes' (Wells and Bandoni, 2001) and 'homobasidiomycetes' (Hibbett and Thorn, 2001) are also employed for keeping continuity with the traditional system.

Data on individual taxa are compiled from published articles. The delimitation of aquatic hyphomycetes is not sharp against other hyphomycetes occurring in water; hence some terrestrial taxa known to occur on decaying plant debris in water or appearing in foam are also incorporated. Their names are in parentheses. The ecologically different group of aeroaquatic hyphomycetes with air-trapping propagules is not dealt with. Marine taxa are considered only when there are freshwater members in the same genera.

Arrangement and Explanation of Entries

In the two first columns, teleomorph and the corresponding anamorph names are given. The reproductive structures other than conidia have previously not been 'tested' from the point of homo- or heterogeneity in the classes, orders or families housing holomorphic and ana-holomorphic aquatic hyphomycete taxa. Accordingly, the next two columns are for conidial ontogeny and its two phases: conidial initiation and proliferation of conidiogenous cells with continuing conidiogenesis.

Hennebert and Sutton (1994) presented detailed parameters for describing individual phases of conidiogenesis. Some of them were adopted here.

(1) Conidium origin and proliferation of conidiogenous cells with respect to production of conidia:
Blastic: With discrete conidial initiation, i.e. the initial growth of conidial primordium is restricted to a very small area of the conidiogenous cell wall.
   (a) Polyblastic/sequential: Several conidia appear on the conidiogenous cell within a short time, each from its own conidiogenous locus.
   (b) Blastic/phialidic: Conidia are repeatedly formed from the same conidiogenous locus.

Table 1. Arrangement of aquatic hyphomycete holomorphs or ana-holomorphs according to their affinity to Asco- and Basidiomycota. The crucial stages of conidiogenesis are displayed as well

ASCOMYCOTA
(LOCULOASCOMYCETES)
Dothideomycetes, Pleosporales, Massarinaceae

Teleomorph	Anamorph	Conid. init./conidiog. cls	Secess.	Shape	Synana	Androm	Ref. for teleo/ana or affinity
Massarina sp.	Anguillospora longissima	thalloblastic/percurrent	rhexo	sigmoid	–	pycnid.	Willoughby & Archer 1973
Massarina sp.	Clavariopsis aquatica	thalloblastic/percurrent	schizo	tetraradiate	–	pycnid.	Webster & Descals 1979
Massarina aquatica	Tumularia aquatica	thalloblastic/percurrent	schizo	fusoid	–	pycnid.	Webster 1965
(Massarina tetraploa)	(Tetraploa aristata)	thalloblastic/determinate	schizo	tetraradiate	–	–	Scheuer 1991
Affinity to **Pleosporales**							
–	Lemonniera pseudofloscula	blastic/phialidic	schizo	tetraradiate	–	–	Campbell et al. 2002a
Affinity to **Dothideales**							
–	Anguillospora rubescens	thalloblastic/percurrent	schizo	sigmoid	–	–	Belliveau & Bärlocher 2005
(DISCOMYCETES)							
Leotiomycetes, Helotiales, Dermateaceae							
Mollisia uda	Anguillospora crassa	thalloblastic/percurrent	schizo	sigmoid	–	h-phial.	Webster 1961
Mollisia casaresiae	Casaresia sphagnorum	thalloblastic/determinate	rhexo	branched	–	h-phial.	Webster et al. 1993
Mollisia sp.	Filosporella sp.	thalloblastic/percurrent	schizo	sigmoid	–	h-phial.	Webster & Descals 1979
Leotiomycetes, Helotiales, Helotiaceae							
Hymenoscyphus imberbis	Anguillospora fustiformis	thalloblastic/percurrent	schizo	straight	–	–	Marvanová & Descals 1996
Hymenoscyphus foliicola	Dimorphospora foliicola	polyblastic/sequential	schizo	reniform	–	h-phial.	Abdullah et al. 1981
Hymenoscyphus splendens	Tricladium splendens	thalloblastic/percurrent	schizo	tricladioid	–	h-phial.	Abdullah et al. 1981
Hymenoscyphus africanus	Geniculospora grandis	thalloblastic/percurrent	schizo	tetraradiate	–	–	Descals et al. 1984
Hymenoscyphus malawiensis	Unnamed anamorph	thalloblastic/determinate	schizo	straight	–	–	Fisher & Spooner 1987
Hymenoscyphus varicosporoides	Varicosporium sp. or Tricladium sp.	thalloblastic/determinate	schizo	branched	–	–	Tubaki 1966 Sivichai et al. 2003
Ombrophila tetracladia	Articulospora tetracladia	polyblastic/sequential	schizo	tetraradiate	–	–	Abdullah et al. 1981
Cudoniella indica	Tricladium indicum	thalloblastic/determinate	schizo	tricladioid	–	–	Webster et al. 1995
Leotiomycetes, Helotiales, Leotiaceae							
Pezoloma sp.	Anguillospora furtiva	thalloblastic/percurrent	schizo	sigmoid	–	?	Descals et al. 1998

(Contd.)

Teleomorph	Anamorph	Conid. init./conidiog. cls	Secess.	Shape	Synana	Androm	Ref. for teleo/ana or affinity
Leotiomycetes, Helotiales, Hyaloscyphaceae							
Hydrocina chaetocladia	Tricladium chaetocladium	thalloblastic/percurrent	schizo	tricladioid	–	h-phial.	Webster et al. 1991
Aff. to Hyaloscyphaceae	Mycoarthris corallina	thalloblastic /sympodial	schizo	arthroconid.	–	–	Marvanová et al. 2002
Leotiomycetes, Helotiales, Vibrisseaceae							
Vibrissea flavovirens	Anavirga dendromorpha	thalloblastic/sympodial	schizo	branched	phial.	–	Hamad & Webster 1987
Leotiomycetes, Helotiales, Loramycetaceae							
Loramyces juncicola	Anguillospora sp.	thalloblastic/determinate?	schizo	sigmoid	–	–	Digby & Goos 1987
Leotiomycetes, Rhytismatales, Rhytismataceae							
Coccomyces sp.	Tricladiopsis flagelliformis	thalloblastic/sympodial	schizo	tricladioid	–	–	Descals & Webster 1982
Affinity to **Leotiomycetes Helotiales**							
–	Anguillospors madiocris	thalloblastic/perecurrent	schizo	sigmoid	–	pycnid	Belliveau & Bärlocher 2005
Affinity to **Leotiomycetes**							
–	Lemonniera aquatica	blastic/phialidic	schizo	tetraradiate	–	–	Campbell et al 2002a
–	Lemonniera centrosphaera	blastic/phialidic	schizo	tetraradiate	–	–	Campbell et al 2002a
–	Lemonniera cornuta	blastic/phialidic	schizo	tetraradiate	–	–	Campbell et al 2002a
–	Lemonniera filiformis	blastic/phialidic	schizo	tetraradiate	–	–	Campbell et al 2002a
–	Lemonniera terrestris	blastic/phialidic	schizo	tetraradiate	–	–	Campbell et al 2002a
–	Margaritispora aquatica	blastic/phialidic	schizo	glob./fusoid	–	–	Campbell et al 2002a
–	Alatospora acuminata	blastic/phialidic	schizo	sigmoid	–	–	Belliveau & Bärlocher 2005
–	Anguillospora filiformis	thalloblastic/percurrent	schizo	sigoid	–	–	Belliveau & Bärlocher 2005
–	Goniopila monticola	polyblastic/sympodial	schizo	globosea	–	h-phial.	Belliveau & Bärlocher 2005
–	Tetrachaetum elegans	thalloblastic/percurrent	rhexo	tetraradiate	–	–	Belliveau & Bärlocher 2005
–	Tetrachaetum angulatum	thalloblastic/percurrent	schizo	tricladioid	–	–	Belliveau & Bärlocher 2005
–	Tricladium patulum	thalloblastic/percurrent	schizo	tricladioid	–	–	Belliveau & Bärlocher 2005
Affinity to **Leotiomycetes, Erysiphales** and **Onygenales**							
–	Tetracladium apiense	thalloblastic/sympodial	schizo	branched	–	–	Nikolcheva & Bärlocher 2002
–	Tetracladium furcatum	thalloblastic/sympodial	schizo	branched	–	–	Nikolcheva & Bärlocher 2002
–	Tetracladium marchalianum	thalloblastic/sympodial	schizo	branched	–	–	Nikolcheva & Bärlocher 2002
–	Tetracladium maxilliforme	thalloblastic/sympodial	schizo	branched	–	–	Nikolcheva & Bärlocher 2002
–	Tetracladium. setigerum	thalloblastic/sympodial	schizo	branched	–	–	Nikolcheva & Bärlocher 2002
Orbiliomycetes. *Orbiliales, Orbiliaceae*							

(Contd.)

Teleomorph	Anamorph	Conid. init./conidiog. cls	Secess.	Shape	Synana	Androm	Ref. for teleo/ana or affinity
Orbilia junci	Dwayaangam junci	thalloblastic/determinate?	schizo	dichotom.	–	–	Kohlmeyer et al. 1998
Orbilia sp.	Anguillospora rosea	thalloblastic/percurrent	schizo	sigmoid	thallob.	–	Descals et al. 1998
Orbilia cf. luteorubella	(Anguillospora sp.)	thalloblastic/determinate?	schizo	sigmoid	thallob?	–	Pfister 1997
Orbilia trinacriifera	Trinacrium sp.	thalloblastic/sympodial	schizo	dichotom.	–	–	Matsushima 1995
Orbilia alnea	Dicranidion sp. ´fragile´	thalloblastic/sympodial	schizo	dichotom.	–	–	Pfister 1997
Orbilia tricellularia	Tricellula inaequalis	polyblastic	schizo	triradiate			Matsushima 2001
Orbilia delicatula	Dicranidion sp. ´gracile´	thalloblastic/sympodial	schizo	dichotom.	–	–	Pfister 1997
Aff. to Orbiliaceae	Dactylellina appendiculata	thalloblastic/sympodial	schizo	basiverticil.	–	–	Scholler et al. 1999
Pezizomycetes, Pezizales, Pyronemataceae							
Miladina lechithina	Actinosporella megalospora	thalloblastic/retrogressive	schizo	tetraradiate	–	–	Descals & Webster 1978
(PYRENOMYCETES)							
Sordariomycetes, Hypocreales, Nectriaceae							
Nectria lugdunensis	Heliscus lugdunensis	blastic/phialidic	schizo	clove-shap.	phial.	–	Webster 1959
Nectria curta	Flagellospora curta	blastic/phialidic	schizo	sigmoid	phial.	–	Webster 1993
Nectria penicillioides	Flagellospora penicillioides	blastic/phialidic	schizo	sigmoid	phial.	–	Ranzoni 1956
Neonectria discophora var. discophora	(Cylindrocarpon ianthothele var. majus)	blastic/phialidic	schizo	sigmoid	–	–	Mantiri et al. 2001 Brayford et al. 2004
Cosmospora purtonii	Fusarium aquaeductuum var. aquaeductuum	blastic/phialidic	schizo	falcate	–	–	Rossman et al. 1999
Cosmospora episphaeria	Fusarium aquaeductuum var. medium	blastic/phialidic	schizo	falcate	phial.	–	Rossman et al. 1999
Affinity to **Sordariomycetes**, Sordariales							
–	Lunulospora curvula	thalloblastic/sympodial	rhexo?	sigmoid	–	–	Belliveau & Bärlocher 2005
Sordariomycetes, Halosphaeriales, Halosphaeriaceae							
Corollospora luteola	Sigmoidea luteola	thalloblastic/progressive	schizo	lunate	–	–	Nakagiri & Tubaki 1982
Corollospora pulchella	Clavatospora bulbosa	thalloblastic/determinate	schizo	tetraradiate	–	–	Kohlmeyer et al. 1967
Sordariomycetes, Lulworthiales, Lulworthiaceae							
Lindra obtusa	Anguillospora marina	thalloblastic/percurrent	schizo	sigmoid	–	–	Nakagiri & Tubaki 1983
BASIDIOMYCOTA (HETEROBASIDIOMYCETES) **Urediniomycetes,** Microbotryomycetidae incertae sedis: Camptobasidiaceae							

(Contd.)

Teleomorph	Anamorph	Conid. init./conidiog. cls	Secess.	Shape	Synana	Androm	Ref. for teleo/ana or affinity
Camptobasidium hydrophilum	Crucella subtilis	thalloblastic/sympodial	schizo	tetraradiate	–	–	Marvanová & Suberkropp 1990
Urediniomycetes incertae sedis: Classiculales, Classiculaceae							
Classicula fluitans	Naiadella fluitans	thalloblastic/sympodial	schizo	acroverticil.	–	–	Bauer et al. 2003
Aff. to Classiculacee	Jaculispora submersa	thalloblastic/sympodial	schizo	acroverticil.	–	–	Bauer et al. 2003
Urediniomycetes incertae sedis: Atractiellaes, Phleogenaceae							
Helicogloea angustispora	Infundibura adhaerens	thalloblastic/determinate?	schizo	fusoid	–	–	Kirschner 2004
(HOMOBASIDIOMYCETES) ***Hymenomycetes**, Hymenomycetidae, Polyporales, Sistotremataceae*							
Sistotrema hamatum	Ingoldiella hamata	thalloblastic/determinate	rhexo	tetraradiate/ tricladioid	–	–	Nawawi & Webster 1982
Hymenomycetes, Hymenomycetidae, Polyporales, Atheliaceae							
Fibulomyces crucelliger	Taeniospora descalsii	thalloblastic/sympodial	schizo	cruciate	–	–	Marvanová & Stalpers 1987
Fibulomyces sp.	Taeniospora gracilis var. enecta	thalloblastic/sympodial	schizo	tricladioid	–	–	Marvanová & Stalpers 1987

Thalloblastic: With integrated conidial initiation, i.e. the growth of conidial primordium occurs from an area as wide as the conidiogenous cell (most aquatic hyphomycetes).

(a) Thalloblastic/percurrent: Growth of new conidia through the detachment scar

(b) Thalloblastic/sympodial: Growth of new conidia below the detachment scar

(c) Thalloblastic/retrogressive: New conidia are gradually transformed in basipetal sequence from the series of cells below the first conidium by swelling of the conidial body and polar growth of conidial branches

(d) Thalloblastic/determinate: No growth after secession of the first conidium, no further conidia produced. It may also be an artefact due to cultural conditions or lack of observations

(2) Secession (liberation) of conidia from conidiogenous cells:

Schizolytic: By splitting of the basal (delimiting) conidial septum.

Rhexolytic: By breaking of the periclinal wall of the cell below the basal conidial septum.

(3) Conidial shape: For simplicity the numerous variants were categorised as follows:

Sigmoid: Unbranched, long, curved

Straight: Unbranched, long

Branched: Complex branching, not tetraradiate or cruciate, not tricladioid, not acro- or basiverticillate

Tetraradiate: With four branches not in one plane, radiating from a centre

Cruciate: Like tetraradiate, but all arms in one plane

Acroverticillate: With fusoid axis and typically 3 (usually thin) branches below the apex or two paired median branches or bent axis

Basiverticillate: With fusoid axis and 3 branches near its base

Tricladioid: *Tricladium*-like. An axis with 2 to 3 alternate primary branches

For the rest of shapes self-explaining terms are used.

The next two columns are devoted to the eventual presence of synanamorph or andromorph. The difference between them is explained in the first paragraph of the Introduction. For the synanamorph only the type of conidiogenous cell is given, the hyphomycetous character of the state is assumed. For the andromorphs the prefix 'h-' means hyphomycetous. The non-hyphomycetous andromorphs are referred to as pycnidial. The details of the conidiogenesis in pycnids are not given, in aquatic hyphomycetes they are mostly poorly known.

In the last column, there is the reference to the author(s) who first published the respective teleomorph/anamorph connection or the affinity of the ana-holomorph to asco- or basidiomycetes.

Results and Discussion

At present, the aquatic hyphomycete holomorphs and several ana-holomorphs are classified in two phylla of fungi. In the phyllum Ascomycota, they can be attributed to 13 families spread over 8 orders which belong to 5 classes, all housed in the subphyllum Pezizomycotina (cf. also Marvanová, 2002). Affinities of several ana-holomorphs were ascertained only at class or order level. The five classes are represented in the three traditional major groups roughly distinguished by ascoma and ascus morphology as follows:

In **loculoascomycetes**, only Dothideomycetes have aquatic hyphomycete members.

In **discomycetes**, aquatic hyphomycetes are housed in three classes: Pezizomycetes, Leotiomycetes and Orbiliomycetes

In **pyrenomycetes** only the class Sordariomycetes hosts representatives of aquatic hyphomycetes.

In ascomycetous ana-holomorphs the sequencing of the 18S and 28S and of the ITS1-5.8S-ITS2 regions of ribosomal or mitochondrial DNA genes also did not reveal affinity outside Pezizomycotina.

In Basidiomycota, aquatic hyphomycete holomorphs are encountered in the class Urediniomycetes and in the subclass Hymenomycetidae of Hymenomycetes.

Within some taxonomic groups, the morphological similarity of some anamorphic reproductive structures or the presence of some phases of the life cycle is in accordance with phylogenetic relationships. In the holomorphs with established teleomorph/anamorph connections (species to species), the conidial initiation is mostly of the thalloblastic type throughout the orders. The leotiomycetan ana-holomorphic *Tetracladium* as well as *Anguillospora filiformis, Tetrachaetum elegans* and two species of Tricladium with their thalloblastic conidiogenesis fit well in this scheme. *Lemonniera* spp. with the blastic-phialidic conidiogenesis break down this picture: one species appears in Pleosporales and five other species and *Margaritispora* are placed in Leotiomycetes, in both cases contrasting with other taxa of Pleosporales and Leotiomycetes.

For the freshwater Nectriaceae the blastic-phialidic conidiogenesis is typical.

From the point of view of phylogenetic relationhips, the two frequent kinds of proliferation connected with thalloblastic initiation, i.e. percurrent and sympodial, are varying. This weakens the arguments for their importance for delimitation of genera. Similarly, the two modes of conidial secession, schizolytic and rhexolytic, although often considered decisive for generic delimitation, seem to be phylogenetically not informative.

The evolution of conidial shapes is unclear. It is not possible to say which shapes are linked to basal (ancestral) groups of fungi, e.g. the sigmoid conidia appear here in four classes of Ascomycota and in two of Basidiomycota. Similarly, branched conidia are encountered in three classes of Ascomycota and in two of Basidiomycota. They are relatively frequent in Helotiales: tetraradiate, tricladioid or complexly branched, but do not appear in Sordariomycetes (if we exclude *Heliscus lugdunensis*, whose short conical conidial outgrowths cannot be considered typical branches). Typical tetraradiate and tricladioid shapes are so far not present in Orbiliomycetes.

Staurosporous and scolecosporous taxa appear also in terrestrial hyphomycetes and coelomycetes. According to Descals (1985) and Kendrick (2003) the sigmoid and branched conidia are in fact transformed hyphae. In aquatic hyphomycetes these shapes are believed to be an adaptation to the life in lotic milieu. In the deeper moist strata in the leaf litter carpet they may facilitate movement in thin layers of water usually present on the surfaces (Bandoni and Koske, 1974). Their advantage in dry terrestrial habitats is unclear, but perhaps the better contact with the surface may also here contribute to more successful colonisation.

Nectriaceae is the only family with hyphomycetous phialidic synanamorphs both in terrestrial as well as aquatic members.

Andromorphs are encountered in some Dothideomycetes (pycnidial) and in Leotiomycetes (mostly hyphomycetous, in one case pycnidial).

Notes on Teleomorphic Classes, Orders and Families Housing Aquatic Hyphomycete Anamorphs

Ascomycota, Pezizomycotina

LOCULOASCOMYCETES

They are characterized by dark pseudothecia with bitunicate asci and pseudoparaphyses. According to Eriksson et al. (2001) in the modern system they are represented in two classes, Dothideomycetes and Chaetothyriomycetes.

Dothideomycetes

This class is considered poorly supported in phylogenetic analyses. Aquatic hyphomycete teleomorphs appear in a single order Pleosporales and a single family Massarinaceae. Four anamorphic species are traditionally linked with *Massarina*. They form colonies in dark grey or brown colours. They are uniform in conidiogenesis, which is thalloblastic, and also in percurrent proliferation of conidiogenous cells ('determinate' in connection with proliferation of a thalloblastic conidiogenous cell may mean lack of observation), but heterogeneous in conidial secession and in conidial shapes. The first three taxa with aquatic anamorphs have pycnidial andromorphs. Some *Massarina* spp. of the 'terrestrial' group have coelomycetous anamorphs typically with nonseptate minute conidia formed on phialidic conidiogenous cells in pycnids (Sivanesan, 1984). Hypothetically, these might have changed their function along with the adaptation of some taxa to water environment.

According to Scheuer (1991) the *Massarina* species having aquatic anamorphs differ from the type species of the genus. *Massarina* spp. were recently studied and revised by several authors (e.g. Hyde, 1995; Aptroot, 1998; Hyde and Aptroot, 1998; Tanaka and Harada, 2003) and their heterogeneity pointed out. Liew et al. (2000), in a study based on sequencing 18S rDNA showed close relationships between some members of the genus *Massarina* and *Lophiostoma*. In a subsequent study (Liew et al. 2002) sequenced the ITS1 and ITS2 region of 5.8S rDNA of a larger sample of *Massarina* species and of some members of closely related genera. They found that *Massarina* species with broadly fusiform to ellipsoidal ascospores clustered with *M. eburnea*, the type species of the genus, whereas species with narrowly fusiform ascospores formed a monophyletic clade with some members of *Lophiostoma*. Eriksson and Hawksworth (2003) resurrected the family Massarinaceae for *Massarina eburnea* and some related taxa which formed a clade with 100% BT support in a parsimony analysis of SSU rDNA sequences.

Massarina spp. of all three aquatic anamorphs have narrowly fusiform ascospores (Sivanesan, 1984) and very probably should be classified in *Lophiostoma,* outside Massarinaceae. The same may be true of *Massarina tetraploa* with a terrestrial staurosporous anamorph *Tetraploa aristata.* Unfortunately none of these taxa were included in the above molecular studies.

In loculoascomycetes most of the terrestrial anamorphs are dematiaceous hyphomycetes or coelomycetes. In Pleosporales the major anamorphs are in the genera *Helminthosporium* and its segregates and in *Alternaria,* where the conidiogenesis is tretic (the conidial primordium is connected with the conidiogenous cell through a channel in its thickened cell wall). However, the variation of conidial shapes in Pleosporales is great: there are tricladioid conidia in *Perisporiopsis*, tri- or tetraradiate in *Batistinula*, complexly branched in *Seuratia,* scoilecoid in *Asteromassaria* and in *Leptosphaeria*, multiradiate in *Pleomassaria*, scopiform or tetraradiate in *Podonectria*, helicoid conidia in *Tubeufia* (Sivanesan, 1984) and *Spirosphaera*-like with close affinity to *Anguillospora longissima* (Voglmayr, 2004).

Ana-holomorphic Taxa with Affinity to Pleosporales and Dothideales

The relationships of *Lemonniera pseudofloscula* to loculoascomycetes revealed by analyses of the 28S rDNA (Campbell et al. 2002a) is striking, because the only dematiaceous structures are the small sclerotia in *Lemonniera* and the conidiogenesis is phialidic. It is difficult to find any morphological or other feature reflecting the heterogeneity of *Lemonniera* whose other members were placed in Leotiomycetes. All described species of *Lemonniera* have similar colonies, conidiogenous cells and conidia. In contrast, the affinity of *Anguillospora rubescens* to Dothideomycetes, based on analysis of the 18S rDNA (Belliveau and Bärlocher, 2005) seems in accordance with the morphological similarity of this species with *A. longissima*, already mentioned in the protologue (Gulis and Mrvanová, 1999).

Discomycetes

This is a heterogeneous group of fungi with open, cup-shaped ascomata and unitunicate asci (Pfister and Kimbrough, 2001). The molecular phylogenetic studies using DNA sequence data showed that the orders are sometimes only distantly related. The affinity of Erysiphales to discomycetes based on sequencing the

18S rDNA gene remains unresolved (Saenz et al. 1994). Gargas and Taylor (1995) in their study including ten representative species of typical discomycetan orders (SSU rDNA analysis) found that the discomycetes in the traditional sense were paraphyletic. The anamorphs are of various kinds, very often pycnidial, but also hyphomycetous, both with phialidic, blastic, thalloblastic or thallic conidiogenesis (Hennebert and Bellemère, 1979).

Aquatic hyphomycete anamorphs living in aquatic or humid habitats are known in three classes: Leotiomycetes, Orbiliomycetes and Pezizomycetes. A question may be raised, why are they so frequently anamorphic discomycetes? In temperate climate, it is possible to speculate that the tiny fleshy apothecia with the hymenium exposed to air do not withstand frosts and so the anamorps ensure overwintering, at least for teleomorphs developing on leaves. Conidia can be found even in streams under ice cover (Müller-Haeckel, 1977). On the other hand, discomycetes often occupy wood, where they may survive as mycelia for several seasons (Shearer, 1992).

Leotiomycetes

This class according to Eriksson et al. (2001) is not supported by high bootstrap values. Aquatic hyphomycete teleomorphs are represented in two orders: *Helotiales* (in six families) and *Rhytismatales* (in one family). Helotiales seem to be heterogeneous.

Helotiales

Dermateaceae hosts species with dark excipular tissues of textura angularis. According to Baral (in Hawksworth, 1994, p. 405) Dermateaceae may comprise two groups, the *Dermea/Pezicula* complex and the *Mollisia/Pyrenopeziza* complex. The latter may be more close to Hyaloscyphaceae. Terrestrial anamorphs in Dermateaceae are typically coelomycetous, often with phialidic conidiogenesis, sometimes with falcate or scolecoform conidia (Hennebert and Bellemère, 1979).

There are three *Mollisia* species having aquatic anamorphs, which in pure culture may develop hyphomycetous, Phialophora-like andromorphs. Whereas the conidiogenesis of anamorphs is uniform, the conidial secession is different in *Casaresia*, which differs markedly also in conidial shape, having two pairs of opposite branches on conidial axis and hooked distal conidial ends.

Helotiaceae based on sequencing the partial SSU of rDNA (Gernandt et al. 2001) is polyphyletic. *Hymenoscyphus* has the highest number of species with aquatic anamorphs described (6), and one aeroaquatic, helicosporous (not incorporated here). Controversy exists around its homogeneity: according to Kohn (1993), *Hymenoscyphus* is a "wastebasket" and should be reassessed after studies by means of developmental as well as molecular taxonomy methods. In contrast, Pfister and Kimbrough (2001) consider it a monophyletic unit. Very recently, Zhang and Zhuang (2004) on the basis of sequence analysis of ITS1-5.8S-ITS2 region of rDNA recognized two clades within *Hymenoscyphus*: a monophyletic one around *H. fructigenus* (lectotype of *Hymenoscyphus*) and another with Ericaceae-associated spp. and *Calycina herbarum*. *Hymenoscyphus ericae* was recombined in a new genus *Rhizoscyphus*.

Hymenoscyphus is heterogeneous also with respect to the aquatic anamorph characters: The prevailing conidiogenesis is thalloblastic percurrent or determinate, in one species it is polyblastic. Among terrestrial anamorphs of *Hymenoscyphus*, Kendrick and DiCosmo (1979) list only *Idriella*, which has sympodial conidiogenous cells and small lunate conidia.

Articulospora tetracladia is a very variable species as regards the conidial shape. Several very similar taxa have been distinguished and formally described, but later considered synonymous (Khan, 1983). However, they may not have the same teleomorph: *Ombrophila tetracladia*, originally described as *Hymenoscyphus tetracladius* teleomorph of *Articulospora tetracladia* (Abdullah et al. 1981) was accompanied by a drawing illustrating *Articulospora angulata* (Tubaki, 1957). Terrestrial coelomycetous anamorph

Acleistia with cupulate conidioma and minute hyaline ellipsoid conidia borne on phialidic conidiogenous cells (Kirk et al. 2001) is reported for *Ombrophila* by Kendrick and DiCosmo (1979).

Tricladium indicum was described as anamorphic fungus by Sati and Tiwari (1992). Webster et al. (1995) collected a discomycete specimen in South Africa, which in culture, derived from ascospores, produced an anamorph comparable to *Tricladium indicum*. They classified the teleomorph as *Cudoniella indica*. Sivichai et al. (2003) collected, in Thailand, a discomycete which they identified as *Hymenoscyphus varicosporoides* described by Tubaki (1966) with an anamorph classified in *Varicosporium*. In contrast, Sivichai et al. (2003) tentatively assigned their anamorph to *Tricladium* similar to *T. indicum,* but differring from it in the width of conidial elements, which was only 5-7.5 µm. There is one more ana-holomorphic species with similar large conidia, but also with narrower elements: *Tricladium marylandicum* (Crane, 1968). Thorough comparison of ex-type or authentic cultures of these fungi and an approach employing methods of molecular taxonomy will be desirable for resolving affinities among these taxa.

Leotiaceae is now understood in a very narrow sense, with only five genera (Eriksson et al. 2001). Its accommodation within Helotiales was doubted by Gernandt et al. (2001), but no formal relocation has been done. *Anguillospora furtiva* is the only aquatic anamorph pertaining to this family. The teleomorph for *A. furtiva,* now classified as *Pezoloma* sp. was first published as *Rutstroemia* sp. by Webster and Descals (1979). The latter genus is now classified in a small family Rutstroemiaceae.

A. furtiva is morphologically and culturally very similar to *Anguillospora crassa* and there is some overlap in conidial size. According to Belliveau and Baerlocher (2005, in press), these two taxa (along with *Anguillospora fustiformis* and *A. mediocris*) have identical sequences of the nSSU rDNA. However, *A. crassa* has the teleomorph in *Mollisia* and *A. fustiformis* is connected with *Hymenoscyphus*. No terrestrial anamorph of *Pezoloma* is mentioned in Kendrick and DiCosmo (1979).

Hyaloscyphaceae were characterized as heterogeneous by Huhtinen (in Hawksworth 1994, p. 407). Superficial similarity of *Hydrocina* apothecia to *Psilachnum* or *Calycina* has been mentioned by Webster et al. (1991). The aquatic anamorph *Tricladium chaetocladium* is in all aspects entirely dissimilar to *Mycoarthris corallina,* whose relationships to Hyaloscyphaceae was revealed on the basis of SSU rDNA and ITS 1 and 2 sequencing (Marvanová et al. 2002). The latter has whitish colonies and forms conidiogenous hyphae breaking down into arthroconidia, on sympodial conidiogenous cells.

Vibrisseaceae probably form a monophyletic lineage within Helotiales (Pfister and Kimbrough, 2001). In *Vibrissea flavovirens* (as *Apostemidium torrenticola*) an andromorphic(?) synanamorph *Phialocephala* sp. was reported beside the *Anavirga* state (Hamad and Webster, 1987). *Phialocephala* spp. are terrestrial anamorphs with ascomycetous affinity. Sequence analysis of ITS1-5.8S-ITS2 of the rDNA of some *Phialocephala* spp. (Grünig et al. 2002), revealed their polyphyletic origin, with 4 spp. grouped in a clade within Leotiales (= Helotiales in the sense of Eriksson et al. 2004) and the other 3 spp. in Sordariales, a phylogenetically distant order.

Interestingly, an unnamed synanamorph (probably also andromorphic) similar to *Phialocephala dimorphospora,* was reported in an ana-holomorphic aquatic hyphomycete *Variocladium giganteum* (as *Tricladium giganteum,* Willoughby and Minshall, 1975).

Loramycetaceae is a small family in Helotiales and probably forms a monophyletic lineage (Pfister and Kimbrough, 2001). *Loramyces juncicola* (SSU rDNA, MP tree) has *Phialocephala fortinii,* an endophyte in roots, as a sister taxon (Gernandt et al. 2001).

Rhytismatales

Only one aquatic hyphomycete teleomorph, *Coccomyces* sp., is classified in Rhytismataceae, characterized by opening of ascomata by longitudinal or radial splits. Conidia of its aquatic anamorph (Descals and

Webster, 1982) are similar to those of *Tricladium*. Known terrestrial anamorphs are coelomycetous (Pfister and Kimbrough, 2001). *Coccomyces* has a plant pathogenic terrestrial species *C. hiemalis* with *Septoria*-like conidial state. Other anamorphs can be *Cylindrosporium, Schizothyrella* or *Leptothyrium* (Kendrick and DiCosmo, 1979).

Ana-holomorphic taxa with affinity to Leotiomycetes

The placement of five *Lemonniera* species within Leotiomycetes is supported by cladistic analysis of the 28S rDNA (Campbell et al. 2002a). On the basis of morphological taxonomy *Lemonniera* was considered a homogeneous genus. The individual species have similar colonies with black microsclerotia, conidiophores with penicillate head on a stalk, phialidic conidiogenous cells and tetraradiate conidia with small differences in conidial morphogenesis (Descals et al. 1977). The more surprising is the placement of *L. pseudofloscula* in Dothideomycetes (see above).

It is more or less in accordance with expectations that *Margaritispora aquatica* is phylogenetically very near to most of the *Lemonniera* species. The similarity of colonies and conidiogenous structures is striking. A distinct difference is in the conidial shape. *Alatospora* although dissimilar culturally as well as in conidial morphology, has also philalidic conidiogenous cells. The rest of taxa have thalloblastic/percurrent or thalloblastic/sympodial conidiogenous cells, which fits well in the scheme leotimycetan aquatic anamorphs.

In *Tetracladium,* five species were linked to Leotiomycetes on the basis of sequencing partial regions of the 18S rDNA gene (Nikolcheva and Bärlocher, 2002). They were placed close to *Bulgaria inquinans*, a terrestrial wood inhabiting saprotrophic fungus, now classified in Helotiales in a separate family Bulgariaceae (Eriksson et al. 2004).The genus *Tetracladium* was found monophyletic (MP and NJ, Nikolcheva and Bärlocher, 2002), which is in accordance with its relative morphological as well as cultural homogeneity. Furthermore, analysis of the data placed the *Tetracladium* species closely to Onygenales and Erysiphales, but without any clear pattern. Erysiphales is classified within Leotiomycetes (Eriksson et al. 2004) but Onygenales is an order of Eurotiomycetes.

Orbiliomycetes

This class has a somewhat unstable position among discomycetous classes. Its members are characterised with small pale apothecia with apically truncate asci; the paraphyses usually have inflated apices. In the past it was classified as a family (Orbiliaceae) in Helotiales. Pfister (1997) on the basis of 18S rDNA sequencing found that the family had rather an isolated position in this order. The study of Gernandt et al. (2001), also based on sequencing partial SSU rDNA supported this view. Eriksson et al. (2004) classify *Orbilia* together with *Hyalorbilia* in a separate class Orbiliomycetes. However, Tehler et al. (2003) in their phylogenetic analysis of SSU rDNA place Orbiliomycetes as an unsupported pezizalean sister group to Helvellaceae, Morchellaceae, Tuberaceae etc.

Many members of Orbiliaceae are known as predaceous fungi capturing nematode (Hagedorn and Scholler 1999). Based on analyses of the partial 18S rDNA and the ITS1- 5.8S rDNA- ITS2 region, Scholler et al. (1999) recognized six anamorphic genera belonging to various spp. of *Orbilia*: *Dactylellina* (among others *Dactylellina* (*Laridospora*) *appendiculata*, an aquatic hyphomycete known from waters in warmer climates), *Gamsyella, Arthrobotrys, Drechslerella*, *Dactylella* pp. and *Dicranidion* pp. Pfister (1997) illustrates an *Anguillospora* anamorph of an *Orbilia* tentatively identified as *O. luteorubella,* collected in non-aquatic habitat. However, according to the scale his *Anguillospora* has much thinner conidia and may be different from *A. rosea*. Another anamorph, *Dicranidion* sp. derived from *Orbilia,* was obtained by the same author. The starosporous and scolecosporous anamorphs occurring in water or in humid terrestrial habitats are relatively homogeneous in the characters of conidiogenesis as well as in colony pigmentation. All produce pale colonies, the conidiogenesis is mostly thalloblastic, conidiogenous cells are sympodial except for *Anguillospora rosea* and *Tricellula inaequalis*; the conidial secession is schizolytic and conidia are dichotomously branched in 3 of 8 taxa of this group.

Pezizomycetes

There is a single teleomorphic taxon with aquatic hyphomycete anamorph classified in the order Pezizales, family Pyronemataceae: *Miladina lechithina*. Some authors consider Pezizales as basal (ancestral) to other apothecial ascomycetes (Gargas and Taylor, 1995). Based on SSU rDNA analysis, Pezizales was found paraphyletic: Pezizaceae appeared monophyletic, whereas the other taxa of Pezizales clustered with Orbiliales (Tehler et al. 2003).

The members of Pyronemataceae have apothecial or cleistothecial ascomata, often brightly coloured due to carotenoid pigments. They are saprotrophic on wood or soil, some occur on burnt ground.

Acladium, Hansfordia, Oedocephalum and *Scytalidium* are mentioned as terrestrial anamorphs of some members of Pyronemataceae (Kendrick and DiCosmo, 1979). Sutton and Hennebert (1994) add *Botrytis* and *Dichobotrys*. The freshwater anamorph *Actinosporella megalospora* of *Miladina lechithina* with huge spores has a unique position among these anamorphs. The only similarity can be seen in the robustness of conidiogenous structures and in dichotomous conidiophore branching in *Dichobotrys* and *Actinosporella*. But conidiogenesis in *Actinosporella* is retrogressive and thalloblastic, whereas it is progressive and polyblastic in Dichobotrys. The possibility of heterogeneity of Pyronemataceae was admitted by Kimbrough and Curry (1986). Kimbrough and Gibson (1990) found wide variation of the septal plugs at the base of asci in members of this family. Affinity of *Miladina* to other members of the family is not known.

PYRENOMYCETES

Pyrenomycetes may be characterised by perithecial ascomata and unitunicate asci with apical apparatus. According to Eriksson et al. (2001), the class Sordariomycetes is usually well supported in SSU rDNA analyses.

Sordariomycetes

In this class only one order, Hypocreales and one family, Nectriaceae, host holomorphic freshwater hyphomycetes. Nearly all terrestrial anamorphic states of Hypocreales have phialidic conidiogenesis (Sutton and Hennebert, 1994; Rossman et al. 1999). This is also true for Nectriaceae including *'Nectria'* spp. with aquatic anamorphs. As was mentioned above, the main characters of conidiogenesis are in accordance with those known in terrestrial Nectriaceae. The presence of microconidial synanamorphs and the absence of andromorphs are also common characters in Nectriaceae.

Rossman et al. (1999), in their extensive studies based on morphology of the teleomorphs, KOH reaction of ascomatal wall and links to anamorph genera, segregated several genera from the genus *Nectria*, e.g. *Cosmospora* for some slow-growing *Fusarium* spp. among them and also for those found often in water. Unfortunately, no chemical reaction was probably tested on the *Nectria* spp. with aquatic anamorphs in *Flagellospora* and *Heliscus*.

Mantiri et al. (2001) in their study based on mitochondrial rDNA sequences segregated *Nectria discophora* with *Cylindrocarpon ianthothele* var. *majus* anamorph often isolated from water environment, into the genus *Neonectria*. This was confirmed in a subsequent study of *Neonectria mammoidea* group and its *Cylindrocarpon* anamorphs by Brayford et al. (2004). In contrast, according to the results of ITS and beta-tubulin gene analysis (Halleen et al. 2004), the *Neonectria mammoidea* group appears closely related to the newly described anamorphic *Campylocarpon* spp. and both seem only remotely related with the other *Neonectria* groups.

Unfortunately, neither the *'Nectria'* teleomorphs of the aquatic genera *Heliscus* and *Flagellospora* nor these anamorphs themselves were included in these studies of hypocrealean fungi. However, O'Donnell (1993) in his study of *Fusarium* spp. and their relatives based on LSU rDNA phylogenetic analysis included also three isolates (only two shown) of *Nectria lugdunensis* and one of *N. penicillioides*. All were classified in Nectriaceae. The *Nectria lugdunensis* ex-type isolate appeared as a sister taxon to *Nectria mauritiicola* (syn. of *Corallomycetella repens* acc. to Rossmann et al. 1999); one of the other two isolates clustered with *Nectria coccinea* and *N. galligena* (both with *Cylindrocarpon* anamorph) segregated later into *Neonectria* (Rossman et al. 1999). The second isolate clustered with *Cosmospora aurantiicola* with *Fusarium larvarum* anamorph. *Nectria penicillioides* appears as a rather weakly supported sister taxon of *Calonectria pyrochroa* with a *Cylindrocladium* anamorph. *Fusarium aquaeductuum* var. *medium* as well as var. *aquaeductuum* are considered anamorphic *Cosmospora purtonii* and *C. episphaeria,* respectively (Rossman et al. 1999; Mantiri et al. 2001).

Nectria in the narrow sense of Rossman (1993) and Rossman et al. (1999) should have anamorphs of the genus *Tubercularia* or *Tubercularia*-like (sporodochial or pycnidial, orange, in pure culture often *Acremonium*-like), which do not appear in any of the aquatic representatives. The above controversial results indicate that there is a need for further phylogenetic studies of '*Nectria*' with aquatic anamorphs.

The two other orders, Halosphaeriales and Lullworthiales, contain aquatic holomorphic taxa occurring in marine as well as in brackish waters. Neither *Anguillospora marina* nor *Sigmoidea luteola* have so far been incorporated in a phylogenetic study. *Sigmoidea luteola* is morphologically very different from its freshwater anamorphic counterparts and very probably will be segregated. *Clavatospora bulbosa* was subjected to cladistic analysis of the 28S rRNA (Campbell et al. 2002b) and its relationships to *Corollospora pulchella* confirmed.

Ana-holomorphic taxa with affinity to Sordariomycetes, Sordariales

Sordariales according to Spatafora and Blackwell (1994) are not monophyletic. On the basis of SSU rDNA analysis they were able to reveal two sublades: one grouping with taxa having brightly coloured perithecia Hypocreales and Clavicipitales (classified in the subclass Hypocreomycetidae in Eriksson et al. 2001), the other with dark-perithecial taxa Diaporthales and Ophiostomatales (classified in the subclass Sordariomycetidae in Eriksson et al. 2001). *Lunulospora curvula* clusters with dark-perithecial genera *Chaetomium* and *Sordaria* in the study based on SSU rDNA analysis (Belliveau and Bärlocher, 2005). There are no congruent characters present in *Lunulospora* and anamorphs of *Sordaria* spp.

Basidiomycota

Basidiomycotan teleomorphs having aquatic anamorph are rather rare. It may be speculated that the role of the conidial states, at least in temperate climate, is similar like that for discomycetes: the aquatic conidial representatives are so far restricted to Hymenomycetes and Urediniomycetes with tiny unprotected layers of basidia, probably not capable to survive low temperatures.

HETEROBASIDIOMYCETES

Heterobasidiomycetous fungi is a very heterogeneous and polyphyletic group, which has been intensively studied in the last few decades (Wells and Bandoni, 2001). Traditionally they were characterised by transversely or longitudinally divided basidia, variation in septal pore morphology and presence of intracellular structures unique in fungi, but this concept appeared untenable if confronted with new results of ultrastructural and molecular studies. Based on nucleotide sequence analyses of 18S rRNA, Swann and Taylor (1995) proposed three basal classes of basidiomycetes: Urediniomycetes, Ustilaginomycetes and Hymenomycetes. In Hymenomycetes they recognized two subclasses: Tremellomycetidae for dimorphic basidiomycetes typically with yeast phase in their life cycle and Hymenomycetidae without yeast state.

Sampaio and Bauer (2003) characterise Urediniomycetes as having simple septal pores with electron-dense microbodies without parenthesomes, disc-like spindle pole bodies and typically transversely septate basidia. The major cell wall carbohydrate component is mannose; xylose is absent. Besides Microbotryomycetidae (Swann et al. 1999), the same authors (Swann et al. 2001) distinguished two other subclasses in Urediniomycetes: Agaricostilbomycetidae (mostly dimorphic) and Urediniomycetidae mostly monomorphic.

Urediniomycetes

The aquatic taxa are rather isolated in the class; the taxonomic distance between *Classicula* and other urediniomycetan members is at the order level and *Camptobasidium* cannot be placed at present in any order of Microbotryomycetidae (Sampaio and Bauer, 2003). This may be due to the fact that the ultrastructural characters, e.g. septal pore apparatus and other cytoskeletal features, important for classification, are not known in these and many other species. The order Classiculales was proposed on the basis of the combination of simple septal pores associated with microbodies and tremelloid haustorial cells present in *Classicula fluitans* and *Jaculispora submersa* (Bauer et al. 2003). The morphologically very similar ana-holomorphic taxon *Jaculispora submersa* described by Hudson and Ingold (1960) from Jamaica without recognising its basidiomycetous affinity, forms together with *Classicula* a statistically well supported clade (SSU rRNA, Bauer et al. 2003). Unfortunately no ex-type culture was established for *Jaculispora*; the study of Bauer et al. was carried out with a somewhat morphologically different strain isolated in Japan (Matsushima, 1987).

Infundibura adhaerens, whose conidia were encountered on leaves in a pond in Great Britain (Goos and Descals, 1985) and in stream foam in Portugal (Marvanová, unpubl.), was recently connected to *Helicogloea angustispora* (Kirschner, 2004) on the basis of a culture obtained from basidiospores. The affinity to *Helicogloea* is supported by analysis of partial 26S RNA gene sequences (Kirschner, 2004).

Undoubtedly there are more heterobasidiomycetous fungi in waters, but their conidia mostly do not have septa with clamps and so when the dicaryotic status of the cells or the apparent trait of conidial germination accompanied with 'protoplasm migration' are not observed, their possible affinity to this group escapes attention.

Camptobasidium as well as *Classicula* anamorphs have, in common, the capability of biotrophism; both are at least facultative mycoparasites. *Crucella* was observed to produce hyphae coiling around conidia of other aquatic hyphomycetes in water (Marvanová and Suberkropp, 1990), *Naiadella* forms haustorial branches on mycelia, but the host remains unknown (Marvanová and Bandoni, 1987, Bauer et al 2003). Urediniomycetan affinity and mycobiotrophic character may hold true also for *Anguillomyces acadiensis* and *Nodulospora inconstans* described from Canadian streams, which showed better and longer viability when cultivated in dual cultures with other fungi. Germination was accompanies with 'protoplasm migration' (Marvanová and Bärlocher, 2000).

Hypothetically, mycoparasitism may have been a fundamental factor in the evolution of the basidiomycetes (Weiss et al. 2004). Do some taxa need certain substances from the host for the right development of the basidiomycetous phase?

Homobasidiomycetes

Hymenomycetes

Aquatic holomorphs are so far known only in two families of the order Polyporales: Atheliaceae (*Fibulomyces/Taeniospora*) and Sistotremataceae (*Sistotrema/Ingoldiella*). Unlike the aquatic urediniomycetan anamorphs, the conidia here bear clamped septa and no 'protoplasm migration' during germination occurs. Also here we know ana-holomorphic representatives occurring in water, e.g. *Fibulotaeniella canadensis* (Marvanová and Bärlocher, 1988), or *Arcispora bisagittaria* (Marvanová and Bärlocher, 1998), which according to the morphology are presumed members of Polyporales. This holds true also for *Fibulomyces crucelliger, Fibulomyces* sp. and *Sistotrema hamatum*, whose affinity to other species of these genera is unknown. Originally, the teleomorph for *Taeniospora gracilis* was identified as *Leptosporomyces galzinii*, another member of Atheliaceae (Nawawi et al. 1977). *Fibulomyces* and *Leptosporomyces* are difficult to distinguish from each other (K. Hjortstam, pers. communication).

Conclusions

In any case, it should be borne in mind, that this compilation reflects the 'ad interim' status of our knowledge on the theme. The jig-saw puzzle called 'aquatic-hyphomycetes-who-are-they' is not near to being solved. Along with the phylogenetic heterogeneity there seem to be trophic heterogeneity as well. Although the anamorphic phases are typically saprotrophic, a capability of other 'diet' is also present. The orbiliomycetan anamorphs often belong to predacious fungi: their mycelia may develop various structures for capturing nematodes or other invertebrates. Are these fungi dependent on this trophic, peculiarity ensured by the anamorphic phase?

Some ana-holomorphic aquatic hyphomycetes are at least distantly related to Erysiphales (Nikolcheva and Bärlocher, 2002) – does it support the hypothesis that plant parasites invaded the aquatic habitats together with their hosts (Shearer, 1993)? Indeed, there are a few hyphomycete taxa found on plants as well as in water: *Gyoerffyella entomobryoides* found in wounded tissue of *Rosa* (Boerema and Arx, 1964), *Mycocentrospora acerina* pathogenic to *Acer* seedlings (Braun, 1995) and often occurring in streams, *Fusarium* and. *Cylindrocarpon* spp., frequently seen on submerged leaves (Bärlocher and Kendrick, 1974).

Aquatic hyphomycetes are systematically studied with classical methods based on morphology and ontogeny only for 60 years and there is not enough knowledge accumulated about individual taxa. Some species were described on the basis of a single isolate and since then not obtained again. In few others the protologue is vague and type material not existing or unavailable—such taxa must be treated as doubtful. Even the common species are not well known, sometimes there are subtle details in the morphology, which are commonly considered as strain variation, but in several cases a thorough pure culture study led to recognition of a new taxon, e.g. *Pleuropedium multiseptatum*, segregated from *Pleuropedium tricladioides* on the basis of different conidiogenesis (Marvanová and Descals, 1996), or *Tetracladium palmatum,* undiscovered for long time among many morphological variants of the common *T. marchalianum* (Roldán et al. 1989). Small morphological differences exist between conspecific isolates from different continents, e.g. *Heliscus submersus* strains from tropical or subtropical climates *vs* strains from Portugal (Marvanová et al. 2003), *Lunulospora curvula* from Europe vs. strains from India (Ingold and Webster, 1973) or from New Zealand (Aimer and Segedin, 1985). Are these strains only different phenotypes or are we dealing with various genotypes? Though a few studies on intraspecific genetic variation, based on large amount of isolates of a single species exist, these isolates are from a geographically small area and do not give information on the species variation on the world wide scale (Peláez et al. 1996; Charcosset and Gardes, 1999).

The type specimens of many taxa of aquatic hyphomycetes are permanent slides with conidia and conidiogenous structures; few living ex-type cultures are available. The dried samples of pure cultures which may be also used for DNA isolation and sequencing are sometimes contaminated with other fungi and hence useless (personal experience). Authentic isolates from authors of the taxa or from group specialists are often abandoned in molecular studies and unfortunately, even common ana-holomorphic aquatic hyphomycetes are represented by relatively few isolates in large culture collections (Marvanová, 1997). For many anamorphic genera isolates of the full sets of described species are not available (Nikolcheva and Bärlocher, 2002),

In connection with the contemporary paucity of taxonomists (Hawksworth and Mouchacca, 1994; Kerrigan, 2003), studies aimed at classification based on pure cultures are rare and rather scattered and incidental; there is no research centre or group devoted to studies of taxonomy of aquatic hyphomycetes from various aspects (morphology, ontogeny, cell ultrastucture, cell wall composition, gene sequencing), accumulating and preserving strains for this purpose. Information on these fungi appears mostly in ecological and biodiversity studies. The seeming easiness of their identification on the basis of detached conidia is in its effect counterproductive: it does not bring up the need to work with pure cultures and so the actual knowledge about individual species may be partly based on unreliable identifications. There are several species whose detached conidia are practically indistinguishable from those of other, sometimes taxonomically distant species (Marvanová, 1997).

There is little or no collaboration between ascomycete and basidiomycete molecular taxonomists on one side and aquatic hyphomycete taxonomists on the other side. Several large studies (e.g. Rossman et al. 1999; Aptroot, 1998), which concern also genera containing aquatic hyphomycete teleomorphs have not included aquatic hyphomycete isolates, or the results appeared controversial, e.g. linking each of two strains of *Heliscus lugdunensis* to a different genus in Hypocreales (O´Donnel, 1993).

Cautions for Anamorph/Teleomorph Relationships Studies

Collecting teleomorphs in nature, isolating the asco-or basidiospores and obtaining the anamorph should be preferred, though it may be less productive. Mating experiments may enhance the chance of success when we start from the anamorph and try to obtain the teleomorph in vitro. However, the teleomorphic states obtained in cultures on agar may not develop characters necessary for reliable identification and this may lead to difficulties with classification (see under Helotiaceae, Leotiaceae and Hymenomycetes).

Where possible, ex-type cultures or herbarium type specimens or at least cultures from reliable source should be used for molecular studies. Even cultures deposited in large culture collections may be misidentified or mislabelled (Trappe, 2004 and personal experience), hence a revision by an appropriate specialist should be ensured.

The data from the GenBank should be taken with cautions, because some may be based on misidentified specimens (Hawksworth, 2004).

It should be stressed, that a very good documentation be kept about cultures used for molecular studies, so that in case of work with misidentified isolate it can be traced back. Sequenced private isolates should be deposited in a culture collection to be at disposal for future studies. In publications, the provenance of the cultures with the strain number and acronym of the collection where the fungus is maintained should be given.

Also, it is desirable to give some morphological data on the strains processed, which will allow to recognise eventual intraspecific variation of the taxon. It will be ideal to sequence more isolates of one species in order to reveal possible cryptic lineages, which can be then eventually defined morphologically. More collaboration between dicaryomycete and aquatic hyphomycete specialists will be advantageous for both sides.

References

Abdullah, S.K., Descals, E. and Webster, J. (1981) Teleomorphs of three aquatic hyphomycetes. Transactions of the British mycological Society 77: 475–483.

Aimer, R. and Segedin, B.P. (1985) Some aquatic hyphomycetes from New Zealand streams. New Zealand Journal of Botany 23: 273–299.

Aptroot, A. (1998) A world revision of *Massarina* (Ascomycota). Nova Hedwigia 66: 89–162.

Bandoni, R.J. and Koske, R.E. (1974) Monolayers and microbial dispersal. Science 183: 1079–1081.

Bauer, R., Begerow, D., Oberwinkler, F. and Marvanová, L. (2003). *Classicula*: the teleomorph of *Naiadella fluitans*. Mycologia 99: 756–764.

Bärlocher, F. and Kendrick, B. (1974) Dynamics of the fungal population on leaves in a stream. Journal of Ecology 62: 761–791.

Belliveau, M.J-R. and Bärlocher, F. (2005) Molecular evidence confirms multiple origins of aquatic hyphomycetes. Mycological Research 109 (in press).

Baschien, C. (2003) Development and evaluation of rRNA targetted in situ probes and phylogenetic relationships of freshwater fungi. PhD Thesis. Technical University, Berlin, 225 pp.

Boerema, G.H. and Arx, J.A. von. (1964) Ein neuer zur Gattung *Ingoldia* gehörender Pilz. Nova Hedwigia 8: 297–300.

Braun, U. (1995) A Monograph of *Cercosporella, Ramularia* and Allied Genera Phytopathogenic Hyphomycetes). Vol. I. IHW Verlag, Eching.

Brayford, D., Honda, B.M., Mantiri, F.R. and Samuels, G.J. (2004) *Neonectria* and *Cylindrocarpon*: The *Nectria mammoidea* group and species lacking microconidia. Mycologia 96: 572–597.

Campbell, J., Shearer, C.A. and Marvanová, L. (2002a) Evolutionary relationships between aquatic anamorphs and teleomorphs: *Lemonniera*. The 7th International Mycological Congress, Oslo 11–17 Aug. 2002. Book of Abstracts, No. 193.

Campbell, J., Shearer, C.A., Mitchell, J.I. and Eaton, R.A. (2002b) *Corollospora* revisited: a molecular approach. In: Fungi in marine Environments, (eds.), Hyde, K.D. and Vrijmoed, L.L.P. Fungal Diversity Press, Hong Kong, 15–33.

Charcosset, J.-Y. and Gardes, M. (1999) Infraspecific genetic diversity and substrate preference in the aquatic hyphomycete *Tetrachaetum elegans*. Mycological Research 100: 736–742.

Crane, J.L. (1968) Freshwater hyphomycetes of the Northern Appalachian Highland including New England and three Coastal Plain states. American Journal of Botany 55: 996–1002.

Crous, P.W., Aptroot, A., Ji-Chuang Kang, Braun, U. and Wingfield, M.J. (2000) The genus Mycosphaerella and its anamorphs. In: Molecules, Morphology and Classification: Towards Monophyletic Genera in the Ascomycetes, (eds.), Seifert, K.A., Gams, W., Crous, P.W. and Samuels, G.J. Studies in Mycology 45, Centraalbureau voor Schimmelcultures, Baarn/Delft, 107–121.

Descals, E. (1985) Conidia as modified hyphae. Proceedings of the Indian Academy of Sciences (Plant Sciences) 94: 209–227.

Descals, E. and Webster, J. (1978) *Miladina lechithina* (Pezizales), the ascigerous state of *Actinospora megalospora*. Transactions of the British mycological Society 70: 466–472.

Descals, E. and Webster, J. (1982) Taxonomic studies on aquatic hyphomycetes. III. Some new species and a new combination. Transactions of the British mycological Society 78: 405–437.

Descals, E., Fisher, P.J. and Webster, J. (1984) The *Hymenoscyphus* teleomorph of *Geniculospora grandis*. Transactions of the British mycological Society 83: 541–546.

Descals, E., Marvanová, L. and Webster, J. (1999) New taxa and combinations of aquatic hyphomycetes. Canadian Journal of Botany 76: 1647–1659.

Descals, E, Webster, J. and Dyko, B.J. (1977) Taxonomic studies on aquatic hyphomycetes. I. *Lemonniera* de Wildeman. Transactions of the British mycological Society 69: 89–109.

De Wildeman, É. (1893) Notes Mycologiques. Annales de la Société Belge de Microscopie 17: 35–68.

De Wildeman, É. (1894) Notes Mycologiques. Annales de la Société Belge de Microscopie 18: 135–161.

De Wildeman, É. (1895) Notes Mycologiques. Annales de la Société Belge de Microscopie 19: 193–206.

Digby, S. and Goos, R.D. (1987) Morphology, development and taxonomy of *Loramyces*. Mycologia 79: 821–831.

Eriksson, O.E., Baral, H.-O., Currah, R.S., Hansen, K., Kurtzman, C.P., Rambold, G. and Laessoe, T. (2001) Outline of Ascomycota – 2001. http//www.umu.se/myconet/M7.html

Eriksson, O.E., Baral, H.-O., Currah, R.S., Hansen, K., Kurtzman, C.P., Rambold, G. and Laessoe, T. (2004) Outline of Ascomycota – 2004. http//www.umu.se/myconet/curr/current.html

Eriksson, O.E. and Hawksworth, D.L. (2003) *Saccharicola*, a new genus for two *Leptosphaeria* species on sugar cane. Mycologia 95: 426–433.

Fisher, P.J. and Spooner, B. (1987) Two new ascomycetes from Malawi. Transactions of the British mycological Society 88: 47–54.

Gams, W. (2000) *Phialophora* and some similar morphologically little-differentiated anamorphs of divergent ascomycetes. In: Molecules, Morphology and Classification: Towards Monophyletic Genera in the Ascomycetes (eds.), Seifert, K.A., Gams, W., Crous, P.W. and Samuels, G.J. Studies in Mycology 45, Centraalbureau voor Schimmelcultures, Baarn/Delft, 187–199.

Gams, W. and Nirenberg, H.I. (1989) A contribution to the generic definition of *Fusarium*. Mycotaxon 35: 407–416.

Gargas, A. and and Taylor, J.W. (1995) Phylogeny of Discomycetes and early radiation of the apothecial Ascomycotina inferred from SSU rDNA sequence data. Experimental Mycology 19: 7–15.

Gernandt, D.S., Platt, J.L., Stone, J.K. and Spatafora, J.W. (2001) Phylogenetics of Helotiales and Rhytismatales based on partial small subunit nuclear ribosomal DNA sequences. Mycologia 93: 915–933.

Goos, R. and Descals, E. (1985) Occurrence of *Infundibura adhaerens* in aquatic habitats in Britain. Transactions of the British mycological Society 84: 170–171.

Greuter, W., McNeil, J., Barrie, F.R., Burdet, H.M., Demoulin, V., Filgueiras, T.S., Nicholson, D.H., Silva, P.C., Skog, J.E., Trehane, P., Turland, N.J. and Hawksworth, D.L., Eds. (2000) International Code of Botanical Nomenclature (Saint Louis Code). Koeltz Scientific Books, Königstein.

Grünig, C.R., Sieber, T.N., Rogers, S.O. and Holdenrieder, O. (2002) Genetic variability among strains of *Phialocephala fortinii* and phylogenetic analysis of the genus *Phialocephala* based on rDNA ITS sequence comparison. Canadian Journal of Botany 80: 1239–1249.

Gulis, V. and Marvanová, L. (1999) Three new scolecosporous hyphomycetes from water in Belarus. Mycotaxon 72: 237–250.

Gulis, V., Marvanová, L. and Descals, E. (2005) An illustrated key to the common temperate species of aquatic hyphomycetes. In: Methods to Study Litter Decomposition: A Practical Guide, (eds.), Graça, M., Bärlocher, F. and Gessner, M.O., Springer, Dordrecht, 153–167.

Halleen, F., Schroers, H.-J., Groenewald, J.Z. and Crous, P.W. (2004) Novel species of *Cylindrocarpon* (*Neonectria*) and *Campylocarpon* gen. nov., associated with black foot disease of grapevines (*Vitis* spp.). Studies in Mycology 50: 431–455.

Hamad, S.R. and Webster, J. (1987) *Anavirga dendromorpha*, anamorph of *Apostemidium torrenticola*. Sydowia 40: 60–64.

Hagedorn, G. and Scholler, M. (1999) A reevaluation of predatory orbiliaceous fungi. I. Phylogenetic analysis using rDNA sequence data. Sydowia 51: 27–48.

Hawksworth, D.L. (1994) Ascomycete Systematics: Problems and Perspectives in the Nineties. Plenum Press, New York and London.

Hawksworth, D.L. (2004) 'Misidentifications' in fungal DNA sequence databanks. New Phytologist 161: 13–15.

Hawksworth, D.L. and Mouchacca, J. (1994) Ascomycete systematics in the nineties. In: Ascomycete Systematics: Problems and Perspectives in the Nineties. Ed., Hawksworth, D.L., Plenum Press, New York and London, 3–11.

Hawksworth, D.L., Sutton, B.C. and Ainsworth, G.C. (1983) Ainsworth and Bisby´s Dictionary of the Fungi. 7th Ed. Commonwealth Mycological Institute, Kew.

Hennebert, G.L. (1987) Pleoanamorphy and its nomenclatural problem. In: Pleomorphic Fungi: the Diversity and its Taxonomic Implication, (ed.), Sugiyama J. Elsevier, Amsterdam, 263–290.

Hennebert, G. L. and Bellemère, A. (1979) Les formes conidiennes des Discomycètes. Essai taxonomique. Revue de Mycologie 43: 259–315.

Hennebert, G. L. and Sutton, B.C. (1994) Unitary parameters in conidiogenesis. In: Ascomycete Systematics. Problems and Perspectives in the Nineties, (ed.), Hawksworth, D.L. Plenum Press, New York and London, 65–76.

Hibbett, D.S. and Thorn, R.G. (2001) Basidiomycota: Homobasidiomycetes. In: The Mycota, Volume VII, Systematics and Evolution, Part B, (eds.), McLaughlin, D.J., Mc Laughlin, E.G. and Lemke, P.A. Springer-Verlag, Berlin-Heidelberg, 121–168.

Hudson, H.J. and Ingold, C.T. (1960) Aquatic hyphomycetes from Jamaica. Transactions of the British mycological Society 43: 469–478.

Hyde, K.D. (1995) The genus *Massarina* with a description of *M. eburnea* and an annotated list of *Massarina* names. Mycological Research 99: 291–296.

Hyde, K.D. and Aptroot, A. (1998) Tropical freshwater species of the genera *Massarina* and *Lophiostoma*. Nova Hedwigia 66: 489–502.

Ingold, C.T. (1942) Aquatic hyphomycetes of decaying alder leaves. Transactions of the British mycological Society 25: 339–417

Ingold, C.T. (1975a) An illustrated guide to aquatic and waterborne hyphomycetes (Fungi Imperfecti) with notes on their biology. Freshwater Biological Association Scientific Publications 30: 1–96

Ingold, C.T. (1975b) Hooker Lecture 1974. Convergent evolution in aquatic fungi: the tetraradiate spore. Biological Journal of the Linnean Society 7: 1–25.

Ingold, C.T. and Webster J. (1973) Some aquatic hyphomycetes from India. Kavaka 1: 5–9.

Kendrick, B. (1989) "Subdivision Deuteromycotina"—a fungal chimera. Sydowia 41: 8–14.

Kendrick, B. (2003) Analysis of morphogenesis in hyphomycetes: new characters derived from considering some conidiophores and conidia as condensed hyphal systems. Canadian Journal of Botany 81: 75–100.

Kendrick, B. and DiCosmo, F. (1979) Teleomorph-anamorph connections in ascomycetes. In: The Whole Fungus Vol. I., (ed.), Kendrick, B., National Museums of Canada and Kananaskis Foundation, Ottawa, 283–359.

Kerrigan, R.W. (2003) New taxa: is one collection sufficient for publication? Inoculum 54(6): 5–6.

Khan, M.A. (1983). Conidial dimorphism in *Articulospora tetracladia*. Transactions of the British mycological Society 80: 173–175.

Kimbrough, J.W. and Curry, K.J. (1986) Septal structures in apothecial tissues of the tribe Aleurieae in the Pyronemataceae (Pezizales, Discomycetes). Mycologia 78: 407–417.

Kimbrough, J.W. and Gibson, J.L. (1990) Ultrastructural and cytological observations of apothecial tissues of *Geopyxis carbonaria* (Pezizales, Ascomycetes). Canadian Journal of Botany 68: 243–257.

Kirk, P.M., Cannon, P.F., David, J.C. and Stalpers, J.A. (2001) Dictionary of the Fungi. 9th Ed. CAB International, Wallingford.

Kirschner, R. (2004) Sporodochial anamorphs of species of *Helicogloea*. In: Frontiers of Basidiomycote Mycology, (eds.), Agerer, R., Piepenbring, M. and Blanz, P., IHW-Verlag, Eching, 165–178.

Kohlmeyer, J., Baral, H.-O. and Volkmann-Kohlmeyer, B. (1998) Fungi on *Juncus roemerianus*. 10. A. new *Orbilia* with ingoldian anamorph. Mycologia 90: 303–309.

Kohlmeyer, J., Schmidt, I. and Nair, N.B. (1967) Eine neue *Corollospora* (Ascomycetes) aus dem Indischen Ozean und der Ostsee. Berichte der Deutschen Botanischen Gesellschaft 80: 98–102.

Kohn, L.M. (1993). What do we need to know about discomycetous anamorphs? In: The Fungal Holomorph: Mitotic, Meiotic and Pleomorphic Speciation in Fungal Systematics, (eds.), Reynolds, D.R. and Taylor, J.W., CAB International, Wallingford, 129–139.

Liew, E.C.Y., Aptroot, A. and Hyde, K.D. (2000) Phylogenetic significance of the pseudoparaphyses in loculoascomycete taxonomy. Molecular Phylogenetics and Evolution 16: 392–402.

Liew, E.C.Y., Aptroot, A. and Hyde, K.D. (2002) An evaluation of the monophyly of *Massarina* based on ribosomal DNA sequences. Mycologia 94: 803–813.

Mantiri, F.R., Samuels, G.J., Rahe, J.E. and Honda, B.M. (2001) Phylogenetic relationships in *Neonectria* species having *Cylindrocarpon* anamorphs inferred from mitochondrial ribosomal DNA sequences. Canadian Journal of Botany 79: 334–340.

Marvanová, L.(1997) Freshwater hyphomycetes: a survey with remarks on tropical taxa. In: Tropical Mycology, (eds.), Janardhanan, K.K., Rajendran, C., Natarajan, K. and K. Hawksworth, D.L., Science Publishers, Inc., Enfield, U.S.A., 169–226.

Marvanová, L. (2002) Aquatic hyphomycetes—emerging outlines of their classification in the fungal system. 3rd International Meeting on "Plant Litter Processing in Freshwater". Sept. 7–12, 2002. Szentendre, Hungary. Abstract, p. 11.

Marvanová, L. and Bandoni, R.J. (1987) *Naiadella fluitans*, gen. et sp. nov.: a conidial basidiomycete. Mycologia 79: 578–586.

Marvanová, L. and Bärlocher, F. (1988) Hyphomycetes from Canadian streams. I. Basidiomycetous anamorphs. Mycotaxon 32: 339–351.

Marvanová, L. and Bärlocher, F. (1998) Hyphomycetes from Canadian streams III. *Arcispora bisagittaria* gen. et sp. nov. Mycologia 90: 531–536.

Marvanová, L. and Bärlocher, F. (2000) Hyphomycetes from Canadian streams. V. Two new conidial basidiomycetes. Mycotaxon 75: 409–423.

Marvanová, L. and Descals, E. (1996) Hyphomycetes from streams: new taxa and new combination. Mycotaxon 60: 455–469.

Marvanová, L. and Stalpers, J.A. (1987) The genus *Taeniospora* and its teleomorphs. Transactions of the British mycological Society 89: 489–498.

Marvanová, L. and Suberkropp, K. (1990) *Camptobasidium hydrophilum* and its anamorph, *Crucella subtilis*: a new heterobasidiomycete from streams. Mycologia 82: 208–217.

Marvanová, L., Landvik, S., Fisher, P.J., Moss, S.T. and Ainsworth, A.M. (2002) A new fungus with arthroconidia from foam. Nova Hedwigia 75: 255–269.

Marvanová, L., Pascoal, C. and Cássio, F. (2003) New and rare hyphomycetes from streams of Northwest Portugal. Part I. Cryptogamie, Mycologie 24: 339–358.

Matsushima, T. (1987) Matsushima Mycological Memoirs No. 5. Matsushima Fungus Collection, Kobe.

Matsushima, T. (1995) Matsushima Mycological Memoirs No. 8. Matsushima Fungus Collection, Kobe.

Matsushima, T. (2001) Matsushima Mycological Memoirs 10. Matsushima Fungus Collection, Kobe. CD–ROM.

Müller-Haeckel, A. (1977) Annual production of hyphomycete conidia in a subarctic stream. Oikos 29: 396–397.

Nakagiri, A. and Tubaki, K. (1982) A new marine ascomycete and its anamorph from Japan. Transactions of the Mycological Society of Japan 23: 101–110.

Nakagiri, A. and Tubaki, K. (1983) *Lindra obtusa*, a new marine ascomycete and its *Anguillospora* anamorph. Mycologia 75: 487–497.

Nawawi, A. and Webster, J. (1982) *Sistotrema hamatum* sp. nov., the teleomorph of *Ingoldiella hamata*. Transactions of the British mycological Society 78: 287–291.

Nawawi, A., Descals, E. and Webster, J. (1977) *Leptosporomyces galzinii*, the basidial state of a clamped branched conidium from fresh water. Transactions of the British mycological Society 68: 31–36

Nikolcheva, L.G. and Bärlocher, F. (2002) Phylogeny of *Tetracladium* based on 18S rDNA. Czech Mycology 53: 285–295.

O'Donnell, K. (1993) *Fusarium* and its near relatives. In: The Fungal Holomorph: Mitotic, Meiotic and Pleomorphic Speciation in Fungal Systematics, (eds.), Reynolds, D.R. and Taylor, J.W., CAB International, Wallingford, 225–233.

Parbery, D.G. (1996) Spermatial states of fungi are andromorphs. Mycological Research 100: 1400.

Peláez, F., Platas, G., Collado, J. and Díez, M.T. (1996) Infraspecific variation in two species of aquatic hyphomycetes assessed by RAPD analysis. Mycological Research 100: 831–837.

Pfister, D.H. (1997) Castor, Pollux and life histories of fungi. Mycologia 89: 1–23.

Pfister, D.H. and Kimbrough, J.W. (2001) Discomycetes. In: The Mycota VII, Systematics and Evolution, Part A, (eds.), McLaughlin, D.J., Mc Laughlin, E.G. and Lemke, P.A. Springer-Verlag, Berlin-Heidelberg, 258–281.

Ranzoni, F.V.(1956) The perfect stage of *Flagellospora penicillioides*. American Journal of Botany 43: 13–17.

Read, S.J., Moss, S.T. and Jones, E.B.G. (1992a) Germination and development of attachment structures by conidia of aquatic Hyphomycetes: light microscope studies. Canadian Journal of Botany 70: 831–837.

Read, S.J., Moss, S.T. and Jones, E.B.G. (1992b) Germination and development of attachment structures by conidia of aquatic Hyphomycetes: a scanning electron microscope study. Canadian Journal of Botany 70: 838–845.

Roldán, A., Descals, E. and Honrubia, M. (1989) Pure culture studies on *Tetracladium*. Mycological Research 93: 452–465.

Rossman, A.Y. (1993) Holomorphic hypocrealean fungi. In: The Fungal Holomorph: Mitotic, Meiotic and Pleomorphic Speciation in Fungal Systematics, (eds.), Reynolds, D.R. and Taylor, J.W., CAB International, Wallingford, 153–160.

Rossman, A.Y., Samuels, G.J., Rogerson, C.T. and Lowen, R. (1999) Genera of Bionectriaceae, Hypocreaceae and Nectriaceae (Hypocreales, Ascomycetes). Studies in Mycology 42, Centraalbureau voor Schimmelcultures, Baarn/Delft..

Saenz, G.S., Taylor, J.W. and Gargas, A. (1994) 18S rDNA gene sequences and supraordinal classification of the Erysiphales. Mycologia 86: 212–216.

Sampaio, J.P. and Bauer, R. (2003) The classification of dimorphic basidiomycetes. http://www.crem.fct.unl.pt/dimorphic_basidiomycetes/Papers/DMBclassification/classification.htm

Sati, S.C. and Tiwari, N. (1992) A new species of *Tricladium* from Kumaon Himalaya, India. Mycological Research 96: 229–232.

Scheuer, Ch. (1991) *Massarina tetraploa* sp. nov., the teleomorph of *Tetraploa aristata*. Mycological Research 95: 126–128.

Scholler, M., Hagedorn, G. and Rubner, A. (1999) A reevaluation of predatory orbiliaceous fungi. II. A new generic concept. Sydowia 51: 89–113.

Seifert, K. A. and Gams, W. (2001) The taxonomy of anamorphic fungi. In The Mycota VII, Systematics and Evolution, Part A, McLaughlin, D.J., McLaughlin, E.G. and Lemke, P.A. Eds., Springer-Verlag Berlin-Heidelberg, 307–347.

Shearer, C.A. (1992) The role of woody debris. In The Ecology of Aquatic Hyphomycetes, Bärlocher F., Ed., Springer-Verlag Berlin-Heidelberg, 77–98.

Shearer, C.A. (1993) The freshwater Ascomycetes. Nova Hedwigia 56: 1–33.

Sivanesan, A. (1984) The Bitunicate Ascomycetes and their Anamorphs. J. Cramer, Vaduz.

Sivichai, S. and Jones, E.B.G. (2003) Teleomorphic-anamorphic connections of freshwater fungi. In: Freshwater Mycology, (eds.), Tsui, C.K.M. and Hyde, K.D. Fungal Diversity Press, Hong Kong, 259–274.

Sivichai, S., Jones, E.B.G. and Hywel-Jones, N.L. (2003) Lignicolous freshwater Ascomycota from Thailand: *Hymenoscyphus varicosporoides* and its *Tricladium* anamorph. Mycologia 95: 340–346.

Spatafora, J.W. and Blackwell, M. (1994) Cladistic analysis of partial, ssrDNA sequences among unitunicate perithecial ascomycetes and its implications on the evolution of centrum development. In: Ascomycete Systematics. Problems and Perspectives in the Nineties, Haowksworth, D.L. (Ed), Plenum Press, New York and London, 233–242.

Sutton, B.C. and Hennebert, G.L. (1994) Interconnections amongst anamorphs and their possible contribution to ascomycete systematics. In: Ascomycete Systematics. Problems and Perspectives in the Nineties. (ed.), Hawksworth, D.L., Plenum Press, New York and London, 77–100.

Swann, E.C. and Taylor, J.W. (1995) Phylogenetic perspectives on basidiomycete systematics: evidence from the 18S rRNA gene. Canadian Journal of Botany 73: 649–1433.

Swann, E.C., Frieders, E.M. and McLaughlin, D.J. (1999) *Microbotryum, Kriegeria* and the changing paradigm in basidiomycete classification. Mycologia 91: 51–66.

Swann, E.C., Frieders, E.M. and McLaughlin, D.J. (2001) Urediniomycetes. . In: The Mycota VII, Systematics and Evolution, Part B, (eds.), McLaughlin, D.J., Mc Laughlin, E.G.and Lemke, P.A. Springer-Verlag Berlin-Heidelberg, 37–56.

Tanaka, K. and Harada, Y. (2003) Pleosporales in Japan. The genus *Massarina*. Mycoscience 44: 173–185.

Tehler, A., Little, D.P. and Farris, J.S. (2003) The full-length phylogenetic tree from 1551 ribosomal sequences of chitinous fungi, *Fungi*. Mycological Research 107: 901–916.

Trappe, J. (2004) The ways of Herbaria: a cautionary note for users of herbarium collections. Inoculum 55(3): 3–4.

Tubaki, K. (1957) Studies on the aquatic hyphomycetes. (III) Aquatic group. Bulletin of the National Science Museum, Tokyo 3: 249–268.

Tubaki, K. (1966) An undescribed species of *Hymenoscyphus*, a perfect stage of *Varicosporium*. Transactions of the British mycological Society 49: 345–349.

Voglmayr, H. (2004) *Spirosphaera cupreorufescens* sp. nov., a rare aeroaquatic fungus. Studies in Mycology 50: 221–228.

Webster, J. (1959) *Nectria lugdunensis* sp. nov., the perfect state of *Heliscus lugdunensis*. Transactions of the British mycological Society 42: 322–327.

Webster, J. (1961) The *Mollisia* perfect state of *Anguillospora crassa*. Transactions of the British mycological Society 44: 559–564.

Webster, J. (1965) The perfect state of *Pyricularia aquatica*. Transactions of the British mycological Society 48: 449–452.

Webster, J. (1987) Convergent evolution and the functional significance of spore shape in aquatic and semi-aquatic fungi. In: Evolutionary Biology of the Fungi. Symposium of the British Mycological Society, April 1986, Bristol, (eds.), Rayner, A.D.M., Brasier, C.M. and Moore, D., Cambridge University Press, Cambridge, 191–201.

Webster, J. (1993) *Nectria curta* sp. nov. (Ascomycetes, Hypocreales) an aquatic fungus and its *Flagellospora* anamorph. Nova Hedwigia 56: 455–464.

Webster, J. and Descals, E. (1979) The teleomorphs of waterborne Hyphomycetes from fresh water. In: The Whole Fungus. Vol. 2. (ed.), Kendrick B., National Museums of Canada and Kananaskis Foundation, Ottawa, 419–451.

Webster, J., Eicker, A. and Spooner, B.M. (1995) *Cudoniella indica* sp. nov. (Ascomycetes, Leotiales), the teleomorph of *Tricladium indicum*, an aquatic fungus isolated from a South African river. Nova Hedwigia 60: 493–498.

Webster, J., Scheuer, Ch. and Om-Kalthoom Khattab, S. (1991) *Hydrocina chaetocladia* gen. et sp. nov., the teleomorph of *Tricladium chaetocladium*. Nova Hedwigia 52: 65–72.

Webster, J., Shearer, C.A. and Spooner, B.M. (1993) *Mollisisa casaresiae* (Ascomycetes) the teleomorph of *Casaresia sphagnorum*, an aquatic fungus. Nova Hedwigia 57: 483–487.

Weiss, M., Bauer, R. and Begerow, D. (2004) Spotlights on heterobasidiomycetes. In: Frontiers of Basidiomycote Mycology, (eds.), Agerer, R., Piependring, M. and Blanz, P. IHW-Verlag Eching, 7–48.

Wells, K. and Bandoni, R.J. (2001) Heterobasidiomycetes. In: The Mycota VII, Systematics and Evolution, Part B, (eds.), McLaughlin D.J., Mc Laughlin E.G.and Lemke P.A. Springer-Verlag Berlin-Heidelberg, 85–120.

Willoughby, L.G. and Archer, J.F. (1973) The fungal spora of a freshwater stream and its colonization pattern on wood. Freshwater Biology 3: 219–239.

Willoughby, L.G. and Minshall, G.W. (1975) Further observations on *Tricladium giganteum*. Transactions of the British mycological Society 65: 77–82.

Zhang, Yan-Hui and Zhuang, Wen Ying. (2004) Phylogenetic relationships of some members in the genus *Hymenoscyphus* (Ascomycota, Helotiales). Nova Hedwigia 78: 475–484.

10. Recent Trends in Phylogeny and Classification of Lichen-Forming Ascomycetes

H. Thorsten Lumbsch
Department of Botany, The Field Museum of Natural History,
1400 S. Lake Shore Drive, Chicago, IL 60605, USA

SUMMARY This article presents recent progress in our understanding of the evolution of lichen-forming fungi. It illustrates the importance of molecular markers for modern systematic research in these fungi using examples from recently published studies. The focus is on phylogenetic studies aimed at clarifying relationships at three taxonomic levels. The first part discusses higher-level phylogeny of euascomycetes and the integration of lichen-forming fungi into the modern classification of Fungi, while the second part deals with the evaluation of morphological concepts of families and genera using molecular data sets. Phylogenetic studies investigating the circumscription and detection of reproductively isolated populations, i.e. species, is also described.

Introduction

Lichen-forming fungi are a large and diverse group of fungi that includes more than 13000 described species. These fungi occur in all ecosystems on all continents and are the dominating organisms in extreme environments, including polar or alpine vegetations. They are able to grow on different substrates, including bare soil, rocks, barks of trees, wood, shells of barnacles and leaf surfaces. The fungi that form lichens establish stable and self-supporting symbiotic relationships with one or more photosynthetic partners. The photosynthetic partner, also called photobiont, may belong to different groups of algae or cyanobacteria. Most commonly the photobiont belongs to the green algae, and among these the two genera *Trebouxia* and to lesser extent *Trentepohlia* are the most common symbiotic partners. A number of fungi utilize cyanobacteria as symbiotic partners, which are able to fix atmospheric nitrogen. The genus *Nostoc* is the most common cyanobacterium in lichen symbioses. The fungal partner of the symbiosis is referred to as mycobiont. The mycobiont is the exhabitant in lichen symbiosis (Hawksworth, 1988). The large majority of mycobionts belong to the Ascomycota, while some genera in the Basidiomycota also form lichen symbioses. The latter are not dealt with in this review. Interested readers are referred to a recent review on lichen-forming basidiomycetes by Oberwinkler (2001). The present review concentrates on recent developments in our understanding of the phylogeny of lichen-forming ascomycetes and consequences for their classification. I have given more thorough discussions of the morphological and chemical characters of the different groups of lichen-forming ascomycetes and initial molecular studies on the phylogeny of these fungi elsewhere, which are not reiterated here (Lumbsch, 1998, 2000; Ott and Lumbsch, 2001). An overview of the different classes including lichenized ascomycetes is given by Grube and Winka (2002). The variety of fungal-algal interactions is described by Hawksworth (1988) and a current classification of Ascomycota, including lichen-forming fungi can be found in Eriksson et al. (2003).

The introduction of the polymerase chain reaction and the subsequent ease to obtain DNA sequences provides a new perspective for phylogenetic research that has not been imaginable before. As in other groups of organisms, phylogenetic studies employing molecular data in lichen-forming ascomycetes have steadily increased during the last ten years. Sequence data from ubiquitous genes are now routinely used to study phylogenetic relationships among lichen-forming and related ascomycetes and have been used to address issues at different taxonomic levels. The sheer magnitude of new papers appearing each year makes it impossible to review the literature entirely. Instead I will try to describe and discuss current trends in the study of the evolution and classification of lichen-forming fungi at three major taxonomic levels using specific examples. In the first part studies at higher phylogenetic level will be discussed, which illustrate the integration of the classification of lichenized ascomycetes into the system of Fungi. In the second part research at the family and generic level and the evaluation of classical concepts using molecular data will be discussed, while in the third part recent trends in our understanding of recognition and circumscription of species will be illustrated.

This article concentrates on the systematics of lichen-forming fungi and utilization of molecular data to resolve phylogenetic problems. Methodology will not be discussed. Applications of PCR in studies of lichen-forming fungi were reviewed by Crespo et al. (1998), Cubero and Crespo (2002) and Arup (2002). Initial molecular studies on lichen-forming fungi mainly used sequence data of the nuclear SSU rDNA (e.g. Gargas and Taylor, 1995, Gargas et al. 1995). However, the development of new fungal specific primers for additional genes (e.g. Döring et al. 2000; Gardes and Bruns, 1993; Liu et al. 1999; Myllys et al. 2002; Printzen, 2002; Printzen and Ekman, 2003), has increased dramatically the array of possible genes that can be utilized for specific studies.

Molecular data are not only employed to study phylogeny and systematics of lichenized fungi, but have also been used to address a number of biological phenomena in lichens. These questions are beyond the scope of this review and only some examples are mentioned here to illustrate the power of molecular data to resolve biological issues in lichens. Most spectacular were attempts to address the question of the occurrence of lichen symbiosis in the phylogeny of fungi. While a study by Gargas et al. (1995) identified multiple origins of lichen symbioses in fungi, Lutzoni et al. (2001) suggested that some groups of fungi, including the economically important Eurotiomycetes and Chaetothyriomycetes, are derived from lichenized ancestors. Further, they showed that from the existing data, it appeared likely that the loss of lichenization was more common than its formation. Consequently, they concluded that lichen symbiosis occurred earlier in the evolution of Ascomycota than previously assumed.

Aspects of co-evolution, symbiotic specificity and phylogeny of the photosynthetic partners were studied in several papers (e.g. Beck et al. 2002; Kroken and Taylor, 2000; Lohtander et al. 2003; Piercey-Normore and DePriest, 2001; Rikkinen et al. 2002; Tibell, 2001a; Wirtz et al. 2003). The influence of the symbiotic lifestyle on evolutionary rates of its symbionts has been studied using omphalinoid fungi as a model. These fungi belong to the Basidiomycota. The evolutionary rates of lichenized omphalinoid fungi were compared to those of non-lichenized fungi (Lutzoni and Pagel, 1997) and to those of the inhabiting photosynthetic partner (Zoller and Lutzoni, 2003). The evolutionary rates in lichenized fungi were found to be significantly higher in both cases.

The strong influence of the interaction of the symbiotic partners on the morphogenesis of lichens has been demonstrated by the presence of photomorphs in Lobariaceae and Stictaceae (James and Henssen, 1976). In these lichens, a lichen fungus can form symbioses with a green algal partner or a cyanobacterium. In some cases, such as the genus *Sticta*, the two photomorphs are morphologically very distinct, forming a foliose lichen in the presence of a green algal photobiont and fruticose thallus in the presence of a cyanobacterial photobiont. In other genera, such as *Pseudocyphellaria*, these photomorphs may be morphologically similar, but are usually regarded as different species. The detailed morphological analysis by James and Henssen (1976), however, suggests that a single fungal species is involved in these photomorphs. The conclusions of these authors based on morphological observations is supported by molecular evidence (Armaleo and Clerc, 1991; Stenroos et al. 2003).

A fascinating and newly developing research area within lichen-forming fungi is the exploration of mating systems. Since cultured mycobionts do not form sexual stages and consequently no ascomata, mating systems are difficult to study. Zoller et al. (1999) and Kroken and Taylor (2001a) used indirect evidence in their sequence data to suggest outcrossing. In studies devoted to mating systems in lichenized fungi, Murtagh et al. (1999, 2000), Dyer et al. (2001a) and Honegger et al. (2004) investigated the breeding systems with RAPD-PCR DNA fingerprinting using axenically cultured single spore isolates. Both, self-fertilizing (homothallic) and heterothallic species were found. It will be interesting to see the studies on mating systems extended towards the analysis of the mating genes themselves in the near future.

Ribosomal DNA in Ascomycota, especially in lichen-forming species, is rich in introns that were identified as group I and spliceosomal introns (Cubero et al. 2000; DePriest and Been, 1992). Their evolution and spread has been the subject of numerous studies (e.g. Bhattacharya et al. 2000, 2001; Cubero et al. 2000; Ivanova et al. 1998).

Lichen-forming fungi are well known for the production of sometimes large number of secondary metabolites. While the biological roles of these substances has been widely discussed (e.g. Rikkinen, 1995) and their taxonomic utility is undisputed, the genes that are responsible for the biosynthesis of these substances have largely been neglected. Only recently studies on the phylogeny of polyketide synthase genes has given new insights into the phylogeny of secondary metabolism in lichens (Grube and Blaha, 2003; Miao et al. 2001).

Molecular data are certainly extremely useful in addressing a number of questions. However, it should be pointed out that sequence data from GenBank must be used with caution. Bridge et al. (2003) showed that up to 20% of sequences deposited in GenBank are erroneous. We hope that the steady stream of new sequences entering this database will simplify the detection of erroneous sequences.

Beside the immense increase in use of molecular data for phylogenetic studies in lichen-forming fungi, several studies have successfully employed morphological and chemical data to infer phylogenetic estimates (e.g. Dennetière and Péroni, 1998; Ferraro et al. 2001; Grube, 1998; Hyvönen et al. 1995; Kärnefelt et al. 1992; Louwhoff and Crisp, 2000; Lücking and Grube, 2002; Lumbsch and Tehler, 1998; Nordin and Mattsson, 2001; Scheidegger et al. 2001; Tehler, 1990, 1993, 1994; Tehler and Egea, 1997). Phylogenetic studies based on morphological and chemical data remain to be important as all of these characters potentially provide important phylogenetic information that should not be neglected (Lumbsch, 2004). Morphological and chemical characters may serve as a framework that can be evaluated with molecular data and analyzed in combined analyses.

Progress in Understanding Higher-level Phylogeny and the Integration of Lichen-Forming Ascomycetes in Fungi System

The higher-level classification of euascomycetes to which all lichenized ascomycetes belong has been unstable for a long time. This culminated in the avoidance of all supraordinal ranks in the classification by Eriksson and Hawksworth (1993). In the classical period of the 19th and early 20th century, groups were distinguished on the basis of ascoma morphology, resulting in a schematic distinction of classes, such as Discomycetes with apothecioid species or Pyrenomycetes with perithecioid taxa. This classification, however, was too coarse and did not allow proper placement of taxa with intermediate ascoma-types. Consequently, other characters were used for the circumscription of major clades of euascomycetes, such as the ascoma development (Nannfeldt, 1932) and ascus-type (Luttrell, 1955). However, these classifications also became unstable with the growing knowledge of the morphological diversity of these organisms. Phylogenetic studies employing nuclear SSU rDNA sequences provided support for a combination of ascoma-types, ascoma development, and ascus-type characters to circumscribe monophyletic groups, even though the individual characters were homoplasious (Berbee and Taylor, 1992, 1995; Gargas and Taylor, 1995; Spatafora,

1995). This resulted in a supraordinal classification with the description of several classes of closely related orders (Eriksson and Winka, 1997). Among these were the mainly lichenized Arthoniomycetes and Lecanoromycetes and the ascolocular fungi, classified into Chaetothyriomycetes and Dothideomycetes, which also include lichen-forming taxa. Tehler et al. (2000, 2003) conducted large-scale analyses of nuclear SSU rDNA data and showed that this data set alone is insufficient to resolve higher-level phylogeny of euascomycetes with confidence. Additional genes have since been added for the elucidation of the phylogeny of higher-level phylogeny of euascomycetes and used in single and combined analyses (Bhattacharya et al. 2000; Landvik et al. 2001; Lindemuth et al. 2001; Liu et al. 1999; Lumbsch et al. 2000, 2002; Lutzoni et al. 2001).

The analysis of the additional data sets basically confirmed the classification based on nuclear SSU rDNA. In concatenated analyses some poorly supported classes, such as Dothideomycetes and Lecanoromycetes, gained significant support, and some relationships among classes, such as the sister-group relationship of Chaetothyriomycetes and Eurotiomycetes, were strongly supported. However, most relationships between classes remained poorly supported. Some differences, however, were also found. Analyses of RPB2 sequences revealed loculoascomycetes as a monophyletic group (Liu et al. 1999, Liu and Hall, 2004), while ribosomal DNA consistently showed them as polyphyletic (Berbee, 1996; Lindemuth et al. 2001). Reeb et al. (2004) studied combined data sets of a) nuclear SSU and LSU rDNA and b) in addition RPB2 sequences. They suggested several changes to the classification of Eriksson and Winka (1997). In their study the Arthoniomycetes and Dothideomycetes were classified as subclasses of Sordariomycetes, although they only formed a monophyletic clade in the three gene analyses that included only one or two representatives of these two classes and did not form a clade in other analyses (e.g. Liu and Hall, 2004; Lumbsch et al. 2002). The Chaetothyriomycetes and Eurotiomycetes were found as sister-groups in both analyses and hence classified as subclasses within one class Eurotiomycetes, sensu lato, although they did not form a clade in the RPB2 analysis by Liu and Hall (2004).

The Lichinales are a group of mainly small crustose cyanophilous lichens that have either prototunicate or unitunicate asci (Henssen, 1979). Their centers of distribution are the semi-arid to arid regions of the subtropics. The relationships with other groups of ascomycetes are uncertain. Henssen and Jahns (1974) and Poelt (1973) placed them as a suborder of the Lecanorales, and later distinguished them as a separate order with uncertain relationship. Tehler (1996) regarded the Lichinales as a basal group of euascomycetes due to the prototunicate ascus of most Lichinales. In studies employing SSU rDNA sequences, Schultz et al. (2001) and Schultz and Büdel (2003) could demonstrate that the species currently placed in this order form a monophyletic group, but the relationship remained unresolved. In further studies employing additional molecular data sets, the isolated position of Lichinales was supported (Lutzoni et al. 2001; Miadlikowska and Lutzoni, 2004; Reeb et al. 2004). The Lichinales appeared as a basal group to the Lecanoromycetes, Chaetothyriomycetes and Eurotiomycetes. Consequently, Reeb et al. (2004) proposed to regard the order as a separate class Lichinomycetes. The relationships of this class, however, remain uncertain.

Reeb et al. (2004) studied the phylogenetic position and evolution of the polyspored Acarosporaceae. They restricted the family to six genera and found, using a likelihood-based method of character reconstruction, that true polyspory with more than 100 spores per ascus may have originated at least five times during the evolution of ascomycetes. The Acarosporaceae were found to be basal in Lecanoromycetes in the three-gene analysis, but without support. A new subclass Acarosporomycetidae was described in the Lecanoromycetes to accommodate the six genera belonging in Acarosporaceae along with a new subclass Ostropomycetidae to include families that were already regarded as being closely related by Kauff and Lutzoni (2002). The phylogenetic relationships of Lecanoromycetes was studied by Lumbsch et al. (2004). They found a strongly supported monophyletic Lecanoromycetes with two major clades. Clade 1 agrees largely with Ostropomycetidae sensu Reeb et al. (2004), with the exception that the two orders Agyriales (Lumbsch et al. 2001a) and Pertusariales did not cluster here, but in clade 2. Clade 2 which included Lecanoromycetidae sensu Reeb et al. (2004) and the two previously mentioned orders.

Within the large order Lecanorales, the cyanophilous taxa classified in the suborder Peltigerineae (or Peltigerales) has currently received some attention. Wiklund and Wedin (2003) and Miadlikowska and Lutzoni (2004) studied the phylogeny of this group with largely similar results, but different conclusions regarding the translation into a classification. The two studies agree in showing that Peltigerineae is monophyletic and includes the bulk of lecanoralean lichens with a cyanobacterium as primary photobiont that form two major clades within the group. Further, both studies suggested that Cladoniineae and Teloschistineae are nested within Lecanorineae. While Wiklund and Wedin (2003) did not suggest any nomenclatural changes, Miadlikowska and Lutzoni (2004) proposed numerous changes to the current classification, including the resurrection of the order Peltigerales and the description of two suborders within this order based on the two major clades found in the analysis. It should be noted that the latter classification leaves a number of taxa unclassified (e.g. *Porpidia* would be placed outside Lecanorales).

Evaluation of Family and Generic Concepts Using Molecular Data

In the classical period, the distinction of families in lichenized fungi has been very coarse and based on simple, easy to recognize characters or correlations of very few characters, such as growth form and spore characters (e.g. colour or septation). These traditional circumscriptions were long identified as producing unnatural groupings and consequently other characters, such as exciple- or ascus-types and combination of characters were used to circumscribe families. Especially after the seminal work of Hafellner (1984), the ascus-type was regarded as a primary character to circumscribe families, although some morphological studies questioned the use of this character by some groups (Lumbsch, 1997; Lutzoni and Brodo, 1995; Printzen, 1995). Classifications based solely on ascus-types were rejected in some molecular studies (e.g. Buschbom and Mueller, 2004; Ekman and Wedin, 2000; Lumbsch et al. 2001), suggesting that the ascus-type may be a poor indicator of phylogenetic relationships, if not correlated with independent characters. Several traditionally circumscribed families were shown to be poly- or paraphyletic (Andersen and Ekman, 2004; Ekman, 2001; Ekman and Jørgensen, 2002; Mattsson and Wedin, 1999). Further, growth form was demonstrated to be of no importance at the family and even generic level (e.g. Ekman, 2001; Schmitt et al. 2001; Stenroos and dePriest, 1998; Wedin et al. 1999).

Our understanding of the phylogeny of most families of lichenized ascomycetes is rather poor and suffers from a general lack of phylogenetic estimates based on morphological characters that may serve as a framework for molecular studies to test these hypotheses. The Physciaceae and Roccellaceae are two notable exceptions. The Physciaceae have been the subject of two morphologically based phylogenetic studies (Nordin and Mattsson, 2001; Scheidegger et al. 2001) and are well studied by a group of different groups of researchers. Grube and Arup (2001) studied the phylogeny of the Physciaceae using molecular data and evaluated characters used for distinction of major clades within this family and Lohtander et al. (2000) studied the phylogeny of Fennoscandian species of that family. Molecular data were also used to support the status of *Diplotomma* as an independent genus within the Physciaceae (Molina et al. 2002). The latter genus has often been regarded as a synonym of *Buellia* (e.g. Purvis et al. 1992). Surprisingly, lichenized fungi that were formerly classified in the Caliciaceae belong to the Physciaceae as well (Wedin et al. 2000a, 2002). The Caliciaceae belong to a group of calicioid fungi and lichens that were traditionally classified within a separate order Caliciales. The main character to cirumscribe this group was the passive spore dispersal, coupled with prototunicate asci and the production of a spore mass at the top of the ascoma that is called mazaedium. Molecular data suggest that this complex of characters correlated with the passive spore dispersal has evolved several times independently during the evolution of ascomycetes. The Physciaceae were found to include two major clades (Wedin et al. 2002), one including the Caliciaceae.

Helms et al. (2003) supported this finding, but proposed to accept the two major clades as different families.

The Roccellaceae have been studied intensely based on morphological characters thanks mainly to efforts by Anders Tehler and Grube (Grube, 1998; Tehler, 1990, 1994, 1995). Tehler's group also employed molecular data to evaluate the morphologically based phylogenies (Myllys et al. 1998, 1999; Tehler, 1995). The family was found to be monophyletic and relationships among some groups could be clarified. Incongruences with morphologically inferred phylogenies were discussed.

The phylogenetic position and evolution of several other families, including Gomphillaceae (Lücking et al. 2004), Lobariaceae (Thomas et al. 2002), Parmeliaceae (Mattsson and Wedin, 1998; Wedin et al. 1999), Pertusariaceae (Lumbsch and Schmitt, 2001; Schmitt and Lumbsch, 2004), and Umbilicariaceae (Ivanova et al. 1999), was studied using molecular data. Cladoniiform lichens were shown to be polyphyletic by Stenroos and dePriest (1998) and subsequently the phylogeny of genera showing similar growth forms was elucidated in several publications (Stenroos et al. 2002; Wedin and Döring, 1999; Wedin et al. 2000).

An interesting case is the genus *Lepraria* that is only known in sterile condition. The thallus in this genus is entirely composed of soredia giving it a leprose appearance. Although no sexual reproduction is known in this group, speciation seems to be as abundant as in any sexual group (Ekman and Tønsberg, 2002). A further group of lichens that are only known in sterile condition are the fruticose genera *Siphula* and *Thamnolia*. These were studied by Platt and Spatafora (2000) who demonstrated that they are closely related to the Icmadophilaceae.

The placement of small genera that have unique characters, which can be used to readily distinguish them, but make recognition of phylogenetic relationships difficult, has been the focus of some papers (Ivanova and Hafellner, 2002; LaGreca and Lumbsch, 2001; Poulsen et al. 2001; Schmitt et al. 2003a; Tibell 2001, 2003).

Due to the nature of binomial nomenclature, the rank of genus has a prominent position in the classification of organisms. Change of placement of a species at generic level involves the change of a name. This is especially troublesome, since ranks above species are arbitrary and the generic concept differs even within different groups of ascomycetes (Lumbsch, 2002). Several studies focused at the distinction of monophyletic groups at the generic level, especially in the Lecanoraceae, Parmeliaceae, and Teloschistaceae. The latter family is a good example for incongruence of morphological classifications and results of molecularly inferred phylogenies (Gaya et al. 2003; Kasalicky et al. 2000; Søchting and Lutzoni, 2004). The traditional generic concept appears highly artificial, since none of the accepted three major genera was monophyletic in the molecular analyses. Thallus morphology was employed as a major character and is shown to be of limited value. However, Søchting and Lutzoni (2003) could demonstrate some correlation with the occurrence of secondary metabolites, suggesting that chemistry is more suitable to identify clades in the Teloschistaceae than thallus morphology. The importance of secondary metabolite chemistry for the recognition of monophyletic groups is also supported in other groups, such as Pertusariaceae (Schmitt and Lumbsch, 2004) and Lecanoraceae. Martin Grube's group in Graz studied the latter family. They found that genera and subgenera based on thallus morphology were not monophyletic (Arup and Grube, 1998, 2000), while some clades characterized by their secondary metabolite chemistry were monophyletic (Arup and Grube, 1998; Grube et al. 2004).

The Parmeliaceae is one of the largest families of lichen-forming fungi. Two large groups that had been classified into the large genera *Cetraria* and *Parmelia* in traditional classifications dominate it. During the last three decades, the generic concepts in lichenized ascomycetes changed dramatically and the Parmeliaceae are a prominent example of this. Morphological and chemical characters were employed to segregate numerous groups at generic level within cetrarioid and parmelioid lichens (Elix, 1993; Kärnefelt, 1998). The acceptance for some of these segregates was not universal. Molecular data were used to evaluate numerous of these generic segregates in cetrarioid taxa (Kärnefelt and Thell, 2000, 2001; Saag et al. 2002;

Thell, 1998, 1999; Thell et al. 1998, 2000, 2002, 2002a) and parmelioid genera (Blanco et al. 2004; Crespo and Cubero, 1998; Crespo et al. 1999, 2001; Guzow-Krzeminska and Wegrzyn, 2003).

The phylogeny within genera was studied using molecular data and helped to clarify the relationships of species groups and character evolution in several genera (e.g. Cubero et al. 2004; Heidmarsson, 2000; Ihlen and Ekman, 2002; Kainz and Rambold, 2004; Lohtander et al. 2002 ; Martín et al. 2000; Miadlikowska and Lutzoni, 2000; Mohr et al. 2004; Ohmura, 2002; Schmitt et al. 2003; Stenroos et al. 2002).

Concept of Lichen Species in Light of Molecular Evidence

Studies on the circumscription of species in lichen-forming fungi suggest that we poorly understand how to recognize and circumscribe species in this group of organisms. Grube and Kroken (2000) reviewed the species concepts in lichenized fungi and suggested to apply the phylogenetic species concept based on a gene genealogical approach as outlined by Taylor et al. (2000) to lichenized fungi. The first study to consequently use this approach on lichen-forming fungi was that by Kroken and Taylor (2001), who found several cryptic species in the genus *Letharia*. Cryptic species were subsequently found in further unrelated genera, such as *Parmelia* (Crespo et al. 2002; Molina et al. 2004), or *Sphaerophorus* (Högnabba and Wedin, 2003), or the non-lichenized calicioid genus *Mycocalicium* (de los Angeles et al. 2001).

In some taxa, an unexpectedly high amount of sequence variation was found (Dyer and Murtagh, 2001; Martín et al. 2000; Murtagh et al. 2002), without separation of distinct clades. Other previously distinguished species were shown to be conspecific (Articus et al. 2002; Lohtander et al. 1998; Myllys et al. 1999a; Ott et al. 2004), especially in the so-called species pairs. The occurrence of vegetative propagules that ensure simultaneous dispersal of both symbiotic partners has been used as a character to distinguish otherwise similar or identical species. Two major types of these amphigenous diaspores are found in lichens: isidia that are corticated propagules on the thallus and soralia that are powdery structures including both symbiotic partners. Morphologically similar taxa that differ in their dispersal strategies are usually referred as species pairs in lichenology (Poelt, 1970), with a fertile primary species and a sterile secondary species. This concept has been extensively debated (Mattsson and Lumbsch, 1989; Tehler, 1982). Recently published studies suggest that the presence of soralia may not indicate isolated populations. In a study on the species pair *Usnea florida* (fertile) and *U. subfloridana* (sorediate), Articus et al. (2002) demonstrated that both morphotypes do not form monophyletic groups and thus the two previously distinguished species were regarded as conspecific. Similar results were found in studies on species pairs in Roccellaceae (Lohtander et al. 1998; Myllys et al. 1999a).

The taxonomic use of secondary metabolite chemistry at the species level has been much disputed in lichenology (e.g. Hawksworth, 1976; Lumbsch, 1998a). Tehler and Källersjö (2001) studied the case of the two well-known lichens *Parmeliopsis ambigua* and *P. hyperopta* that are mainly distinguished by their cortical chemistry. The two chemotypes formed distinct clades in their analyses, suggesting that two isolated species are involved.

In some notoriously difficult species complexes that include several morphotypes, molecular characters helped to evaluate the use of characters to distinguish species and thus to clarify the species concept in these groups (e.g. Goffinet et al. 2003; McCune et al. 2002; McDonald et al. 2003; Miadlikowska et al. 2003). The genetic variability of widely disjunct taxa was studied for bipolar species (Murtagh et al. 2002; Myllys et al. 2003), and disjuncts between western North America and Europe (Arnerup et al. 2004; Högberg et al. 2002; Printzen and Ekman, 2002; Printzen et al. 2003). On a different scale Zoller et al. (1999) and Walser et al. (2003) studied genetic variation and dispersal abilities of the threatened *Lobaria pulmonaria* in Switzerland.

Conclusions

This article gives a very brief overview of the current trends in our understanding of the evolution of lichen-forming fungi. Molecular characters have immensely extended the array of questions to be addressed by lichenologists and are expected to widen it even further. Increasingly biological questions not directly linked to phylogenies will be addressed using molecular data, but systematic questions will continue to play an important part of lichenological research. The growing body of knowledge both in numbers of sequences from species and the number of species sequenced will gradually improve our understanding of the evolution of these fascinating organisms. Sequences of the whole genome from model organisms in lichens will further accelerate the progress in research. However, numerous phylogenetic studies using molecular data clearly show that critical re-evaluation of morphological and chemical characters previously employed is necessary. Therefore a profound knowledge of both morphological and molecular techniques is absolutely necessary as a basis for young lichenologists entering this exciting field of research.

Acknowledgements

Imke Schmitt (Chicago) is thanked for helpful comments on the manuscript.

References

Andersen, H.L. and Ekman, S. (2004) Phylogeny of the Micareaceae inferred from nrSSU DNA sequences. Lichenologist 36: 27–35.

Armaleo, D. and Clerc, P. (1991) Lichen chimeras: DNA analysis suggests that one fungus forms two morphotypes. Experimental Mycology 15: 1–10.

Arnerup, J., Högberg, N. and Thor, G. (2004) Phylogenetic analysis of multiple loci reveal the population structure within Letharia in the Caucasus and Morocco. Mycological Research 108: 311–316.

Articus, K., Mattsson, J-E., Tibell, L., Grube, M. and Wedin, M. (2002) Ribosomal DNA and beta-tubulin data do not support the separation of the lichens *Usnea florida* and *U. subfloridana* as distinct species. Mycological Research 106: 412–418.

Arup, U. (2002) PCR techniques and automated sequencing in lichens. In: "Protocols in lichenology". (eds), Kranner, I., Beckett, R.P. and Varma, A.K. Springer, Berlin. 392–411.

Arup, U. and Grube, M. (1998) Molecular systematics of *Lecanora* subgenus *Placodium.* Lichenologist 30: 415–425.

Arup, U. and Grube, M. (2000) Is *Rhizoplaca* (Lecanorales, lichenized Ascomycota) a monophyletic genus? Canadian Journal of Botany 78: 318–327.

Beck, A., Kasalicky, T. and Rambold, G. (2002) Myco-photobiontal selection in a Mediterranean cryptogam community with *Fulgensia fulgida.* New Phytologist 153: 317–326.

Berbee, M.L. (1996) Loculoascomycete origins and evolution of filamentous ascomycete morphology based on 18S rRNA gene sequence data. Molecular Biology and Evolution 13: 462–470.

Berbee, M.L. and Taylor, J.W. (1992) Two Ascomycete classes based on fruiting-body characters and ribosomal DNA sequence. Molecular Biology and Evolution 9: 278–284.

Berbee, M.L. and Taylor, J.W. (1995) From 18S ribosomal sequence data to evolution of morphology among the fungi. Canadian Journal of Botany 73 (Suppl. 1): S677–S683.

Bhattacharya, D., Lutzoni, F., Reeb, V., Simon, D., Nason, J. and Fernandez, F (2000) Widespread occurrence of spliceosomal introns in the rDNA genes of ascomycetes. Molecular Biology and Evolution 17: 1971–1984.

Bhattacharya, D., Friedl, T. and Helms, G. (2001) Vertical evolution and intragenic spread of lichen-fungal group I introns. Journal of Molecular Evolution 55: 74–84.

Blanco, O., Crespo, A., Divakar, P.K., Esslinger, T.L., Hawksworth, D.L. and Lumbsch, H.T. (2004) *Melanelixia* and *Melanohalea*, two new genera segregated from *Melanelia* (Parmeliaceae) based on molecular and morphological data. Mycological Research, 108: 873–884.

Bridge, P.D., Roberts, P.J., Spooner, B.M. and Panchal, G. (2003) On the unreliability of published DNA sequences. New Phytologist 160: 43–48.

Buschbom, J. and Mueller, G. (2004) Resolving evolutionary relationships in the lichen-forming genus *Porpidia* and related allies (Porpidiaceae, Ascomycota). Molecular Phylogenetics and Evolution 32: 66–82.

Crespo, A. and Cubero, O.F. (1998) A molecular approach to the circumscription and evaluation of some genera segregated from *Parmelia* s. lat. Lichenologist 30: 369–380.

Crespo, A., Cubero, O.F. and Grube, M. (1998) PCR applications in studies of lichen-forming fungi. In: "Applications of PCR in Mycology". (eds), Bridge, P.D., Arora, D.K., Reddy, C.A. and Elander, R.P. CAB International, Wallingford, 85–105.

Crespo, A., Gavilán, R., Elix, J.A. and Gutiérrez, G. (1999) A comparison of morphological, chemical and molecular characters in some parmelioid genera. Lichenologist 31: 451–460.

Crespo, A., Blanco, O. and Hawksworth, D.L. (2001) The potential of mitochondrial DNA for establishing phylogeny and stabilising generic concepts in the parmelioid lichens. Taxon 50: 807–819.

Crespo, A., Molina, M.C., Blanco, O., Schroeter, B., Sancho, L.G., Hawksworth, D.L. (2002) rDNA ITS and beta-tubulin gene sequence analyses reveal two monophyletic groups within the cosmopolitan lichen *Parmelia saxatilis*. Mycological Research 106: 788–795.

Cubero, O. and Crespo, A. (2002) Isolation of nucleic acids from lichens. In: "Protocols in lichenology". (eds). Kranner, I., Beckett, R.P. and Varma, A.K. Springer, Berlin. 381–391.

Cubero, O.F., Bridge, P.D., Crespo, A. (2000) Terminal-sequence conservation identified spliceosomal introns in ascomycete 18S RNA genes. Molecular Biology and Evolution 17: 751–756.

Cubero, O.F., Crespo, A., Esslinger, T.L. and Lumbsch, H.T. (2004) Molecular phylogeny of the genus *Physconia* (Ascomycota, Lecanorales) inferred from a Bayesian analysis of nuclear ITS rDNA sequences. Mycological Research 108: 498–505.

Döring, H., Clerc, P., Grube, M. and Wedin, M. (2000) Mycobiont-specific PCR primers for the amplification of nuclear ITS and LSU rDNA from lichenized ascomycetes. Lichenologist 32: 200–204.

de Los Angeles Vinuesa, M., Sanches-Puelles, J.M. and Tibell, L. (2001) Intraspecific variation in *Mycocalicium subtile* (Mycocaliciaceae) elucidated by morphology and the sequences of the ITS1-5.8S-ITS2 region of rDNA. Mycological Research 105: 323–330.

Dennetière, B., Péroni, J. (1998) Approche phylogénétique des Gomphillaceae. Cryptogamie, Bryologie-Lichenolgie 19: 105–121.

dePriest, P.T. and Been, M.D. (1992) Numerous group I introns with variable distributions in the ribosomal DNA of a lichen fungus. Journal of Molecular Biology 228: 315–321.

Dyer, P.S. and Murtagh, G.J. (2001) Variation in the ribosomal ITS-sequence of the lichens *Buellia frigida* and *Xanthoria elegans* from the Vestfold Hills, eastern Antarctica. Lichenologist 33: 151–159.

Dyer, P.S., Murtagh, G.J. and Crittenden, P.D. (2001a) Use of RAPD-PCR DNA fingerprinting and vegetative incompatibility tests to investigate genetic variation within lichen-forming fungi. Symbiosis 31: 213–229.

Ekman, S. (2001) Molecular phylogeny of the Bacidiaceae (Lecanorales, lichenized Ascomycota). Mycological Research 105: 783–797.

Ekman, S. and Jørgensen, P.M. (2002) Towards a molecular phylogeny for the lichen family Pannariaceae (Lecanorales, Ascomycota). Canadian Journal of Botany 80: 625–634.

Ekman, S. and Tønsberg, T. (2002) Most species of *Lepraria* and *Leproloma* form a monophyletic group closely related to *Stereocaulon*. Mycological Research 106: 1262–1276.

Ekman, S. and Wedin, M. (2000) The phylogeny of the families Lecanoraceae and Bacidiaceae (lichenized Ascomycota) inferred from nuclear SSU rDNA sequences. Plant Biology 2: 350–360.

Elix, J.A. (1993) Progress in the generic delimitation of *Parmelia* sensu lato lichens (Ascomycotina: Parmeliaceae) and a synoptic key to the Parmeliaceae. Bryologist 96: 359–383.

Eriksson, O.E. and Hawksworth, D.L. (1993) Outline of the ascomycetes–1993. Systema Ascomycetum 12: 51–257.

Eriksson, O.E. and Winka, K. (1997) Supraordinal taxa of Ascomycota. Myconet 1: 1–16.

Eriksson, O.E., Baral, H.-O., Currah, R.S., Hansen, K., Kurtzman, C.P., Rambold, G. and Laessoe, T. (2003) Outline of Ascomycota - 2003. Myconet 9: 1–89.

Ferraro, L.I., Lücking, R. and Sérusiaux, E. (2001) A world monograph of the lichen genus *Gyalectidium* (Gomphillaceae). Botanical Journal of the Linnean Society 137: 311–345.

Gardes, M. and Bruns, T.D. (1993) ITS primers with enhanced specificity for basidiomycetes. Application to the identification of mycorrhizae and rust. Molecular Ecology 2: 113–118.

Gargas, A. and Taylor, J.W. (1995) Phylogeny of Discomycetes and early radiations of the apothecial Ascomycotina inferred from SSU rDNA sequence data. Experimental Mycology 19: 7–15.

Gargas, A., DePriest, P.T., Grube, M. and Tehler, A. (1995) Multiple origins of lichen symbioses in fungi suggested by SSU rDNA phylogeny. Science 268: 1492–1495.

Gaya, E., Lutzoni, F., Zoller, S. and Navarro-Rosines, P. (2003) Phylogenetic study of *Fulgensia* and allied *Caloplaca* and *Xanthoria* species (Teloschistaceae, lichen-forming Ascomycota). American Journal of Botany 90: 1095–1103.

Goffinet, B., Miadlikowska, J. and Goward, T. (2003) Phylogenetic inferences based on nrDNA sequences support five morphospecies within the *Peltigera didactyla* complex (lichenized Ascomycota). Bryologist 106: 349–364.

Grube, M. (1998) Classification and phylogeny in the Arthoniales (lichenized Ascomycetes). Bryologist 101: 377–391.

Grube, M. and Arup, U. (2001) Molecular and morphological evolution in the Physciaceae (Lecanorales, lichenized Ascomycotina), with special emphasis on the genus *Rinodina*. Lichenologist 33: 63–72.

Grube, M. and Blaha, J. (2003) On the phylogeny of some polyketide synthase genes in the lichenized genus *Lecanora*. Mycological Research 107: 1419–1426.

Grube, M. and Kroken, S. (2000) Molecular approaches and the concept of species and species complexes in lichenized fungi. Mycological Research 104: 1284–1294.

Grube, M. and Winka, K. (2002) Progress in understanding the evolution and classification of lichenized ascomycetes. *Mycologist* 16: 67–76.

Grube, M., Baloch, E. and Arup, U. (2004) A phylogenetic study of the *Lecanora rupicola* group (Lecanoraceae, Ascomycota). Mycological Research 108: 506–514.

Guzow-Krzeminska, B. and Wegrzyn, G. (2003) A preliminary study on the phylogeny of the genus *Melanelia* using nuclear large subunit ribosomal DNA sequences. Lichenologist 35: 83–86.

Hafellner, J. (1984) Studien in Richtung einer natürlicheren Gliederung der Sammelfamilien Lecanoraceae und Lecideaceae. In: Beiträge zur Lichenologie. Festschrift J. Poelt. Beiheft zur Nova Hedwigia 79., (eds), Hertel, H. and Oberwinkler, F. J. Cramer, Vaduz, 241–371.

Hawksworth, D.L. (1976) Lichen chemotaxonomy. In: Lichenology: Progress and Problems. (eds), Brown, D.H., Hawksworth, D.L. and Bailey, R.H. Academic Press, London, 139–184.

Hawksworth, D.L. (1988) The variety of fungal-algal symbioses, their evolutionary significance, and the nature of lichens. Botanical Journal of the Linnean Society 96: 3–20.

Heidmarsson, S. (2000) Taxonomy and phylogeny of *Dermatocarpon* (Verrucariales, lichenized Ascomycotina) with Special emphasis on the Nordic Species. Comprehensive Summaries of Uppsala Dissertations from the Faculty of Science and Technology, 555, Uppsala University, Uppsala, Sweden. 25 pp.

Helms, G., Friedl, T. and Rambold, G. (2003) Phylogenetic relationships of the Physciaceae inferred from rDNA sequence data and selected phenotypic characters. Mycologia 95: 1078–1099.

Henssen, A. (1979) Problematik der Gattungsabgrenzung bei den Lichinaceen. Berichte der Deutschen Botanischen Gesellschaft 92: 483–506.

Henssen, A. and Jahns, H.M. (1974) Lichens. Thieme, Stuttgart.

Högberg, N., Kroken, S., Thor, G. and Taylor, J.W. (2002) Reproductive mode and genetic variation suggest a North American origin of European *Letharia vulpina*. Molecular Ecology 11: 1191–1196.

Hognabba, F. and Wedin, M. (2003) Molecular phylogeny of the *Sphaerophorus globosus* species complex. Cladistics 19: 224–232.

Honegger, R., Zippler, U., Gansner, H. and Scherrer, S. (2004) Mating systems in the genus *Xanthoria* (lichen-forming ascomycetes). Mycological Research 108: 480–488.

Hyvönen, J., Ahti, T., Stenroos, S. and Gowan, S.P. (1995) Genus *Cladina* and the section *Unciales* of the genus *Cladonia* (Cladoniaceae, lichenized Ascomycotina), a preliminary phylogenetic analysis. Journal of The Hattori Botanical Laboratory 78: 243–253.

Ihlen, P.G. and Ekman, S. (2002) Outline of phylogeny and character evolution in *Rhizocarpon* (Rhizocarpaceae, lichenized Ascomycota) based on nuclear ITS and mitochondrial SSU ribosomal DNA sequences. Botanical Journal of the Linnean Society 77: 535–546.

Ivanova, N.V., DePriest, P.T., Bobrova, V.K. and Troitsky, A.V. (1998) Group I introns in the fungal 18S rDNA of the lichen family Umbilicariaceae. Doklady Akademii Nauk SSSR 363: 400–404.

Ivanova, N.V., DePriest, P.T., Bobrova, V.K. and Troitsky, A.V. (1999) Phylogenetic analysis of the lichen family Umbilicariaceae based on nuclear ITS1 and ITS2 rDNA sequences. Lichenologist 31: 477–489.

Ivanova, N.V. and Hafellner, J. (2002) Searching for the correct placement of *Megaspora* by use of ITS1, 5.8S and IS2 rDNA sequence data. In: Progress and Problems in Lichenology at the Turn of the Millenium. Bibliotheca Lichenologica, (eds), Llimona, X., Lumbsch, H.T. and Ott, S. J. Cramer, Berlin, Stuttgart, 113–122.

James, P.W. and Henssen, A. (1976) The morphological and taxonomic significance of cephalodia. In: Lichenology: Progress and problems. (eds), Brown, D.H., Hawksworth, D.L. and Bailey, R.H., Academic Press, London. 27–77.

Kainz, C. and Rambold, G. (2004) A phylogenetic study of the lichen genus *Protoblastenia* (Lecanorales, Psoraceae) in Central Europe. In: Contributions to Lichenology: Festschrift in honour of Hannes Hertel. Bibliotheca Lichenologica, (eds), Döbbeler, P. and Rambold, G. J. Cramer, Berlin, 267–299.

Kärnefelt, I. (1998) Teloschistales and Parmeliaceae—a review of the present problems and challenges in lichen systematics at different taxonomic levels. Cryptogamie, Bryologie-Lichenolgie 19: 93–104.

Kärnefelt, E.I. and Thell, A. (2000) On the systematic position of the genus *Cetrariella* (Parmeliaceae: Ascomycotina) indicated by ITS rDNA data. In: New Aspects in Cryptogamic Research. Contributions in Honour of Ludger Kappen. Bibliotheca Lichenologica, (eds), Schroeter, B., Schlensog, M. and Green, T.G.A., J. Cramer, Berlin, Stuttgart, 27–32.

Kärnefelt, E.I. and Thell, A. (2001) Delimitation of the lichen genus *Tuckermannopsis* Gyeln. (Ascomycotina Parmeliaceae) based on morphology and DNA sequences. In: Lichenological Contributions in Honour of Jack Elix. Bibliotheca Lichenologica, (eds), McCarthy, P.M., Kantvilas, G. and Louwhoff, S.H.J.J. J. Cramer, Berlin, Stuttgart, 193–209.

Kärnefelt, I., Mattsson, J-E. and Thell, A. (1992) Evolution and phylogeny of cetrarioid lichens. Plant Systematics and Evolution 183: 113–160.

Kasalicky, T., Döring, H., Rambold, G. and Wedin, M. (2000) A comparison of ITS and LSU nrDNA phylogenies of *Fulgensia* (Teloschistaceae, Lecanorales), a genus of lichenised ascomycetes. Canadian Journal of Botany 78: 1580–1589.

Kauff, F. and Lutzoni, F. (2002) Phylogeny of the Gyalectales and Ostropales (Ascomycota, Fungi): among and within other relationships. Molecular Phylogenetics and Evolution 25: 138–156.

Kroken, S. and Taylor, J.W. (2000 [2001]) Phylogenetic species, reproductive mode, and specificity of the green alga *Trebouxia* forming lichens with the fungal genus *Letharia*. Bryologist 103: 645–660.

Kroken, S. and Taylor, J.W. (2001) A gene genealogical approach to recognize phylogenetic species boundaries in the lichenized fungus *Letharia*. Mycologia 93: 38–53.

Kroken, S. and Taylor, J.W. (2001a) Outcrossing and recombination in the lichenized fungus *Letharia*. Fungal Genetics and Biology 34: 83–92.

LaGreca, S. and Lumbsch, H.T. (2001) No evidence from rDNA ITS sequence data for a placement of *Ramalinora* in the Ramalinaceae. Lichenologist 33: 172–176.

Landvik, S., Eriksson, O.E. and Berbee, M.L. (2001) *Neolecta* – a fungal dinosaur? Evidence from beta-tubulin amino acid sequences. Mycologia 93: 1151–1163.

Lindemuth, R., Wirtz, N. and Lumbsch, H.T. (2001) Phylogenetic analysis of nuclear and mitochondrial sequence data supports that loculoascomycetes (Ascomycota) are not monophyletic. Mycological Research 105: 1176–1181.

Liu, Y.J. and Hall, B.D. (2004) Body plan evolution of ascomycetes, as inferred from an RNA polymerase II phylogeny. Proceedings of the National Academy of Sciences USA 101: 4507–4512.

Liu, Y.J., Whelen, S. and Hall, B.D. (1999) Phylogenetic relationships among ascomycetes: evidence from an RNA polymerase II subunit. Molecular Biology and Evolution 16: 1799–1808.

Lohtander, K., Källersjö, M. and Tehler, A. (1998) Dispersal strategies in *Roccellina capensis* (Arthoniales). Lichenologist 30: 341–350.

Lohtander, K., Källersjö, M., Moberg, R. and Tehler, A. (2000) The family Physciaceae in Fennoscandia: phylogeny inferred from ITS sequences. Mycologia 92: 728–735.

Lohtander, K., Oksanen, I. and Rikkinen, J. (2002) A phylogenetic study of *Nephroma* (lichen-forming Ascomycota). Mycological Research 106: 777–787.

Lohtander, K., Oksanen, I. and Rikkinen, J. (2003) Genetic diversity of green algal and cyanobacterial photobionts in *Nephroma* (Peltigerales). Lichenologist 35: 325–339.

Louwhoff, S.H.J.J. and Crisp, M.D. (2000) Phylogenetic analysis of *Parmotrema* (Parmeliaceae: lichenized Ascomycotina). Bryologist 103: 541–554.
Lücking, R. and Grube, M. (2002) Facultative parasitism and reproductive strategies in *Chroodiscus* (Ascomycota, Ostropales). Stapfia 80: 267–292.
Lücking, R., Stuart, B.L. and Lumbsch, H.T. (2004) Phylogenetic relationships of Gomphillaceae and Asterothyriaceae: evidence from a combined Bayesian analysis of nuclear and mitochondrial sequences. Mycologia 96: 283–294.
Lumbsch, H. T. (1997) A comparison of the ascoma ontogeny supports inclusion of the Eigleraceae in the Hymeneliaceae (Lecanorales). Bryologist 100: 180–192.
Lumbsch, H.T. (1998) Taxonomic use of metabolic data in lichen-forming fungi. In: Chemical Fungal Taxonomy. (eds), Frisvad, J.C., Bridge, P.D. and Arora, D.K., Marcel Dekker Inc., New York, 345–387.
Lumbsch, H.T. (1998a) The use of metabolic data in lichenology at the species and subspecific levels. Lichenologist 30: 357–367.
Lumbsch, H.T. (2000) Phylogeny of filamentous ascomycetes. Naturwissenschaften 87: 335–342.
Lumbsch, H.T. (2002) How objective are genera in filamentous ascomycetes? Perspectives in Plant Ecology, Evolution and Systematics 5: 91–101.
Lumbsch, H. T. (2004) Comparison of homoplasy supports that characters have no apriori value in phylogenetic analyses of lichenized ascomycetes. Bibliotheca Lichenologia 88: 427–431.
Lumbsch, H.T. and Schmitt, I. (2001) Molecular data suggest that the lichen genus *Pertusaria* is not monophyletic. Lichenologist 33: 161–170.
Lumbsch, H.T. and Tehler, A. (1998) A cladistic analysis of the genus *Diploschistes* (Ascomycotina, Thelotremataceae). Bryologist 101: 398–403.
Lumbsch, H.T., Lindemuth, R. and Schmitt, I. (2000) Evolution of filamentous ascomycetes inferred from LSU rDNA sequence data. Plant Biology 2: 525–529.
Lumbsch, H.T., Schmitt, I., Döring, H. and Wedin, M. (2001) ITS sequence data suggest variability of ascus types and support ontogenetic characters as phylogenetic discriminators in the Agyriales (Ascomycota). Mycological Research 105: 265–274.
Lumbsch, H.T., Schmitt, I., Döring, H. and Wedin, M. (2001a) Molecular systematics supports the recognition of an additional order of Ascomycota: the Agyriales. Mycological Research 105: 16–23.
Lumbsch, H.T., Wirtz, N., Lindemuth, R. and Schmitt, I. (2002) Higher level phylogenetic relationships of euascomycetes (Pezizomycotina) inferred from a combined analysis of nuclear and mitochondrial sequence data. Mycological Progress 1: 57–70.
Lumbsch, H.T., Schmitt, I., Palice, Z., Wiklund, E., Ekman, S. and Wedin, M. (2004) Supraordinal phylogenetic relationships of Lecanoromycetes based on a Bayesian analysis of combined nuclear and mitochondrial sequences. Molecular Phylogenetics and Evolution 31: 822–832.
Luttrell, E.S. (1955) The ascostromatic Ascomycetes. Mycologia 47: 511–532.
Lutzoni, F.M. and Brodo, I.M. (1995) A generic redelimitation of the *Ionaspis-Hymenelia* complex (lichenized Ascomycotina). Systematic Botany 20: 224–258.
Lutzoni, F., Pagel, M. and Reeb, V. (2001) Major fungal lineages are derived from lichen symbiotic ancestors. Nature 411: 937–940.
Lutzoni, F.M. and Pagel, M. (1997) Accelerated evolution as a consequence of transitions to mutualism. Proceedings of the National Academy of Sciences USA 94: 11422–11427.
Martín, M.P., Winka, K., Llimona, X. and Lumbsch, H.T. (2000) Evaluation of morphological variation in the lichen *Diploschistes ocellatus* (Ascomycota, Ostropales): evidence from nuclear rDNA ITS sequence data. Plant Biology. 2: 571–577.
Mattsson, J.-E. and Lumbsch, H.T. (1989) The use of the species pair concept in lichen taxonomy. Taxon 38: 238–241 (1989).
Mattsson, J.-E. and Wedin, M. (1998) Phylogeny of the Parmeliaceae: DNA data versus morphological data. Lichenologist 30: 463–472.
Mattsson, J.-E. and Wedin, M. (1999) A re-assessment of the family Alectoriaceae. Lichenologist 31: 431–440.
McCune, B., Camacho, F. and Ponzetti, J. (2002) Three new species of *Trapeliopsis* on soil in western North America. Bryologist 105: 78–85.

McDonald, T., Miadlikowska, J. and Lutzoni, F. (2003) The lichen genus *Sticta* in the Great Smoky Mountains: a phylogenetic study of morphologial, chemical, and molecular data. Bryologist 106: 61–79.

Miadlikowska, J. and Lutzoni, F. (2000) Phylogenetic revision of the genus *Peltigera* (lichen-forming Ascomycota) based on morphological, chemical, and large subunit nuclear ribosomal DNA data. International Journal of Plant Sciences 161: 925–958.

Miadlikowska, J. and Lutzoni, F. (2004) Phylogenetic classification of peltigeralean fungi (Peltigerales, Ascomycota) based on ribosomal RNA small and large subunits. American Journal of Botany 91: 449–464.

Miadlikowska, J., Lutzoni, F., Goward, T., Zoller, S. and Posada, D. (2003) New approach to an old problem: incorporating signal from gap-rich regions of ITS and rDNA large subunit into phylogenetic analyses to resolve the *Peltigera canina* species complex. Mycologia 95: 1181–1203.

Miao, V., Coëffet-LeGal, M.-F., Brown, D., Sinnemann, S., Donaldson, G. and Davies, J. (2001) Genetic approaches to harvesting lichen products. Trends in Biotechnology 19: 349–355.

Mohr, F., Ekman, S. and Heegaard, E. (2004) Evolution and taxonomy of the marine *Collemopsidium* species (lichenized Ascomycota) in north-west Europe. Mycological Research 108: 515–532.

Molina, M.C., Crespo, A., Blanco, O., Hladun, N. and Hawksworth, D.L. (2002) Molecular phylogeny and status of *Diploicia* and *Diplotomma*, with observations on *Diploicia subcanescens* and *Diplotomma rivas-martinezii*. Lichenologist 34: 509–519.

Molina, M.C., Crespo, A., Blanco, O., Lumbsch, H.T. and Hawksworth, D.L. (2004) Phylogenetic relationships and species concepts in *Parmelia* s.str. (Parmeliaceae) inferred from nuclear ITS rDNA and beta-tubulin sequences. Lichenologist 36: 37–54.

Murtagh, G.J., Dyer, P.S., McClure, P.C. and Crittenden, P.D. (1999) Use of randomly amplified polymorphic DNA markers as a tool to study variation in lichen-forming fungi. Lichenologist 31: 257–267.

Murtagh, G.J., Dyer, P.S. and Crittenden, P.D. (2000) Sex and the single lichen. Nature 404: 564.

Murtagh, G.J., Dyer, P.S., Furneaux, P.A., Crittenden, P.D. (2002) Molecular and physiological diversity in the bipolar lichen-forming fungus *Xanthoria elegans*. Mycological Research 106: 1277–1286.

Myllys, L., Källersjö, M. and Tehler, A. (1998) A comparison of SSU rDNA data and morphological data in Arthoniales (Euascomycetes) phylogeny. Bryologist 101: 70–85.

Myllys, L., Lohtander, K., Källersjö, M. and Tehler, A. (1999) Applicability of ITS data in Roccellaceae (Arthoniales, Euascomycetes) phylogeny. Lichenologist 31: 461–476.

Myllys, L., Lohtander, K., Källersjö, M. and Tehler, A. (1999a) Sequence insertions and ITS data provide congruent information on *Roccella canariensis* and *R. tuberculata* (Arthoniales, Euascomycetes) phylogeny. Molecular Phylogenetics and Evolution 12: 295–309.

Myllys, L., Stenroos, S. and Thell, A. (2002) New genes for phylogenetic studies of lichenized fungi: glyceraldehyde-3-phosphate dehydrogenase and beta-tubulin genes. Lichenologist 34: 237–246.

Myllys, L., Stenroos, S., Thell, A. and Ahti, T. (2003) Phylogeny of bipolar *Cladonia arbuscula* and *Cladonia mitis* (Lecanorales, Euascomycetes). Molecular Phylogenetics and Evolution 27: 58–69.

Nannfeldt, J.A. (1932) Studien über die Morphologie und Systematik der nicht-lichenisierten inoperculaten Discomyceten. Nova Acta Regiae Societatis Scientiarum Upsaliensis, ser. IV 8 (2): 1–368.

Nordin, A. and Mattsson, J.E. (2001) Phylogenetic reconstruction of character development in Physciaceae. Lichenologist 33: 3–23.

Oberwinkler, F. (2001) Basidiolichens. In: The Mycota. Vol. IX. Fungal Associations. (ed.), Hock, B Springer-Verlag, Berlin, Heidelberg, New York, 211–225.

Ohmura, Y. (2002) Phylogenetic evaluation of infrageneric groups of the genus *Usnea* based on ITS regions in rDNA. Journal of The Hattori Botanical Laboratory 92: 231–243.

Ott, S. and Lumbsch, H.T. (2001) Morphology and phylogeny of ascomycete lichens. In: The Mycota. IX. Fungal Associations. (eds.), Hock, B. and Esser, K., Springer-Verlag, Berlin, Heidelberg, pp. 189–210.

Ott, S., Brinkmann, M., Wirtz, N. and Lumbsch, H.T. (2004) Mitochondrial and nuclear ribosomal DNA data do not support the separation of the Antarctic lichens *Umbilicaria kappenii* and *Umbilicaria antarctica* as distinct species. Lichenologist 36: 227–234.

Piercey-Normore, M.D. and DePriest, P.T. (2001) Algal switching among lichen symbionts. American Journal of Botany 88: 1490–1498.

Platt, J.L. and Spatafora, J.W. (2000) Evolutionary relationships of nonsexual lichenized fungi: molecular phylogenetic hypotheses for the genera *Siphula* and *Thamnolia* from SSU and LSU rDNA. Mycologia 92: 475–487.

Poelt, J. (1970) Das Konzept der Artenpaare bei den Flechten. Deutsche Botanische Gesellschaft, neue Folge 4: 187–198.

Poelt, J. (1973) Classification. In: The Lichens, (eds), Ahmadjian, V. and Hale M.E., Academic, New York. 599–632.

Poulsen, R.S., Schmitt, I., Søchting, U. and Lumbsch, H.T. (2001) Molecular and morphological studies on the subantarctic genus *Orceolina* (Agyriaceae). Lichenologist 33: 323–329.

Printzen, C. (1995) Die Flechtengattung *Biatora* in Europa. Bibliotheca Lichenologica, 60, J. Cramer, Berlin, Stuttgart. pp.275.

Printzen, C. (2002) Fungal specific primers for PCR-amplification of mitochondrial LSU in lichens. Molecular Ecology Notes 2: 130–132.

Printzen, C. and Ekman, S. (2002) Genetic variability and its geographical distribution in the widely disjunct *Cavernularia hultenii*. Lichenologist 34: 101–111.

Printzen, C. and Ekman, S. (2003) Local population subdivision in the lichen *Cladonia subcervicornis* as revealed by mitochondrial cytochrome oxidase subunit 1 intron sequences. Mycologia 95: 399–406.

Printzen, C., Ekman, S. and Tønsberg. T. (2003) Phylogeography of *Cavernularia hultenii*: evidence of slow genetic drift in a widely disjunct lichen. Molecular Ecology 12: 1473–1486.

Purvis, O.W., Coppins, B.J., Hawksworth, D.L., James, P.W. and Moore, D.M. (eds.) (1992) The Lichen Flora of Great Britain and Ireland. Natural History Museum Publications and British Lichen Society, London. 710 pp.

Reeb, V., Lutzoni, F. and Roux, C. (2004) Contribution of RPB2 to multilocus phylogenetic studies of the euascomycetes (Pezizomycotina, Fungi) with special on the Acarosporaceae and the evolution of polyspory. Molecular Phylogenetics and Evolution 32: 1036–1060.

Rikkinen, J. (1995) What's behind the pretty colours? A study on the photobiology of lichens. Bryobrothera 4: 1–239.

Rikkinen, J., Oksanen, I. and Lohtander, K. (2002) Lichen guilds share related cyanobacterial symbionts. Science 297: 357.

Saag, A., Randlane, T., Thell, A. and Obermayer, W. (2002) Phylogenetic analysis of cetrarioid lichens with globose ascospores. Proceedings of the Estonian Academy of Sciences, Biology Ecology. 51: 103–123.

Scheidegger, C., Mayrhofer, H., Moberg, R. and Tehler, A. (2001) Evolutionary trends in the Physciaceae. Lichenologist 33: 25–45.

Schmitt, I. and Lumbsch, H.T. (2004) Molecular phylogeny of the Pertusariaceae supports secondary chemistry as an important systematic character set in lichen-forming ascomycetes. Molecular Phylogenetics and Evolution 33: 43–55.

Schmitt, I., Messuti, M.I., Feige, G.B. and Lumbsch, H.T. (2001) Molecular data support rejection of the generic concept in the Coccotremataceae (Ascomycota). Lichenologist 33: 315–321.

Schmitt, I., Lumbsch, H.T. and Søchting, U. (2003) Phylogeny of the lichen genus *Placopsis* and its allies based on Bayesian analyses of nuclear and mitochondrial sequences. Mycologia 95: 827–835.

Schmitt, I., Martín, M.P., Türk, R. and Lumbsch, H.T. (2003a) Phylogenetic position of the genera *Varicellaria, Melanaria* and *Thamnochrolechia* (Pertusariales). In: Lichenological Contributions in Honour of G.B. Feige. Bibliotheca Lichenologica, (ed), Jensen, M. J., Cramer, Berlin, Stuttgart, 147–154.

Schultz, M. and Büdel, B. (2003) On the systematic position of the lichen genus *Heppia*. Lichenologist 35: 151–156.

Schultz, M., Arendholz, W.-R. and Büdel, B. (2001) Origin and evolution of the lichenized Ascomycete order Lichinales: monophyly and systematic relationships inferred from ascus, fruiting body and SSU rDNA evolution. Plant Biology 3: 116–123.

Søchting, U. and Lutzoni, F. (2004) Molecular phylogenetic study at the generic boundary between the lichen-forming fungi *Caloplaca* and *Xanthoria* (Ascomycota, Teloschistaceae). Mycological Research 107: 1266–1276.

Spatafora, J.W. (1995) Ascomal evolution of filamentous ascomycetes: evidence from molecular data. Canadian Journal of Botany. 73 (Suppl. 1): S811–S815.

Stenroos, S.K. and DePriest, P.T. (1998) SSU rDNA phylogeny of cladoniiform lichens. American Journal of Botany 85: 1548–1559.

Stenroos, S., Hyvönen, J., Myllys, L., Thell, A. and Ahti, T. (2002) Phylogeny of the genus *Cladonia* s.lat. (Cladoniaceae, Ascomycetes) inferred from molecular, morphological, and chemical data. Cladistics 18: 237–278.

Stenroos, S., Stocker-Wörgötter, E., Yoshimura, I., Myllys, L., Thell, A. and Hyvönen, J. (2003) Culture experiments and DNA sequence data confirm the identity of *Lobaria* photomorphs. Canadian Journal of Botany. 81: 232–247.

Taylor, J.W., Jacobsen, D.J., Kroken, S., Kasuga, T., Geiser, D., Hibbett, D.S. and Fisher, M.C. (2000) Species recognition and species concepts in fungi. Fungal Genetics and Biology 29: 21–32.

Tehler, A. (1982) The species pair concept in lichenology. Taxon 31: 708–714.

Tehler, A. (1990) A new approach to the phylogeny of euascomycetes with a cladistic outline of Arthoniales focussing on Roccellaceae. Canadian Journal of Botany 68: 2458–2492.

Tehler, A. (1993) The genus *Schismatomma* (Arthoniales, Euascomycetidae). Opera Botanica 118: 1–38.

Tehler, A. (1994) The phylogeny of *Roccellina* (Roccellaceae), revisited. Acta Botanica Fennica 150: 185–195.

Tehler, A. (1995) Arthoniales phylogeny as indicated by morphological and rDNA sequence data. Cryptogamic Botany 5: 82–97.

Tehler, A. (1996) Systematics, phylogeny and classification. *In*: Nash, T.H. (ed) Lichen Biology. Cambridge Univ, Cambridge. pp. 217–239.

Tehler, A. and Egea, J.M. (1997) The phylogeny of *Lecanactis* (Opegraphaceae). Lichenologist 29: 397–414.

Tehler, A. and Källersjö, M. (2001) *Parmeliopsis ambigua* and *P. hyperopta* (Parmeliaceae): species or chemotypes? Lichenologist 33: 403–408.

Tehler, A., Farris, J.S., Lipscomb, D.L. and Källersjö, M. (2000) Phylogenetic analyses of the fungi based on large rDNA data sets. Mycologia 92: 459–474.

Tehler, A., Little, D.P. and Farris, J.S. (2003) The full-length phylogenetic tree from 1551 ribosomal sequences of chitinous fungi, Fungi. Mycological Research 107: 901–916.

Thell, A. (1998) Phylogenetic relationships of some cetrarioid species in British Columbia with notes on *Tuckermannopsis*. Folia Cryptogamica Estonica 32: 113–122.

Thell, A. (1999) Group I intron versus ITS sequences in phylogeny of cetrarioid lichens. Lichenologist 31: 441–449.

Thell, A., Berbee, M. and Miao, V. (1998) Phylogeny within the genus *Platismatia* based on rDNA ITS sequences (lichenized Ascomycota). Cryptogamie, Bryologie-Lichenologie 19: 307–319.

Thell, A., Stenroos, S. and Myllys, L. (2000) A DNA study of the *Cetraria aculeata* and *C. islandica* groups. Folia Cryptogamica Estonica 36: 95–106.

Thell, A., Feuerer, T., Kärnefelt, I., Myllys, L. and Stenroos, S. (2002) Phylogeny and ecology of *Cetraria obtusata, Coelopogon epiphorellus*, and related taxa (Parmeliaceae, lichenized ascomycetes). Mitteilungen aus dem Institut für allgemeine Botanik in Hamburg 30–32: 283–296.

Thell, A., Stenroos, S., Feuerer, T., Kärnefelt, I., Myllys, L. and Hyvönen, J. (2002a) Phylogeny of cetrarioid lichens (Parmeliaceae) inferred from ITS and beta-tubulin sequences, morphology, anatomy and secondary chemistry. Mycological Progress 1: 335–354.

Thomas, M.A., Ryan, D.J., Farnden, K.J.F. and Galloway, D.J. (2002) Observations on phylogenetic relationships within Lobariaceae Chevall. (Lecanorales, Ascomycota) in New Zealand, based on ITS–5.8s molecular sequence data. In: Progress and Problems in Lichenology at the Turn of the Millenium. Bibliotheca Lichenologica, Llimona, X., (eds), Lumbsch, H.T. and Ott, S.J. Cramer, Berlin, Stuttgart, 123–138.

Tibell, L. (2001) *Cybebe gracilenta* in an ITS, 5.8S rDNA based phylogeny belongs to *Chaenotheca* (Coniocybaceae, lichenized Ascomycetes). Lichenologist 33: 519–525.

Tibell, L. (2001a) Photobiont association and molecular phylogeny of the lichen genus *Chaenotheca*. Bryologist 104: 191–198.

Tibell, L. (2003) *Tholurna dissimilis* and generic delimitations in Caliciaceae inferred from nuclear ITS and LSU rDNA phylogenies (Lecanorales, lichenized ascomycetes). Mycological Research 107: 1403–1418.

Walser, J.-C., Sperisen, C., Soliva, M. and Scheidegger, C. (2003) Fungus-specific microsatellite primers of lichens: application for the assessment of genetic variation on different spatial scales in *Lobaria pulmonaria*. Fungal Genetics and Biology 40: 72–82.

Wedin, M. and Döring, H. (1999) The phylogenetic relationship of the Sphaerophoraceae, *Austropeltum* and *Neophyllis* (lichenized Ascomycota) inferred by SSU rDNA sequences. Mycological Research 103: 1131–1137.

Wedin, M., Döring, H. and Mattsson, J.-E. (1999) A multi-gene study of the phylogenetic relationships of the Parmeliaceae. Mycological Research 103: 1185–1192.

Wedin, M., Döring, H. and Ekman, S. (2000) Molecular phylogeny of the lichen families Cladoniaceae, Sphaerophoraceae, and Stereocaulaceae (Lecanorales, Ascomycotina). Lichenologist 32: 171–187.

Wedin, M., Döring, H., Nordin, A. and Tibell, L. (2000a) Small subunit rDNA phylogeny shows the lichen families Caliciaceae and Physciaceae (Lecanorales, Ascomycotina) to form a monophyletic group. Canadian Journal of Botany 78: 246–254.

Wedin, M., Baloch, E. and Grube, M (2002) Parsimony analyses of mtSSU and nITS rDNA sequences reveal the natural relationships of the lichen families Physciaceae and Caliciaceae. Taxon 51: 655–660.

Wiklund, E. and Wedin, M. (2003) The phylogenetic relationships of the cyanobacterial lichens in the Lecanorales suborder Peltigerineae. Cladistics 19: 419–431.

Wirtz, N., Lumbsch, H.T., Green, T.G.A., Türk, R., Pintado, A., Sancho, L. and Schroeter, B. (2003) Lichen fungi have low cyanobiont selectivity in maritime Antarctica. New Phytologist 160: 177–183.

Zoller, S. and Lutzoni, F. (2003) Slow algae, fast fungi: exceptionally high nucleotide substitution rate differences between lichenized fungi *Omphalina* and their symbiotic green algae *Coccomyxa*. Molecular Phylogenetics and Evolution 29:629–640.

Zoller, S., Lutzoni, F. and Scheidegger, C. (1999) Genetic variation within and among populations of the threatened lichen *Lobaria pulmonaria* in Switzerland and implications for its conservation. Molecular Ecology 8: 2049–2059.

11. Significance of Lichens and Their Secondary Metabolites: A Review

D.K. Upreti and S. Chatterjee
Lichenology Laboratory, National Botanical Research Institute, Lucknow-226001, India

SUMMARY Lichens, the symbiotic phenotypes of nutritionally specialized fungi associated with algae, are an outstandingly successful group, exploiting a wide range of habitats throughout the world and dominating about 8% of terrestrial ecosystems. Lichens have been used by various ethnic groups from the time of early civilization. Irrespective of the advances in medical sciences, the tribal people still utilize these plants. The lichens are utilized for different purposes depending on their nutritive, medicinal, decorative, brewing, distilling, dyeing, cosmetic and perfumery properties. These different uses are substantiated by the complex lichen secondary metabolism, producing about 400 secondary compounds referred as "Lichen Substances".

The Lichen substances are unique as they are unknown in other plant sources. Lichens contain many characteristic phenolic compounds with known antibiotic activity. They grow too slowly to provide large amounts of material for commercial testing. However, several techniques are available by which cultured lichen tissues grow 10-12 times faster than the natural lichens. The medicinal aspect of lichens has turned to new directions after the untiring research work done by Japanese workers, who studied the anti-tumor and anti-HIV activity of lichens. Lichens may be a good potential source of phytochemicals.

In recent years a large amount of information regarding the significance of lichens and their metabolites are available which are discussed in this article. The references on antiviral, antimicrobial, antiproliferative, antimitotic, antioxidant activity of lichens together with cytotoxic effects and other inhibitory activities of certain lichen metabolites are also included in this article.

Introduction

Lichens are symbiotic associations of fungi, green algae and/or cyanobacteria. They have a varied chemistry and produce many polyketide derived compounds such as depsides and depsidones most of which are not known from other groups of plants. Lichens have been appreciated in traditional medicine (Table 1) but their importance has largely been ignored by the modern pharmaceutical industry because of the difficulties in establishing axenic cultures and conditions for rapid growth. These factors preclude their routine use in most conventional screening processes.

Recently molecular genetic techniques using PCR genomic library construction and heterologus expression have provided an alternative approach to begin exploring the diversity of polyketide biosynthetic pathways in lichens. The techniques can be expanded to cover other pathways and be integrated with conventional culture collection based screening to provide a comprehensive search for novel chemical entities in these organisms (Miao et al. 2001).

Lichen metabolites exert a wide variety of biological actions such as anti-microbial, antiviral, antiinflamatory, analgesic, antipyretic, antiproliferative and cytotoxic effects.

El-fattah (1997) studied the biological profile of two octanoic acid derivatives from the ether extract of

Xanthoria stenesi. The derivatives were identified as 3-hydroxy-2,4,6-triethyl octanoic acid (xanthoric acid A) and 3-0-[3′ hydroxy-2,4,6-trimethyl octanoate]2,4,6-trimethyl octanoic acid (xanthoric acid B). In addition, 3-methyl orsellinic acid methyl ester (2,4 dihydroxy-3,6-dimethyl benzoic acid methyl ester), Physcion a 5 α stigmast-7-en-3-beta-ol were isolated and identified on the basis of spectral data. Xanthoric acid B has a wide range of bioactivities such as cytotoxicity, anticonvulsant, antibacterial and antimycotic.

Traditional Uses of Lichens

Lichens were held in high regard by medicinal practitioners in medieval times and their use has persisted to this day. In various Pharmacopoeias, lichens are listed purely on the basis of their folklore medicinal uses. Several species of lichens enjoyed a good position in ancient and traditional systems of Indian medicine like Ayurveda and Unani.

Phamacopoeia Universalis of 1846 (Varita, 1973) lists eleven lichen species viz.: *Cetraria islandica* (L) Ach., *C. nivalis* (L) Ach. (Parmeliaceae); *Cladonia coccifera* (L) Willd. *C. pyxidata* (L.) Hoffm. (Cladoniaceae); *Usnea plicata* A L. Smith (Usneaceae); *Peltigera canina* (L) Willd, *P. venosa* (L) Hoffm. *P. polydactyla* (Neck.) Hoffm., *P. horizontalis* (Huds.) Raumg. (Peltigeraceac); *Lobaria pulmonaria* (L) Hoffm., *Xanthoria parietina* (L) Fr. (Teloschistaceae) and *Evernia prunastri* (L) Ach. (Usneaceae).

Gonzalez-Tejero et al. (1995) emphasized that lichens have been used in traditional medicine since the time of the first Chinese and Egyptian civilizations, but according to Kumar and Upreti (2001) the word "Shipal" is used for algae in Rigveda (6000-4000 BC), a text where the first authentic record of "Oshadi" (medicine) has been described. The medicinal properties attributed to "Shipal" as lichen are in "Avkolva", as mentioned in Athravaveda (1500 BC). Subsequently a number of sanskrit synonyms of lichens for example, Shailaya and Shilapushp have been described in Sushruta Samhita (1000 BC), Charak Samhita (300-200 BC), and several Nighantu (AD 1100-1800). The Sanskrit names of lichens were later identified with several species of the genus Parmelia, such as *Parmelia cirrhata* and *P. perforata*.

(a) The Doctrine of Signatures

The fame of many lichens derives from the Doctrine of Signatures, which is very old in genesis and was raised to an accepted science by Paracelsus but is now discredited. The Doctrine relies on thc concept that like affects like. Hence plants designed by the creator for human beings may be expected to each bear a sign pointing to its special use. Lichens used medicinally in remote areas of the world are described by Llano (1944). *Lobaria pulmonaria* (L) Hoffm., was used to treat lung disease because of its superficial resemblance to lung tissue, *Parmelia sulcata* Taylor, was a remedy for cranial maladies, *Peltigera canina* was used for rabies. (Llano, 1944)

The Seminole Indians in Florida and the Chinese employ various lichens in medicine especially as expectorants. The most commonly used lichen is *Usnea*. In Sweden, the Iceland Moss, *Cetraria islandica* (L) Ach., is widely sold in apothecary shops for treating diabetes, lung diseases and catarrh (Hale, 1967).

The long filaments of *Usnea barbata Wigg*., were used for strengthening after confinement. *Xanthora parietina*, being a yellow lichen, was supposed to cure jaundice, while *Peltigera apthosa* (L.) Willd., the thallus of which is dotted with small wart-like tubercles, was recommended for children who suffered from 'Thrush'. Species of *Evernia*, *Peltigera*, *Parmelia*, *Cladonia*, *Roccella* and *Pertusaria* were used to control fever, diarrhoea, infections, skin diseases, epilepsy, convulsions and as purgatives (Llano, 1944).

(b) Lichens Used in Ayurvedic and Unani Systems of Medicine

Chandra and Singh (1971) provided a detailed description of a crude drug 'chharila' comprising three species of *Parmelia*, *P. perlata* (L) Ach, *P. perforata* (Wulf.) Ach. and *P. cirrhata* Fr. sold in Indian markets. It has astringent, resolvent, laxative and carminative properties and is also supposed to possess aphrodisiac property. It is considered to be useful in dyspepsia, spermatorrhoea, amenorrhoea, calculi, diseases of the blood and heart, stomach disorders, enlarged spleen, bronchitis, bleeding piles, scabies, leprosy, excessive salivation, soreness of throat, tooth ache and pain in general. The smoke of 'chharila' is believed to relieve headache. The powdered drug is applied to wounds, besides being a good cephalic snuff. According to Subrahmanian and Ramakrishnan (1964), *Peltigera canina* contains high amount of the amino acid methionine and is used for liver ailments in various parts of India.

(c) Ethnomedicinal Uses

During the last five decades, exhaustive ethnobotanical work has been carried out by several workers in different parts of the country, with major emphasis on higher plants whereas ethnobotanical aspects of cryptogamic groups such as algae, fungi, lichens and bryophytes have not received adequate attention. However, ethnobotanical data on some lower plant groups are available for countries other than India (Llano, 1956; Zaveneld 1959; Wasson, 1962). Recently, Brij Lal et al. (1985), Brij Lal (1988), Brij Lal and Upreti (1995), Saklani and Upreti (1992), have collected ethnobotanical information on a few species of lichens used by the different tribal and non-tribal communities of India.

Most of the acids present in the lichens used by the tribals have been reported as antibacterial or antiviral by Asahina and Shibata (1954). These are mainly phenolic-carboxylic acids (atranorin, lobaric and salazinic acids), fatty acids (protolichesterinic acid) and triterpenoid (zeorin) derivatives.

Azenha et al. (1998) provided a detailed account of ethnolichenology, biological activities and biochemistry of Amazonian lichens. The ethanol, benzene and acetone extract of *Usnea, Cladonia* and *Parmelia* species were found to exhibit antibacterial activity against *Staphylococcus aureus.*

The lichen species which are used by the different tribal communities in India and other parts of the world are enumerated in Table 1.

Properties of Lichens

More than 50% of lichen species are known to possess antibiotic value. The most promising lichen substance is usnic acid, a broad spectrum antibiotic. It strongly inhibits *Mycobacterium* and is commercially used under the name 'Usno'. This medication is available in Europe and used in treating external burns and superficial wounds. Similar commercial preparations are marketed in Germany as 'Usni plant' and 'Binan' in Russia. The main source of usnic acid for these drugs are the yellow species of *Cladonia* (reindeer mosses). Despite the wealth of evidence demonstrating the antibiotic activity of lichen substances, the mechanism of this activity is almost unknown. Early studies suggested that the common cortical compound, usnic acid interrupts cell division, perhaps through inhibition of DNAase-activity (Marshak and Fager, 1950) or uncoupling of oxidative phosphorylation (Johnson et al. 1950).

Lawrey (1986) described in detail the role of iichen substances as antibiotics, activity against micro-organism in nature and inhibition effects on seed germination of vascular plants. The lichens grow very slowly and are resistant to decay by micro-organisms. Lichen substances can inhibit seed germination of vascular plants and spore germination of mosses and lichens, demonstrating allelopathic role. Despite their availability as food to herbivores, lichens are seldom eaten; lichen substances deter feeding. Recent research done primarily with invertebrate consumers indicates that some lichen compounds are capable of defending lichen thalli from predation.

Perry et al. (1999) studied the bioactivity of 69 lichen extracts of New Zealand lichens. Active extracts were generally from species known to contain phenolic compounds. Bioactivity directed isolation work on *Cladia retipora, Pseudocyphellaria glabra* and *P. hornoeophylla* led to the identification of usnic acid as the main antimicrobial, cytotoxic and antiviral component in these three species.

Table 1

Species (Family)	Local name	Ethnic group	Use
1. In India			
Buellia subsoriroides (Physciaceae)	Maidi	Garhwal herdsmen	A substitute for 'henna'. On whole lichens herdsmen spit and work them (with a small stone) into a paste; the paste is used to make designs on lips and palms (Brij Lal and Upreti, 1995).
Cetrelia collata (Everniastrum cirrhatum, Melanelia infumata, Paramotrema nilgherrense) (Parmeliaceae)	Budu	Gaddi tribe of Kangra valley	Materials for sacrificial fire ('havan') (Singh et al. 2000)
Everniastrum cirrhatum (Parmeliaceae)	Jhavila	Baiga, Bhil, Bhilala, Gond, Korka, Muria of Madhya Pradesh	As spice and flavouring agent for meat and vegetables (Brij Lal et al. 1985)
	Rukhu-ku-jau	Lepchas and Nepalese of Sakyong Valley, North Sikkim	Whole thallus is boiled; after decanting, the boiled lichen is fried and eaten as vegetable (Saklani and Upreti, 1992)
Heterodermia diademata (Physciaceae)	Dhungo-ku-setojhau	Nepalese of Sikkim	Thalli applied on cuts for protecting from wetting and infection (Saklani and Upreti, 1992)
Heterodermia tremulans (Physciaceae)	Jhavila	Baiga, Bhil, Bhilala, Gond, Korka, Muria of Madhya Pradesh	As spice and flavouring agent for meat and vegetables (Brij Lal et al. 1985)
Lobaria pulmonaria (Lobariaceae)		Darjeeling and Sikkim	Hill men use it for curing eczema on the head and cleaning hair. Used for strengthening hair. Remedy for lung troubles, haemorrhages and asthma (Biswas, 1956)
Parmelia abessinica (Parmeliaceae)		Western Andhra	Used as indigenous medicine "Rathipuvvu" (Llano, 1951)
Parmelia kamschadalis (Parmeliaceae)		India	"Bitter, febrifuge, astringent, resolvent, emollient, demulcent, considered useful as a diuretic; also soporific and sedative. Used in diarrhoea, dyspepsia, spermatorrhoea, amenorrhoea, and dysentery. Applied externally as a poultice, applied to renal and lumbar regions, which causes a copious flow of urine. As a liniment it is applied to the head in cases of headaches. Powder is applied to promote healing of wounds (Nadkarni, 1976)

Parmelia sancti-angeli (*Parmotrema sancti-angeli*) (Parmeliaceae)	Jhavila	Gond and Oran	30-50g fresh plant is burnt and the ash is mixed with mustard (*Brassica nigra*) or linseed oil and used for ringworm-like skin disease (Brij Lal and Upreti, 1995)
Parmelia tinctorum (*Parmotrema tinctorum*) (Parmeliaceae)	Jhavila	Baiga, Bhil, Bhilala, Gond, Korka, Muria	As spice and flavouring agent (Brij Lal et al. 1985)
Parmelia spp. (*Parmelia perlata* (L.) Ach., *P. perforata* (Wulf.) Ach., *P. cirrhata* Fr.) (Parmeliaceae)	Chharila	India	Crude drug, extracted from 3 spp. of *Parmelia* is sold in Indian bazaars and used in Ayurvedic and Unani systems of medicine. Powdered drug applied to wounds. Carminative and aphrodisiac; considered to be useful in dyspepsia, spermatorrhoea, amenorrhoea, calcui, diseases of the blood and heart, stomach disorders, enlarged spleen, bronchitis, bleeding piles, scabies, leprosy, excessive salivation, soreness of the throat, toothache, and pain in general. The smoke of 'chharila' is believed to relieve headache (Saklani and Upreti, 1992)
Parmotrema chinense (*Parmelia perlata*) (Parmeliaceae)		India	Used as a poultice, applied to renal and lumbar regions which causes a copious flow of urine. Powder is applied to promote healing of wounds. Bitter, febrifuge, astringent, resolvent, emollient, demulcent, useful as a diuretic; also soporific and sedative used in diarrhoea, dyspepsia, spermatorrhoea, amenorrhoea, and dysentery used as an incense especially to relieve headache (Nadkarni, 1976)
Parmotrema perforatum (*Parmelia perforata*) (Parmeliaceae)		India	Used as a poultice, applied to renal and lumbar regions which causes a copious flow of urine. As a liniment it is applied to head in cases of headaches powder is applied to promote healing of wounds. Bitter, febrifuge, astringent, resolvent, emollient, demulcent, considered useful as a diuretic; also soporific and sedative used in diarrhoea, dyspepsia, spermatorrhoea, amenorrhoea, and dysentery. Used as an incense especially to relieve headache (Nadkarni, 1976)
Peltigera canina (Peltigeraceae)		India, Sikkim	Eaten as a remedy for liver ailments (Saklani and Upreti 1992)
Peltigera polydactyla (Peltigeraceae)	Jhau	Lepchas of Sakyong valley, North Sikkim	Paste of thalli applied to cuts to stop bleeding (Saklani and Upreti, 1992)
Ramalina subcomplanta (Ramalinaceae)	Jhavila	Baiga, Bhil, Bhilala, Gond, Korka, Muria	As spice and flavoring agent (Brij Lal et al. 1985)

(Contd.)

Species (Family)	Local name	Ethnic group	Use
		Darjeeling and Sikkim	It yields an efficacious medicine used in hydrophobia and jaundice (Saklani and Upreti, 1992)
Rimelia reticulata (Parmeliaceae)	Jhavila	Baiga, Bhil, Bhilala, Gond, Korka, Muria of Madhya Pradesh	As spice (Richardson, 1988)
Stereocaulon himalayense (Stereocaulaceae)	Dhungo-ku-jhau	Lepchas of Sikkim	Thalli are pounded and boiled in water and used as antidote for urinary trouble (Sakalani and Upreti, 1992)
Thamnolia vermicularis (Imperfect lichenized fungi)	Chhai dhoop	Bhotias of Nanda Devi Biosphere Reserve	Smoke from burning thalli kills worms (white, 1-2 mm long) in utensils used for preparing and storing butter milk (Upreti and Negi, 1995)
Usnea longissima (Usneaceae)	Syara	Bhotia and Garhwalis of remote areas of Uttarkashi district, Garhwalis Himalayas	Stuffing for pillows and cushions: an ingredient of a poultice for bone setting; as spice, and as staple food for musk deer (Brij Lal and Upreti, 1995)
		Madhya Pradesh, Baiga tribe	Used for treating bone fractures (Brij Lal, 1988)
Usnea sikkimensis (Usneaceae)		Darjeeling and Sikkim	Powdered lichen used to strengthen hair. Remedy for lung troubles, haemorrhage, and asthma (Biswas, 1956)
2. Other Countries			
Bryoria fremontii (Lecanoraceae)	–	Okanagan-Colville Indians, interior British Columbia	Fresh lichen dried, mixed with grease, and rubbed on naval of new born baby to prevent infection (Turner et al. 1980)
	–	Nez Perce Great Basin, Western North America	Lichen is taken internally for upset stomach, indigestion and diarrhoea. Lichen soaked in stream water is cooked in a steam pit overnight cut into loaves and consumed (Turner, 1977)
Bryoria sp. (Lecanoraceae)	Black moss	Atsugewi Northern California	Applied as poultice for reducing swellings. The moss was taken from pine trees, dried, and pounded up. It was then boiled, or sometimes was used dry (Garth, 1953)
Bryoria trichodes (*Alectoria americana*) (Lecanoraceae)	–	Chugach Natives, Alaska	Lichen collected in quantity and piled on a sick person in the steam bath to hold the heat on his body and also used to staunch the blood from a wound (Wennekens, 1985)
Cetraria islandica (L.) Ach. (Parmeliaceae)	Iceland moss	Europe	A decoction was recommended as a tonic for convalescents. A light diet for invalids, and a mild tonic for Phthisis, chronic catarrh, dyspepsia, chronic diarrhoea, and dysentery (Schneider, 1904)

		People of boreal and subarctic region of Norway	Lichen is mixed with flour to make bread or porridge; boiled to extract gelatin (Richardson, 1991)
		Ubaue Valley, France	Decoction is used as pectoral, emollient (Novaretti and Lemordant, 1990)
		Venezia Giulia, Italy	Reconstituent after tuberculosis, anticatarrhal (Lokar and Poldini, 1988)
		Spain	In treatment of catarrh, asthma and to mitigate inflammation (Ahmadjian and Nilsson, 1963)
		Sweden	Used to treat asthma, diabetes, nephritis, lung diseases and wasting diseases. Thought to be effective treatment for whooping cough and colds (Airaksinen et al. 1986)
Cladonia rangiferina (Cladoniaceae)	Reindeer moss	Ojibwa, Great Lakes	New born baby bathed with the water in which lichen had been boiled (Vogel 1970)
Cladonia sp. (Cladoniaceae)	Reindeer moss	Alaska	Boiled juice drunk as medicine for diarrhoea and taken as tea for internal chest pains (Kari, 1987; Smith, 1973)
Cladonia stellaris (*Cladonia alpestris*) (Cladoniaceae)		East central Saskatchewan	Decoction imbibed to expel intestinal worms; dried powdered lichen used in winter for the same purpose (Leighton, 1985)
Cladonia chlorophaea (Cladoniaceae)		Okanagan-Colville Indians, British Columbia	Whole plant boiled, decoction used to wash sores which were slow to heal (Turner et al. 1980)
Cladonia pyxidat (Cladoniaceae)		Europe	Used to treat "whooping-cough" and as a febrifuge (Lindley, 1849)
Dictyonema sericeum (Thelephoraceae)	–	Primitive lowland Amazonian Auca	Powerful Shaman medicine as hallucinogenic (Davis and Yost 1983)
Dictyonema sp. (Thelephoraceae)		Eastern Equador	Reported to cause sterility; put into a child's drink to cause barrenness (Davis and Yost, 1983)
Falvocetraria nivalis (*Cetraria nivalis*) (Parmeliaceae)	Beard of rock	Andeans Province Bautista Saavedra, Peru	Tea drunk as cardio-pulmonary tonic for altitude sickness and heart attacks (Bastien, 1983)
Flavoparmelia caperata (*Parmelia caperata*) (Parmeliaceae)		Tarahumar of Mexico	Dried, crushed, and dusted upon burns (Pennington, 1963)
Hypogymnia physodes (*Parmelia physodes*) (Parmeliaceae)		In Forest Potawatomi, Great Lakes	Cures constipation. Eaten from the trees while in the woods (Smith, 1933)
Letharia vulpina (*Evernia vulpina*) (Parmeliaceae)		Salishan tribes, Okangon and Thompson mountain states	Used as medicine (Teit and Boas, 1928)

(Contd.)

Species (Family)	Local name	Ethnic group	Use
		Blackfoot mountain states	Blackened in fire and rubbed on rash, eczema, and wart sores. Infusion of (*Evernia vulpina*) and marrow was taken for stomach disorders like ulcers (Hellson and Gadd, 1974)
		Interior British Columbia	Boiled stronger solution used to wash external sores and wounds. Boiled and taken in a weak solution for internal problem (Turner et al. 1980)
		In Mendocino County, California	For drying up running sores and relieving accompanying inflammation (Mead, 1972).
Lobaria pulmonaria (Lobariaceae)		British Columbia	Applied to the faces of children and when peeled, reveal a light-coloured 'patchiness' very much like the lichen itself." Considered very sacred and was specially prepared (Turner and Efrat, 1982)
		West coast of Vancouver Island, B.C., Canada	Lichen when growing on hemlock tree used as medicine against coughing up blood (Turner and Efrat, 1982)
Lobaria quercizans (*Sticta glomulifera*) (Lobariaceae)		Menomini, Great Lakes	Grows on many kinds of trees but gathered only from hard maple or hemlock. Eaten as food and medicine. Stored dried. Put into soups. Highly esteemed for its tonic effect on the system and blood. Valued for flavour (Smith, 1923)
Lobaria retigera (Lobariaceae)		China	Used in Chinese medicine (Hu et al. 1980)
Lobaria spp. (Sticta) (Lobariaceae)		British Columbia, Canada	The entire plant of (Sticta), from certain trees only, was boiled and the decoction taken internally for pains in the stomach and externally as an eyewash and poultice (Turner, 1973)
Nephroma arcticum (Nephromataceae)		Alaska	Cooked alone in water and fed to a person in weak condition to make him strong. Reported to have been a very effective medicine (Oswalt, 1957)
Parmelia saxicola (Parmeliaceae)		Northern California	Colic is treated with a tea made from this greenish-gray lichen found growing on stones (Powers, 1874)
Parmelia saxatilis (Parmeliaceae)		China	Used in Chinese medicine (Hu et al. 1980)
		Malung, Dalarna, Sweden	Used to dissolve warts on hands (Ahmadjian and Nilsson, 1963)
Parmotrema perforatum (Parmelia perforata) (Parmeliaceae)		Kabul, Afghanistan	Imported for unspecified medicinal use (Younos et al. 1987)

Peltigera aphthosa (Peltigeraceae)		Sweden	Boiled in milk as a cure for aphthae (Lindley, 1849)
		Alaska	A lichen (probably Peltigera aphthosa) was dried, powdered and applied to the "hurt" for burns and scalds (Emmons et al. 1991)
Peltigera britannica (*Peltigera aphthosa*) (Peltigeraceae)		West coast of Vancouver Island	Used for running sores that were hard to heal. Mashed and made into a poultice and used especially for sores on the leg caused by bruises from walking among rocks. Chewed and eaten as a remedy for tuberculosis (Turner et al. 1983)
Peltigera canina (Peltigeraceae)		Europe	An imaginary remedy for hydrophobia (Lindley, 1849)
Peltigera membranacea (Peltigeraceae)		West coast of Vancouver Island, B.C., Canada	Used as some kind of medicine (Turner and Efrat, 1982)
		West coast of Vancouver Island, Nitinaht Indians	Reported to cure a man who could not urinate (Turner et al. 1983)
Peltigera sp. (Peltigeraceae)		British Columbia	Mixed with spruce pitch and applied as poultice to wounds (Moerman, 1998)
Pertusaria amara (Pertusariaceae)			Used in cases of intermittent fever (Smith, 1921)
Pertusaria communis (Pertusariaceae)		Europe	Used to cure intermittent fever; more useful on men than women (Llano, 1951)
Pseudevernia furfuracea (Parmeliaceae)	Musgo	Spain, Andalucia	Decoction used to treat respiratory diseases (Gonzalez et al. 1995)
		Europe	A 15th century drug "Lichen quercinus virdes." Alongwith *Evernia prunastri* and *Hypogymnia physodes* (Llano, 1951)
Pseudocyphellaria aurata (Lobariaceae)		Madagascar, Ambavaniasy	Used as a tea to treat the indigestion (Scheidegger, 1998)
Ramalina bourgeana (Ramalinaceae)	Flor de Piedra (Stone flower)	Spain, Andelucia	Decoction (one cup daily) used in the areas of Viso and Nijar as a diuretic for the treatment of kidney stones (Gonzalez et al. 1995)
Ramalina thrausta (Ramalinaceae)		Finland	Mixed with beard mosses, that are placed on wounds, skin eruptions, and athlete's foot (Vartia, 1973)
Rimelia reticulata (*Parmelia reticulata*) (Ramalinaceae)		Chihuahua	Common lichen used in preparing a tea to relieve discomfort from kidney disorders or venereal disease. Prepared in late afternoon and set aside for one night before use (Pennington, 1969)

(Contd.)

Species (Family)	Local name	Ethnic group	Use
Roccella babingtonii (Roccellaceae)		Mainland side of Sea of Cortez	Ground plant, mixed with water, was used to bathe a child with a fever". Ground on metate, squeezed through a cloth with a bit of water; resulting liquid placed on a burn or sore. Tea used as remedy for shortness of breath, febrifuge. "ground with castime" (a kind of reddish clay) and water used as a febrifuge and to cure diarrhoea (Felger and Moser, 1985)
		Sicnani, Peru; Peruvian Indians	Two species in Roccellaceae are used one for coughs, one for fever (Llano, 1951)
Stereocaulon paschale (Stereocaulaceae)		China	Used in Chinese medicine (Hu et al. 1980)
Umbilicaria sp. (Umbilicariaceae)		Northern Quebec	Used as a tea by those suffering from tuberculosis (Stevens, 1983)
Umbilicaria esculenta (Umbilicariaceae)		China	Used in Chinese medicine (Hu et al. 1980)
Usnea (Usneaceae)		Europe	A number of companies produce pastilles and pills for sore throats made from [*Cetraria islandica*] or from the beard lichen Usnea (Richardson, 1991)
Usnea africana (Usneaceae)		East Africa, Hewas	Chewed fresh and juice swallowed to treat stomach ache (Kokwaro, 1976)
Usnea articulata (Usneaceae)		Heinemann, Auckland	Absorbent Fiber, Samoa, used for wounds and skin bruises (Brooker et al. 1987)
Usnea barbata (Usneaceae)		U.S.	Indicated in the treatment of fungal infections of the mouth, stomach, intestines, anus, vagina, nose, ear, and skin, as well as "systemic fungal infections." Administered internally, topically, and as a douche (Herbal Resources, 1998)
		Europe	Used to strengthen hair (Llano, 1951)
		Greece	Hippocrates recommended for uterine trouble. Dioscorides recommended "beard moss" for certain diseases peculiar to women (Vartia 1973)
		New Zealand, Maori	Used for bandaging wounds to check bleeding (Brooker et al. 1987)
		Malay Peninsula, Malaysians	The natives of the Malay Peninsula use this plant for colds and strengthening after confinement (Llano, 1944)
Usnea densirostra (Usneaceae)		Uruguay	Probable identification of "yerba de la piedra," which has medicinal uses. (Osorio, 1982).
Usnea diffracta (Usneaceae)		China	Used in Chinese medicine (Hu et al. 1980)

		"Lao-tzu's beard," "pine gauze," "female gauze" described in earliest herbal (500 AD). Picked in 5th lunar month, dried in the shade. Stops sweating, dizziness, cold, pain, phlegm. Benefits urinary tract and stops swelling in female genitalia (Strickmann, 1988)
Usnea filipendula (Usneaceae)	Russia: Primorye and South Sakhalin ("Soviet Far East")	Used in powdered form to treat wounds as antibacterial (Moskalenko, 1986)
Usnea flavescens (Usneaceae)	East Africa, Hewas (Iraqw)	Chewed fresh and juice swallowed to treat stomach ache (Kokwaro, 1976)
Usnea florida (Usneaceae)	South-Central Chile	Infusion taken for diarrhoea (Houghton and Manby, 1985)
Usnea longissima (Usneaceae)	China	Used in Chinese medicine especially as an expectorant and in the treatment of ulcers (Chopra et al. 1958) "Lao-tzu's beard," "pine gauze," "female gauze" described in earliest herbal (500 AD). Picked in 5th lunar month, dried in the shade. Stops sweating, dizziness, cold, pain, phlegm. Benefits urinary tract and stops swelling in female genitalia (Hu et al. 1980)
	West coast of Vancouver Island, Nitinaht Indians	Valued for absorbent qualities — wound dressing (Turner et al. 1983)
Usnea sp. (Usneaceae)	Africa, Mt. Kilimanjaro	Ingredient in (apparently effective) herbal tea given by African guide to relieve altitude sickness (Blackwell, 1990)
	Italy: "Brocon" Pass, Valsugana Valley, Trentino district	Shepherds put Usnea sp. in their shoes to prevent or treat blisters (Zorer, 1998)
	Karen: Doi Inthanon (national park), Chiang Mai Province, Thailand	Used in a bath for women following the birth of a child, to aid parturition and to prevent infection (Wolseley, 1998)
	Makah: Olympic Peninsula, Washington	Used to treat boils (Moerman, 1998)
	Finland	Included in the (beard mosses) that are placed on wounds, skin eruptions, and athlete's foot (Vartia, 1973)
	Sweden, Malung, Dalarna	Used to treat foot blisters (Ahmadjian and Nilsson, 1963)

(Contd.)

Species (Family)	Local name	Ethnic group	Use
		East-central Saskatchewan	Fresh lichen inserted into nostril to stop a nose bleed (Leighton 1985)
Xanthoparmelia chlorchroa (*Parmelia molliuscula*) (Parmeliaceae)		New Mexico	Used as remedy for impetigo (Elmore, 1944)
Xanthoparmelia conspersa (*Parmelia conspersa*) (Parmeliaceae)		Africa	Decoction drunk and powder applied locally for treatment of venereal disease, especially syphilis. Taken internally and applied locally as a snake-bite remedy (Watt and Breyer-Brandwijk, 1962)
Xanthoria parietina (*Parmelia parietina*) (Teloschistaceae)		Europe	Used in intermittent fevers, to cure jaundice (Llano, 1951)
Xanthoria parietina subsp. *ectanea* (Teloschistaceae)	Flor de Piedra, Rompiedra	Spain, Andalucia	Decoction used with wine to treat menstrual complaints, kidney disorders, toothache, various pains. It is one of the ingredients of cough syrup (Gonzalez et al. 1995)

Polysaccharides from Lichens

Olafsdottir and Ingolfsdotter (2001) described the structural characteristics and biological activity of lichen polysaccharides. The lichen polysaccharides are mainly of three different types i.e. β-glucans, α-glucans and galactomannans. In addition, a few complex heteroglycans have recently been described such as thamnolan, a water soluble, immunologically active heteroglycan with a novel rhamnopyranosyl galactofuran type of structure.

Antioxidant Activity of Lichens

Racine et al. (1980) studied the antioxidant properties of wax from Yugoslavian Oakmoss (*Evernia prunastri*). Wax from the lichens was extracted in industrial benzene and by solvents of different polarities. The wax and the extracts were tested for antioxidant activities using (+) limonene as peroxidizable test substrate and were found to have such activity. The extracts are more active than the wax itself. The wax and the extracts contain a small amount of substance with antioxidant activity comparable to that of the usual synthetic antioxidants.

Jayaprakasha and Rao (2000) extracted *Parmotrema stuppeum* with benzene and acetone. Both the extracts were fractioned on 1% oxalic acid impregnated silica gel column to obtain four phenolic compounds. The antioxidant activity of benzene extract, acetone extract and isolated compounds were evaluated in a beta-carotene-linoleate model system. The pure compounds showed moderate antioxidant activity. This is the first report on the isolation and characterisation of compounds from the lichen *Parmotrema stuppeum* as well as on their antioxidant activity.

Gülcin et al. (2002) evaluated the antioxidant activity of aqueous extract of *Cetraria islandica*. The antioxidant activity, reducing power, superoxide anion radical scavenging and free radical scavenging activities were studied. About 50, 100, 250 and 500 μg of the aqueous extract of *Cetraria islandica* showed higher antioxidant activity than 500 μg of α-tocopherol. The results obtained indicate that *Cetraria islandica* is a potential source of natural antioxidant.

Enzyme Inhibitory Activity

Yamamoto et al. (1993) extracted metabolites from cultured lichen tissue and conducted a detailed study as growth inhibitors of gram-positive bacteria (*Propionibacterium acnes, Staphylococcus aureus* and *Bacillus subtilis*), of fungi (*Penicillium italicum, P. digitatum, Trichothecium roseum, Botrytis cinerea, Curvularia* sp. and *Pyricularia oryzae*). Superoxide dismustase (SoD) like activity and inhibition of tyrosinase was seen in some metabolites/extracts.

Inhibition of β-glucosidose was detected in extracts of the lichen, *Umbilicaria esculenta* by Lee and Kim (2000). The extract showed strong inhibition of disaccharide hydrolysing enzymes of mold and mammalian origin, but weak or no inhibition of polysaccharide hydrolysing enzymes except glucoamylase and laminarinase. The inhibitor in the extract was very stable, retaining more than 95% of its activity when treated with heat, acid, alkali, and some hydrolytic enzymes. By NMR spectrometry the purified inhibitor was identified as 1-deoxynojirimycin (1,5-dideoxy-1, 5-imino-D-glucitol), which was known to be produced by *Streptomyces* sp., and the plant *Morus* sp. Extracts from *Parmelia austrosinensis, Parmelia praesorediosa* and an unidentified lichen species, showed glucosidase inhibitory activities similar to *Umbilicaria esculenta*.

Inhibitory activity of usnic acid against plant p-hydroxyphenyl pyruvate dioxygenase was studied by Romagni et al. (2000). Usnic acid exists as α(–) and α(+) enantiomers indicating a alpha or beta projection of the methyl group at position 9b, respectively. The (–)-usnic acid causes a dose-dependent bleaching of the cotyledonary tissues associated with a decrease of both chlorophyll and carotenoids in treated plants whereas no bleaching was observed with the (+) enantiomer. The (–)-usnic acid inhibited protophorphyrinogen oxidase activity (1C50 = 3 μM), but did not lead to protoporphyrin LX accumulation. Bleaching appears to be caused by irreversible inhibition of the enzyme 4-hydroxyphenylpyruvate dioxygenase by (–)-usnic acid (apparent, IC 50 = 50 μM).

Rojas et al. (2000) investigated the inhibitory effect of lichen metabolites on thylakoid electron transport and photophosphorylation in isolated spinach chloroplasts. The Lecanonn, methyl-beta-orchinol carboxylate, methyl orsellinate, orcinol, and methyl haematommate isolated from *Parmotrema tinctorum* caused significant inhibition of the radicle growth and germination of seedlings of *Amaranthus hypochondiriacus* and *Echinochloa crusgalli*. In addition, lecanoric and gyrophoric acid significantly inhibited the light-dependent synthesis of ATP and uncoupled electron transfer on the reducing side of photosystem II in freshly lysed, illuminated spinach chloroplasts.

Antiproliferative and Cytotoxic Activity of Lichens

Leao et al. (1997) observed the cytotoxic effect against HeLa cells of polysaccharides obtained from *Ramalina celastri*. The most active polysaccharides which show anti-tumoral activity are (1→3)-β-D-glucans, branched or not at O-6. Since these structures are sometimes poorly soluble in aqueous media, a-D-glucans and their chemical derivatives, which are more soluble were also studied. The α-D-glucans altered the cellular volume, cytoplasmic densities, and mitosis.

TEM analysis showed cytoplasmic blebbing and the presence of an amorphous electron-dense material that occurs free in the cytoplasm and interior membranes. Enhanced injury caused by the sulphated derivatives was apparent, altering cell adhesion and causing cell aggregation. Nuclear modification such as fragmentaiton and condensation of chromatin under the nuclear envelop, which appeared to be convoluted, suggested the occurrence of cell death by apoptosis.

Kumar and Müller (1999a) studied the inhibitory action of atranorin, diffractaic and protolichensterinic acid isolated from *Parmelia tinctorum and Parmelia nepalensis*. The acids inhibit LTB4 biosynthesis in polymorphonuclear leukocyte while ethyl hematommate (3) and (±)-usnic acid (1) were only weak inhibitors. Methyl beta-orchinocarboxylate and gyrophoric acid were inactive at concentrations upto 60 μM.

Kumar and Müller (1999b) showed the antiproliferative and cytotoxic activity of lichen acids (gyrophoric, usnic and diffractaic acid) on human keratinocyte growth. Different species of the parmelioid lichen (*Parmelia nepalensis = Everniastrum nepalense, Parmelia tinctorum = Parmotrema tinctorum*) were used for isolation of lichen metabolites. The sensitivity of the human keratinocyte cell line HaCaT to the lichen acids was evaluated. The tridepside gyrophoric acid, the dibenzofuran derivative (+) usinic acid and the didepside, diffractaic acid exhibit a potent antiproliferative activity and inhibit cell growth (IC 50 values of 1.7, 2.1, and 2.6 micro M, respectively).

Antimitotic Effects

Cardarelli et al. (1997) described several interesting biological properties of usnic acid. It is an antibiotic and it also seems to exert an antimitotic action. Usnic acid can play a role as an environmental indicator, since its concentration varies according to the presence of toxic agents. A series of tests that have been run on different biological systems such as fungi, yeast, plant cells and neoplastic human cell cultures in order to make a general evaluation of the properties of usnic acid and to highlight any analogy between its effects on phylogenetically distant organisms. The result obtained confirm some of the already known properties of usnic acid and identify concentration ranges that are active against cells from different organisms. Furthermore, at low concentrations, the acid displays a capacity to stimulate cell metabolism in some of the biological systems tested.

Antimicrobial Activity

Lauterwein et al. (1995) tested the lichen secondary metabolites against aerobic and anaerobic microorganisms. Vulpinic acid was less active than usnic acid. Gram-negative rods and fungi were not inhibited by these compounds but *Staphylococcus aureus, Enterococcus faecalis, Enterococcus faecium* and some anaerobic species (*Bacteroides* and *Clostridium* species) were found susceptible.

Fournet et al. (1997) isolated the secondary metabolites from some Chilean lichens and studied their activity against the cutaneous leishmaniasis. Usnic acid, pannarine and 1′-chloropannarine were tested against promastigote forms of three strains of *Leishmania* sp. Pannarine and 1′-chloropannarine exhibited *in vitro* activity at 50 mg/ml and (+) usnic acid at 25 μg/ml. BALB/C mice infected with *Leishmania amazonensis* were treated 4 weeks post infection with (+) usnic acid by subcutaneous or oral routes for 15 days at 25 mg/kg or by five intralesional injections at intervals of 4 days at 25 mg/kg of body weight. The reference drug, N-methyl-glucamine antimonate (glucantime) was administered by subcutaneous injections (regimens of 28 mg of pentavalent antimony) for 15 days. The subcutaneous and oral treatments with (+) usnic acid did not produce any effect but by intraleisional administration a significant effect that reduced by 43. 34%, the weight of the lesion and by 72.28% the parasites loads in infected footpads was observed.

Ingolfsdottir et al. (1998) screened for *in vitro* activity of several common lichen metabolites against *Mycobacterium aurum*, a non-pathogenic organism with a similar sensitivity profile as *M. tuberculosis*. Of the different compounds tested, usnic acid from *Cladonia arbuscula* exhibited the highest activity with an MIC value of 32 mg/ml. Atranorin and lobaric acid, both isolated from *Sterocaulon alpinum*, salazinic acid from *Parmelia saxatilis* and protolichesterinic acid from *Cetraria islandica* all showed MIC values ≥125 μg/ml.

Garcia et al. (1999) studied the antimicrobial activity of some lichen products from South Spain and a very high activity against Gram-positive bacteria has been observed in lichens containing usnic acid.

Immunological Active Substances of Lichens

Olafsdottir et al. (1999) investigated the immunologically active compound, (1→3)-(1→4)-α-D-glucan isolated from *Cetraria islandica*. A polysaccharide, Ci-3 resembling isolichenen except with a much higher degree of polymerization has been isolated from the water extract, as well as from the alkali extract of the lichen *Cetraria islandica* purified using ethanol fractionation, dialysis, ion-exchange chromatography and gel filtration. The immunomodulating activity of Ci-3 was tested in an *in vitro* phagocytosis assay and as anti-complementary, and proved to be active in both tests.

Antiviral Activity

Hirabayashi et al. (1989) studied the inhibitory effect of a lichen polysaccharide sulphate, GE-3-S, on the replication of human immunodeficiency virus (HIV) *in vitro*. A sulphate (GE-3-S) prepared by chlorosulfonic acid treatment of GE-3, a partially acetylated b (1-6) glucan of the lichen *Umbilicaria esculenta*, inhibited a cytopathic effect on HIV and suppressed the HIV-antigen expression in cells. GE-3-S also suppressed the giant cell formation of HIV infected cells and inhibited HIV induced plaque formation by 50% at a dose of 19.5 μg/ml and completely inhibited at 250 μg/ml in cells. GE-3-S has no direct effect on the reverse transcription of HIV.

Cohen et al. (1996) investigated the antiviral activities of some anthraquinones, bianthrones and hypericin derivatives derived from lichens, The antiviral activities were compared by the end point CPE (Viral cytopathic effects) methods and plaque assays. Under optimal conditions of exposure to light, hypericin, 7,7′-dichlorophypericin and 5,7-dichloroemodin exhibited strong inhibitory activity against HSV-1 (herpes simplex virus type) in both assays. Partial inactivation of the virus was shown by emodin, 7-chloro-1-o-methyl emodin; the bianthrones and other anthraquinones were found to be inactive. Antiviral activity appeared to be positively correlated with increasing substitution of chlorine in the anthraquinone structure. In the absence of light, only hypericin and 7,7′-dichloropheric in displayed detectable activity.

Antitumor Activity

Japanese workers have studied the antitumor activity of certain lichen species. The antitumor activity of *Umbilicaria, Lasallia, Cetraria, Cladonia, Parmelia* and *Usnea* depends on the polysaccharide components in lichen species.

Nishikawa et al. (1974) studied the antitumor polysaccharides of six lichen species viz. *Lasallia pensylvania* (Hoffm.) Llano, *Usnea rubescens* Stirton, *Cladonia crispata* (Ach.) Floton., *C. mitis* Sandot., *C. rangiferina* (L.) Wigg. and *C. squamosa* (Scop.) Hoffm. From these, a variety of polysaccharide fractions were prepared and their antitumor effects were tested against the sarcoma-180 implanted in mice. It has been suggested that *Lasallia pensylvanica* and *Usnea rubicunda* have polysaccharides with higher antitumor effect than those of other species.

Takashashi et al. (1974) studied the antitumor active polysaccharides of lichens of stictaceae, especially *Lobaria orientalis, L. isidiosa, L. pseudopulmonaria, L. linita, L. japonica, Sticta gracilis* and *S. wrightii.* The studies revealed that the polysaccharide of *Lobaria orientalis* contained at least two antitumor active glycopeptides, LOF-1 and and LOF-2. The water soluble heteropolysaccharide containing glycopeptides occur in several kinds of stictaceae.

Yamamoto et al. (1995) studied the inhibition of tumor promoter induced Epstein-Barr Virus (EBV) activation through tissue culture and thallus extracts of lichens. *Usnea longissima* thallus and *Cetraria ornata* tissue culture showed strong inhibitory activity. The (±) usnic acid, barbatic acid, diffractaic acid 4-o-demethyl barbatic acid and evernic acid were identified as inhibitors of EBV activation by the extract of *Usnea longissima* thallus. Of these compounds, (±) usnic acid exhibited the highest inhibitory activity (IC 50 = 1.0 μM)

Sone et al. (1996) isolated a heteropolysaccharide and a β-D-glucan from *Gyrophora esculenta* (*Umbilicaria esculenta*) cold water and alkali extracts and studied their antitumor activity. Both the galactomannan and (1→6)-β-D-glucan exhibit moderate inhibitory activities on Sarcoma180.

Antifungal Properties of Lichen

The antifungal properties of lichen substances are not well known and much work remains to be done in this field. It has been discovered that the growth of *Neurospora crassa* is strongly inhibited by usnic acid and haematommic acid, a monocyclic phenol derivative of lichen depsides. Water extracts of lichen inhibit wood decaying fungi and mycorrhiza; but it is unclear how these inhibitors are active (Brown and Mikola, 1974; Henningsson and Lundstrom 1970).

Malicki (1965) and Dawson et al. (1984) found that lichen substances of the terricolous lichen, *Cladonia* leached usnic acid in soil and inhibited the growth of bacterial decomposers.

Land and Lundstrom (1998) tested the inhibitory effects of water extracts from *Nephroma articum* and *Peltigera aphthosa* against various yeasts, moulds, blue-strain fungi and rot fungi. *Nephroma articum* water extracts showed a clear fungicidal effect on a diverse range of fungi while *Peltigera apthosa* had no effect.

Shahi et al. (2000) screened the aqueous extract of *Everniastrum cirrhatum* for antifungal activity against human pathogenic fungi, viz. *Epidermophyton floccosum, Microsporum audouinii, M. nanum, M. canis, M. gypseum, Trichophyton mentagrophytes, T. rubrum, T. violaceum* and *T. tonsurans* and found that 40 μl/ml showed fungistatic action and the fungicidal concentration was found to be 60 μl/ml against human pathogenic fungi. Clinical trials showed 35% recovery and 40% improvement against fungal infections.

The antifungal properties of aqueous extracts of *Heterodermia leucomela* against a number of pathogenic fungi were studied by Shahi et al. (2001). The aqueous extracts showed broad spectrum antifungal properties at 80 μl/ml concentrations against some human and plant pathogens. The minimum fungicidal concentrations against human pathogenic fungi were found to be 60 μl/ml and 80 μl/ml. The plant pathogens tested were inhibited at 80 μl/ml.

Conclusion

In recent years most of the lichen secondary metabolites have been identified, many are unique to lichens and have excellent antimicrobial activity.

The experimental studies of lichens are by and large neglected due to the difficulty in culturing lichens. Natural lichens contain many characteristic phenolic compounds. However, they grow too slowly to provide large amounts of material for commercial testing. Certain techniques for culturing lichen tissue *in vitro* are developed recently. Since cultured tissues grow a dozen times faster than natural lichens, this technical method is very useful and important for industrial purposes.

The culture of whole lichen thalli is still difficult and many problems are unsolved. At present workers have succeeded in reconstruction of new lichen thalli from thallus fragments, soredia, spores and hymenial algae or from two components grown separately and then combined. The mycobiont mostly separated from the parent thallus either by spores or from thallus fragments. The cell-aggregates method from thallus fragments seems to be best because it gives the most reproducible results. The cell-aggregates are composed of algal and fungal cells.

All the secondary metabolites which are so characteristic of lichens are of fungal origin. More than 630 secondary metabolities known from lichens, are unique to these organisms and only a small (50-60) occur in other fungi or higher plants.

The major classes of secondary metabolites in lichens belongs to three pathways. The secondary aliphatic acids, their esters and related derivatives and the polyketide derived aromatic compounds of acetyl-polymalonyl pathway, the disester and trisester terpenes and steroids of the mevalonic acid pathway and the terphenylquinones and pulvinic acid derivatives of the shikimic acid pathway.

Information on a variety of products isolated from lichens, shows a wide range of potentially useful biological activities.

Only after improving the cell and molecular methodologies related to lichen tissue culture can these unique plants hope to be exploited commercially for production of novel medicines.

India is a megadiversity region of the world and is represented by the occurrence of more than 2000 species of lichens comprising c.a. 10% of the world's lichen flora. The varied climatic condition and the higher altitudes of Southern India support the luxuriant growth of a variety of lichen species. The medicinal aspects of lichens has taken new directions due to the untiring research work done by Japanese workers who have studied the antitumor and anti HIV activity of lichens. Extraction of anti HIV products from these plants will definitely attract the attention of scientist to conduct more intensive researches in this field. In the Indian context, so far only a few lichen species have been screened for their biological activity. It can therefore be inferred that there is a lot of scope for carrying out such studies in India and there is a vast treasure of medicine hidden in this small, unique group of plants and a more systematic study will definitely unfold new leads to the discovery of some potential medicines.

References

Ahmadjian, V. and Nilsson, S. (1963). Swedish lichens. Yearbook (American Swedish Historical Foundation).

Asahina, Y. and Shibata, S. (1954). Chemistry of lichen substances. Japan Society for the Promotion of Sciences, Tokyo. pp. 240.

Airaksinen, M.M., Peura, P., Ala-Fossi-Salokangas, L., Antere, S., Lukkarinin, J., Saikkonen, M. and Stenback, F. (1986). Toxicity of plant material used as emergency food during famines in Finland. Journal of Ethnopharmacology 18: 273–296.

Azenha, G., Iturriaga, T., Michelangeli, F.I. and Rodriguez, E. (1998). Ethnolichenology, Biological Activity and Biochemistry of Amazonian Lichen Species. Emanations from the rain forest 1(1): 7–23.

Bastien, J.W. (1983). Pharmacopeia of the Qollahuaya Andeans. Journal of Ethnopharmacology 8: 97–111.

Biswas, K. (1956). Common Medicinal Plants of Darjeeling and the Sikkim Himalayas. 90 pp.

Blackwell, W.H. (1990). Poisonous and Medicinal Plants. Prentice-Hall, Englewood Cliffs, N.J. 103 pp.

Brij Lal (1988). Traditional remedies for bone fracture among the tribals of Madhya Pradesh, India. Aryavaidyan 1(3): 190–195.

Brij Lal, Upreti, D.K. and Kalakoti, B.S. (1985). Ethnobotanical utilization of Lichens by the tribals of Madhya Pradesh. Journal of Economic and Taxonomic Botany 7: 203–204.

Brij Lal, and Upreti, D.K. (1995). Ethnobotanical notes on three Indian lichens. Lichenologist 27(1): 77–79.

Brooker, S.G., Cambie, R.C. and Cooper, R.C. (1987). New Zealand Medicinal Plants. Heinemann, Auckland. 63 pp.

Brown, R.T. and Mikola, P. (1974). The influences of fruticose soil lichens upon the mycorrhizae and seedling growth of forest trees. Acta Forestalia Fennica. 141: 1–23.

Cardarelli, M., Serino, G., Campanella, I., Ercole, P., De Cicco, N.F., Alesiani, O. and Rossiello, F. (1997). Antimitotic effects of usnic acid on different biological systems. Cellular and Molecular Life Sciences 53(8): 667–672.

Chandra, S. and Singh, A. (1971). A lichen crude drug ('Chharila') from India. Journal of Research Indian Medicine 6: 209–215.

Chopra, R.N., Chopra, I.C., Handa, K.L. and Kapur, L.D. (1958). Indigenous Drugs of India. Academic Publishers, Calcutta and New Delhi. 645 pp.

Cohen, P.A., Hudson, J.B. and Towers, G.H. (1996). Antiviral activities of anthraquinones, bianthrones and hypericin derivatives from lichens. Experientia 52(2): 180–183.

Davis, E.W. and Yost, J.A. (1983). Novel hallucinogens from eastern Ecuador. Botanical Museum Leaflets (Peabody Museum) 29 (3): 291–295.

Dawson, H.J., Hrutfiord, B.F. and Ugolini, F.C. (1984). Mobility of lichen compounds from *Cladonia mitis* in arctic soil. Soil Sciences 138: 40–45.

el-Fattah, H.A. (1997). A new bioactive metabolite from *Xanthoria sterineri*. Bollettino Chimico Farmaceutico 136(3): 257–261.

Elmore, F.H. (1944). Ethnobotany of the Navago. U. of New Mexico Press. 16 pp.

Emmons, G.T., de Laguna, F. and Low, J. (1991). The Tlingit Indians. U. of Washington Press, The American Museum of Natural History 362 pp.

Felger, R.S. and Moser, M.B. (1985). People of the Desert and Sea; Ethnobotany of the Seri Indians. University of Arizona Press. 217 pp.

Fournet, A., Ferreira, M.E., Rojas, A.A., Torres, O.S.I.A., Yaluff, G., Quilhot, W., Fernandex, E. and Hidalgo, M.E. (1997). Activity of compounds isolated from Chilean lichens against experimental cutaneous leishmaniasis. Comparative Biochemistry and Physiology: C-Pharmacology, Toxicology and Endocrinology 116(1): 51–54.

Garcia, R.J., Garcia, G.M.D. and Saenz, R.M.T. (1999). Some lichen products have antimicrobial activity. Zeitschrift für Naturforschung 54(7-8): 605–609.

Garth, T.W. (1953). Atsugewi ethnography. Anthropology Records. UC Press. 140 pp.

Gonzalez-Tejero, M.R., Martinez-Lirola, M.J. Casares-Porcel, M. and Molero-Mesa,. J. (1995). Three lichens used in popular medicine in Eastem Andalucia (Spain). Economic Botany 49:96–98.

Gülcin, I., Oktay, M., Küfrevioglu, I. Ö and Aslan, A. (2002). Determination of antioxidant activity of lichen *Cetraria islandica* (L.) *Ach.* Journal of Ethnopharmacology. 79: 325–329.

Hale, M.E. (1967). The Biology of Lichens. Edward Arnold Press, London.

Hellson, J.C., Gadd M. (1974). Ethnobotany of the Blackfoot Indians. National Museum of Man Mercury Series, Canadian Ethnology Service 19: 76.

Henningsson, B. and Lundstrom, H. (1970). The influence of lichens, lichen extracts and usnic acid on wood destroying fungi. Material und Organismen 5: 19–31.

Herbal Resources, Inc. (1998). Spilanthes-Usnea Compound: Anti-Fungal Remedy.

Hirabayashi, K., Susumu, I., Ito, M., Shigea, S., Narui, T., Mort, T. and Shibata, S. (1989). Inhibitory effect of a lichen polysaccharide sulfate, GE-3-S, on the replication of Human Immunodeficiency Virus (HIV) *in vitro.* Chemical and Pharmaceutical Bulletin 37(9): 2410–2012.

Houghton, P.J., Manby, J. (1985). Medicinal plants of the Mapuche. Journal of Ethnopharmacology 13: 89–103.

Hu S.-y., Kong, Y.C. and But, P.P.H. (1980). An Enumeration of the Chinese Materia Medica. The Chinese University Press, Hong Kong. 59 pp.

Ingolfsdottir, K, Chung, G.A., Skulason, V.G., Gissurarson, S.R. and Vilhelmsdottir, M. (1998). Antimycobacterial activity of lichen metabolites *in vitro.* European Journal of Pharmaceutical Sciences 6(2): 141–144.

Jayaprakasha, G.K. and Rao, L.J. (2000). Phenolic constituents from the lichen *Parmotrema stuppeum* (Nyl.) Hale and their antioxidant activity. Zeitschrift für Naturforschung 55(11-12): 1018–1022.

Johnson, R.B., Feldott, G. and Lardy, H.A. (1950). The mode of action of the antibiotic, usnic acid. Archives of Biochemistry 28: 317–323.

Kari, P.R. (1987). Tanaina Plantlore. US Nat. Park Service, Alaska Region 176 pp.

Kokwaro, J.O. (1976). Medicinal Plants of East Africa. East African Literature Bureau, Lampala, Nairobi, Dar es Salaam. 10 pp.

Kumar, K. and Upreti, D.K. (2001). *Parmelia* spp. (lichens) in ancient medicinal plant lore of India. Economic Botany 55(3): 458–459.

Kumar, K.C. and Muller, K. (1999a). Lichen metabolites. 1. Inhibitory action against leukoriene B4 biosynthesis by a non-redox mechanism. Journal of Natural Products 62(6): 817–820.

Kumar, K.C. and Muller, K. (1999b). Lichen metabolites. 2. Antiproliferative and cytotoxic activity of gyrophoric, usnic, and diffractaic acid on human keratinocyte growth. Journal of Natural Products 62(6): 821–823.

Land, C.J. and Lundstrom, H. (1998). Inhibition of fungal growth by water extracts from the lichen *Nephroma arcticum.* Lichenologist 30(3): 259–262

Lauterwein, M., Oethinger, M., Belsner, K., Peters, T. and Marre, R. (1995). *In vitro* activities of the lichen secondary metabolites vulpinic acid, (+)-usnic acid, and (-)-usnic acid against aerobic and anaerobic microorganisms. Antimicrobial Agents and Chemother. 39(11): 2541–2543.

Lawrey, J.D. (1986). Biological use of lichen substances. Bryologist 89: 111–112.

Leao, A.M., Buchi, D.F., Iacomini, M., Gorin, P.A. and Oliveira, M.B. (1997). Cytotoxic effect against HeLa cells of polysaccharides from the lichen *Ramalina celastri.* Journal of Submicroscopic Cytology and Pathology 29(4): 503–509.

Lee, K.A. and Kim, M. S. (2000). Glucosidase inhibitor from *Umbilicaria esculenta.* Canadian Journal of Microbiology 46(11): 1077–81.

Leighton, A.L. (1985). Wild plant use by the Woods Cree (Nihithawak) of East-Central Saskatchewan. National Museum of Man, Mercury Series, Canadian Ethnology Service 101: 19 and 20.

Lindley, J. (1849). Medical and Oeconomical Botany. Bradbury and Evans, London. 18 and 21pp.

Llano, G.A.P. (1944). Lichens: their biological and economic significance. The Botanical Review 10 (1): 1–65.

Llano, G.A. (1951). Economic uses of lichens. Ann. Rep. Smiths. Inst. : 385–422.

Llano, G.A.P. (1956). Utilization of lichens in the Arctic and sub-arctic. Economic Botany. 10(4): 367–392.

Lokar, L.C. and Poldini, L.(1988). Herbal remedies in the traditional medicine of the Venezia Giulia Region (north east Italy). Journal of Ethnopharmacology 22: 231–278.

Malicki, J. (1965). The effect of lichen acids on the soil microorganisms. Part I. The washing down of the acids into the soil. Annals Universit. Marie Curie Sklodowska, Sectio C. Biologia 20: 239–248.

Marshak, A. and Fager, J. (1950). Prevention of nuclear fusion and mitosis and inhibition of deoxyribonuclease by d-usnic acid. Journal of Cellular Comparative Physiology, 35: 317–329.

Mead, G.R. (1972). The ethnobotany of the CA Indians: a compendium of the plants, their users, and their uses. Mus. Anth., Occ. Publ. in Anth., Ethnol. Ser. 30. U. Northern CO, Greeley, CO. p 91.

Miao, V., Coeffect, Le Gal, Le Gal M.F., Brown, D., Sinnemann, S., Donaldson, G. and Davies, J. (2001). Genetic Approaches to harvesting lichen products. Trends in Biotechnology. 19(9): 349–355.

Moerman, D. (1998). Ethnobotany of Native America. www.umb.umich.edu/cgi-bin/herb/.

Moskalenko, S.A. (1986). Preliminary screening of far-eastern ethnomedicinal plants for antibacterial activity. Journal of Ethnopharmacology 15: 321–259.

Nadkarni, K.M. (1976). Indian Materia Medica. (ed), Nadkarni, A.K. pp 922.

Nishikawa, Y., Ohki, K., Takahashi, K., Kurono, G., Fukuoka, F. and Emori, M. (1974). Studies on the water soluble constituents of lichens. II. Antitumour polysaccharides of *Lasallia, Usnea* and *Cladonia* species. Chemical and Pharmaceutical Bulletin. 22(11): 2691–2702.

Novaretti, R. and Lemordant, D. (1990). Plants in the traditional medicine of the Ubaye Valley. Journal of Ethnopharmacology 30: 1–34.

Olafsdottir, E.S. and Ingolfsdotter, K. (2001). Polysaccharides from lichens: structural characteristic and biological activity. Planta Medica. 67(3); 199–208.

Olafsdottir, F.S., Ingolfsdottir, K., Barsett, H., Paulsen, B.S., Jurcic, K. and Wagner, H. (1999). Immunologically active (1→3)–(1→4)-α-D-glucan from *Cetraria islandica.* Phytomedicine 6(1): 33–39.

Osorio, H.S. (1982). Contribution to the lichen flora of Uruguay XVII. The scientific name of the "Yerba de la Piedra". Phytologia 52: 217–220.

Oswalt, WH. (1957). A Western Eskimo ethnobotany. Anthropological Papers of the University of Alaska 6(1): 16–36.

Pennington, C.W. (1963). The Tarahumar of Mexico. University of Utah Press. p. 193.

Pennington, CW. (1969). The Tepehuan of Chihuahua. University of Utah Press. p 188.

Perry, N.B., Benn, H.M., Brennan, N.J., Burgess, E.J., Elliss, G., Galloway, D.J., Lorimer, S.D. and Tangney, R.S. (1999). Antimicrobial, Antiviral and Cytotoxic activity of New Zealand Lichens. Lichenologist 31(6): 627–636.

Powers, S. (1874). Aboriginal botany. Proceedings of the California Academy of Sciences 5: 373–379.

Racine, Ph., Hartmann, V.E. and Tollard D'Audiffret, Y. (1980). Antioxidant properties of wax from Yugoslavian oakmoss (*Evernia prunastri*). International Journal of Cosmetic Science (2):305–313.

Richardson, D.H.S. (1988). Medicinal and other economic aspects of lichens. In: CRC Handbook of Lichenology. Volume III. Galun, M., ed. 93–108.

Richardson D.H.S. (1991). Lichens and man. In: Frontiers in Mycology, (ed), Hawksworth, D.L. Oxford, UK: CABI. 187–210.

Rojas, I.S., Lotina-Hennsen, B. and Mata, R. (2000). Effect of lichen metabolites on thylakoid electron transport and photophosphorylation in isolated spinach chloroplasts. Journal of Natural Products. 63(10): 1396–1399.

Romagni, J.G., Meazza, G., Nanayakkara, N.P. and Dayan, F.E. (2000). The phytotoxic lichen metabolite, usnic acid, is a potent inhibitor of plant *p*-hydroxyphenylpyruvate dioxygenase. FEBS Letters 480(2–3): 301–305.

Saklani, A. and Upreti, D.K. (1992). Folk uses of some lichens in Sikkim. Journal of Ethnopharmacology 37: 229–233.

Scheidegger, C. (1998). Personal communication.

Schneider, A. (1904). A Guide to the Study of Lichens. Knight and Miller, Boston. 24 p.

Shahi, S.K, Shukla, A.C., Upreti, D.K. and Dikshit, A. (2000). Use of lichens as antifungal drugs against superficial fungal infections. Journal of Medicinal and Aromatic Plant Sciences 22 (4A) & 23 (1A): 169–172.

Shahi, S.K., Shukla, A.C., Dikshit, A. and Uperti, D.K. (2001). Broad spectrum antifungal properties of the lichen, *Heterodermia leuocomela.* Lichenologist 33(1): 177–179.

Singh, K.K., Upreti, D.K. and Kaushal Kumar (2000). Ethnobotanical note on some less known Himalaya Lichens. Bulletin British Lichen Society 86: 36–37.

Smith, A.L. (1921). Lichens. Chapter X. Economical and technical. Cambridge Univ. Press. p 408.

Smith, G.W. (1973). Arctic Pharmacognosia. Arctic 26(4): 328–333.

Smith, H.H. (1923). Ethnobotany of the Menomini Indians. Bulletin of the Public Museum of the City of Milwaukee 4 (1): 21–60.

Smith, H.H. (1933). Ethnobotany of the Forest Potawatomi Indians. Bulletin of the Public Museum of the City of Milwaukee 7 (1): 68.

Sone, Y., Isoda-Johmura, M. and Misaki, A. (1996). Isolation and chemical characterization of polysaccharides from Iwatake, *Gyrophora esculenta* Miyoshi. Bioscience, Biotechnology, and Biochemistry. 60(2): 213–215.

Stevens, J. (1983). Traditional Medicine Project. Avataq Cultural Institute 35 pp.

Strickmann, M. (1988). Lichen notes from all over. Personal communication unpubl. notes p. 2.

Subrahmanian, S.S. and Ramakrishnan, S. (1964). Amino acids of *Peltigera cannina*. Current Science 33: 522.

Takashashi, K., Takeda, T., Shibata, S., Inomata, M. and Fukuoka, F. (1974). Polysaccharides of lichens and fungi – VI. Antitumor active polysaccharides of lichens of Stictaceae. Chemical and Pharmaceutical Bulletin 22(2): 404–408.

Teit, J.A. and Boas F. (1928). The Salishan tribes of the western plateaus. Annual Reports of the Bureau of American Ethnology 45: 23–396.

Turner, N.J. (1973). The ethnobotany of the Bella Coola Indians of British Columbia. Syesis 6: 193–220.

Turne, N.J. (1977). Economic importance of black tree lichen (*Bryoria fremontii*) to the Indians of western North America. Economic Botany 31: 461–470.

Turner, N.J and Efra, B.S. (1982). Ethnobotany of the Hesquiat Indians of Vancouver Island. British Columbia Provincial Museum Cultural Recovery Paper No. 2: 26–99.

Turner, N.J., Bouchard, R. and Kennedy, D.I.D. (1980). Ethnobotany of the Okanagan-Colville Indians of British Columbia and Washington. Occasional Papers of the British Provincial Museum No. 21, Province of British Columbia. p. 14–15.

Turner, N.J., Thoma, J. and Carlson, R.T. (1983). Ethnobotany of the Nitinaht Indians of Vancouver Island. British Columbia Provincial Museum Occas. Paper No. 24: 56–165.

Upreti, D.K. and Negi, H.S. (1995). Lichens of Nanda Devi Biosphere Reserve, Uttar Pradesh, India. Journal of Economic and Taxonomic Botany 19(3): 627–636.

Varita, K.O. (1973). Antibiotics in lichens. In: The Lichens, (Eds. Ahmadjian, V. and Hale, M.E.), pp. 547–561. Academic Press, New York and London.

Vogel, V.J. (1970). American Indian Medicine. U. of Oklahoma Press, Norman. 236 pp.

Wasson, R.G. (1962). The hallucinogenic mushrooms of Mexico and Prylocifin: A bibiliography. Botanical Museum Leaflet, Harvard University 20: 25–73.

Watt, J.M., Breyer-Brandwijk, M.R. (1962). The Medicinal and Poisonous Plants of Southeastern and Eastern Africa E. and S. Livingstone, Ltd., Edinburgh and London. 1132 pp.

Wennekens, A.J. (1985). Traditional Plant Usage by Chugach Natives around Prince William Sound and on the Lower Kenai Peninsula, Alaska. M.A. Thesis. University of Alaska, Anchorage. 39 pp.

Yamamoto, Y., Miura, Y., Higuchii, M., Kinoshita, Y. and Yoshimura, Y. (1993). Using lichen tissue cultures in modern biology. Bryologist 96: 384–393.

Wolseley, P. (1998). Personal communication.

Yamamoto, Y., Miura, Y., Kinoshita, Y., Higuchi, M., Yamada, Y., Murakami, A., Ohigashi, H. and Koshimizu, K. (1995). Screening of tissue cultures and thalli of lichens and some of their active constituents for inhibition of tumor promoter-induced Epstein-Barr virus activation. Chemical and Pharmaceutical Bulletin. 43(8): 1388–1390.

Younos, C., Fleurentin, J., Notter, D., Mazars, G., Mortier, F. and Pelt, J.M. (1987). Repertory of drugs and medicinal plants used in traditional medicine of Afghanistan. Journal of Ethnopharmacology 20: 245–290.

Zaveneld, J.S. (1959). The utilization of marine algae in Tropical South and East Asia. Economic Botany 3: 89–131.

Zorer, R. (1998). Personal communication.

12. Diversity and Applications of Endophytic Fungi Isolated from Tropical Plants

João Lúcio Azevedo and Welington Luiz de Araujo
Department of Genetics, Escola Superior de Agricultura "Luiz de Queiroz", Universidade de São Paulo, P.O. Box 83, 13400-970 Piracicaba, São Paulo, Brazil

SUMMARY Endophytes are all microorganisms culturable or not, that inhibit the interior of plant tissues, causing no harm to the host and that do not develop external structures, excluding in this way nodulating bacteria and mycorrhizal fungi, which although are endophytes will not be included in the present chapter. The use of these microorganisms in agriculture has increased consistently during the last years, mainly their using in plant growth promotion or pest control as substitutes to xenobiotic products, preserving the environment. Besides, these microorganisms have been used as an important source for new drugs and biotechnological processes. At least 98 species were already isolated from tropical plants and among these, some have shown potential for production of drugs against cancer as well as to inhibit the bacterial and fungal growth. The diversity and biotechnological potential of endophytic fungi are listed and include toxin production as well as the influence of these compounds on plant and livestock and how their production may be affected by genetic and environmental conditions. As the literature has shown, there is a lack of information on endophytes from tropical hosts, which due to their diversity could host many new fungi species that could provide new pharmaceutical compounds. Having this in mind, we have included an updated and extensive literature in this review, concerning new findings from tropical plants, including the characterization of endophytic fungi from several Amazon trees.

Endophytes: General Aspects

Plant have the ability to interact with different microorganisms such as fungi and bacteria. This allows them to establish themselves in their habitat. The use of microorganisms in agriculture has increased consistently during the last years. A key role has been played by endophytic microorganisms, which naturally or deliberately have been used to promote plant growth or to control pests as substitutes to xenobiotic products, thus preserving the environment

Definitions

First described by de Bary (1866), endophytes are defined as microorganisms that colonize internal plant tissues. Carroll (1986) defined endophytes as asymptomatic microorganisms living inside plants, while Petrini (1991) defined them as microorganisms that inhabit, at least for one period of their life cycle, plant inner tissues without causing any apparent harm to the host. Wilson (1995) gave an aditional definition, considering endophytes, as bacteria and fungi that for all or part of their life cycle invade the tissues of living plants and that cause unapparent and asymptomatic infections to the plant host. Hallmann et al. (1997) described them, as those that may be isolated from surface sterilized plant parts or extracted from inner tissues and cause no damage to the host plant. It has now been shown by molecular techniques that microorganisms that are not culturable on usual media and conditions, can be found inside plants. In the present article a modified version of the definition proposed by Azevedo et al. (2000) will be used: endophytes will be considered to be all microorganisms that are culturable or not, that inhabit the interior of plant tissues, causing no harm to the host and that do not develop external structures, excluding in this way nodulating bacteria and mycorrhizal fungi.

Importance of Endophytic Microorganism

Endophytes were initially considered neutral, not causing benefits nor showing detrimental effects to their hosts. Further studies indicate that, in many cases, they have an important role in host plant protection, acting against predators and pathogens. Landmark reports on the subject started in 1981. From 1981 to 1985, which may be considered a historical period, it was shown that protection of plants against herbivores, can be provided by endophytic microorganisms. For example, the presence endophytes could result in the reduction of insect attacks (Azevedo et al. 2000). These microorganisms have a mutualistic interaction with the host plant, which provides nutrients and protection to the endophytic fungi and receive chemical compounds such as enzymes, alkaloids and antibiotics. These compounds may increase plant resistance against biotic and abiotic stress. These observations resulted in increasing studies of endophytic fungi, mainly for their potential biotechnological value.

Vertical and Horizontal Transmission and Isolation of Endophytes

Although endophytic microorganisms may be isolated from seeds in some plant species, usually they are not vertically transmited through seeds. Colonization of different plant tissues and organs begins by endophyte penetration through natural or artificial openings, such as stomata, root emission zone, injuries caused by agricultural practices and root growth. After penetration, these endophytes colonize the host, being found in all tissues and plant organs, inhabiting the apoplast, plant vessels and in some cases the interior of cells. Inside the host plant or during colonization, endophytes interact with epiphytic, pathogenic and/or other endophytic microorganisms, resulting in an equilibrium of the microbial community associated with the plant. In some cases, when this equilibrium is broken, endophytic microorganisms may be pathogenic (Di Fiori and Del Gallo, 1995).

Endophytic fungi can be detected by direct microscopic examination of plant tissues, but most of the work with endophytes is carried out by isolation of fungi from hosts. Usually the first step is the sterilization of the outer surface of the plant, followed by the transfer of plant fragments to appropriate culture media. After incubation, fungi can be isolated, purified and characterized. According to the objectives of the research and plant host species, several different variations of the basic procedure may be applied (Bacon, 1990; Araujo et al. 2002).

Ubiquity of Endophytic Fungi

Fungal endophytes have been isolated from practically all studied plants. There are reports of endophytes isolated from lichens, moss, ferns, gymnosperms (Petrini, 1986; Petrini et al. 1990; Carroll and Carroll, 1978) and from both monocotyledoneous and dicotyledoneous plants, ranging from wood tree species, such as *Eucalyptus* (Bettucci et al. 1999) to herbaceous crop plants, such as maize (Fisher et al. 1992; Pamphile and Azevedo, 2002) and *Stylosanthes* (Pereira et al. 1993). They also have been isolated from plants growing in several different environments as pastures, forests, mangroves and agricultural fields. There is some degree of host specificity, but some fungi can be found in many different hosts (Isaac, 1992; Azevedo, 1997; 1998). Their occurrence may be affected by organ or tissue age, by the season of the year, by plant genotype, anthropogenic and several other factors. It has been proposed that each plant species has endophytic microorganisms still not classified and with potential biotechnological interest. Since, it is known that the plant biodiversity in tropical and subtropical area is higher than in temperate climates, the biodiversity of endophytic fungi associated inside these plants should be also higher than that observed in plants from temperate regions.

The distinction among endophytes, epiphytes and plant pathogenic fungi is not always clear. There is a gradient among them (Redlin and Carris, 1996; Azevedo et al. 2000). According to Strobel and Daisy (2003), although the symptomless nature of endophytic occupation in plant tissue has been proposed as a symbiotic or mutualistic relationship. The observed biodiversity of endophytes suggests that they may also be aggressive, acting as saprophytes or opportunistic pathogens. Some species of endophytic fungi are taxonomically related to phytopathogens in the same plant host, as observed in *Citrus sinensis*, which is colonized by the endophytic fungi *Guignardia mangiferae* (syn. *G. endophyllicola*). This fungal species is phylogenetically related to *G. citricarpa*, the causal agent of the citrus black spot (Baayen et al. 2002). Mutations may change the relation between plant and fungus, as observed in the phytopathogen *Colletotrichum magna*, which after a single mutation, colonized endophytically the host plant. Suggesting that genetic differences between endophytes and pathogens belonging to the same species may be small (Freeman and Rodriguez, 1993).

This article focusses mainly on tropical endophytic fungi diversity and biotechnological application of these microorganisms. Recent studies have demonstrated that these endophytes can be used as vectors to provide new characteristics with biotechnological interest to the host plant. In this aspect, endophytic fungi can be genetically modified and express heterologous genes. They can be therefore be used to control pathogens, promote plant growth and produce vitamins, amino acids and vaccine inside the host plant. Therefore, it is extremely important to look at endophytes as microorganisms with biotechnological potential besides their biological role. These topics will be discussed in this chapter, including results that have been obtained in studies with endophytic fungi in indigenous and cultivated plant species in Brazil.

Biodiversity of Endophytic Fungi from Tropical Plants

Several authors have stressed the abundance of endophytic fungi in tropical regions but till recently, only scanty information was available about endophytic fungi isolated from the tropical forests. Petrini and Dreyfus (1981) were one of the first to isolate tropical endophytic fungi, from Araceae, Bromeliaceae and Orchidaceae in French Guyana followed by endophytes isolation from Piperaceae and Crassulaceae in Brazil and Colombia (Dreyfuss and Petrini, 1984). Rodrigues and Samuels (1990) isolated, for the first time, endophytic fungi from a tropical palm tree (*Licuala ramasayi*) growing in the rainforest of Queensland, Australia. The same authors isolated endophytic fungi from other Australian and Brazilian palm tree species. In some instances, as was the case of endophytes isolated from açai (*Euterpe oleraceae*) a palm species common in floodplains of the Brazilian Amazon basin, some new fungal species were described. The authors showed the potential of these microorganisms for pharmaceutical purposes (Rodrigues, 1991; Rodrigues and Samuels, 1992; Rodrigues, 1994; 1996). These and related findings increased the interest to investigate endophytes from tropical plants, that have different morphology (Fig. 1) and biotechnological potential. Research in nineties was carried out by groups based mainly in Brazil, India and Central America. In Brazil, besides endophytes isolated from palm trees already mentioned (see Rodrigues, 1996 for a review), endophytic fungi were isolated from banana (*Musa acuminata*), and *Stylosanthes* (Pereira et al. 1993; 1999), maize (*Zea mays*) by Pimentel (2001), soybean (*Glycine max*) by Mendes et al. (2001) and Pimentel (2001), several species of citrus (Gai et al. 2000, Araújo et al. 2001; Araújo et al. 2002a; Glienke-Blanco et al. 2002), cocoa (*Theobroma cacao*) by Rubini (2004), *Anacardium occidentale*, known as cashew tree (Medeiros, 1988; Oliveira, 1999), *Pachystrizus erosus* (Stamford, 1997) Guarana (*Paulinia cupana*) (Guimarães, 1998), cupuaçu (*Theobroma grandiflorum*), *Pueraria phaseoloides* and *Scleria pterota* (Galvão, 1998), graminaceous plants as *Brachiaria brizantha* cv. Marandu, *Brachiaria humidicola Brachiaria decumbens*, *Digitaria insularis*, *Paspalum notatum* and *Rhyncheritrum roseum* (Alquati, 1999), *Vigna*

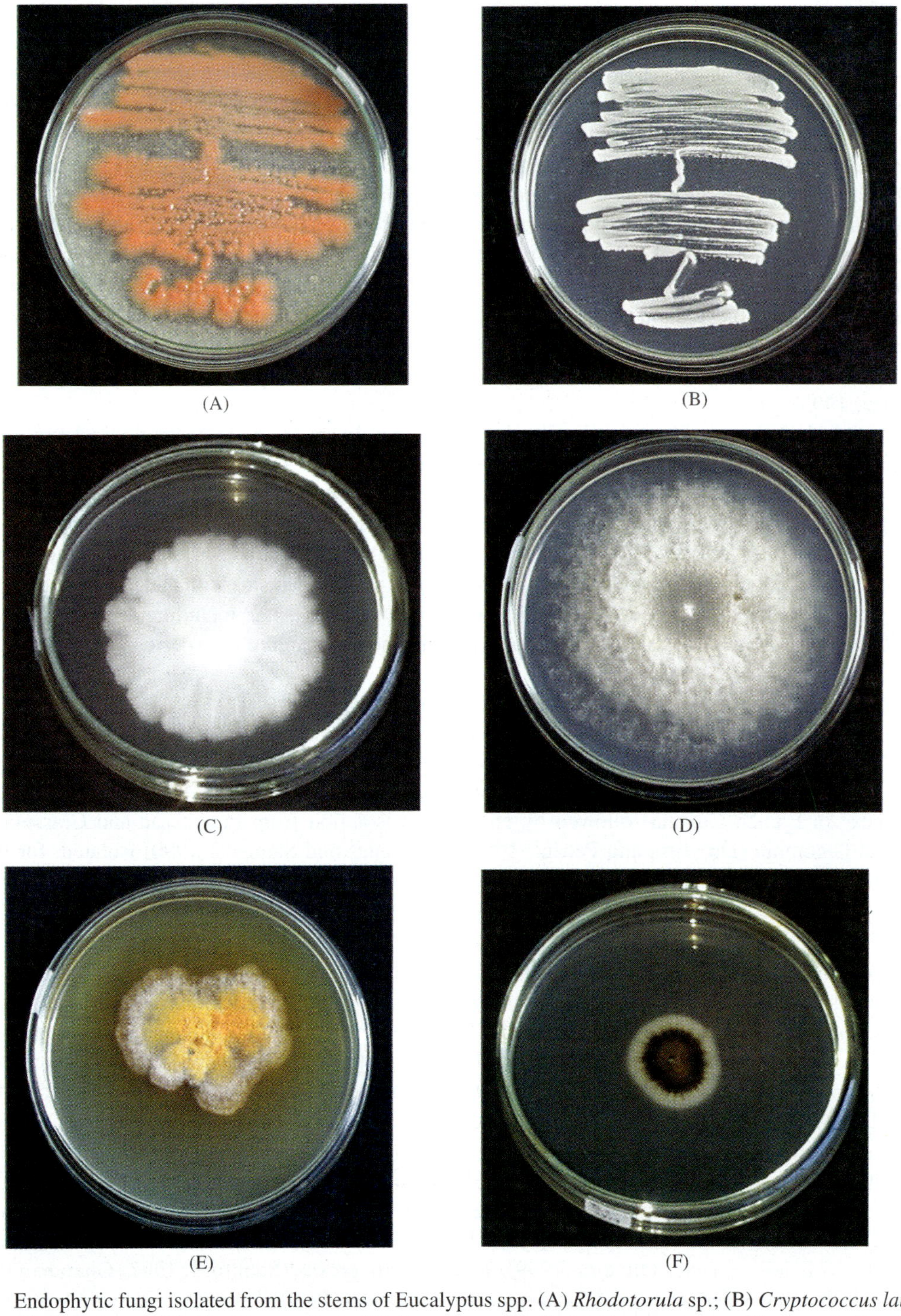

Fig. 1. Endophytic fungi isolated from the stems of Eucalyptus spp. (A) *Rhodotorula* sp.; (B) *Cryptococcus laurentii*; (C) *Xylaria* sp.; (D) *Mycelia sterilia* 1; (E) *Mycelia sterilia* 2; (F) *Hormonema* sp. Pictures kindly given by Aline Romão and Ricardo Yara, ESALQ/University of São Paulo, Brazil.

unguiculata (Rodrigues, 1999), *Copaifera multijuga* (Barbosa, 2001), coffee (*Coffea arabica*) by Yamada (1999), *Ilex paraguaensis* (Penna, 2000; Takeda et al. 2003), the Anacardiaceae *Spondias mombin* from Rio de Janeiro and from the Amazonian state of Pará (Rodrigues and Samuels, 1999; Rodrigues et al. 2000), several other amazonian plants such as *Himanthus sucuuba* and *Paulicorea longiflora* and *Stryctupos cugens* (Magalhães, 2000; Souza, 2001). Studies on tropical endophytes in India include neem (*Azadirachta indica*), a tropical native plant of that country (Rajagopal and Suryanarayanan, 2000), mangrove trees as *Rhizophora apiculata* and *R. mucronata* halophytes from mangrove habitats and the seagrass *Halophyla ovalis* (Suryanarayanam et al. 1998; Suryanarayanam and Kumaresan, 2000; Kumaresan and Suryanarayanan, 2002; Maria and Sridhar, 2003; Devarajan et al. 2002), *Cuscuta reflexa* and its angiosperm hosts (Suryanarayanan et al. 2000), *Ficus benghalensis* (Suryanarayanan and Vijaykrishna, 2001), and *Plumeria rubra* (Suryanarayanan and Thennarasan, 2004). From upper South America, Central America and Mexico, endophytic fungi were isolated from the Compositae *Gynoxis oleifolia* in Ecuador (Fisher et al. 1995) cocoa (*Theobroma cacao*) and from dicotyledoneous neotropical trees in Panama (Arnold et al. 2001; Arnold and Herre, 2003; Arnold et al. 2003); from the Sapotaceae *Manilkara bidentata* (Lodge et al. 1996), from a tropical timber tree (*Guarea guidonia*) and orchids such as *Lepanthes rupestris* and others in Puerto Rico (Gamboa and Bayman, 2001; Bayman et al. 2002; Otero et al. 2002); from maize (*Zea mays*) in Costa Rica (Danielsen and Jensen, 1990); from several tree species in the Iwokrama forest reserve in Guyana (Cannon and Simmons, 2002; Lu et al. 2004); from two species of palms (*Sabal bermudana* and *Livistona chinensis*) from Bermuda (Southcott and Johnson, 1997) and from the weed *Parthenium hysterophorus* in Mexico (Romero et al. 2001). Endophytic fungi were isolated in Brunei Darussalam from australian palms of the genus *Licuala* (Frölich et al. 2000), from the saltbush *Atriplex vesicatoria* in Australia (Cother and Gilbert, 1994) and from banana (*Musa acuminata*) in Thailand (Photita et al. 2001) Australia and Hong-Kong (Brown et al. 1998).

After the increase in the number of plant species evaluated, not only the diversity of endophytic fungi from tropical plants was shown, but also the biotechnological potential that had not been exploited. As shown in Table 1, plants from tropical areas may be colonized by a great diversity of genera and species, including basidiomycetes and ascomycetes, such as *Acremonium* sp., *A. implicatum, Alternaria* sp., *A. alternata, A. zinniae, A. helianthi, Ampullifera* sp., *Ascochyta* sp., *Aspergillus* sp., *A. flavipes, A. glaucus, A. niger, Aureobasidium, A. pullulans, Beauveria* sp., *B. bassiana, Blastomyces* sp., *Botryodiplodia* sp., *Botryotrichum* sp., *Botryosphaeria* sp., *Camarosporium* sp., *Candida famata, Ceratobasidium* sp., *Chaetomium* sp., *Cladosporium* sp., *Colletotrichum* sp., *C. gloeosporioides, Cordana musa, Cryptococcus laurentti, Curvularia* sp., *C. brachyspora, C. lunata, Cylindrocarpon* sp., *Dactylaria* sp., *Deightoniella torulosa, Dreschlera* sp., *Dreschslera halodes, Epicoccum purpuracens, Fusarium* sp., *F. anthophilum, F. avenaceum, F. equiseti, F. moniliforme, F. oxysporum, F. sentectum, F. solani, Fusicoccum* sp., *Geotrichum* sp., *Gliocadium* sp., *G. roseum, Glomerella* sp., *G. cingulata , Guignardia* sp., *G. citricarpa, G. philoprina, G. cocoicola, Humicola* sp., *Lasiodiplodia* sp., *Lasiodiplodia theobromae, Monilochaetes* sp., *Mucor* sp., *Nigrospora oryzae, Nodulisporium* sp., *Paecilomyces* sp., *Papulosporium* sp., *Penicillium* sp., *Penicillium funiculosum, Periconia* sp., *Pestalotia palmarum, Pestalotiopsis* sp., *Pithomyces* sp., *Phialophora* sp., *Phomopsis* sp., *P. anacardii, Phylloslicta* sp., *P. sapotae, Phaeotrichoconis* cf. *crotalariae, Phoma* sp., *Pleurotus* sp., *Pseudeurotium* sp., *Pseudoseptoria* sp., *Pyriculariopsis parasitica, Rhizopus* sp., *Rhodotorula glutinis, Scopulariopsis* sp., *Septochyta* sp., *Sporormiella minima, Syncephalastrum* sp., *Thanatephorus* sp., *Thozetellopsis* sp., *Torula* sp., *Trichoderma* sp., *Ustilago* sp., *Verticillium* sp., *Xylaria* sp. and *X. cubenses*.

Even though it remains difficult to compare earlier and more recent studies on isolated endophytic fungi, certain trends are apparent with predominant fungal types isolated as endophytes.

Table 1. Some endophytic fungi isolated from plants cultivated in tropical countries (for a review till 1996 see Rodrigues and Petrini, 1997

Host plant	Endophytic fungi	References
Anacardium occidentale	*Acremonium, Alternaria* sp., *Colletotrichum gloeosporioides, Fusarium solani, Guignardia* sp., *Lasiodiplodia* sp., *Pestalotia* sp., *Phomopsis* sp., *P. anacardii, Phaeotrichoconis* cf. *crotalariae, Phoma* sp., *Septoria, Stagonospora* sp.	Medeiros (1988), Oliveira (1999) and Freire and Bezerra (2000)
Azadirachta indica	*Fusarium avenaceum*	Rajagopal and Suryanarayanan (2000)
Brachiaria brizantha	*Acremonium implicatum*	Kelemu et al. (2001)
Citrus spp.	*Rhodotorula glutinis, Candida famata* (teleomorph *Debaryomyces hansenii*), *Cryptococcus laurentti, Colletotrichum gloeosporioides, Guignardia citricarpa, Cladosporium* sp.	Gai et al. (2000); Glienke-Blanco et al. (2002); Araújo et al. (2001)
Cuscuta reflexa	*Phomopsis* sp., *Cladosporium* sp., *Colletotrichum* sp., *Acremonium* sp., *Alternaria alternata, A. longissima, Aspergillus flavipes, A. niger, A. terreus, A. versicolor, Aureobasidium, Curvularia lunata, C. tuberculara, Dreschslera halodes, D. Hawaiiensis, Fusarium* sp., *Nigrospora* sp., *Nodulisporum* sp., *Penicillium* sp., *Periconia* sp., *Pithomyces* sp., *Thozetellopsis* sp., *Torula* sp., *Trichoderma* sp., *Botryodiplodia* sp., *Pestalotiopsis* sp., *Phoma* sp., *Phomopsis* sp., *Phyllosticta* sp., *Pseudoseptoria* sp., *Septochyta* sp., *Glomerella* sp., *Guignardia* sp., *Sporormiella minima.*	Suryanarayanan et al. (2000)
Euterpe oleracea	*Colletotrichum gloeosporioides, F. sentectum, F. oxysporum, Pestalotia palmarum, Xylaria cubense*	Rodrigues (1999); Rodrigues and Samuels, (1992)
Ficus benghalensis	*Sporormiella minima, Lasiodipolodia theobromae, Colletotrichum gloeosporioides, Fusicoccum* sp., *Phoma* sp., *Phomopsis* sp., *Phyllosticta* sp., *Aspergillus niger, Aspergillus* sp., *Aureobasidium pullulans, Curvularia lunata, Fusarium* sp., *Gliocladium roseum, Gliocadium* sp., *Paecilomyces* sp., *Penicillium* sp., *Phialophora* sp., *Trichoderma* sp.	Suryanarayanan and Vijaykrishma (2001)
Glycine max	*Ascochyta* sp., *Cladosporium* sp., *Colletotrichum* sp., *Nodulisporium* sp., *Phyllosticta* sp. and *Xylaria* sp.	Pimentel (2001)
Glycine max	*Ampelomyces quercinus, Cochliobolus intermedius, Colletotrichum gloeosporioides, Didymella bryoniae, Eurotium amstelodami, Fusarium denticulatum, F. dlaminii, F. oxysporum* f. sp. *vasinfectum, Macrophomina phaseolina, Nectria haematococca and Peniophora au rantiaca*	Mendes et al. (2001)
Gramineas	*Acremonium* sp., *Fusarium* sp. and *Ustilago* sp.	Alquati (1999)
Guarea guidonia	*Phomopsis, Colletotrichum, Xylaria, Rhizoctonia*-like fungi	Gamboa and Bayman (2001)
Halophytes from 3 dicotiledon families	*Camarosporium, Colletotrichum, Phomopsis, Phyllosticta, Sporormiella minima, Acremonium, Alternaria, Ampullifera, Cladosporium, Fusarium* sp, *Guignardia* sp.	Suryanarayanan and Kumaresan, (2000)

Host plant	Endophytic fungi	References
Himanthus sucuuba	*Colletotrichum* sp., *Guignardia* sp., *Glomerella* sp., *Phomopsis* sp., *Pestalotia* sp. and *Xylaria* sp.	Magalhães (2000)
Ilex paraguariensis	*Acremonium* sp., *Aspergillus* sp., *Colletotrichum* sp., *Fusarium* sp., *Guignardia philoprina*, *Paecilomyces* sp., *Papulosporium* sp., *Penicillium* sp., *Nodulisporium* sp. *Xylaria* sp., *X. cubensi*	Penna (2000) and Takeda et al. (2003)
Lepanthes rupestris	*Guignardia*, *Colletotrichum*, *Xylaria* sp.	Bayman et al. (2002)
Licula sp.	*Xylaria* sp.	Frölich et al. (2000)
Livistona chinwnsis	*Idriella* sp. *and Aspergillus* sp.	Southcott and Johnson (1997)
Mangifera indica	*Alternaria arborescens*, *A. tenuissima*, *Botryosphaeria parva*, *Diaporthe helianthi*, *Eutypella scoparia*, *Glomerella cingulata*, *Guignardia mangiferae*, *Hypoxylon stygium*, *Nodulisporium* sp., *Stenocarpella maydis* and *Xylaria longipes*	Silva (personal communication)
Manilkara bidentata	*Xylaria* cf. *multiplex*, *X.* cf. *adscendens*, *Colletotrichum*, *Pestalotiopsis versicolor*, *Fusarium solanum*, *Phylosticta sapotae*, *Glomerella cingulata*	Lodge et al. (1996)
Musa acuminata	*Xylaria* spp., *Guignardia cocoicola*, *Colletotrichum gloeosporioides*, *Deightoniella torulosa*, *Pyriculariopsis parasitica*, *Dactylaria* sp.	Photita et al. (2001)
Musa sp.	*Alternaria* sp., *Aspergillus* sp., *Cordana musa*, *Curvularia* sp., *Dreschlera* sp., *Epicoccum purpuracens*, *Fusarium* sp., *Glomerella cingulada*, *Humicola* sp., *Nigrospora oryzae*, *Periconia* sp., *Phomopsis* sp., *Phyllosticta* sp., *Trichoderma* sp., *Xylaria* sp.	Photita et al. (1999)
Orchids	*Rhizoctonia*-like fungi, *Ceratobasidium*, *Tulasnella*, *Thanatephorus*	Otero et al. (2002)
Pachystrizus erosus	*Mucor* and *Rhizopus*	Stamford (1997)
Parthenium hysterophorus	*Alternaria zinniae*, *A. helianthi*, *Cylindrocarpon* sp., *Curvularia brachyspora*, *Fusarium* sp., *Nigrospora oryzae*, *Penicillium funiculosum*, *Periconia* sp.	Romero et al. (2001)
Paulicourea longiflora	*Aspergillus niger*, *Colletotrichum* sp., *Guignardia* sp., *Glomerella* sp., *Phomopsis* sp. and *Xylaria* sp.	Souza (2001)
Paullinia cupana	*Guignardia* sp., *Phomopsis* sp., *Glomerella cingulada*, *Xylaria* sp.	Guimarães (1998)
Pueraria phaseoloides	*Glomerella cingulada*, *Humicola* sp., *Xylaria* sp.	Galvão (1998)
Rhizophora apiculata and *R. mucronata*	*Acremonium* sp., *Alternaria* sp., *Aspergillus glaucus*, *Botryotrichum* sp., *Chaetomium* sp., *Cladosporium* sp., *Glomerella* sp., *Nigrospora* sp., *Penicillium* sp., *Phialophora* sp., *Pestalotiopsis* sp., *Phoma* sp., *Phomopsis* sp., *Phyllosticta* sp., *Pseudeurotium* sp., *Sporormiella minima*, *Thielavia* sp., *Trichoderma* sp.	Suryanaraanyan et al. (1998)
Saccharum sp.	*Athelia bombacina*, *Cochliobolus homomorphus*, *C. intermedius*, *C. melinidis*, *Exserohilum mcginnisii*, *E. rostratum*, *Leptosphaeria bicolour* and *Xylaria longipes*	Araújo and Kuklinsky-Sobral, 2004
Scleria pterota	*Curvularia* sp., *Glomerella cingulata* and *Pestalotia* sp.	Galvão (1998)
Spondias mombin	*Colletotrichum gloeosporioides*, *Guignardia* sp., *Phomopsis* sp.	Rodrigues and Samuels (1999)

(Contd.)

Host plant	Endophytic fungi	References
Stryctupos cugens	*Aspergillus niger*, *Colletotrichum* sp., *Guignardia* sp., *Trichoderma* sp.	Souza (2001)
Stylosanthes sp.	*Alternaria* sp., *Glomerella cingulada*, *Glomerella* sp., *Nigrospora oryzae*, *Nodulisporium* sp., *Periconia* sp., *Phomopsis* sp., *Phyllosticta* sp., *Xylaria* spp.	Pereira et al. (1993)
Subal bermudana	*Idriella* sp. and *Aspergillus* sp.	Southcott and Johnson (1997)
Theobroma grandiflorum	*Fusarium* sp., *Glomerella cingulada*, *Guignardia* sp., *Humicola* sp., *Pestalotia* sp.	Galvão (1998)
Theobroma cacao	*Verticillium* sp., *Gliocadium* sp., *Monilochaetes* sp., *Trichoderma* sp., *Geotrichum* sp., *Syncephalastrum* sp., *Fusarium* sp., *Acremonium* sp., *Xylaria* sp., *Cladosporium* sp., *Blastomyces* sp., *Pleurotus* sp., *Botryosphaeria* sp.	Rubini et al. (2005)
Vigna unguiculata	*Aspergillus* sp., *Fusarium* sp., *F. equiseti*, *F. solani*, *F. sentectum*, *F. oxysporum*, *F. anthophilum*, *F. moniliforme*, *Penicillium* sp.	Rodrigues (1999)
Zea mays	*Acremonium* sp., *Aspergillus* sp., *Beauveria* sp., *B. bassiana*, *Chaetomium* sp., *Colletotrichum* sp., *Curvularia* sp., *Dreschlera* sp., *Fusarium* sp., *F. moniliforme*, *Glomerella cingulada*, *Humicola* sp., *Paecilomyces* sp., *Penicillium* spp., *Periconia* sp., *Phomopsis* sp., *Nigrospora oryzae*, *Rhizopus* sp., *Scopulariopsis* sp., *Syncephalastrum* sp., *Trichoderma* sp., *Xylaria* sp.	Pimentel (2001); Pamphile (1997); Pamphile and Azevedo (2002); Silva (1997)

Biodiversity Evaluated by Molecular Methods

Of the thousand myriads of ecosystems on earth, those having the greatest biodiversity seem to be ones also presenting the greatest number of endophytes and revealing the greatest biodiversity of microorganisms (Strobel and Daisy, 2003). According to these authors, tropical rainforests are a remarkable example of the ecosystems where biological diversity corelates with chemical diversity due to the constant and active chemical innovation. This suggests that endophytes inhabiting rainforests and tropical cultivated plants could be a source of novel molecular structures and biologically active compounds. The rational strategies for collection of plants for endophyte isolation are as follows: (a) plants from unique environmental settings, especially those with an unusual biology, and possessing novel strategies for survival; (b) plants that have an ethnobotanical history (used by indigenous peoples) that are related to the specific uses or applications of interest are selected for study; (c) plants that are endemic, having an unusual longevity, or that have occupied a certain ancient land mass, are also more likely to lodge endophytes with active natural products; (d) plants growing in areas of great biodiversity also have the propects of harboring endophytes with great biodiversity (Strobel and Daisy, 2003).

As already mentioned, data indicate that tropical plant hosts contain a great diversity of endophytic microorganisms, some of them not yet classified and probably belonging to a new genera and species. According to Hawksworth (2001), from the estimate number of 1.5 millions of existing fungal species, less than 5% have been described and the undiscovered ones may be found in the tropics possibly the majority as endophytes. The same conclusion is reached by Hyde (2001) and by Arnold et al. (2000). On the other hand, Cannon and Simmons (2002) and Suryanarayanan et al. (2002, 2003) suggest that this is not always the case, at least in some tropical environments. Ways to detect diversity intraspecies and interspecies have been facilitated by the use of molecular markers. At first, isozymes were used to detect diversity among tropical isolated fungi as, for instance, by Rodrigues et al (1993), who analysed many isolates and species of the endophytic genus *Xylaria*, obtained from the palm tree *E. oleraceae*, to show high degree of diversity among them. Oliveira (1999) used eletrophoretic methods to detect variability in endophytes isolated from cashew trees. With the introduction of molecular procedures based on DNA markers, several authors used these methods to evaluate the diversity among endophytes isolated from tropical hosts. Pamphile (1997) analysed via RAPD 21 endophytic *Fusarium* isolated from maize. The results showed high degree of diversity among them and also it was shown that they could be grouped according to the maize populations from which they were isolated. Variability was also studied through RAPD analysis in more than 100 endophytic and pathogenic isolates from the genus *Guignardia* (Glienke, 1995; Glienke-Blanco, 1999; Glienke et al. 2002). They were divided in two distinct groups and one of the bands detected by RAPD was found exclusively from DNA of pathogenic isolates. This band was cloned and sequenced and a pair of primers was synthesized based on the obtained sequence and used to discriminate pathogenic isolates from endophytic ones. Galvão (1998), Guimarães (1998) also used RAPD technique to analyse variability among endophytic isolates from amazonian trees. Pandey et al. (2003) studied different endophytic species of *Phyllosticta* isolated in India, using ITS-RFLP. They found no differences within species concluding that they can live in different hosts and habitats. Analysis by RAPD and ISSR-PCR from a collection of *Colletotrichum* strains derived from 12 species of plant hosts from Guyana have shown that no strains were genetically identical; indicating that diversity in tropical forests may be high (Lu et al. 2004).

Useful Biological Activities of Endophytic Microorganisms

Endophytic microorganisms have many biological functions, which may interfere in plant development or may be used for biotechnological purposes (Table 2). These functions may be related to biological control of pests, herbivores and diseases, plant growth promotion and production of metabolites, such as alkaloids,

Table 2. Examples of endophytic microorganisms-plant interaction (for more details see Azevedo, 1998 and Azevedo et al. 2000)

	Type of interaction	Reference
1.	Biological control of insects	
	Endophytic fungi: *Beauveria bassiana*	Wagner and Lewis (2000)
	Host plant: Maize	
	Effect: Protection against *Ostrinia nubilalis*	
2.	Protection against herbivores	
	Endophytic fungi: *Acremonium* sp.	White and Cole (1985)
	Host plant: *Lolium* sp. and others	
	Effect: Production of toxins against lamb and cow	
3.	Physiological effects	
	Endophytic fungi: *Piriformospora indica*	Varma et al. (1999)
	Host plant: Maize	
	Effect: Plant growth promotion	
4.	Production of drugs by endophytic fungi	
	Endophytic fungi: *Taxomyces andreanea*	Stierle et al. (1993)
	Compound: Anticancer, Taxol	
	Endophytic fungi: *Fusarium subglutinans*	Lee et al. (1995)
	Compound: Imunosupressive, Subglutinol	
5.	Disease control	
	Endophytic fungi: *Fusarium oxysporum*	Benhamou et al. (2002)
	Host plant: Cucumber	
	Effect: Control of *Pythium ultimum*	

antibiotics, immunosupressives and anticancer drugs. Furthermore, synthesis of toxic compounds by plants against herbivores may be stimulated by the presence of enzymes or compounds produced by endophytic microorganisms. In this case, this stimulus increases the plants survival by reduction of herbivory. In other cases, plants that are known to produce toxic compounds, are in fact, hosts of endophytes that are the real producers of toxic alkaloids responsible for intoxication in wild animals, causing severe effects, as observed, for instance, in cervids (Wolfe et al. 1998).

Although these fungi increase the survival of the host plant, from the economic point of view, they result in serious losses, since they are responsible for intoxication in ruminants, reducing their production or resulting in the death of livestock. However, the benefits caused by endophytes in fungi-plant interaction are higher than the losses, and they are becoming an important tool in modern and sustainable agriculture.

Endophytes and Pharmaceutical Products

Endophytes produce certain phytochemicals originally characteristic of the host. This may be explained by horizontal gene transfer between the endophyte and the host and could occur during the evolutionary process. Studies carried out by Stierle et al. (1993) showed that the fungus *Taxomyces andreanea*, found as an endophyte of *Taxus brevifolia*, is able to produce a complex diterpen, called taxol, used as an antitumoral. Furthermore, another endophytic fungi, *Pestalotiopsis microspora* isolated from *T. wallachiana* also produced taxol (Strobel et al. 1996). Taxol is normally obtained from the host plant, but these findings showed that microorganisms could also produce this drug, possibly allowing a new and more efficient method to produce such compounds. Futhermore, this would reduce the need to harvest slow growing and possibly rare plants, preserving the world's ever diminishing biodiversity.

Table 3 shows some pharmaceutical substances produced by endophytic microorganisms. Some of these natural products have high commercial value and are produced in low amounts by the host plant. Produced by endophytes, the metabolite pathway can be better studied, and key genes involved with the synthesis of the compounds may be found. Antibiotics and antifungal substances have been evaluated by researchers worldwide (Mandala et al. 1997; Strobel et al. 1999; Rodrigues et al. 2000).

Table 3. Chemical compounds produced by endophytic fungi

Chemical compounds	Endophytes	Biological activity	Reference
Leucinostatin A	*Acremonium* sp.	Antitumoral and antifungic	Strobel and Hess (1999)
Lolitrems, peramin and paxilin	*A. lolii*	Mycotoxins	Rowan (1993)
Isocoumarins	*Conoplea elegantula*	Control of *Choristoneura fumiferana*	Findlay et al. (1995)
Criptocandin	*Cryptosporiopsis* cf. *quercina*	Antifungal	Strobel et al. (1999)
Criptocin	*C.* cf. *quercina*	Antifungal	Li et al. (2000)
Citonic acid A and B	*Cytonaema* sp.	*Citomegalovirus protease* inhibitors	Guo et al. 2000
Subglutinol A and B	*Fusarium subglutinans*	Immunosupressive	Lee et al. 1995
Taxol	*Pestalotiopsis microspora*	Antitumoral	Strobel et al. 1996
Taxol	*Stegolerium kukenani*	Antitumoral	Strobel et al. 2001
Taxol	*Taxomyces andreanea*	Antitumoral	Stierle et al. 1993

Endophytic Microorganisms and Biological Control

Chemical compounds have been extensively used in agriculture causing serious damage to the environment. Biological control of disease and pests could be a partial solution for these problems. Taking into account that endophytic microorganisms colonize an ecological niche similar to that of phytopathogens, especially vascular wilt pathogens, this might favor endophytes as candidates to biocontrol agents. Intensive work on biocontrol agents has shown that endophytic microorganisms isolated from surface disinfected plant tissues exhibit potential as biocontrol agents against microbial pathogens, insects and nematodes (Carroll, 1986; Azevedo et al. 2000).

The endophytic fungi *Neotyphodium* (sin. *Acremonium*) and *Fusarium* have been used as biocontrol agents due their good performance in this area (Tables 4 and 5). The most important mechanism for biocontrol is the Induced Systemic Resistance (ISR) that is started by active penetration of the endophyte. In this case, the host plant is induced to produce resistance proteins, phenolic compounds or changing the plant morphology by increasing glucan or lignin formation. Plants propagated by seeds or by in vitro culture could be inoculated with endophytes and transferred to the field, allowing the introduced colonization by endophytic microorganisms, and increasing plant protection.

Table 4. Endophytic fungi described as potential agent of biological control of diseases caused by fungi and nematodes

Endphytes	Pathogen	Host plant	Reference
Colletotrichum magna	*C. magna*	*Citrullus lanatus* and *Cucumis sativus*	Redman et al. (1999)
Fusarium spp.	*Radopholus similis* (nematóide)	Banana	Pocasangre et al. (2000)
Neotyphodium spp.	*Drechslera sorokiniana*; *Lolium* sp.; *Fusarium graminearum*; F. *culmorum*; *Rhizoctonia cerealis* and *Gaeumannomyces graminis*	*Triticum* spp. and *Festuca arundinacea*	Tunali et al. (2000)

Table 5. Endophytic fungi described as potential agent for biological control of insect-pest

Endophytes	Insect pest	Reference
Acremonium sp.	*Parapediasia teterrella*	Koga et al. (1997)
	Blissus leucopterus hirtus	Saha et al. (1987)
A. alternatum	*Plutella xylostella*	Raps and Vidal (1998)
A. coenophialum	*Parapediasia teterrella*	Koga et al. (1997)
A. lolii	*Microctonus hyperodae*	Barker and Addison (1996)
A. strictum	*Trialeurodes vaporariorum*	Vidal (1996)
Balansia cyperi	*Spodoptera frugiperda*	Clay et al. (1985)
Beauveria bassiana	*Ostrinia nubilalis*	Wagner and Lewis (2000)
Cladosporium shaerosperum	*Adelges abietis*	Lasota et al. (1983)
Discula quercina	*Besbiscus mirabilis*	Wilson and Carroll (1997)
Epichloe sylvatica	*S. frugiperda*	Brem and Leuchtmann (2001)
Neotyphodium spp.	*Blissus leucopterus*	Richmond and Shetlar (2000)
N. lolii	*Listronotus bonariensis*	Goldson et al. (2000)
Rhabdocline parkeri	*Contarinia* sp.	Sherwoodpike et al. (1986)

Fast growing endophytic species may be used as biocontrol agents, as they are able to colonize the specific site and inhibit, by competition or antibiosis, the establishment of the pathogen. In tomato, *Fusarium oxysporum* can be controlled by avirulent *Fusarium oxysporum*, which colonize endophytically the host plant. The endophytic microorganism may attack the pathogen cells by use of lytic enzymes, such as chitinases and proteases. However, in this case the contact between pathogen and endophyte is necessary. Although, chemotaxis may occur, the contact beween these microorganism occurs in random decreasing the biocontrol efficiency.

The use of endophytes in biocontrol of fungal and bacterial disease has been consistently described. However, few studies have been focused on viral diseases, just because they colonize different niches: fungi and bacteria colonize preferentially the intercellular space while viruses are intracellularly colonizing, but both virus and endophytes depend on the plant metabolism to obtain nutrients.

Some endophytic fungi produce, in association with the host plant, alkaloids that may have insecticidal and nemoticidal activity, protecting the host plant. However in some instances the pest control is indirect. Environmental factors should be considered in this interaction, because it has been observed that in low levels of nutrients for the host plant, the endophytic fungi *Neotyphodium coenophialum* protects the host plant against *Spodoptera frugiperda*. However, at high nutritional levels the reduction of the herbivory was not observed.

It is known that some fungi designated as entomopathogens are able to cause diseases in insects. They are often deliberately used for biological control of agriculture insect-pests. Bing and lewis (1991; 1992a; 1992b) described *Beauveria bassiana*, a entomopathogenic fungus which is a natural endophyte found in maize and is able to control the European corn bore *Ostrinia nubilalis*. Studies carried out in Brazil also show that important crops such as maize and soybean have a variety of endophytic fungi that include some entomopathogenic fungi as the genera *Cladosporium, Paecilomyces* and *Beauveria*. Some of the *Beauveria* described from maize are possibly from other species as *B. amorpha* (Pimentel, 2001). These findings emphasize the importance of endophytic fungi acting as natural controllers of insect/pests, which cause severe damages to cultivated crops in tropical countries.

The interaction between endophytes and insects is complex and depends on many environmental conditions. The insect is able to recognize the tissue of plant colonized by some endophyte, and is rejected by the insect. This shows that the practical application of the endophytes in biological control is promising, but the complexity of the interaction endophyte-pest-plant should be completely understood, allowing the reproducibility of the biocontrol agent in field conditions. Other examples of endophytes controlling insects can be found in the review of Azevedo et al. (2000).

Plant Growth Promotion

Endophytic fungi can affect plant growth directly or indirectly. The direct promotion of plant growth, for the most part, entails providing the plant with compounds that are synthesized by the fungi for facilitating the uptake of certain nutrients from the environment. The indirect promotion of plant growth occurs when endophytic fungi decrease or prevent the deleterious effects of one or more phytopathogenic organisms.

The endophytic fungi *Piriformospora indica* (Hymenomycetes, Basidiomycota) colonizes endophytically the roots of many host plants, promoting the plant growth (Varma et al. 1999). The authors suggested that *P. indica* is a good candidate to improve commercial plant production and might be especially useful in agroforestry and flori-horticulture applications. Research on the cellular and molecular basis of fungi-plant interactions should contribute to the understanding of the beneficial associations between plants and fungi. Besides, the interaction between the plant genotype and endophytic community, other events may be involved in this process, such as epiphytic and rhizospheric communities.

Plant growth regulators, such as ethylene, auxins, gibberelin or cytokinins, produced by some isolates of endophytic fungi may also be considered to be causal agents for altering plant growth and development. *Colletotrichum gloeosporioides* and *Fusarium oxysporum*, for instance, are generally regarded as being pathogenic fungi in many plant species, but some strains may also be isolated as endophytes, being found within the roots of some plants, such as soybean and citrus. These species are able to produce and excrete auxins in the culture medium, suggesting that during plant fungi interaction the excretion of this plant regulator could be responsible for high development of the host plant.

Final Considerations

Taking in mind that there are approximately 300,000 plant species, and that the endophytes inhabiting these plants are a source of bioactive and chemically novel compounds with potential for exploitation in pharmaceutical, agricultural and industrial areas, the evaluation of the diversity of these microorganisms, mainly in tropical plants, should be considered. Probably, the major problem concern the discovery of natural products from endophytic fungi from tropical plants is the rapid reduction of the forests in tropical areas, which will result in the loss of many new fungi species with biotechnological potential still not evaluated.

The intimate relationship between endophytic microorganisms and their hosts involves co-evolutionary processes and may influence the physiology of the plant and also interfere with the presence of other endophytes. This endophytes-plant interaction may have been naturally selected during long climatic changes thus allowing a great genetic variability in endophyte populations that opens perspectives for the discovery of improved or new enzymes, drugs, other products with new and useful properties. Thus, the loss of information about endophytes due to the decrease of tropical plants could result in an enhanced use of xenobiotic compounds and consequently increase of environmental damage. Countries that are responsible for the great animal, plant and microbial diversity found in our planet, located mainly in tropical regions, should develop strategies to preserve plant and microbial species including endophytes, and exploit the biotechnological potential present in the endophytic community associated with their hosts.

References

Alquati, S.A.B. (1999). *Paracoccidioides brasiliensis* não ocorre na forma endofítica em gramíneas da região endêmica de Botucatu, SP, Brasil, Ms. Thesis, Botucatu, UNESP, pp. 115.

Araújo, W.L. and Kuklinsky-Sobral, J. (2004). Microbial communities of transgenic sugarcane. In: 8th International Symposium on the Biosafety of Genetically Modified Organisms, Montpellier, França, pp. 211–215.

Araujo, W.L., Lima, A.O.S., Azevedo, J.L., Kuklinsky-Sobral, J. and Lacava, P.T. (2002a). Manual: Isolamento de microrganismos endofíticos. ESALQ/USP, Piracicaba.

Araújo, W.L., Saridakis, H.O., Barroso, P.A.V., Aguilar-Vildoso, C.I. and Azevedo, J.L. (2001). Variability and interactions between endophytic bacteria and fungi isolated from leaf tissues of citrus rootstocks. Canadian Journal of Microbiology 47: 229–236.

Arnold, A.E. and Herre, E.A. (2003). Canopy cover and leaf age affect colonization by tropical fungal endophytes: ecological pattern and process in *Theobroma cacao* (Malvaceae). Mycologia 95: 388–398.

Arnold, A., Maynard, Z. and Gilbert. G.S. (2001). Fungal endophytes in dicotyledoneous neotropical trees: patterns of abundance and diversity. Mycological Research 105: 1502–1507.

Arnold, A.E., Maynard, Z., Gilbert, G.S., Coley, P.D. and Kursar, T. A. (2000). Are tropical fungal endophytes hyperdiverse? Ecological Letters 3: 267–274.

Arnold, A.E., Mejia, L.C., Kyllo, D., Rojas, E.I., Maynard, Z., Robbins, N. and Herre, E.A. (2003). Fungal endophytes limit pathogen damage in a tropical tree. Proceedings of the National Academy of Sciences USA 100: 15649–15654.

Azevedo, J.L. (1997). Endophytic fungi and their roles in tropical plants. In: Progress in Microbial Ecology, (eds),

Martins, T., Sato, A., Tiedje, J. M., Hagler, L.C. M., Dobereiner, J. and Sanchez I. Sociedade Brasileira de Microbiologia, São Paulo, pp. 279–287.

Azevedo, J.L. (1998). Biodiversidade microbiana e potencial biotecnológico. In: Ecologia Microbiana, (eds), Melo, I.S. and Azevedo, J.L. Embrapa-CNPMA, Jaguariúna, pp. 117–137.

Azevedo, J.L., Maccheroni Jr., W., Pereira, J.O. and Araújo, W.L. (2000). Endophytic microorganisms: a review on insect control and recent advances on tropical plants. EJB: Electronic Journal of Biotechnology [online]. 3: 40–65.

Baayen, R.P., Bonants, P.J.M., Verkley, G., Carroll, G.C., Van Der Aa, H.A., De Weerdt, M., Van Brouwershaven, I.R., Schutte, G.C., Maccheroni Jr., W., Glienke-Blanco, C., Azevedo J. L. (2002). Nonpathogenic isolates of the Citrus Black Spot fungus, *Guignardia citricarpa*, identified as a cosmopolitan endophyte of woody plants, *G. mangiferae (Phyllosticta capitalensis)*. Phytopathology 92: 464–477.

Bacon, C.W. (1990). Isolation, culture and maintenance of endophytic fungi of grasses. In : Isolation of Biotechnological Organisms from Nature. (ed), Labeda, D.P. McGraw-Hill Publishing Co., New York, pp. 259–282.

Barbosa, L.C.A. (2001). Enzimas de interesse biotecnológico de fungos endofíticos de *Copaifera multijuga*. MS. Thesis, Universidade de Brasília, Brazil, pp. 106.

Barker, G.M. and Addison, P.J. (1996). Influence of Clavicipitaceous endophyte infection in ryegrass on development of the parasitoid *Microctonus hyperodae* Loan (Hymenoptera: Braconidae) in *Listronotus bonariensis* (Kuschel) (Coleoptera: Curculionidae). Biological Control 3: 281–287.

Bayman, P., Gonzales, E.J., Fumero, J.J. and Tremblay, R.L. (2002). Are fungi necessary? How fungicides affect growth and survival of the orchid *Lepanthes rupestris* in the field. Journal of Ecology 90: 1002–1008.

Benhamou, N., Garand, C. and Goulet, A. (2002). Ability of nonpathogenic *Fusarium oxysporum* Strain Fo47 to induce resistance against *Pythium ultimum* infection in Cucumber. Applied and Environmental Microbiology 68 (8): 4044–4060.

Bettuci, L., Alonso, R. and Tiscornia, S. (1999). *Endophytic mycobiota* of healthy twigs and the assemblage of species associated with twig lesions of *Eucalyptus globosus* and *E. grandis* in Uruguay. Mycological Research 103: 468–472.

Bing, L.A. and Lewis, L.C. (1991). Suppression of *Ostrinia nubilalis* (Hübner) (Lepidoptera: Pyralidae) by endophytic *Beauveria bassiana* (Balsamo) Vuillemin. Environmental Entomology 20: 1207–1211.

Bing, L.A. and Lewis, L.C. (1992a). Temporal relationships between Zea mays, Ostrinia nubilalis (Hübner) (Lep.: Pyralidae) and endophytic *Beauveria bassiana* (Balsamo) Vuillemin. Entomophaga 37: 525–536.

Bing, L.A. and Lewis, L.C. (1992b). Endophytic *Beauveria bassiana* (Balsamo) Vuillemin in corn: the influence of plant growth stage and *Ostrinia nubilalis* (Hübner). Biocontrol Science and Technology 2: 39–47.

Brem, D. and Leuchtmann, A. (2001). Epichloe grass endophytes increase herbivore resistance in the woodland grass *Brachypodium sylvaticum*. Oecologia, 126: 522–530.

Brown, K.B., Hyde, K.D. and Guest, D.I. (1998). Preliminary studies on endophytic fungal communities of *Musa acuminata* species complex in Hong Kong and Australia. Fungal Diversity 1: 27–51.

Cannon, P.F. and Simmons, C.M. (2002). Diversity and host preference of leaf endophytic fungi in the Iwokrama Forest Reserve, Guyana. Mycologia 94: 210–220.

Carroll, G. (1986). Fungal associates of woody plants as insect antagonists in leaves and stems. In: Microbial mediation of plant–herbivore interactions, (eds), Barbosa, P., Krischik, V.A. and Jones, C.G. John Wiley and Sons, New York, pp. 253–271.

Carroll, G. and Carroll, F.E. (1978). Studies on the incidence of coniferous needle endophytes in the Pacific Northwest. Canadian Journal of Botany 56: 3034–3043.

Clay, K., Hardy, T.N. and Hammond Jr., A.M. (1985). Fungal endophytes of Cyperus and their effect on the insect herbivore. American Journal of Botany 72:1284–1289.

Cother, E.J. and Gilbert, R.L. (1994). The endophytic mycoflora of bladder saltbush (*Atriplex vesicatoria* Hew. Ex Benth.) and its possible role in the plant periodic decline. Proceedings of the Linnean Society of New South Wales 114: 149–169.

Danielsen, S. and Jensen, D.F. (1990). Fungal endophytes from stalks of tropical maize and grasses: isolation, identification and screening for antagonism against *Fusarium verticillioides* in maize stalks. Biocontrol Science and Technology 9: 545–553.

Devarajan, P.T., Suryanarayanan, T.S. and Geetha, V. (2002). Endophytic fungi associated with the tropical seagrass *Halophila ovalis* (Hydrocharitaceae). Indian Journal of Maritime Sciences 31: 73–74.

De Bary, A. (1866). Morphologie Physiologie der Pilze, Flechten, und Myxomyceten, Vol. II. Holmeister´s Handbook of Physiological Botany, Leipzig.

Di Fiori, S. and Del Gallo, M. (1995). Endophytic bacteria: their possible role in the host plants. In: Azospirillum VI and related microorganisms (eds), Fendrik, I., Del Gallo, M., Vanderleyden, J. and de Zamaroczy, M., Springer-Velag, Berlin, pp. 169–187.

Dreyfuss, M. and Petrini, O. (1984). Further investigations on the occurrence and distribution of endophytic fungi in tropical plants. Botanica Helvetica 94: 33–40.

Findlay, J.A., Buthelezi, S., Lavoie, R., Pena-Rodriguez, L. and Miller, J.D. (1995). Bioactive isocoumarins and related metabolites from conifer endophytes. Journal of Natural Products 58: 1759–1766.

Fisher, P.J., Petrini, L.E. and Sutton, B.C. (1995). A study of fungal endophytes from leaves, stem and roots of *Gynoxis oleifolia*, Muchler (Compositae) from Ecuador. Nova Hedwigia 60: 589–594.

Fisher, P.J., Petrini, O. and Lappin-Scott, H.M. (1992). The distribution of some fungi and bacteria endophytes in maize (*Zea mays*, L.). New Phytologist 122: 299–305.

Freeman, S. and Rodriguez, R.J. (1993). Genetic conversion of a fungal plant pathogen to a nonpathogenic, endophytic mutualist. Science 260: 75–78.

Freire, C.O. and Bezerra, J.L. (2001). Foliar endophytic fungi of Ceara State (Brazil): A preliminary study. Summa Phytopathologica 27: 304–308.

Frölich, J., Hyde, K.D. and Petrini, O. (2000). Endophytic fungi associated with palms. Mycological Research 104: 1202–1212.

Gai, C.S., Maccheroni Jr., W., Araujo, W.L., Azevedo, J.L. (2000). Microscopia eletrônica de varredura para localização de leveduras endofiticas em citros (*Citrus sinensis*) e analise de variabilidade entre os isolados através de marcadores de RAPD. In: Simpósio Internacional de Iniciação Cientifica da Universidade de São Paulo, Piracicaba, SP, Brazil.

Galvão, R.M.S. (1998). Variabilidade genética detectada por RAPD em Glomerella cingulata, um dos fungos endofíticos mais freqüentes, isolados em *Theobroma grandiflorum*, *Pueraria phaseoloides*e, *Scleria pterota*, MS. Thesis, FUA/UFSCAR, Manaus, Amazonas, Brazil, pp.151.

Gamboa, M.A. and Bayman, P. (2001). Communities of endophytic fungi in leaves of a tropical timber tree (*Guarea guidonia*) Meliaceae. Biotropica 33: 352–360.

Glienke, C. (1995). Variabilidade genética no fungo endofítico Gignardia citricarpa Kiely, detectada por RAPD. Ms. Thesis, Universidade federal do Paraná, Curitiba, Paraná, Brazil, pp. 115.

Glienke-Blanco, C. (1999). *Guignardia citricarpa*: Kiely: Análise genética, cariotípica e interação com o hsopedeiro. Ph.D. Thesis. ESALQ/Universidade de São Paulo, Piracicaba, São Paulo, Brazil, pp. 200.

Glienke-Blanco, C., Aguilar-Vildoso, C.I., Vieira, M.I.C., Barroso, P.A.V. and Azevedo, J.L. (2002). Genetic variability in the endophytic fungus *Guignardia citricarpa* isolated from citrus plants. Genetics and Molecular Biology 25: 251–255.

Goldson, S.L., Proffitt, J.R., Fletcher, L.R. and Baird, D.B. (2000) Multitrophic interaction between the ryegrass Lolium perenne, its endophyte *Neotyphodium lolii*, the weevil pest *Listronotus bonariensis*, and its parasitoid *Microctonus hyperodae*. New Zealand Journal of Agricultural Research 43: 227–233.

Guimarães, V.C. (1998) Isolamento de fungos endofíticos do hospedeiro *Paullinia cupana* H.B.K. var. sorbilis (Mart.) Ducke e análise da variabilidade genética, detectada por marcadores RAPD no endófito *Glomerella cingulata*. MS. Thesis, FUA/UFSCAR, Manaus, pp.123.

Guo, B., Dai, J.R., Ng, S., Huang, Y., Leong, C., Ong, W. and Carte, B.K. (2000) Cytonic acids A and B: novel tridepside inhbitors of hCMV protease from the endophytic fungus *Cytonaema* species, Journal of Natural Products 63: 602–604.

Hallmann, J., Quadt-Hallmann, A., Mahaffee, W.F. and Kloepper, J.W. (1997). Bacterial endophytes in agriculture crops. Canadian Journal Microbiology 43: 895–914.

Hawksworth, D.L. (2001). The magnitude of fungal diversity: the 1.5 million species estimate revisted. Mycological Research 105: 1422–1432.

Hyde, K.D. (2001). Where are the missing fungi? Does Hong Kong have any answers? *Mycological Research* 105: 1514–1518.

Isaac, S. (1992). Fungal-plant interactions, Chapman and Hall, London, pp. 418.

Kelemu, S., White Jr., J.F., Muñoz, F. and Takayama, Y. (2001). An endophyte of the tropical forage grass *Brachiaria brizantha*: Isolating, identifying, and characterizing the fungus, and determining its antimycotic properties. Canadian Journal of Microbiology 47: 55–62.

Koga, H., Hirai, Y., Kanda, K., Tsukiboshi, T. and Uematsu, T. (1997). Successive transmission of resistance to bluegrass webworm to perennial ryegrass and tall fescue plants by artificial inoculation with *Acremonium* endophytes, JARQ 31: 109–115.

Kumaresan, V. and Suryanarayanan, T.S. (2002). Endophytic assemblages in young, mature and senescent leaves of *Rhizophora apiculata*: evidence for the role of endophytes in mangrove litter degradation. Fungal Diversity 9: 81–91.

Lasota , J.A., Waldvogel, M.G. and Shetlar, D.J. (1983). Fungus found in galls of *Adelges abietis* (L.) (Homoptera: Adelgidae): identification within tree distribution and possible impact on insect survival. Environmental Entomology 12: 245–246.

Lee, J.C., Lobkovsky, E., Pliam, N.B., Strobel, G. and Clardy, J. (1995). Subglutinol-A and subglutinol-B-immunosuppressive compounds from the endophytic fungus *Fusarium subglutinans*. Journal of Organic Chemistry 60: 7076–7077.

Li, J.Y., Strobel, G., Harper, J., Lobkovsky, E. and Clardy, J. (2000). Cryptocin, a potent tetramic antimycotic from the endophytic fungus *Cryptosporiopsis* cf. *quercina*. Organic Letters 23: 767–770.

Lodge, D.J., Fisher, P.J. and Sutton, B.C. (1996). Endophytic fungi of Manilkara bidentata leaves in Puerto Rico, Mycologia 88: 733–738.

Lu, G.Z., Cannon, P.F., Reid, A. and Simmons, C.M. (2004). Diversity and molecular relationships of endophytic *Colletotrichum* isolated from the Iwokraama forest reserve Guyana, Mycological Research 53–63.

Magalhães, A.A.S. (2000). Isolamento e variabilidade genética detectada por RAPD do fungo endofítico *Guignardia* sp. *de Himatanthus sucuuba* Spruce (Wood) Apocynaceae. Ms. Thesis, FUA/UFSCAR, Manaus, Brazil, pp130.

Mandala, S.M., Thornton, R.A., Rosenbach, M., Milligan, J., Garcia-Calvo, M., Bull, H.G., Kurtz, M.B. (1997). Khafrefungin, a novel inhibitor of sphingolipid synthesis. Journal of Biological Chemistry 272: 32709–32714.

Maria, G.L. and Sridhar, K.R. (2003). Endophytic fungal assemblage of two halophytes from west coast mangrove habitats, India. Czech Mycology 55: 241–251.

Medeiros, S.A.F. (1988). Microflora da folhagem do cajueiro *Anacardium occidentale* (L.) e controle biológico do agente da antracnose *Colletotrichum gloeosporioides* (Penz) Sacc. "in vitro". Ms. Thesis, UFRPe, Recife, Recife, Brazil, pp106.

Mendes, R., Kuklinsky-Sobral, J., Geraldi, I.O., Araujo, W.L., Azevedo, J.L., Pizzirani-Kleiner, A.A. (2001). Monitoring soybean endophytic fungal community associated with glyphosate. In: XXI Congresso Brasileiro de Microbiologia, Foz do Iguaçú. Sociedade Brasileira de Microbiologia, São Paulo, Brazil, pp. 242.

Oliveira, F.C.O. (1999). Fungos endofíticos de folhas de cajueiro, *Anacardium occidentale* L. *Propriedades antagônicas* a *Colletotrichum gloeosporioides* (Penzig) Sacc. Avaliação enzimática através de eletroforese e substratos específicos. Ms. Thesis, UFRPe, Recife, Brazil, 79p.

Otero, J.T., Ackerman, J.D. and Bayman, P. (2002). Diversity and host specificity of endophyte Rhizoctonia-like fungi from tropical orchids. American Journal of Botany 89: 1852–1858.

Pamphile, J.A. (1997). Variabilidade, transformação genética e transposons em linhagens endofíticas de *Fusarium moniliforme* isoladas de milho (Zea mays L.). Ph.D. Thesis, ESALQ/USP, Piracicaba, Brazil, 199p.

Pamphile, J.A. and Azevedo, J.L. (2002). Molecular characterization of endophytic strains of *Fusarium verticillioides* (=*Fusarium moniliforme*) from maize (*Zea mays* L.). World Journal of Microbiology and Biotechnology 18: 391–396.

Pandey, A.K., Reddy, M.S. and Suryanarayanan, T.S. (2003). ITS-RFLP and ITS sequence analysis of a foliar endophytic *Phyllosticta* from different tropical trees. Mycological Research 107: 439–444.

Penna, E.B.S. (2000). Microrganismos endofíticos em erva-mate (*Ilex paraguariensis* St. Hil.) e variabilidade genética em *Phylosticta* sp. por RAPD. Ms. Thesis, UFPR, Curitiba, Brazil, pp. 123.

Pereira, J.O., Azevedo, J.L. and Petrini, O. (1993). Endophytic fungi of *Stylosanthes*: A first report. Mycologia 85: 362–364.

Pereira, J.O., Carneiro-Vieira, M.L. and Azevedo, J.L. (1999). Endophytic fungi from *Musa acuminata* and their reintroduction into axenic plants. World Journal of Microbiology and Biotechnology 15: 37–40.

Petrini, O. (1986). Taxonomy of endophytic fungi in aerial plant tissues. In: Microbiology of the Phyllosphere, (eds), Fokkema, N.J. and Van den Heuvel, J., Cambridge University Press, Cambridge, pp. 175-187.

Petrini, O. (1991). Fungal endophyte of tree leaves. In: Microbial Ecology of Leaves. (Eds), Andrews, J. and Hirano, S.S. Spring-Verlag, New York, pp. 179–197.

Petrini, O. and Dreyfuss, M.M. (1981). Endophytische pilze in epiphytischen Araceae, Bromeliaceae and Orchidaceae. Sydowia 34: 135–148.

Petrini, O., Hake, U. and Dreyfuss, M.M. (1990). An analysis of fungal communities from fruticose lichens. Mycologia 82: 444–451.

Photita, W., Lumyong, S., Lumyong, P. and Hyde, K.D. (2001). Endophytic fungi of wild banana (*Musa acuminata*) at Doi Suthep Pui National Park, Thailand, Mycological Research 105: 1508–1513.

Pimentel, I.C. (2001). Fungos endofíticos do milho (*Zea mays* L.) e da soja (*Glycine max* L.) Merrril e seu potencial valor biotecnológico no controle de pragas agrícolas. Ph.D. Thesis, UFPr, Curitiba, Brazil, pp. 154.

Pocasangre, L., Sikora, R.A., Vilich, V. and Schuster, R.P. (2000). Survey of banana endophytic fungi from central America and screening for biological control of the burrowing nematode (*Radopholus similis*). Infomusa 9: 3–5.

Raps, A. and Vidal, S. (1998). Indirect effects of an unspecialized endophytic fungus on specialized plant-herbivorous insect interactions. Oecologia 114: 541–547.

Rajagopal, K. and Suryanarayanam, T.S. (2000). Isolation of endophytic fungi from leaves of neem (*Azadirachta indica* A. *juss*.). Current Science 78: 1375–1378.

Redlin, S.C. and Carris, L.M. (1996). Endophytic fungi in grasses and woody plants. The American Phytopathological Society Press, St. Paul, pp. 223.

Redman, R.S., Freeman, S., Clifton, D.R., Morrel, J., Brown, G. and Rodriguez, R.J. (1999). Biochemical analysis of plant protection afforded by a nonpathogenic endophytic mutant of *Colletotrichum magna*. Plant Physiology 119: 795–804.

Richmond, D.S. and Shetlar, D.J. (2000). Hairy chinch bug (Hemiptera: Lygaeidae) damage, population density, and movement in relation to the incidence of perennial ryegrass infected by Neotyphodium endophytes. Journal of Economic Entomology 93: 1167–1172.

Rodrigues, A.A.C. (1999). Fungos endofíticos de sementes de caupi, *Vigna uncuiculata* (L.) Walp. e diferenciação morfológica, patogênica e enzimática de espécies de *Fusarium*. Ms. Thesis, UFPe, Recife, Brazil, pp. 87.

Rodrigues, K.F. (1991). Fungos endofíticos em *Euterpe oleraceae* Mart. com ênfase em Xylariaceae. Boletim do Museu Paraense Emilio Goeldi Série Botânica 7: 429–439.

Rodrigues, K.F. (1994). The foliar fungal endophytes of the amazon palm *Euterpe oleraceae*, Mycologia 86: 376–385.

Rodrigues, K.F. (1996). Fungal endophytes of palms. In: Endophytic fungi in grasses and woody plants. (eds), Redlin S.C. and Carris, L.M. American Phytopathological Society, St. Paul, pp. 1221–1132.

Rodrigues, K.F. (1999). The foliar fungal endophytes of the amazonian palm *Euterpe oleraceae*. Mycologia 86: 376–385.

Rodrigues, K.F., Leuchtmann, A. and Petrini, O. (1993). Endophytes species of *Xylaria*: cultural and isozymic studies. Sydowia 45: 116–138.

Rodrigues, K.F. and Petrini, O. (1997). Biodiversity of endophytic fungi in tropical regions. In Biodiversity of tropical microfungi. (ed), Hyde, K.D. Hong Kong University Press, Hong Kong, pp. 57–69.

Rodrigues, K.F. and Samuels, G.J. (1990). Preliminary study of endophytic fungi in a tropical palm. Mycological Research 94: 827–830.

Rodrigues, K.F. and Samuels, G.J. (1999). Fungal endophytes of *Spondias mombin* leaves in Brazil. Journal of Basic Microbiology 39: 131–135.

Rodrigues, K.F. and Samuels, G.J. (1992). *Idriella* species endophytic in palms, Mycotaxon 43: 271–276.

Rodrigues, K.F., Hesse, M. and Werner, C. (2000). Antimicrobial activities of secondary metabolites produces by fungi from *Spodias mombin*. Journal of Basic Microbiology 40: 261–267.

Romero, A., Carrion, G. and Rico-Gray, V. (2001). Fungal latent pathogens and endophytes from leaves of *Parthenium hysterophorus* (Asteracea). Fungal Diversity 7: 81–87.

Rowan, D.D. (1993). Lolitrems, peramine and paxilline: mycotoxins of the ryegrass/endophyte interaction, Agriculture, Ecosystems and Environment 44: 103–122.

Rubini, M.R. (2004). Microbiota endofítica de cacaueiro (*Theobroma cacao* L.) e o controle biológico de *Crinipellis perniciosa*. Ms. Thesis. Universidade de Caxias do Sul, Brazil, pp. 100.

Saha, D.C., Johnson-Cicalese, J.M., Halisky, P.M., Heemstra, M.I. and Funk, C.R. (1987). Occurrence and significance of endophytic fungi in the fine fescues. Plant Disease 71: 1021–1024.

Sherwoodpike, M., Stone, J.K. and Carroll, G.C. (1986). *Rhabdocline parkeri* a ubiquitous foliar endophyte of Douglas fir. Canadian Journal of Botany 64: 1849–1855.

Silva, A.C. (1997). Isolamento de fungos endofíticos de milho (*Zea mays*, L.). MS. Thesis, ESALQ/USP, Piracicaba, Brazil, pp. 56.

Southcott, K.A. and Johnson, J.A. (1997). Isolation of endophytes from two species of palm from Bermuda. Canadian Journal of Microbiology 43: 789–792.

Souza, A.Q.L. (2001). Fungos endofíticos de plantas tóxicas da Amazônia, *Palicourea longiflora* (aubl.) Rich e *Strychnos cogens* Bentham. Ms. Thesis, FUA/UFSCAR, Manaus, Brazil, pp. 102.

Stamford, T.L.M. (1997). Isolamento e identificação de microrganismos endofíticos, seleção e caracterização de actinomicetos produtores de enzimas amilolíticas. Ph.D. Thesis, UFPe, Recife, Brazil, pp. 110.

Stierle, A., Strobel, G. and Stierle, D. (1993). Taxol and taxane production by *Taxomyces andreanae* an endophytic fungus of Pacific yew. Science 260: 214–216.

Strobel, G. and Daisy, B. (2003). Bioprospecting for microbial endophytes and their natural products. Microbiology and Molecular Biology Reviews 67: 491–502.

Strobel, G.A. and Hess, W.M. (1999). Glucosylation of the peptide leucinostatin A, produced by an endophytic fungus of European yew, may protect the host from leucinostatin toxicity. Chem Biology 4: 529–536.

Strobel, G., Hess, W.M., Baird, G., Ford, E., Li, J.Y. and Sidhu, R.S. (2001). *Stegolerium kukenani* gen, et sp nov na endophytic taxol producing fungus from the Roraima and Kukenan tepuis of Venezuela, Mycotaxon 78: 353–361.

Strobel, G., Yang, X., Sears, J., Kramer, R., Sidhu, R.S. and Hess, W.M. (1996). Taxol from *Pestalotiopsis microspora*, an endophytic fungus of *Taxus wallachiana*, Microbiology-UK 142: 435–440.

Strobel, G.A., Miller, R.V., Martinez-Miller, C., Condron, M.M., Teplow, D.B. and Hess, W.M. (1999). Cryptocandin, a potent antimycotic from the endophytic fungus *Cryptosporiopsis* cf. *quercina*. Microbiology-UK 145: 1919–1926.

Suryanarayanan, T.S. and Kumaresan, V. (2000). Endophytic fungi of some halophytes from a estuarine mangrove forest. Mycological Research 104: 1465–1467.

Suryanarayanan, T.S. Kumaresan, V. and Johnson, J.A. (1998). Foliar fungal endophytes from two species of the mangrove *Rhizophora*. Canadian Journal of Microbiology, 44: 1003–1006.

Suryanarayanan, T.S., Murali, T.S. and Venkatesan, G. (2002). Occurrence and distribution of fungal endophytes in tropical forests across a rainfall gradient. Canadian Journal of Botany 80: 818–826.

Suryanarayanan, T.S., Sethilarasu, G. and Muruuganandam, C. (2000). Endophytic fungi from *Cuscuta reflexa* and its host plants. Fungal Diversity 4: 117–123.

Suryanarayanan, T.S. and Thennarasan, S. (2004). Temporal variation in endophyte assemblages of *Plumeria rubra* leaves. Fungal Diversity 15: 195–202.

Suryanarayanan, T.S., Venkatesan, G. and Murali, T.S. (2003). Endophytic fungal communities in leaves of tropical forest trees: diversity and distribution patterns. Current Sciences 85: 489–493.

Suryanarayanam, T.S. and Vijaykrishna, D. (2001). Fungal endophytes of aerial roots of *Ficus benghalensis*. Fungal diversity 8: 155–161.

Takeda, I., Guerrero, R. and Bettucci, L. (2003). Endophytic fungi of twigs and leaves from *Ilex paraguariensis* in Brazil. Sydowia, 55: 372–380.

Tunali, B., Marshall, D., Royo, C., Nachit, M.M., Fonzo. N.D.I. and Araus, J.L. (2000). Antagonistic effect of endophytes against several root-rot pathogens of wheat. In: Durum wheat improvement in the Mediterranean region: new challenges. Proceedings of a Seminar, Zaragoza, Spain 40: 381–386.

Varma, A., Verma, S., Sudha, Sahay, N., Bütehorn, B., Franken, P. (1999). *Piriformospora indica*, a cultivable plant-growth-promoting root endophyte. Applied and Environmental Microbiology 65: 2741–2744.

Vidal, S. (1996). Changes in suitability of tomato for whiteflies mediated by a non-pathogenic endophytic fungus. Entomologia Experimentalis et Applicata 80: 272–274.

Wagner, B.L. and Lewis, L.C. (2000). Colonization of corn, *Zea mays*, by the entomopathogenic fungus *Beauveria bassiana*, Applied and Environmental Microbiology 66: 3468–3473.

White, J.F. and Cole, G.T. (1985). Endophyte-host associations in forage grasses. 3. *In vitro* inhibition of fungi by *Acremonium coenophialum*, Mycologia 77: 487–489.

Wilson, D. (1995). Endophyte: the evolution of a term and clarification of its use and definition. Oikos 73: 274–276.

Wilson, D. and Carrol, G.C. (1997). Avoidance of high-endophyte space by gall-forming insects. Ecology 2153–2163.

Wolfe, B.A., Bush, M., Monfort, S.L., Mumford, S.L., Pessier, A. and Montali, R.J. (1998). Abdominal lipomatosis attributed to tall fescue toxicosis in deer. Journal of American Veterinary Medical Association 213: 1783–1786.

Yamada, C.M. (1999). Detecção de microrganismos endofíticos em frutos de café. MS. Thesis, Universidade Federal de Viçosa, Brazil, 56 p.

13. Natural Products from Marine Fungi

Adam Piorko
Department of Chemistry, Saint Mary's University, Halifax, Nova Scotia, B3H 3C3 Canada

SUMMARY The last six years have seen a significant increase in research activity in the area of natural products from marine fungi. Over a hundred papers have been published on the isolation of marine fungal products from a variety of fungal strains recovered from different marine environments and on related issues as well. The vast majority of reports described the isolation, structure elucidation, and biological activity of such metabolites. While several metabolites with interesting structures and a wide range of biological activities were described, other aspects such as the analytical methods for identification and quantification of natural products in samples, and biotechnological implications of current knowledge were also reported. This contribution summarizes the recently published results and identifies the areas of current and future interest in marine fungal natural product research.

Introduction

For many years, the study of natural products from marine fungi has been a marginal area of research. A number of papers reviewing marine natural products in the previous decade did not mention any research on marine fungi at all (Carte, 1996; Munro et al. 1999; Faulkner, 2000; Blunden, 2001; Proksch et al. 2002). This situation has changed dramatically in the last few years and, currently, natural products from marine fungi command increasing attention from researchers. The apparent reason for this change is a need for new leads in the search for drugs to treat human diseases, notably in the area of infectious diseases caused by increasing numbers of drug-resistant microorganisms and for cancer (Pinner et al. 1996; Munro et al. 1999; WHO 1996, 1997, 1999; Moore, 2002).

For the purpose of this contribution, marine fungi are defined as an ecological group rather than a taxonomic group according to Kohlmeyer and Kohlmeyer (1979). The natural products, which were obtained from the fungal strains (including lower marine fungi, see Fell and Newell, 1998), isolated from the broadly defined marine environment (Verbist et al. 2000) are presented here.

Marine fungal natural products have been studied for over fifty years. Initial reports have described predominantly metabolites including amino acids, choline, fatty acids, carbohydrates, sterols (also known as primary metabolites), and the enzymatic activities of marine fungi. These earlier reports were summed up by Kirk et al. (1974). Minor and species specific natural products from marine fungi (also known as secondary metabolites) were the sole topic of only a few reviews by Miller (1986) for lignicolous fungi, Liberra and Lindequist (1995), Biabani and Laatsch (1998), Le and Lin (2001), Jensen and Fenical (2002) and most recently, Bugni and Ireland (2004). The most comprehensive review of all natural products from marine fungi (covering primary and secondary metabolites, their biological activities, and fungal enzymes) was presented by Verbist and coworkers (2000). Natural products reported from marine fungi were described along with those isolated from other marine microorganims in several reviews (for example: König, 1992; Fenical and Jensen, 1993; Kobayashi and Ishibashi, 1993; Davidson, 1995; König and Wright, 1996;

Pietra, 1997; Bernan et al. 1997, Jensen and Fenical, 2000) and as part of a chapter in the recent comprehensive natural product chemistry compilation (Kobayashi and Ishibashi, 1999). Among the periodical reports on natural products from all marine microorganisms, the reviews by late J.D. Faulkner (2002) and, currently, by Blunt et al. (2003, 2004), are among the most useful sources of information on new developments in this area. The proceedings and abstracts of the international conferences and symposia on mycology and related areas such as marine biotechnology, marine corrosion and fouling, etc., provide additional insight into current research activities around the globe.

Practical approaches to the screening for natural products from marine fungi have recently been presented by Miller (2000).

This contribution attempts to present important recent developments in the study of natural products isolated from marine fungi and their implications for future research.

Enzymes

The commercial use of enzymes, including those isolated from terrestrial fungi, entails applications covering a wide range of processes and products ranging from manufacturing chemicals and foodstuffs to use as additives in products (Lowe, 1992). Studies on enzymatic activity frequently appear in literature, however, only one enzyme has been isolated from marine fungi so far (Burtseva et al. 2003).

Qiu et al. (2001) identified genes coding for Δ4 and Δ5 fatty acid desaturases in *Thraustochytrium* sp. and used a PCR-based cloning strategy to express the Fad4 gene in *Saccharomyces cervisiae* and *Brassica juncea*. Their study demonstrated that this desaturase is indeed involved in the biosynthesis of unusually high levels of DHA [docosahexaenoic acid, 22:6(n-3)] and EPA [eicosapentaenoic acid, 20:5(n-3)] by *Thraustochytrium* sp., and that a large-scale DHA production in genetically manipulated oilseed crops is possible.

Mohapatra et al. (1998) characterized an amylase from *Mucor* sp. isolated from the sponge. Suresh and Chandrasekaran (1999) studied chitinolytic activity of fungal isolates from marine sediments of India and assessed the impact of fermentation process parameters on chitinase production for the hyperproducing culture of a strain of *Beauveria bassiana*. The extracellular cellulolytic activity of some mangrove fungi from different sources was studied extensively and results presented in series of papers by Pointing et al. (1998, 1999), and Pointing (1999). The glycosylhydrolase activity of 90 marine fungal strains and amylase activity of 245 strains were studied by Burtseva et al. (2003) and Frolova et al. (2002), respectively. Degradation of the polyhydroxybutyric acid by fungi under simulated deep-sea pressure was studied by Gonda et al. (2000). Pointing (2000) presented both qualitative and quantitative assays used in the study of lignocellulolytic enzymes while Molitoris (2000) summarized the methods for evaluation of different enzymatic activities of marine fungi.

Marine Fungi: Metabolites

The number of published papers that describe the isolation of natural products from marine fungi has increased dramatically in six years and now exceeds the number of papers from the previous 50 years, which were reviewed by Biabani and Laatsch (1998) and Verbist et al. (2000). As mentioned earlier this increase in research activity may be attributed primarily to the need for new drugs. A number of government agencies and industrial sponsors fund this research effort in some countries and, indeed, the majority of contributions come from the systematic studies undertaken in Japan, United States, Germany, and China (including Hong Kong). In fact, approximately 70% of all reports originated in these four countries. Some of the reported screening programs involved an enormous number of samples and examples of older reports by Cuomo et al. (1995) in which some 1500 marine and 1450 terrestrial fungi were screened for antimicrobial activity, or screening of 80,000 fermentations (including possibly fungal and other microorganisms) for antifungal antibiotics reported by Albaugh et al. (1998) illustrate the point. Usually, these programs utilize fast screening, bioassay-guided methods that allow for the detection of specific activities and a high throughput of samples. A number of bioassays were developed specifically as tools for such screening, as for example the bioassay to detect substances with antimitotic and antifungal activity (Kobayashi et al. 1996, Namikoshi et al. 2004).

A larger number of marine fungi were screened in the last six years. Currently screening programs frequently address related issues such as the source, abundance, and variety of isolates. Sponga et al. (1999) isolated 395 fungi from marine sediments and sponges of the Mediterranean Sea, Red Sea, and Atlantic Ocean and found that marine sources are richer in fungal strains than soil samples. The majority of isolated strains belonged to the terrestrial genera, and the frequency of antimicrobial activities displayed by marine fungi was the same as for terrestrial strains. The antibacterial activity of 227 marine isolates of fungi from diverse Venezuelan sources including lagoons and mangroves was assessed and the natural products content of polar extracts of mycelia was analyzed using HPLC by Christophersen et al. (1999). As a significant number of extracts exhibited antibacterial activity, authors further studied 61 strains of *Penicillium citrinum* and found the natural content of extracts to be similar. The authors did not find any correlation between the ecology or identity of the host organism and the antibacterial activity of the *P. citrinum* isolate. In addition, they found no evidence that the presence of natural products in the extracts was related to competitive or antagonistic interactions between the fungal species. Malmstrøm et al. (2000) studied the metabolite characteristics of 68 *Penicillium* species from Venezuela. They found that all isolates of *P. citrinum* (25), *P. sizovae* (11), *P. corylophilum* (10), *P. sumatrense* (15), as well as 16 out of 18 isolates of *P. steckii* consistently produced the same, species-specific types of natural products. Höller et al. (2000) studied the diversity and biological activity of metabolites of fungi isolated from 16 marine sponges from the Pacific Ocean (Australia), Caribbean Sea (Dominican Republic), North Sea (Germany), Mediterranean Sea (Malta), and Atlantic Ocean (Canary Islands). Out of a total of 681 isolates, 92 showed bioactive metabolites in tests detecting the antifungal, antialgàl and antibacterial activities as well as cytotoxicity and inhibition of two enzymes, reverse transcriptase and tyrosine kinase. An unspecified number of fungal isolates from tropical and sub-tropical coral reefs was screened by Namikoshi et al. (2000b) for cytotoxicity, antifungal, and antimitotic activities. The authors found that marine-derived fungi produced metabolites that are identical to those found from different genera of terrestrial fungi. Diversity of marine fungi found in sediments from the Sea of Japan and production of bioactive metabolites by isolated strains was studied by Khudyakova et al. (2000). The authors found that 57% of fungal isolates, out of a sample of 442 species found in sediments, produced bioactive metabolites when the hemolytical activity of mycelial extracts was used as an indicator. For *Chaetomium spiculipilium* isolate, such an activity was exhibited by the fraction containing extracellular fatty acids and pigments. Nine fungal isolates from Fiji were studied for production of antimicrobial metabolites by Bugni and Ireland (2004). They found that most of the fungi exhibited dependence of metabolite production on the artificial seawater (ASW) concentration, with the production at a maximum for 20–60 % ASW and these results are similar to those reported earlier by Masuma et al. (2001). Several higher marine fungi collected from the shores of Prince Edward Island in Canada were screened for their antibacterial activity and new metabolites. A few of the extracts exhibited marginal activity and some already known compounds such as fatty acids, sterols, and other small molecules were recovered from these extracts (MacKenzie et al. 1998, MacKenzie et al. 2004). Nielsen et al. (1999) reported that 6 isolates of *Emericella unguis* exhibited the same HPLC metabolic profile. For each of these 6 isolates, guisinol and dechloronidulin were found among metabolites, while they were absent in the HPLC profiles of other *Emericella* species. This observation is of chemotaxonomic significance. Vala et al. (2000) examined thirteen strains of facultative marine fungi for the presence of siderophores located on the external surface of mycelia. Ten of the strains were found to produce carboxylate and/or hydroxamate siderophores in the iron-deficient media.

Those metabolites that were not included in reviews either by Verbist et al. (2000) or Biabani and Laatsch (1998) are presented in Tables 1 to 6. These compounds are arranged according to the source of a fungal isolate, using the arrangement of Biabani and Laatsch (1998). A few details along with a reference are provided for each product. Structures of some new metabolites are only slightly different from the metabolites known from other sources. The placement of the compound in a particular chemical group, according to the structure, is in many cases tentative, as there is no generally accepted classification scheme and in many cases biosynthetic origin has not been studied. Interestingly, while comparing the number of metabolites reported from different species and the sources of fungal isolates, Bugni and Ireland (2004) concluded that *Penicillium* and *Aspergillum* strains produced the highest number of isolated metabolites, and that the sponge-derived fungi were the most prolific source of metabolites.

Some of the compounds presented in Tables 1 to 6 deserve additional comments because of their potential for further study and eventual development as pharmacological agents. A few of them have already been studied and the mechanism of their action at a molecular level is at least partially understood. The most promising compounds were presented in the annual reports on marine pharmacology, which currently cover the literature up to and including the year 2000 (recent reports: Mayer, 1999; Mayer and Lehmann, 2000, 2001; Mayer and Hamann, 2002, 2004; Mayer and Gustafson, 2003). Included in these reviews are antitumor and cytotoxic aspergillamide A (Toske et al. 1998), gymnastatins (Amagata et al. 1998a), penostatins (Iwamoto et al. 1999), dankasterone (Amagata et al. 1999), mangicols (Renner et al. 2000), and N-methylsansalvamide (Cueto et al. 2000). The mechanisms of action have not been determined for these compounds as yet.

The mechanisms of action for two compounds, both belonging to the class of depsipeptides, have been elucidated. Reports indicate that thiocoraline, isolated from a *Micromonospora* sp. (Baz et al. 1997; Romero et al. 1997) is a DNA-polymerase α inhibitor (Erba et al. 1999), while sansalvamide A (Belofsky et al. 1999) inhibits topoisomerase-catalyzed DNA relaxation (Hwang et al. 1999). Thiocoraline is reported to be cytotoxic. A potential antihelminthic compound, nafuredin, which acts by inhibiting NADH-fumarate reductase of anaerobic organisms, was isolated by Ui et al. (2001), while mechanism of its action was established by Omura et al. (2001).

Studies on the mechanism of action of three metabolites which were included in reviews by Biabani and Laatsch (1998) and Verbist et al. (2000) revealed that gliotoxin inhibits mycobacterial detoxification enzyme mycothiol-S-conjugate amidase (Nicholas et al. 2003), epolactaene selectively inhibits mammalian DNA polymerase α and β and human topoisomerase II (Kakeya et al. 1997, Mizushina et al. 2000), while tryprostatin A arrests the cell cycle progression by inhibiting the microtubule assembly (Kondoh et al. 1998, Usui et al. 1998).

Several new metabolites with cytotoxic or antitumor activity have been reported. These compounds include humicolone (Laurent et al. 2002), virescenosides (Afiyatullov et al. 2000), aspergilloxide (Cueto et al. 2002), aigialomycin D (Isaka et al. 2002), spiroxins (McDonald et al. 1999), neomangicols (Renner et al. 1998), fusaperazines (Usami et al. 2002), leptosins (Yamada et al. 2002b, 2004), scytalidamides (Tan et al. 2003), varitriol (Malmstrøm et al. 2002a), penochalasins (Iwamoto et al. 2001), gymnasterons (Amagata et al. 1998b), evariquinone (Bringmann et al. 2003), gliocladins and glioperazine (Usami et al. 2004), modified verticillins (Son et al. 1999) and phomopsidin (Namikoshi et al. 1997, 2000b; Kobayashi et al. 2003). Their mode of action have not yet been reported. Phomopsidin was reported to exhibit antimitotic properties thus its mode of action may be similar to that of epolactaene. In general, metabolites exhibiting cytotoxic and antitumor properties may be found across the entire spectrum of chemical structures. Interestingly, Namikoshi et al. (2002) reported that a mixture of common fatty acids exhibited antimitotic activity, which may be eventually manifested in the antitumor or antifungal activity of such mixtures. Similarly, weak antibacterial activitities of fractions containing fatty acids and sterols were reported by Khudyakova et al. (2000) and MacKenzie et al. (2004).

Table 1. Marine fungi isolated from sand, sediments, sea, microbial mat, and from unidentified sources—selected natural products

Fungus/Source/Location Reference	Compound(s) chemical group	Biological activity
Aspergillus sp. Sediment, salt lake, Bahama Toske et al. (1998)	Aspergillamide A Peptide	Cytotoxic
Aspergillus sp. Unidentified, Bahama Cueto et al. (2002)	Aspergilloxide Sesterterpene	Cytotoxic
Aspergillus ochraceus Sediment, Sea of Japan Dai et al. (2001)	Circumdatin G Alkaloid	No data
Bispora? Sediment, Caribbean Miljkovic et al. (2001)	Scorpinone Polyketide	No data
Fusarium sp. Sea, Japan Sakaki et al. (2002)	Neurosporaxanthin β-D glucopyranoside Carotenoid glycosyl ester	No data
Halorosellinia oceanica Ocean soil, Thailand Chinworrungsee et al. (2001), Chinworrungsee et al. (2002)	Halorosellinic acid Terpene	Antimalarial
Hypoxylon oceanicum Unknown, South China Sea Li et al. (2001)	Benzopyrones, sterols Polyketides, sterols	No data
Hypoxolon sp. Unknown, South China Sea Le et al. (1999)	Fatty acids Fatty acids	No data
Keissleriella sp. Sediment, Yellow Sea Liu et al. (2002)	Dihydronaphthalenone derivative Isocoumarins Polyketides	Antifungal
Penicillium sp. Sand, South China Sea Lin et al. (2000)	Penicillazine Alkaloid	No data
Penicillium sp. Unknown, South China Sea Shao et al. (1999)	386-A Polyketide	No data
Stilbella aciculosa Sediment, Sea of Japan Kuznetsova et al. (2001)	Fusidic acid Sterol	No data
Zopfiella latipes Soil, India (Indian Ocean) Daferner et al. (2002)	Zopfiellamides A, B Mixed	Antimicrobial
Unknown FOM-8108 Sand, Japan Uchida et al. (2001)	Chlorogentisylquinone Polyketide	Enzyme inhibitor[a]
Unknown 2516 Unknown, South China Sea Yin et al. (2002)	Cyclic tetra- and dipeptides Cyclic peptides	Enzyme inhibitor[a]
Phoma sp. Marine microbial mat, Bahama Liu et al. (2003)	Phomoxin, phomoxide Polyketides	No activity

[a]Selective inhibitor of a neutral sphingomyelinase.

Table 2. Marine fungi isolated from driftwood and decayed wood—selected natural products

Fungus/Source/Location Reference	Compound(s) chemical group	Biological activity
Acremonium neo-caledoniae New Caledonia Laurent et al. (2000)	Verrol-4-acetate Terpenoid	Weak cytotoxic
Aspergillus tamarii Okinawa I., Japan Tsuda et al. (2003)	Speradine A Alkaloid	Enzyme inhibitor[a] Antibacterial
Corollospora maritima Helgoland, North Sea Liberra et al. (1998)	Corollosporine Polyketide	Mild antibacterial
Corollospora pulchella Peleliu (Japan) Alvi et al. (1998)	Pulchellalactam Mixed	Enzyme inhibitor[b]
Fusarium sp. Bahama (mangrove habitat) Renner et al. (1998)	Neomangicol A Neomangicol B Terpenoids	Cytotoxic Cytotoxic, antibacterial
Fusarium sp. Bahama (mangrove habitat) Renner et al. (2000)	Mangicols A and C Mangicols B, D-G Terpenoids	Anti-inflammatory Weak cytotoxic
Humicola grisea New Caledonia Laurent et al. (2002)	Humicolone Polyketide	Cytotoxic
Hypoxylon croceum Florida (mangrove habitat) Daferner et al. (1999)	Hypoxysordarin Terpenoid	Antifungal
Kirschsteiniothelia maritima Submerged wood, Baltic Sea Kusnick et al. (2002)	Ascochital, ascochitine Polyketides	Antibacterial
Myrothecium roridum Palau (Japan) Namikoshi et al. (2001)	12,13-Deoxyroridin E Macrocyclic (mixed)	Weak cytotoxic
Remispora maritima unknown Kong and Carter (2003)	Remisporine A Polyketide	No antimicrobial activity
Verruculina enalia Bahama (Salt Lake) Lin et al. (2002c)	Ennalins A and B Polyketides	No data

[a]Inhibitor of Ca-ATPase and histone deacetylase.
[b]Specific inhibitor of protein CD45 tyrosine phosphatase.

Table 3. Marine fungi from mangroves and other plants—selected natural products

Fungus/Source/Location Reference	Compound(s) chemical group	Biological activity
Aigialus parvus	Aigialomycin D	Cytotoxic, antiplasmodial
Mangrove, unreported	Aigialomycins A, B, C, E	No activity
Isaka et al. (2002)	Macrocyclic	
Eutypa sp.	Eutypoid-A, fatty acids	No data
Mangrove, South China Sea	Mixed, fatty acids	
Lin et al. (2002a)		
Halosarpheia sp. #732	Enniatin G	Antitumor
Mangrove, South China Sea	Enniatins B, B_4	
Lin et al. (2002b)		
Hypoxylon oceanicum	15G256ι, ω	Possibly antifungal
Mangrove; China	Macrocyclic polyesters	
	15G256o,α-2,β-2,ν,π	No data
Schlingmann et al. (2002)	Linear polyesters	
Julella avicenniae	FA, FA esters	No data
Mangrove; China	Fatty acids	
Lin et al. (1998)		
Phomopsis sp.	Phomopsidin	Potent antimitotic
Mangrove, Pohnpei	Mixed	
Namikoshi et al. (1997, 2000b),		
Kobayashi et al. (2003)		
Unknown endophytic	1893A and B	No data
Mangrove, South China Sea	Polyketides	
Chen et al. (2003)		
Unknown endophytic	Avicennins A and B	No data
Mangrove, South China Sea	Polyketides	
Lin et al. (2001c)		
Unknown endophytic No. 1403	Griseofulvin	No data
Mangrove, South China Sea	Anthracenediones	
Jiang et al. (2000)	Polyketides	
Unknown endophytic No. 2533	Avicennin A	No data
Mangrove, South China Sea	Polyketide	
Wan et al. (2001)		
Acremonium striatisporum	Virescenosides M, N	Cytotoxic
Holothurian, Sea of Japan	Terpenic glycosides	
Afiyatullov et al. (2000)		
Acremonium striatisporum	Virescenosides O-Q	Cytotoxic
Holothurian, Sea of Japan	Terpenic glycosides	
Afiyatullov et al. (2002)		
Fusarium sp.	Sansalvamide A	Cytotoxic, antiviral
Sea grass, Bahamas	Depsipeptide	Enzyme inhibitor[a]
Belofsky et al. (1999); mode of		
action – Hwang et al. (1999)		
Gliocladium roseum	Roselipins 1A-B; 2A-B	Enzyme inhibitors[b]
Seaweed, Micronesia	FA derivatives	
Omura et al. (1999);		
Tomoda et al. (1999);		
Tabata et al. (1999)		

(Contd.)

Scytalidum sp. Sea grass, Bahamas Rowley et al. (2003)	Halovirs A-E Pentadepsipeptides	Potent antiviral[c]
Thraustochytrium globosum Sea grass, Bahamas Jenkins et al. (1999)	Thraustochytrosides A-C FA derivatives	No data
Xylaria sp. Seeds of angiosperm tree, China Lin et al. (2001d)	Xyloketal A Xyloketals B, C, D, E Polyketides	Enzyme inhibitor[d] No data
Xylaria sp. Seeds of angiosperm tree, China Lin et al. (2001e)	Xyloallenolide Cyclic peptide (mixed)	No data
Unknown M3 Laver, Japan Byun et al. (2003)	Dioxopiperazine deriv. Dioxopiperazine	Antifungal

[a]Selective inhibitor of a poxvirus topoisomerase.
[b]Inhibitors of diacylglycerol acyltransferase.
[c]Active against herpes simplex viruses 1 and 2.
[d]Inhibitor of acetylcholinesterase.

Table 4. Marine fungi isolated from algae—selected natural products

Fungus/Source/Location Reference	Compound(s) chemical group	Biological activity
Acremonium sp. Brown alga, Spain, Mediterranean Abdel-Lateff et al. (2002)	Acremonin A and glucoside Polyketides (mixed)	Potent antioxidants
Ascochyta salicorniae Green alga, North Sea Osterhage et al. (2000a)	Ascosalipyrrolidinones Ascosalipyrone Polyketide (mixed)	Antiplasmodial Antimicrobial
Aspergillus parasiticus Red alga, Sea of Japan, Korea Son et al. (2002)	Parasitenone Polyketide	Free radical scavenger
Aspergillus sp. Brown alga, Philippines Lorenz et al. (1998)	Mactanamide Dioxopiperazine	Fungistatic
Cladosporium sphaerospermum Brown alga, Philippines Kuznetsova et al. (1998)	Sterols	No data
Drechslera dematioidea Red alga, Spain, Mediterranean Osterhage et al. (2002a)	Isosativenetriol Drechslerins A-G Terpenoids	Antiplasmodial Antifungal
Fusarium chlamydosporium Alga, Japan Usami et al. (2002)	Fusaperazines A, B Dioxopiperazines	Cytotoxic
Fusarium sp. Alga, Scotland Jiang et al. (2002)	JM47 Cyclic tetrapeptide	No activity
Fusarium sp. Alga, US Virgin Islands Cueto et al. (2000)	N-methylsansalvamide Cyclic depsipeptide	No data

(Contd.)

Fungus/Source/Location Reference	Compound(s) chemical group	Biological activity
Leptosphaeria sp. Alga, unknown Yamada et al. (2002b)	Leptosins M_1, N, N_1 Leptosin M Dioxopiperazines	Cytotoxic Cytotoxic, enz. inhibitor[a]
Leptosphaeria sp. Alga, unknown Yamada et al. (2004)	Leptosins O, P Leptosins R, S Dioxopiperazines	Cytotoxic
Penicillium sp. Alga, unknown Iwamoto et al. (1999)	Penostatins A-E Polyketides	Cytotoxic
Penicillium sp. Alga, unknown Iwamoto et al. (2001)	Penochalasins D-H Macrocycles	Cytotoxic
Penicillium sp. Green alga, Caribbean Son et al. (1999)	Dideoxyverticillin A Deoxyverticillin A Dioxopiperazines	Potent cytotoxic Potent cytotoxic
Pestalotia sp. Brown alga, Bahamas Cueto et al. (2001)	Pestalone Mixed	Potent antibiotic Cytotoxic
Phoma tropica Brown alga, Azores Osterhage et al. (2002b)	5-hydroxyramulosin Polyketide	No data
Scytalidium sp. Alga, Bahamas Tan et al. (2003)	Scytalidamides A, B Heptapeptides	Cytotoxic
Wardomyces anomalus Green alga, Baltic Sea Abdel-Lateff et al. (2003)	Polyhydroxyxanthones Polyketides	Antioxidants
Unknown K063 Red alga, Okinawa I., Japan Komatsu et al. (2001)	Dictyonamide A Peptides	Enzyme inhibitor[b]
Unknown K063 Red alga, Okinawa I., Japan Shigemori et al. (1999)	Seragakinone A Polyketide	Antibacterial
Unknown CNC-159 Green alga, unknown Jenkins et al. (1998)	Solanapyrones E – G Terpenoids	Antialgal
Unknown MF001 Green alga, Sea of Japan, Korea Lee et al. (1999)	Ergosteryl myristate Sterol ester	No data

[a]Inhibitor of protein kinases PTK and CaMKIII; specific inhibitor of topoisomerase II.
[b]Inhibitor of cyclin-dependent kinase 4.

Table 5. Marine fungi isolated from sponges—selected natural products

Fungus/Source/Location Reference	Compound(s) chemical group	Biological activity
Aspergillus niger Palau Island Ui et al. (2001), Omura et al. (2001)	Nafuredin Polyketide	Antihelminthic
Aspergillus niger Florida, USA Varoglu and Crews, (2000)	Asperic acid Terpenoid	No activity
Aspergillus ostianus Pohnpei Namikoshi et al. (2003)	Asperlactones Aspyrone Chlorinated polyketides	Antibacterial
Aspergillus versicolor Bali, Indonesia Lin et al. (2001a)	Aspergiones A-F Polyketides	No data
Aspergillus versicolor Bali, Indonesia Lin et al. (2001b)	Aspergillone and derivs. Polyketides	No data
Asteromyces cruciatus North Sea, Germany Höller et al. (1999a)	Furancarbaldehyde Mixed	Antibacterial
Cladosporium herbarum Indonesia Jadulco et al. (2001)	Pandangolides 3, 4 Macrocyclic	No activity
Cladosporium herbarum Indonesia Jadulco et al. (2002)	Herbarins A, B Polyketides	Insecticidal
Coniothyrum sp. Dominican Republic Höller et al. (1999b)	Phenylbutanone deriv. Polyketide	Potent antimicrobial
Curvularia lunata Indonesia Jadulco et al. (2002)	Lunatin Polyketide	Antibacterial
Drechslera hawaiiensis Bali, Indonesia Edrada et al. (2000)	Spiciferones A, B Spiciferol Polyketides	No data
Emericella variecolor Caribbean Sea Malmstrøm et al. (2002a)	Varitriol Varixanthone Polyketides (mixed)	Cytotoxic Antimicrobial
Emericella variecolor Elba, Italy Bringmann et al. (2003)	Evariquinone Isoemericellin Polyketide (mixed)	Cytotoxic No activity
Gymnascella dankaliensis Japan Amagata et al. (1998b)	Gymnasterones A, B Sterols	Cytotoxic
Gymnascella dankaliensis Japan Amagata et al. (1999)	Dankasterone Sterol	Potent cytotoxic

(Contd.)

Fungus/Source/Location Reference	Compound(s) chemical group	Biological activity
Gymnascella dankaliensis	Gymnastatins A-C	Potent cytotoxic
Japan	Gymnastatins D, E	Weak cytotoxic
Amagata et al. (1998a)	Mixed	
Hortaea werneckii	Hortein	No activity
Mediterranean Sea, France	Polyketide	
Brauers et al. (2001)		
Microsphaeropsis sp.	Microsphaeropsisin	Potent antimicrobial
North Sea, Germany	Terpenoid	
Höller et al. (1999b)		
Microsphaeropsis sp.	Betaenone derivs.	
Mediterranean Sea, France	Tetrahydroanthraquinones	Enzyme inhibitors[a]
Brauers et al. (2000)	Polyketides	
Microsphaeropsis sp.	Microsphaerones A, B	No significant activity
Mediterranean Sea, France	Polyketides	
Wang et al. (2002)		
Monochaetia sp.	Substituted phthalide	No activity
Dominican Republic	Polyketide	
Höller et al. (2000)		
Paecilomyces sp.	Deoxynortrichoharzin	No cytotoxicity
Fiji I.	Polyketide	
Rahbaek et al. (1998)		
Penicillium cf. montanense	Xestodecalatones A-C	B-antifungal
Bali Sea, Indonesia	Polyketides	
Edrada et al. (2002)		
Phoma sp.	YM-202204	Potent antifungal
Okinawa I., Japan	Polyketide	
Nagai et al. (2002)		
Trichoderma longibrachiatum	Epoxysorbicillinol	No data
Indonesia	Polyketide	
Sperry et al. (1998)		
Ulocladium botrytis	Ulocladol	Enzyme inhibitor[b]
Dominican Republic	Polyketide	
Höller et al. (1999a)		
Unknown I96S215	Isocladospolide	No antibiotic activity
Indonesia	Pandangolides 1,2	No antibiotic activity
Smith et al. (2000)	Polyketide/macrolides	
Unknown	Hirsutanols A-D, deriv.	A-antimicrobial
Indonesia	Terpenes	
Wang et al. (1998)		
Unknown TUF98F139	Fatty acids, mixture	Antimitotic
Palau I., Japan	Fatty acids	
Namikoshi et al. (2002)		

[a]Inhibitor of PKC-ε, CDK4, and EGF receptor protein kinases.
[b]Inhibitor of tyrosine kinase ($p56^{lck}$).

Table 6. Marine fungi isolated from fauna—selected natural products

Fungus/Source/Location Reference	Compound(s) chemical group	Biological activity
Acremonium sp. Tunicate, Bahama Belofsky et al. (2000)	Oxepinamides A-C Fumiquinazolines H-I Alkaloids	Potent anti-inflammatory Weak antifungal
Aspergillus niger Ascidian, Fiji Bugni et al. (2000)	Yanuthones and derivatives Mixed	Weak antibacterial
Emericella unguis Mollusc and medusa, Venezuela Nielsen et al. (1999)	Guisinol Depside	Antibacterial
Emericella unguis Medusa, Venezuela Malmstrøm, (1999)	Unguisins A-B Cyclic heptapeptide	Antibacterial
Emericella unguis Mollusc, Venezuela Malmstrøm et al. (2002b)	Unguisin C Depside Cyclic heptapeptide	No data
Epicoccum purpurascens Jellyfish, North Sea Wright et al. (2003)	Epicoccamide Mixed	No activity
Gliocladium roseum Sea hare, Japan Usami et al. (2004)	Gliocladins A-C, glioperazine Dioxo- and trioxopiperazines	Cytotoxic
Hyphomycetes sp. Zoanthid, Japan Suenaga et al. (2000)	Kasarin Amino acid derivative	Potent antibacterial Weak cytotoxic
Micromonospora sp. Soft coral, Mozambique Romero et al. (1997); Baz et al. (1997); mode of action – Erba et al. (1999)	Thiocoraline Depsipeptide	Potent cytotoxic Potent antimicrobial
Paecilomyces sp. Coral, Yap I., Japan Namikoshi et al. (2000a); Namikoshi et al. (2000b)	Paecilospirone Polyketide	Weak antibiotic
Penicillium sp. Bivalve, Okinawa Kagata et al. (2000)	Coruscol A Carbohydrate derivative	No data
Penicillium sp. Bivalve, Okinawa Komatsu et al. (2000)	Sculezonones A-B Polyketide	No data
Penicillium steckii Unidentified tunicate, Venezuela Malmstrøm et al. (2000)	Tanzawaic acids E-F Polyketides	No data
Periconia byssoides Sea hare, Japan Yamada et al. (2001)	Macrosphelides E, F, G, I Macrosphelide H Macrocycles	Cell-adhesion inhibitors
Periconia byssoides Sea hare, Japan Yamada et al. (2002a)	Macrosphelide L Macrocycles	Cell-adhesion inhibitor
Unknown LL-37H248 Soft coral, BC, Canada McDonald et al. (1999)	Spiroxins A-E Polyketides	Potent cytotoxic Weak antibacterial

Potent antifungal and/or antibacterial compounds with an undetermined mechanism of action include kasarin (Suenaga et al. 2000), zopfiellamides (Daferner et al. 2002), microsphaeropsisin and phenylbutanone derivative (Höller et al. 1999b), ascochital (Kusnick et al. 2002), ascosalipyrone (Osterhage et al. 2000a), varixanthone (Malmstrøm et al. 2002a), neomangicols (Renner et al. 1998), YM-202204 (Nagai et al. 2002), and hypoxysordarin (Daferner et al. 1999). It is probable that hypoxysordarin inhibits protein synthesis as this mechanism of antimicrobial action was reported for the other compounds of the sordarin group (Justice et al. 1998). A few natural products that exhibit moderate or weak antimicrobial properties are also included in Tables 1 to 6.

Recent reports indicate the mode of action for some of the isolated metabolites, and identify the enzymes targeted by metabolites. Chlorogentisylquinone (Uchida et al. 2001) inhibits neutral sphingomyelases 1 and 2, while activity of acidic sphingomyelinase remains unaffected. A study by Alvi et al. (1998) reported that pulchellactam inhibited a CD45 protein tyrosine phosphatase. Inhibitory activity of speradine A, targetting Ca-ATPase and histone deacetylase, was observed by Tsuda et al. (2003). Xyloketal A, one of the five xyloketals isolated by Lin et al. (2001d), was found to inhibit acetylcholinesterase. Roselipins were found by Tomoda et al. (1999) to inhibit diacylglycerol acyltransferase in *de novo* biosynthesis of triglycerides. A broad range of biological activities, including specific inhibition of two protein kinases PTK and CaMKIII, and of another enzyme, topoisomerase II, by leptosin M, was reported by Yamada et al. (2002b). Another kinase, cyclin-dependent kinase 4, was inhibited by dictyonamide A isolated by Komatsu et al. (2001). Tyrosine kinase ($p56^{lck}$) inhibitor, ulocladol, was isolated and assessed by Höller et al. (1999a). Four inhibitors of protein kinases were isolated by Brauers et al. (2000) and they were found to inhibit PKC-ε, CDK4, and EGF receptor tyrosine kinases.

At a cellular level, Helmholtz et al. (1999) reported that a mixture of glycolipids from *Nia vibrissa* inhibits binding of LPS endotoxin to a human endotoxin receptor. Also, according to the study of Yamada et al. (2001), macrosphelides E-H inhibited the adhesion of human leukemia cells to human-umbilical-vein endothelial cells. A series of peptides, halovirs A-E, was isolated by Rowley et al. (2003) and tested against herpes simplex viruses 1 and 2. These are the only recent marine fungal products, besides the sansalvamide A (Belofsky et al. 1999; Hwang et al. 1999), for which the antiviral activity has been reported so far. Other biological acivities of recent metabolites are indicated in Tables 1 to 6.

It has been noted by many authors that modifications of culture conditions may result in changes in the complement or amount of metabolites produced by a fungus. Cueto et al. (2001) reported that a new antibiotic, pestalone, was produced by *Pestalotia* sp. only when the fungus was co-cultured with the marine bacterium (strain CNJ-328). It appears to be the first report documenting the production of a natural product by marine fungus under challenge. Pestalone is a relatively simple polysubstituted benzophenone and has moderate cytotoxicity against human tumor cells. More importantly, however, pestalone exhibits potent antibacterial activity against methicillin-resistant *Staphylococcus aureus* and vancomycin-resistant *Enterococcus faecium* and these findings offer hope that more new products active against antibiotic-resistant strains of bacteria may be discovered among the marine fungal metabolites.

From the point of view of chemical structures there are a few interesting observations. All major classes of metabolites are represented among the newly reported products (Fig. 1). Most of the new compounds have structures that are not very much different from those reported earlier from various organisms and microorganisms, and only a few structures deserve mention here. Scorpinone (Miljkovic et al. 2001) features an azaanthraquinone system, which is relatively rare among the fungal metabolites. Kasarin structure (Suenaga et al. 2000) contains the rare (naturally occurring) monocyclic β-lactam ring of azetinone. Penicillazine (Lin et al. 2000) is unique as it contains (4*H*)-5,6-dihydro-1,2-oxazine ring system in the structure, which has no precedent among the natural products known. Similarly the spiro[chroman-2,1′(3′*H*)-isobenzofuran] skeleton of paecilospirone is unprecedented among natural products (Namikoshi et al. 2000a,b).

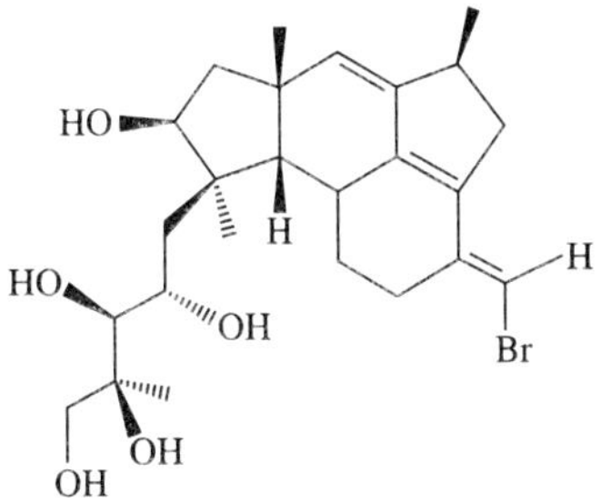

Neomangicol B

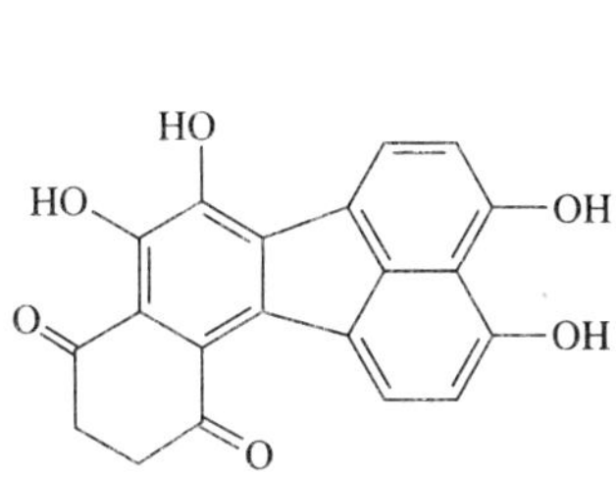

Hortein

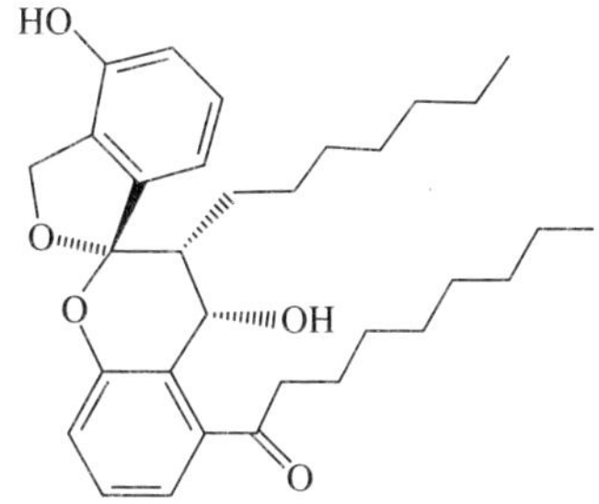

Paecilospirone

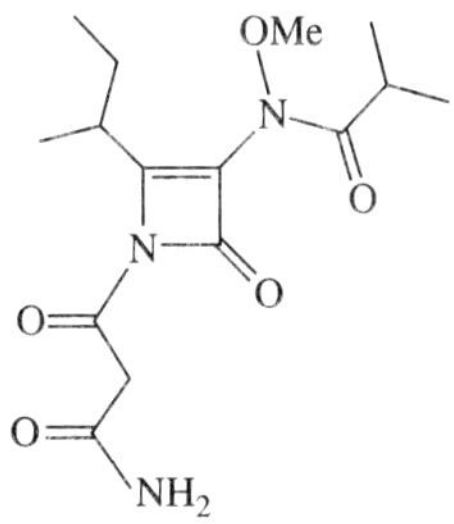

Kasarin

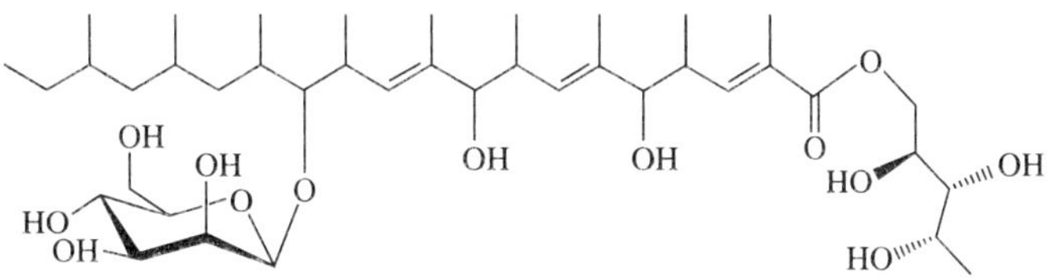

Roselipin 1A

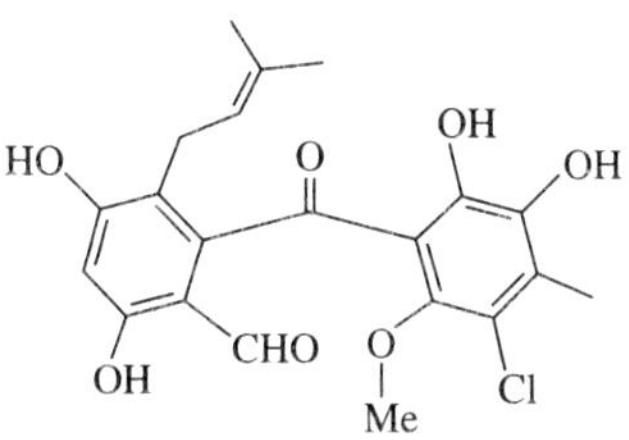

Pestalone

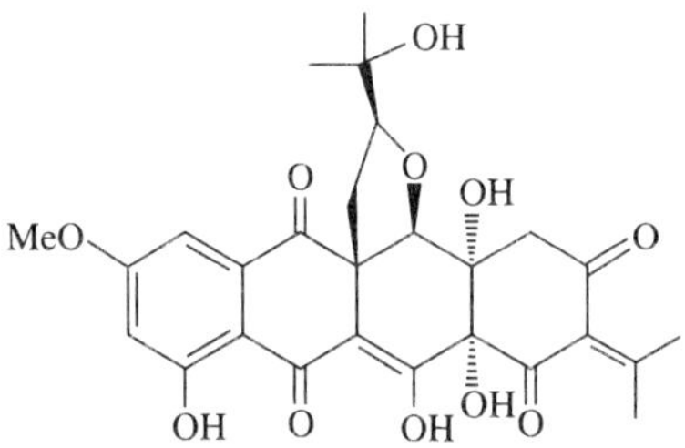

Seragakinone A

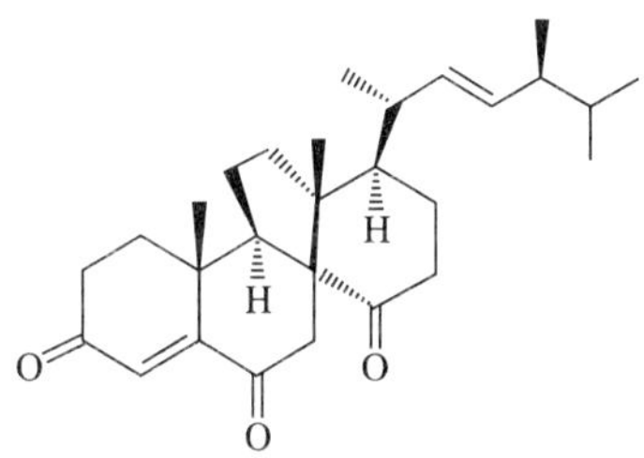

Dankasterone

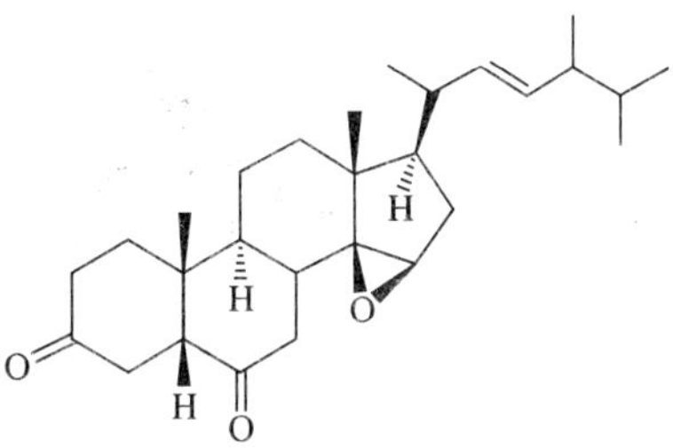

Gymnasterone B

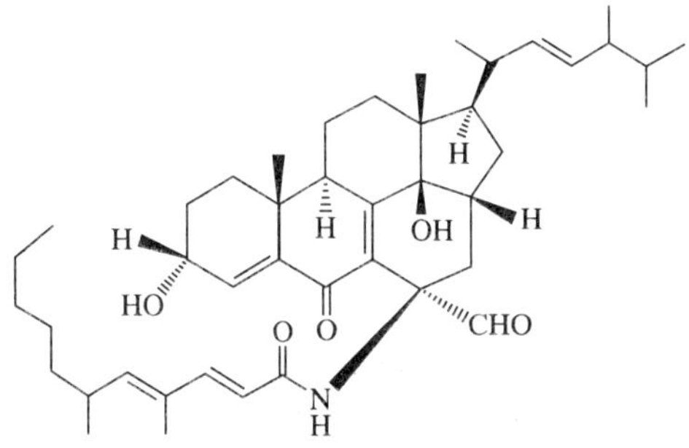

Gymnasterone A

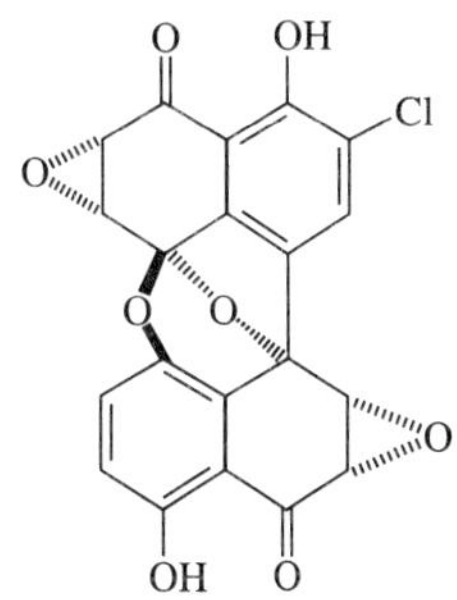

Spiroxin A

Fig. 1

Neomangicols (Renner et al. 1998) and mangicols (Renner et al. 2000) contain a hitherto unreported carbon skeleton, which appears to be a new class of rearranged sesterterpenes. Roselipins (Omura et al. 1999; Tomoda et al. 1999; Tabata et al. 1999) have a structure composed of a unique, highly methylated, unsaturated hydroxyacid, modified with a mannose or mannose acetate and esterified with arabinitol. Seragakinone A (Shigemori et al. 1999) is a pentacyclic anthracycline-related metabolite, containing a hydroxyisopropyl-substituted tetrahydrofuran ring fused to ring B of the anthracycline system. Gymnasterones A and B (Amagata et al. 1998b) have five and four rings, respectively with unusually modified sterol structures. Dankasterone (Amagata et al. 1999) has a keto-sterol structure with a different location of the five-member D-ring and an attachment of a chain, C20-C27, to the C-ring of the system. Hortein structure (Brauers et al. 2001) contains an oxidized acenaphtho[1′,2′:7,8]naphthalene ring system, which has not been found previously among natural products. Finally, spiroxins (McDonald et al. 1999) posses an unusual octacyclic ring system based upon a *bis*-naphthalene unit that resembles the structure of preussomerins.

Fungi from Hypersaline Environments and Salt Lakes

Facultative fungi have recently been found in the hypersaline environments of salterns of Southern Europe. Some of the strains, after isolation and culturing, were studied for production of mycotoxins by Tepšič et al. (1997). They found that strains of one of the dominant species, the highly pathogenic *Aspergillus fumigatus*, produced mycotoxins such as verruculogen, fumigaclavine A and fumitremorgins B and C, and that the composition of mycotoxin mixtures was different under the standard and saltern conditions of growth. Méjanelle et al. (2000) reported on the composition of the lipid fraction from different fungal species. This fraction contained an array of sterols and their complement varied among species, although ergosterol dominated in all cases. Some other sterols were recognized as intermediates in the biosynthesis of ergosterol in the study (Méjanelle et al. (2001). The composition of the sterol fraction allows for recognition of some fungal species, and, because of differences in the fungal, algal, and zooplankton sterol composition, it may serve as a specific fungal biomarker. Liu et al. (2003) reported on the isolation of new polyketide compounds, phomoxin and phomoxide and of the known eupenoxide from a culture of *Phoma* (strain CNC-651) which was recovered from a microbial mat collected from a hypersaline pond located in the Bahamas. The fungus *Verruculina enalia* isolated from a similar saline environment yielded polyketides ennalins A and B and three known diketopiperazines (Lin et al. 2002b). The effect of seawater concentration on growth and antimicrobial metabolite production of several strains of marine fungi was studied by Masuma et al. (2001) and Bugni and Ireland (2004).

Biotechnological Aspects

Research that has approached a possible commercial application is the use of thraustochytrids as a source of the long-chain n-3 polyunsaturated fatty acids (PUFA) production for nutritional purpose. Earlier efforts in this area are summed up by Lewis et al. (1999, 2000) who have reported extraction methods, allowing for efficient recovery of PUFA from cultured thraustochytrids. Optimization of PUFA production by thraustochytrids by the choice of the strains, variation in the growth medium composition and time of cultivation was reported by Bowles et al. (1999). A recent study by Fan et al. (2001) reports the utilization of food processing waste (soymilk) for the growth of thraustochytrids isolated from the mangrove environment. Interestingly, it concludes that mangrove strains produce low levels of EPA in general, and lower levels of DHA on soymilk-supplemented medium than cultures grown on glucose yeast extract medium. As a corollary, Lewis et al. (2001) studied the squalene and sterol content of thraustochytrid cultures grown under the same conditions, since the presence of these lipid components may adversely affect the nutritional value of isolated PUFA products. The use of thraustochytrid biomass as a source of n-3 PUFA in aquaculture resulted in enhanced PUFA content of crustaceans and fish (Jaritkhuan, 2002). A study on cultivation of *Nia vibrissa,* aimed at optimization conditions for the continuous synthesis of bioactive natural products was completed by Helmholz et al. (1999).

Analytical Aspects

Aquaculture farms are exposed to dynamic conditions of marine habitat, and the presence of fungi in farming areas has raised some concerns about the safety of harvested seafood. Farmed organisms, most notably shellfish, may accumulate the toxins produced by these fungi (Sallenave-Namont et al. 2000). Therefore the presence of fungi in farming areas must be monitored continuously and methods for expeditious and unambiguous analysis of the identity and concentrations of known toxins produced by these fungi must be established. Sallenave et al. (1999) studied bioaccumulation of *Trichoderma koningii* toxins by mussels and confirmed that this process does occur with the highest concentration being found in the digestive gland of the mussels. Similarly, Grovel et al. (2003) reported that gliotoxin, produced by a local *Aspergillus fumigatus* strain, is bioaccumulated mainly in the meat of blue mussels. Rapid tests for identification of gliotoxin using ion trap MS^n and combined tests using LC/MS and biological tests for identification of toxic peptaibols produced by *Trichoderma koningii* have been developed by Grovel et al. (2002) and Landreau et al. (2002), respectively.

Similarly, analytical methods for studies of drug metabolism *in vivo* are being developed. HPLC with fluorescence detection method for bioanalysis of thiocoraline, a depsipeptide anticancer drug derived from the marine fungus *Micromonospora* sp., in plasma samples was developed by Sparidans et al. (1999).

Analysis of the natural product complement has long been used as a valuable taxonomic tool. Recently, Osterhage et al. (2000b) reported on their comparative study of metabolites from 26 terrestrial and 16 marine *Phoma* species. The results that were obtained using HPLC-DAD and HPLC-MS methods indicate that there is a significant difference in the metabolite content between marine-derived and terrestrial strains. Similarly, the taxonomic importance of metabolite profiles was demonstrated by Malmstrøm et al. (2000) and Nielsen et al. (1999).

Future Trends

Recent WHO yearly reports paint a grim picture of human suffering and the prohibitive economic costs to treat many diseases, such as cancers, viral diseases, malaria, infectious diseases caused by microbes and the spread of antibiotic-resistant strains. There is a considerable pressure to fight these diseases worldwide. Several factors are considered in planning strategies of action and one of the solutions sought is an introduction of new and selective pharmaceutical agents. An examination of natural products isolated from a wide range of unexplored environments, still remains one of the major sources of new leads for the pharmaceutical industry. According to the analysis of sources for new drugs in the United States between 1981–2002, completed by Newman et al. (2003), the combinatorial chemistry technology, which has been heavily favored by the pharmaceutical industry in recent years, has not produced the expected results as of yet. In fact, the number of new active substances has declined in the last few years. The authors concluded that de-emphasizing and scaling down the natural product research by the industry is to blame for this situation. Similar conclusions have been reached by Butler (2004) in his recent analysis of the importance of natural product research in drug discovery. The industry has reduced the numbers of scientists working on the new natural products even though it may be evident that the natural products play a dominant role in discovery of leads for the development of new drugs. In his assessment of the research efforts on the study of biologically active compounds from marine organisms Blunden (2001) concluded that, from the point of view of new medicinal agents discovery, the results are disappointing. His statement includes marine fungal products. This situation is exacerbated by the fact that a significant proportion of isolated compounds have already been reported from other sources. So far, as the analysis of Munro et al. (1999)

indicates, marine invertebrates seem to be the best source of compounds with significant cytotoxic activity. These facts underscore an urgent need for more intensive research effort directed toward the study of fungal metabolites.

It has been long postulated by many authors that ecological considerations should be applied in the choice of strains for examination (see Gloer, 1997, for terrestrial fungi; Burgess et al. 1999, for bacteria; Strobel, 2003, for terrestrial microbial endophytes; Engel et al. 2002, for marine microbes, *etc.*). The authors of an increasing number of recent papers reviewed here actually applied this principle and the groups of strains isolated from different organisms or from a defined environment were studied. It should be stressed that the recent reports assessing the correlation between the environment and the presence of metabolites or on increase in their concentration have brought conflicting conclusions (see Sponga et al. 1999; Christophersen et al. 1999; Tepšič et al. 1997; Osterhage et al. 2000b, Masuma et al. 2001; Bugni and Ireland, 2004). This issue, as demonstrated by the case of induction of pestalone biosynthesis by a fungus, which was subjected to bacterial challenge (Cueto et al. 2001), requires further study. In a related area, Schulz et al. (1999) indacatied that some of the natural products isolated from terrestrial strains inhabiting extreme environments may differ significantly from those found under more uniform conditions. Therefore, the study of strains inhabiting unusual environments such as salterns, salt marshes, intertidal and mangrove environments, deep sea, underwater vents, etc., may result in the isolation of new and interesting metabolites. Another related area of interest is the precursor-directed biosynthesis of metabolites using fungi, which has been successfully applied in the production of cyclic heptapeptide unguisin D by Malmstrøm et al. (2002).

Immediate attention is required for a detailed assessment of requirements for culturing different fungi in order to obtain replicable results from the same strain, as well as studies of unexplored phyla such as oomycetes, labyrinthulids, chytridiomycetes, or fungi difficult to culture. The results of such studies may lead to the development of innovative biotechnologies. Another underdeveloped area entails the compilation of databases of analytical profiles for fungi commonly found in aquaculture areas. This would be useful for toxicological analyses as well as monitoring safety of produced seafood.

Isolation of new natural products from marine organisms and microorganisms is currently the major direction in natural product research. In this context, broadly defined marine fungi offer a significant promise as the source of new compounds. With a surge of interest in the marine fungal product research, which has occurred in recent years, it is reasonable to expect that new products with interesting biological activities will be discovered.

Conclusions

The brief review of developments in the area of marine fugal natural products indicates clearly a significant increase in research efforts which has occurred in the past six years. The sheer number of publications found in the literature for the years 1998–2004 substantiates this observation. The reports originated from many countries, thus demonstrating that a meaningful, although still inadequate number of scientists, actively participated in the studies. The research completed in four countries dominates this area and such a situation seems to result from the proper organization of research activities, availability of substantial funding from both government and industrial sponsors, and concentration of resources and personnel.

As far as the outcome of the current research effort is concerned, it appears to be impressive as many new compounds with interesting biological activities were reported. Some of the compounds will certainly be further evaluated and eventually either developed directly or used as leads in a search for agents with an improved spectrum of biological action. The results present many alternative sources of already known natural products. Isolation of new compounds with cytotoxic, antimicrobial, antiviral, antifouling, and other biological activities has been accomplished. Of particular interest is a new collection of compounds exhibiting selective activity against some antibiotic-resistant strains of bacteria, against some lines of cancer cells, and selective antagonists of some enzymes, which raises hope for more discoveries with significant impact in future.

From the point of view of mycology and ecology, however, there are reasons for concern. In many laboratories fungi are just one of many marine organisms and microorganisms under study. The dominant part of research, based upon a fast screening model, is undoubtedly profit-oriented, and this marginalizes knowledge-driven research. Some papers on marine fungal metabolites are rushed for publication and even the identity of the fungal isolate is unknown at the time of publication. It was noted by Pomponi (1999) that assays used in high-throughput drug discovery programs provide no information about the ecological position of screened organisms and their relationships with co-habiting organisms. Similarly, König and Wright (1996), Gloer (1997) and Hyde et al. (1998) indicated that the accelerated loss of biodiversity places an element of urgency on the study of fungi, and these comments remain valid today. Fortunately, there are also reports, which may still be considered in the knowledge-driven category and they are able to maintain records about marine habitats.

Acknowledgments

It is my pleasure to thank Dr. Thomas Rand (Department of Biology, Saint Mary's University), for critically reading the manuscript.

References

Abdel-Lateff, A., Klemke, C., König, G.M. and Wright, A.D. (2003) Two new xanthone derivatives from the algicolous marine fungus *Wardomyces anomalus*. Journal of Natural Products 66: 706–708.

Abdel-Lateff, A., König, G.M., Fisch, K.M., Höller, U., Jones, P.G. and Wright, A.D. (2002) New antioxidant hydroquinone derivatives from the algicolous marine fungus *Acremonium* sp. Journal of Natural Products 65: 1605–1611.

Afiyatullov, S.S., Kalinovsky, A.I., Kuznetsova, T.A., Isakov, V.V., Pivkin, M.V., Dmitrenok, P.S. and Elyakov, G.B. (2002) New diterpene glycosides of the fungus *Acremonium striatisporum* isolated from a sea cucumber. Journal of Natural Products 65: 641–644.

Afiyatullov, S.S., Kuznetsova, T.A., Isakov, V.V., Pivkin, M.V., Prokof'eva, N.G. and Elyakov, G.B. (2000) New diterpenic altrosides of the fungus *Acremonium striatisporum* isolated from a sea cucumber. Journal of Natural Products 63: 848–850.

Albaugh, D., Albert, G., Bradford, P., Cotter, V., Froyd, J., Gaughran, J., Kirsch, D.R., Lai, M., Rehnig, A., Sieverding, E. and Silverman, S. (1998) Cell wall active antifungal compounds produced by the marine fungus *Hypoxylon oceanicum* LL-15G256. III. Biological properties of 15G256γ. Journal of Antibiotics 51(3): 317–322.

Alvi, K.A., Casey, A. and Nair, B.G. (1998) Pulchellalactam, a CD45 protein tyrosine phosphatase inhibitor from the marine fungus *Corollospora pulchella*. Journal of Antibiotics 51(5): 515–517.

Amagata, T., Doi, M., Ohta, T., Minoura, K. and Numata, A. (1998a) Absolute stereostructures of novel cytotoxic metabolites, gymnastatins A-E, from *Gymnascella* species separated from a *Halichondria* sponge. Journal of the Chemical Society, Perkin Transactions 1, 3585–3599.

Amagata, T., Doi, M., Tohgo, M., Minoura, K and Numata, A. (1999) Dankasterone, a new class of cytotoxic steroid produced by a *Gymnascella* species from a marine sponge. Journal of the Chemical Society, Chemical Communications 1321–1322.

Amagata, T., Minoura, K. and Numata, A. (1998b) Gymnasterones, novel cytotoxic metabolites produced by a fungal strain from a sponge. Tetrahedron Letters 39: 3773–3774.

Baz, J.P., Canedo, L.M., Fernandez Puentes, J.L. and Silva Elipe, M.V. (1997) Thiocoraline, a new depsipeptide with antitumor activity produced by a marine *Micromonospora*. II. Physico-chemical properties and structure determination. Journal of Antibiotics 50(9): 738–741.

Belofsky, G.N., Anguera, M., Jensen, P.R., Fenical, W. and Köck, M. (2000) Oxepinamides A-C and fumiquinazolines H-I: Bioactive metabolites from a marine isolate of a fungus of the genus *Acremonium*. Chemistry – A European Journal 6(8): 1355–1360.

Belofsky, G.N., Jensen, P.R. and Fenical, W. (1999) Sansalvamide: A new cytotoxic cyclic depsipeptide produced by a marine fungus of the genus *Fusarium*. Tetrahedron Letters 40: 2913–2916.

Bernan, V.S., Greenstein, M. and Maiese, W.M. (1997) Marine microorganisms as a source of new natural products. Advances in Applied Microbiology 43: 57–90.

Biabani, M.A.F. and Laatsch, H. (1998) Advances in chemical studies on low-molecular weight metabolites of marine fungi. Journal für Praktische Chemie 340: 589–607.

Blunden, G. (2001) Biologically active compounds from marine organisms. Phytotherapy Research 15: 89–94.

Blunt, J.W., Copp, B.R., Munro, M.H.G., Northcote, P.T. and Prinsep, M.R. (2003) Marine natural products. Natural Product Reports 20: 1–48.

Blunt, J.W., Copp, B.R., Munro, M.H.G., Northcote, P.T. and Prinsep, M.R. (2004) Marine natural products. Natural Product Reports 21: 1–49.

Bowles, R.D., Hunt, A.E., Bremer, G.B., Duchars, M.G. and Eaton, R.A. (1999) Long-chain n-3 polyunsaturated fatty acid production by members of the marine protistan group the thraustochytrids: Screening of isolates and optimization of docosahexaenoic acid production. Journal of Biotechnology 70: 193–202.

Brauers, G., Ebel, R., Edrada, R.A., Wray, V., Berg, A., Gräfe, U. and Proksch, P. (2001) Hortein, a new natural product from the fungus *Hortaea werneckii* associated with the sponge *Aplysina aerophoba*. Journal of Natural Products 64: 651–652.

Brauers, G., Edrada, R.A., Ebel, R., Proksch, P., Wray, V., Berg, A., Gräfe, U., Schächtele, C., Totzke, F., Finkenzeller, G., Marme, D., Kraus, J., Münchbach, M., Michel, M., Bringman, G. and Schaumann, K. (2000) Anthraquinones and betaenone derivatives from the sponge-associated fungus *Microsphaeropsis* species: Novel inhibitors of protein kinases. Journal of Natural Products 63(6): 739–745.

Bringmann, G., Lang, G., Steffens S., Gınther, E. and Schaumann, K. (2003) Evariquinone, isoemericellin, and stromemycin from a sponge derived strain of the fungus *Emericella variecolor*. Phytochemistry 63: 437–443.

Bugni, T.S., Abbanat, D., Bernan, V.S., Maiese, W.M., Greenstein, M., Van Wagoner, R.M. and Ireland, C.M. (2000) Yanuthones: Novel metabolites from a marine isolate of *Aspergillus niger*. Journal of Organic Chemistry 65: 7195–7200.

Bugni, T.S. and Ireland, C.M. (2004) Marine-derived fungi: A chemically and biologically diverse group of microorganisms. Natural Product Reports 21: 143–163.

Burgess, J.G., Jordan, E.M., Bregu, M., Mearns-Spragg, A. and Boyd, K.G. (1999) Microbial antagonism: a neglected avenue of natural products research. Journal of Biotechnology 70: 27–32.

Burtseva, Y.V., Verigina, N.S., Sova, V.V. Pivkin, M.V. and Zvyagintseva, T.N. (2003) Filamentous marine fungi as producers of O-glycosylhydrolases: β-1,3-glucanase from *Chaetomium indicum*. Marine Biotechnology 5: 349–359.

Butler, M.S. (2004) The role of natural product chemistry in drug discovery. Journal of Natural Products 67: 2141–2153.

Byun, H-G., Zhang, H., Mochizuki, M., Adachi, K., Shizuri, Y., Lee, W-J. and Kim, S-K. (2003) Novel antifungal diketopiperazine from marine fungus. Journal of Antibiotics 56(2): 102–106.

Carte, B.K. (1996) Biomedical potential of marine natural products. BioScience 46 (4): 271–286.

Chen, G., Lin, Y., Wen, L., Vrijmoed, L.L.P. and Gareth Jones, E.B. (2003) Two new metabolites of a marine endophytic fungus (No. 1893) from an estuarine mangrove on the South China Sea coast. Tetrahedron 59: 4907–4909.

Chinworrungsee, M., Kittakoop, P., Isaka, M., Chanphen, R., Tanticharoen, M. and Thebtaranonth, Y. (2002) Halorosellins A and B, unique isocoumarin glucosides from the marine fungus *Halorosellinia oceanica*. Journal of the Chemical Society, Perkin Transactions 1, 2473–2476.

Chinworrungsee, M., Kittakoop, P., Isaka, M., Rungrod, A., Tanticharoen, M. and Thebtaranonth, Y. (2001) Antimalarial harosellinic acid from the marine fungus *Halorosellinia oceanica*. Bioorganic and Medicinal Chemistry Letters 11: 1965–1969.

Christophersen, C., Crescente, O., Frisvad, J.C., Gram, L., Nielsen, J., Nielsen, P.H. and Rahbaek, L. (1999) Antibacterial activity of marine-derived fungi. Mycopathologia 143: 135–138.

Cueto, M., Jensen, P.R. and Fenical, W. (2000) N-Methylsansalvamide, a cytotoxic cyclic depsipeptide from a marine fungus of the genus *Fusarium*. Phytochemistry 55: 223–226.

Cueto, M., Jensen, P.R. and Fenical, W. (2002) Aspergilloxide, a novel sesterterpene epoxide from a marine-derived fungus of the genus *Aspergillus*. Organic Letters 4: 1583–1585.

Cueto, M., Jensen, P.R., Kauffmann, C., Fenical, W., Lobkovsky, E. and Clardy, J. (2001) Pestalone, a new antibiotic produced by a marine fungus in response to bacterial challenge. Journal of Natural Products 64: 1444–1446.

Cuomo, V., Palomba, I., Perretti, A., Guerriero, A., D'Ambrosio, M. and Pietro, F. (1995) Antimicrobial activities from marine fungi. Journal of Marine Biotechnology 2: 199–204.

Daferner, M., Anke, T. and Sterner, O. (2002) Zopfiellamides A and B, antimicrobial pyrrolidinone derivatives from the marine fungus *Zopfiella latipes*. Tetrahedron 58: 7781–7784.

Daferner, M., Mensch, S., Anke, T. and Sterner, O. (1999) Hypoxysordarin, a new sordarin derivative from *Hypoxolon croceum*. Zeitschrift für Naturforschung 54c: 474–480.

Dai, J-R., Carté, B.K., Sidebotteom, P.J., Lee, Sek Yew A., Ng, S-B., Huang, Y. and Butler, M.S. (2001) Circumdatin G, a new alkaloid from the fungus *Aspergillus ochraceus*. Journal of Natural Products 64: 125–126.

Davidson, B.S. (1995) New dimensions in natural products research: Cultured marine microorganisms. Current Opinion in Biotechnology 6: 284–291.

Edrada, R.A., Heubes, M., Brauers, G., Wray, V., Berg, A., Gräfe, U., Wohlfarth, M., Mühlbacher, J., Schaumann, K., Sudarsono, Bringmann, G. and Proksch, P. (2002) Online analysis of xestodecalactones A-C, novel bioactive metabolites from the fungus *Penicillium cf. montanense* and their subsequent isolation from the sponge *Xestospongia exigua*. Journal of Natural Products 65: 1598–1604.

Edrada, R.A., Wray, V., Berg, A., Gräfe, U., Sudarsono, Brauers G. and Proksch P. (2000) Novel spiciferone derivatives from the fungus *Drechslera hawaiiensis* isolated from the marine sponge *Callyspongia aerizusa*. Zeitschrift für Naturforschung 55c: 218–221.

Engel, S., Jensen, P.R. and Fenical, W. (2002) Chemical ecology of marine microbial defense. Journal of Chemical Ecology 28(10): 1971–1985.

Erba, E., Bergamashi, D., Ronzoni, S., Faretta, M., Taverna, S., Bonfanti, M., Catapano, C.V., Faircloth, G., Jimeno, J. and D'Incalci, M. (1999) Mode of action of thiocoraline, a natural marine compound with anti-tumor activity. British Journal of Cancer 80: 971–980.

Fan, K.W., Chen, F., Jones, E.B.G. and Vrijmoed, L.L.P. (2001) Eicosapentaenoic and docosahexaenoic acids production by and okara-utilizing potential of thraustochytrids. Journal of Industrial Microbiology and Biotechnology 27: 199–202.

Faulkner, D.J. (2000) Highlights of marine natural products chemistry (1972–1999) Natural Product Reports 17: 1–6.

Faulkner, D.J. (2002) Marine natural products. Natural Product Reports 19: 1–48 (and previous reports in this journal by the same author).

Fell, J.W. and Newell, S.Y. (1998) Biochemical and molecular methods for the study of marine fungi. In: Molecular Approaches to the Study of the Ocean, (ed), Cooksey, K.E. Chapman and Hall, London, 259–283.

Fenical W. and Jensen P.R. (1993) Marine microorganisms: A new biomedical resource. In: Marine Biotechnology, Volume 1, Pharmaceutical and Bioactive Natural Products, (eds), Attaway, D.H. and Zaborsky, O.R. Plenum Press, New York and London, 419–457.

Frolova, G.M., Sil'chenko, A.S., Pivkin, M.V. and Mikhailov, V.V. (2002) Amylases of the fungus *Aspergillus flavipes* associated with *Fucus evanescens*. Prikladnaya Biokhimiya i Mikrobiologiya 38(2): 155–160.

Gloer, J.B. (1997) Applications of fungal ecology in the search for new bioactive natural products. In: The Mycota, IV, Environmental and Microbial Relationships, (eds), Wicklow, D.T. and Söderström, B.E. Springer-Verlag, Berlin 249–268.

Gonda, K.E., Jendrossek, D. and Molitoris, H.P. (2000) Fungal degradation of the thermoplastic polymer poly-β-hydroxybutyric acid (PHB) under simulated deep sea pressure. Hydrobiologia 426: 173–183.

Grovel, O., Pouchus, Y.F., Robiou du Pont, T., Montagu, M., Amzil, Z. and Verbist, J-F. (2002) Ion trap for identification of gliotoxin, as the cytotoxic factor of a marine strain of *Aspergillus fumigatus* Fresenius. Journal of Microbiological Methods 48: 171–179.

Grovel, O., Pouchus, Y.F. and Verbist, J-F. (2003) Accumulation of gliotoxin, a cytotoxic mycotoxin from *Aspergillus fumigatus*, in blue mussel (*Mytilus edulis*). Toxicon 42: 297–300.

Helmholz, H., Etoundi, P. and Lindequist, E. (1999) Cultivation of the marine basidiomycete *Nia vibrissa* (Moore & Meyers), Journal of Biotechnology 70: 203–206.

Höller, U., König, G.M. and Wright, A.D. (1999a) A new tyrosine kinase inhibitor from a marine isolate of *Ulocladium botrytis* and new metabolites from the marine fungi *Asteromyces cruciatus* and *Varicosporina ramulosa*. European Journal of Organic Chemistry 2949–2955.

Höller, U., König, G.M. and Wright, A.D. (1999b) Three new metabolites from marine-derived fungi of the genera *Coniothyrium* and *Microsphaeropsis*. Journal of Natural Products 62: 114–118.

Höller, U., Wright, A.D., Matthee, G.F., Konig, G.M., Draeger, S., Aust., H-J. and Schulz, B. (2000) Fungi from marine sponges: Diversity, biological activity and secondary metabolites. Mycological Research 104: 1354–1365.

Hwang, Y., Rowley, D., Rhodes, D., Gertsch, J., Fenical, W. and Bushman, F. (1999) Mechanism of inhibition of a poxvirus topoisomerase by the marine natural product sansalvamide A. Molecular Pharmacology 55: 1049–1053.

Hyde, K.D., Jones, E.B.G., Leaño, E.M., Pointing, S.B., Poonyth, A.D. and Vrijmoed, L.L.P., (1998) Role of fungi in marine ecosystems. Biodiversity and Conservation 7: 1147–1161.

Isaka, M., Suyarnsestakorn, C., Tanticharoen, M., Kongsaeree, P. and Thebtaranonth, Y. (2002) Aigialomycins A - E, new resorcyclic macrolides from the marine mangrove fungus *Aigialus parvus*. Journal of Organic Chemistry 67: 1561–1566.

Iwamoto, C., Minoura, K., Oka, T., Ohta, T., Hagishita, S. and Numata, A. (1999) Absolute stereostructures of novel cytotoxic metabolites, penostatins A - E, from a *Penicillium* species separated from an *Enteromorpha* alga. Tetrahedron 55: 14353–14368.

Iwamoto, C., Yamada, T., Ito, Y., Minoura, K. and Numata, A. (2001) Cytotoxic cytochalasans from a *Penicillium* species separated from a marine alga. Tetrahedron 57: 2997–3004.

Jadulco, R., Brauers, G., Edrada, R.A., Ebel, R., Wray, V., Sudarsono, and Proksch, P. (2002) New metabolites from sponge-derived fungi *Curvularia lunata* and *Cladosporium herbarum*. Journal of Natural Products 65: 730–733.

Jadulco, R., Proksch, P., Wray, V., Sudarsono, Berg, A. and Gräfe, U. (2001) New macrolides and furan carboxylic acid derivative from the sponge-derived fungus *Cladosporium herbarum*. Journal of Natural Products 64: 527–530.

Jaritkhuan, S. (2002) Thraustochytrids: A new alternative source of fatty acids for aquaculture. In: Fungi in Marine Environments, (ed), Hyde K.D. Fungal Diversity Research Series, Hong Kong, 7: 345–357.

Jenkins, K.M., Jensen, P.R. and Fenical, W. (1999) Thraustochytrosides A-C: New glycosphingolipids from a unique marine protist, *Thraustochytrium globosum*. Tetrahedron Letters 40: 7637–7640.

Jenkins, K.M., Toske, S.G., Jensen, P.R. and Fenical, W. (1998) Solanapyrones E-G, antialgal metabolites produced by a marine fungus. Phytochemistry 49(8): 2299–2304.

Jensen, P.R. and Fenical, W. (2000) Marine microorganisms and drug discovery: Current status and future potential. In: Drugs from the Sea. (ed), Fusetani N. Karger, Basel, 6–29.

Jensen, P.R. and Fenical, W. (2002). Secondary metabolites from marine fungi. In: Fungi in Marine Environments. (ed), Hyde, K.D. Fungal Diversity Research Series, Hong Kong, 7: 293–315.

Jiang, G., Lin, Y., Zhou, S. Vrijmoed, L.L.P. and Jones, E.B.G. (2000) Studies on the secondary metabolites of mangrove fungus No. 1403 from the South China Sea. Acta Scientiarum Naturalium Universitas Sunyatseni 39(6): 68–72.

Jiang, Z., Barret, M-O., Boyd, K.G., Adams, D.R., Boyd, A.S.F. and Burgess, J.G. (2002) JM47, a cyclic tetrapeptide HC-toxin analogue from a marine *Fusarium* species. Phytochemistry 60: 33–38.

Justice, M.C., Hsu, M.J., Tse, B., Ku, T., Balkovec, J., Schmatz, D. and Nielsen, J. (1998) Elongation Factor 2 as a novel target for selective inhibition of fungal protein synthesis. Journal of Biological Chemistry 273: 3148–3151.

Kagata, T., Shigemori, H., Mikami, Y. and Kobayashi, J. (2000) Coruscol A, a new metabolite from the marine-derived fungus *Penicillium* species. Journal of Natural Products 63: 886–887.

Kakeya, H., Onozawa, C., Sato, M., Arai, K. and Osada, H. (1997) Neuritogenic effect of epolactaene derivatives on human neuroblastoma cells which lack high-affinity nerve growth factor receptors. Journal of Medicinal Chemistry 40: 391–394.

Khudyakova, V. Yu., Pivkin, M.V., Kuznetsova, T.A. and Svetashev, V.I. (2000) Marine fungi in sediments of the Sea of Japan and their biologically active metabolites. Mikrobiologiya 69: 722–726.

Kirk, P.W., Jr., Catalfomo, P., Block, J.H and Constantine, G.H., Jr. (1974) Metabolites of higher marine fungi and their possible ecological significance. Veröffentlichungen. Institut für Meeresforschung in Bremerhaven, Supplement 5: 509–518.

Kobayashi, J. and Ishibashi, M. (1993) Bioactive metabolites of symbiotic marine organisms. Chemical Reviews 93: 1753–1769.

Kobayashi, J. and Ishibashi, M. (1999) Marine natural products and marine chemical ecology. In: Comprehensive Natural Product Chemistry, (eds), Barton, D., Nakanishi, K. and Meth-Cohn, O. Elsevier, Amsterdam, Vol. 8, Miscellaneous Natural Products including Marine Natural Products, Pheromones, Plant Hormones, and Aspects of Ecology, Mori K. vol. Ed., 415–649.

Kobayashi, H., Meguro, S., Yoshimoto, T. and Namikoshi, M. (2003) Absolute structure, biosynthesis, and anti-microtubule activity of phomopsidin, isolated from a marine-derived fungus *Phomopsis* sp. Tetrahedron 59: 455–459.

Kobayashi, H., Namikoshi, M., Yoshimoto, T. and Yokochi, T. (1996) A screening method for antimitotic and antifungal substances using conidia of *Pyricularia oryzae*, modification and application to tropical marine fungi. Journal of Antibiotics 49(9): 873–879.

Kohlmeyer, J. and Kohlmeyer, E. (1979) Marine Mycology: The Higher Fungi. Academic Press, London.

Komatsu, K., Shigemori, H. and Kobayashi, J. (2001) Dictyonamides A and B, new peptides from marine-derived fungus. Journal of Organic Chemistry 66: 6189–6192.

Komatsu, K., Shigemori, H., Mikami, Y. and Kobayashi, Y. (2000) Sculezonones A and B, two metabolites possessing a phenalenone skeleton from a marine-derived fungus *Penicillium* species. Journal of Natural Products 63: 408–409.

Kondoh, M., Usui, T., Mayumi, T. and Osada, H. (1998) Effects of tryprostatin derivatives on microtubule assembly *in vitro* and *in situ*. Journal of Antibiotics 51: 801–804.

Kong, F. and Carter, G.T. (2003) Remisporine B, a novel dimeric chromenone derived from spontaneous Diels-Alder reaction of remisporine A. Tetrahedron Letters 44: 3119–3122.

König, G.M. (1992) Meeresorganismen als Quelle pharmazeutisch bedeutsamer Naturstoffe. Deutsche Apotheker Zeitung 132(14): 673–683.

König, G.M. and Wright, A.D. (1996) Marine natural products research: Current directions and future potential. Planta Medica 62: 193–211.

Kusnick, C., Jansen, R., Liberra, K. and Lindequist, U. (2002) Ascochital, a new metabolite from the marine ascomycete *Kirschsteiniothelia maritime*. Pharmazie 57: 510–512.

Kuznetsova, T.A., Afiyatullov, S.S., Denisenko, V.A., Pivkin, M.V. and Elyakov, G.B. (1998) Sterols from a marine isolate of the fungus *Cladosporium sphaerospermum* Penz.. Biochemical Systematics and Ecology 26: 365–366.

Kuznetsova, T.A., Smetanina, O.F., Afiyatullov, S.S., Pivkin, M.V., Denisenko, V.A. and Elyakov, G.B. (2001) The identification of fusidic acid, a steroidal antibiotic from marine isolate of the fungus *Stibella aciculosa*. Biochemical Systematics and Ecology 29: 873–874.

Landreau, A., Pouchus, Y.F., Sallenave-Naumont, C., Biard, J-F., Boumard, M-C., Robiou du Pont, T., Mondeguer, F., Goulard, C. and Verbist, J-F. (2002) Combined use of LC/MS and a biological test for rapid identification of marine mycotoxins produced by *Trichoderma koningii*. Journal of Microbiological Methods 48: 181–184.

Laurent, D., Guella, G., Mancini, I., Roquebert, M-F., Farinole, F. and Pietra, F. (2002) A new cytotoxic tetralone derivative from *Humicola grisea*, a filamentous fungus from wood in the southeastern lagoon of New Caledonia. Tetrahedron 58: 9163–9167.

Laurent, D., Guella, G., Roquebert, M-F., Farinole, F., Mancini, I. and Pietra, F. (2000) Cytotoxins, mycotoxins and drugs from a new deuteromycete, *Acremonium neo-caledoniae*, from the southwestern lagoon of New Caledonia. Planta Medica 66: 63–66.

Le, Z. and Lin, Y. (2001) Chemical study on marine fungi. Journal of East China Geological Institute 24(1): 48–51.

Le, C., Lin, Y., Jiang, G., Zhou, S., Guan, L. and Luo, W. (1999) Studies on the volatile metabolites of marine fungus *Hypoxolon* sp. Chinese Journal of Marine Drugs, 18(2): 12–14.

Le, Z., Lin, Y., Jiang, G., Zhou, S. and Vrijmoed, L.L.P. (1998) Metabolites of mangrove fungus *Julella avicenniae* from the South China Sea. Acta Scientiarum Naturalium Universitas Sunyatseni 37(4): 119–120.

Lee, D.I., Choi, J.S., Yang, M-R., Lee, W.K., Kim, D-S., Choi, H.D. and Son, B.W. (1999) Ergosteryl myristate, a new ergosterol derivative from unidentified marine algicolous fungus. Natural Products Sciences 5: 93–96.

Lewis, T.E., Nichols, P.D. and McMeekin, T.A. (1999) The biotechnological potential of thraustochytrids. Marine Biotechnology 1: 580–587.

Lewis, T.E., Nichols, P.D. and McMeekin, T.A. (2000) Evaluation of extraction methods for recovery of fatty acids from lipid-producing microheterotrophs. Journal of Microbiological Methods 43: 107–116.

Lewis, T.E., Nichols, P.D. and McMeekin, T.A. (2001) Sterol and squalene content of a docosahexaenoic-acid-producing thraustochytrid: Influence of culture age, temperature, and dissolved oxygen. Marine Biotechnology 3: 439–447.

Li, H., Lin, Y., Wang, L., Zhou, S. and Vrijmoed, L.L.P. (2001) Metabolites of the marine fungus *Hypoxolon oceanicum* from the South China Sea. Acta Scientiarum Naturalium Universitas Sunyatseni 40(4): 70–72.

Liberra, K., Jansen, R. and Lindequist, U. (1998) Corollosporine, a new phthalide derivative from the marine fungus *Corollospora maritima* Werderm. 1069. Pharmazie 53: 578–582.

Liberra, K. and Lindequist, U. (1995) Marine fungi - a prolific resource of biologically active products?, Pharmazie 50(9): 583–588.

Lin, W.H., Fu, H.Z., Li, J. and Proksch, P. (2001a) Novel chromone derivatives from marine fungus *Aspergillus versicolor* isolated from the sponge *Xestospongia exigua*. Chinese Chemical Letters 12(3): 235–238.

Lin, W.H., Li, J., Fu, H.Z. and Proksch, P. (2001b) Four novel hydropyranoindeno-derivatives from marine fungus *Aspergillus versicolor*. Chinese Chemical Letters 12(5): 435–438.

Lin, Y., Li, H., Jiang, G., Zhou, S., Vrijmoed, L.L.P. and Jones, E.B.G. (2002a) A novel γ-lactone, eutypoid-A and other metabolites from marine fungus *Eutypa* sp. (#424) from the South China Sea. Indian Journal of Chemistry, Section B: Organic Chemistry including Medicinal Chemistry 41: 1542–1544.

Lin, Y., Shao, Z., Jiang, G., Zhou, S., Cai, J., Vrijmoed, L.L.P. and Jones, E.B.G. (2000) Penicillazine, a unique quinolone derivative with 4H-5,6-dihydro-1,2-oxazine ring system from the marine fungus *Penicillium* sp. (strain #386) from the South China Sea. Tetrahedron 56: 9607–9609.

Lin, Y., Wan, J., Zhou, S. and Jones E.B.G. (2001c) New isocoumarins from the mangrove endophytic fungus #2533, Chemistry Journal on Internet 3(7): 30–35. (http://www.chemistrymag.org/cji/2001/037030ne.htm; accessed October 10, 2003).

Lin, Y., Wang, J., Wu, X., Zhou, S., Vrijmoed, L.L.P. and Jones, E.B.G. (2002b) A novel compound enniatin G from the mangrove fungus *Halosarpheia* sp. (strain #732) from the South China Sea. Australian Journal of Chemistry 55: 225–227.

Lin, Y., Wu, X., Deng, Z., Wang, J., Zhou, S. Vrijmoed, L.L.P. and Jones, E.B.G. (2002c) The metabolites of the mangrove fungus *Verruculina enalia* No. 2606 from a salt lake in the Bahamas. Phytochemistry 59: 469–471.

Lin, Y., Wu, X., Feng, S., Jiang, G., Luo, J., Zhou, S., Vrijmoed, L.L.P., Jones, E.B.G., Krohn, K., Steingröver, K. and Zsila, F. (2001d) Five unique compounds: Xyloketals from mangrove fungus *Xylaria* sp. from the South China Sea coast. Journal of Organic Chemistry 66: 6252–6256.

Lin, Y., Wu, X., Feng, S., Jiang, G., Zhou, S., Vrijmoed, L.L.P. and Jones, E.B. G. (2001e) A novel N-cinnamoylcyclopeptide containing an allenic ether from the fungus *Xylaria* sp. (strain # 2508) from the South China Sea, Tetrahedron Letters 42: 449–451.

Liu, C.H., Meng, J.C., Zou, W.X., Huang, L.l., Tang, H.Q. and Tan, R.X. (2002) Antifungal metabolite with a new carbon skeleton from *Keissleriella* sp. YS4108, a marine filamentous fungus. Planta Medica 68: 363–365.

Liu, Z., Jensen, P.R. and Fenical, W. (2003) A cyclic carbonate and related polyketides from a marine-derived fungus of the genus *Phoma*. Phytochemistry 64: 571–574.

Lorenz, P., Jensen, P.R. and Fenical, W. (1998) Mactanamide, a new fungistatic diketopiperazine produced by a marine *Aspergillus* sp. Natural Product Letters 12: 55–60.

Lowe, D.A. (1992) Fungal enzymes. In: Handbook of Applied Mycology. (eds), Arora, D.K., Elander, R.P. and Mukerji, K.G. Marcel Dekker, Inc., New York, Vol. 4: Fungal Biotechnology, 681–706.

MacKenzie, S.E., Gurusamy, G.S., Piorko, A., Strongman, D.B, Hu, T., and Wright, J.L.C. (2004) Isolation of sterols from the marine fungus *Corollospora lacera*, Canadian Journal of Microbiology, 50(12): 1069–1072.

MacKenzie, S.E., Gurusamy, G.S., Taboski, M.A., Strongman, D.B, Hu, T., Wright, J.L.C. and Piorko, A. (1998) Unpublished results.

Malmstrøm, J. (1999) Unguisins A and B: New cyclic peptides from the marine-derived fungus *Emericella unguis*. Journal of Natural Products 62: 787–789.

Malmstrøm, J., Christopherson, C., Barrero, A.F., Oltra, J.E., Justicia, J. and Rosales, A. (2002a) Bioactive metabolites from a marine-derived strain of the fungus *Emericella variecolor*. Journal of Natural Products 65: 364–367.

Malmstrøm, J., Christopherson, C. and Frisvad, J.C. (2000) Secondary metabolites characteristc of *Penicillium citrinum*, *Penicillium steckii* and related species. Phytochemistry 54: 301–309.

Malmstrøm, J., Ryager, A., Anthoni, U. and Nielsen, P.H. (2002b) Unguisin C, a GABA-containing cyclic peptide from the fungus *Emericella unguis*. Phytochemistry 60: 869–872.

Masuma, R., Yamaguchi, Y., Noumi, M., Omura, S. and Namikoshi, M. (2001) Effect of sea water concentration on hyphal growth and antimicrobial metabolite production in marine fungi. Mycoscience 42: 455–459.

Mayer, A.M.S. (1999) Marine pharmacology in 1998: Antitumor and cytotoxic compounds. The Pharmacologist 41(4): 159–164.

Mayer, A.M.S. and Gustafson, K.R. (2003) Marine pharmacology in 2000: antitumor and cytotoxic compounds. International Journal of Cancer 105: 291–299.

Mayer, A.M.S. and Hamann, M.T. (2002) Marine pharmacology in 1999: Compounds with antibacterial, anticoagulant, antifungal, anthelmintic, anti-inflammatory, antiplatelet, antiprotozoal, and antiviral activities affecting the cardiovascular, endocrine, immune, and nervous systems; and other miscellaneous mechanisms of action. Comparative Biochemistry and Physiology Part C 132: 315–339.

Mayer, A.M.S. and Hamann, M.T. (2004) Marine pharmacology in 2000: Marine compounds with antibacterial, anticoagulant, antifungal, anti-inflammatory, antimalarial, antiplatelet, antituberculosis, and antiviral activities; affecting the cardiovascular, immune, and nervous systems and other miscellaneous mechanisms of action. Marine Biotechnology 6: 37–52.

Mayer, A.M.S. and Lehmann, V.K.B. (2000) Marine pharmacology in 1998: Marine compounds with antibacterial, anticoagulant, antifungal, anti-inflammatory, anthelmintic, antiprotozoal, and antiviral activities; with actions on the cardiovascular, endocrine, immune, and nervous systems; and other miscellaneous mechanisms of action. The Pharmacologist 42(2): 62–69.

Mayer, A.M.S. and Lehmann, V.K.B. (2001) Marine pharmacology in 1998: Antitumor and cytotoxic compounds. Anticancer Research 21: 2489–2500.

McDonald, L.A., Abbanat, D.R., Barbieri, L.R., Bernan, V.S., Discafani, C.M., Greenstein, M., Janota, K., Korshalla, J.D., Lassota, P., Tischler, M. and Carter, G.T. (1999) Spiroxins, DNA cleaving antitumor antibiotics from a marine-derived fungus. Tetrahedron Letters 40: 2489–2492.

Méjanelle, L., Lòpez, J.F., Gunde-Cimerman, N. and Grimalt, J.O. (2000) Sterols of melanized fungi from hypersaline environments. Organic Geochemistry, 31: 1031–1040.

Méjanelle, L., Lòpez, J.F., Gunde-Cimerman, N. and Grimalt, J.O. (2001) Ergosterol biosynthesis in novel melanized fungi from hypersaline environment. Journal of Lipid Research 42: 352–358.

Miljkovic, A., Mantle, P.G., Williams, D.J. and Rassing, B. (2001) Scorpinone: A new natural azaanthraquinone produced by a *Bispora*-like tropical fungus. Journal of Natural Products 64: 1251–1253.

Miller, J.D. (1986) Secondary metabolites in lignicolous marine fungi. In: The Biology of Marine Fungi (ed), Moss S.T. Cambridge University Press, Cambridge, 61–67.

Miller, J.D. (2000) Screening for secondary metabolites. In: Marine Mycology. A Practical Approach, (eds), Hyde, K.D. and Pointing, S.B. Fungal Diversity Press, Hong Kong, 158–171.

Mizushina, Y., Kobayashi, S., Kuramochi, K., Nagata, S., Sugawara, F. and Sakaguchi, K. (2000) Epolactaene, a novel neuritogenic compound in human neuroblastoma cells, selectively inhibits the activities of mammalian DNA polymerases and human DNA topoisomerase II. Biochemical and Biophysical Research Communications 273: 784–788.

Mohapatra, B.R., Banerjee, U.C. and Bapuji, M. (1998) Characterization of a fungal amylase from *Mucor* sp. associated with the marine sponge *Spirastrella* sp. Journal of Biotechnology 60: 113–117.

Molitoris, H.P. (2000) Methods for determination of enzymatic activities of marine fungi. Czech Mycology 52(2): 97–124.

Moore, B.S. (2002) The interface of natural product chemistry and biology. In: Marine Biotechnology in the Twenty-First Century – Problems, Promise, and Products, Committee on Biotechnology. National Academy Press, Washington, 61–64.

Munro, M.H.G., Blunt, J.W., Dumdei, E.J., Hickford, S.J.H, Lill, R.E., Li, S., Battershill, C.N. and Duckworth, A.R. (1999) The discovery and development of marine compounds with pharmaceutical potential. Journal of Biotechnology 70: 15–25.

Nagai, K., Kamigiri, K., Matsumoto, H., Kawano, Y., Yamaoka, M., Shimoi, H., Watanabe, M. and Suzuki, K. (2002) YM-202204, a new antifungal antibiotic produced by marine fungus *Phoma* sp. Journal of Antibiotics 55(12): 1036–1041.

Namikoshi, M., Akano, K., Meguro, S., Kasuga, I., Mine, Y., Takahashi, T. and Kobayashi, H. (2001) A new macrocyclic trichothecene, 12,13-deoxyroridin E, produced by the marine-derived fungus *Myrothecium roridum* collected in Palau. Journal of Natural Products 64: 396–398.

Namikoshi, M., Kobayashi, H., Liu, H.W., Yao, X.S. and Zhang, H. (2004) Antifungal and antimitotic substances discovered by the bioassay using conidia of *Pyricularia oryzae*. Drug Design Reviews – Online 1(3): 257–271.

Namikoshi, M., Kobayashi, H., Yoshimoto, T. and Hosoya, T. (1997) Phomopsidin, a new inhibitor of microtubule assembly produced by *Phomopsis* sp. isolated from coral reef in Pohnpei. Journal of Antibiotics 50(10): 890–892.

Namikoshi, M., Kobayashi, H., Yoshimoto, T. and Meguro, S. (2000a) Paecilospirone, a unique spiro[chroman-2,1'(3'*H*)-isobenzofuran] derivative isolated from tropical marine fungus *Paecilomyces* sp. Chemistry Letters 308–309.

Namikoshi, M., Kobayashi, H., Yoshimoto, T., Meguro, S. and Akano, K. (2000b) Isolation and characterizetion of bioactive metabolites from marine-derived filamentous fungi collected from tropical and sub-tropical coral reefs. Chemical and Pharmaceutical Bulletin 48(10): 1452–1457.

Namikoshi, M., Negishi, R., Nagai, H., Dmitrenok, A. and Kobayashi, H. (2003) Three new chlorine containing antibiotics from a marine-derived fungus *Aspergillus ostianus* collected in Pohnpei. Journal of Antibiotics 56: 755–761.

Namikoshi, M., Suzuki, S., Meguro, S., Kobayashi, H., Mine, Y. and Kasuga, I. (2002) Inhibitors of microtubule assembly produced by the marine fungus strain TUF 98F139 collected in Palau. Journal of Tokyo University of Fisheries 88: 1–6.

Newman, D.J., Cragg, G.M. and Snader, K.M. (2003) Natural products as sources of new drugs over the period 1981-2002. Journal of Natural Products 66: 1022–1037.

Nicholas, G.M., Eckman, L.L., Newton, G.L., Fahey, R.C., Ray, S. and Bewley, C.A. (2003) Inhibition and kinetics of *Mycobacterium tuberculosis* and *Mycobacterium smegmatis* mycothiol-S-conjugate amidase by natural products. Bioorganic and Medicinal Chemistry 11: 601–608.

Nielsen, J., Nielsen, P.H. and Frisvad, J.C. (1999) Fungal depside, guisinol, from a marine derived strain of *Emericella unguis*. Phytochemistry 50: 263–265.

Omura, S., Miyadera, H., Ui, H., Shiomi, K., Yamaguchi, Y., Masuma, R., Nagamitsu, T., Takano, D., Sunazuka, T., Harder, A., Kölbl, H., Namikoshi, M., Miyoshi, H., Sakamoto, K. and Kita, K. (2001) An anthelmintic compound, nafuredin, shows selective inhibition of complex I in helminth mitochondria. Proceedings of the National Academy of Science USA, 98(1): 60–62.

Omura, S., Tomoda, H., Tabata, N. and Ohyama, Y. (1999) Roselipins, novel fungal metabolites having a highly methylated fatty acid modified with a mannose and an arabinitol. Journal of Antibiotics 52(6): 586–589.

Osterhage, C., Kaminsky, R., König, G.M. and Wright, A.D. (2000a) Ascosalipyrrolidinone A, an antimicrobial alkaloid, from the obligate marine fungus *Ascochyta salicorniae*. Journal of Organic Chemistry 65: 6412–6417.

Osterhage, C., König, G.M., Höller, U. and Wright, A.D. (2002a) Rare sesquiterpenes from the algicolous fungus *Drechslera dematioidea*. Journal of Natural Products 65: 306–313.

Osterhage, C., König, G.M., Jones, P.G. and Wright, A.D. (2002b) 5-Hydroxyramulosin, a new natural product produced by *Phoma tropica*, a marine-derived fungus isolated from the alga *Fucus spiralis*. Planta Medica 68: 1052–1054.

Osterhage, C., Schwibbe, M., König, G.M. and Wright, A.D. (2000b) Differences between marine and terrestrial *Phoma* species as determined by HPLC-DAD and HPLC-MS. Phytochemical Analysis 11: 288–294.

Pietra, F. (1997) Secondary metabolites from marine microorganisms: Bacteria, protozoa, algae and fungi. Achievements and prospects. Natural Product Reports 14: 453–464.

Pinner, R.W., Teutsch, S.M., Simonsen, L., Klug, L.A., Graber, J.M., Clarke, M.J. and Berkelman, R.L. (1996) Trends in infectious diseases mortality in the United States. Journal of the American Medical Association 275(3): 189–193.

Pointing, S.B. (1999) Qualitative methods for determining lignocellulolytic enzyme production by tropical fungi. Fungal Diversity 2: 17–33.

Pointing, S.B. (2000) Lignocellulolytic enzyme assays. In: Marine Mycology. A Practical Approach, (eds), Hyde K.D. and Pointing S.B. Fungal Diversity Press, Hong Kong, 137–157.

Pointing, S.B., Buswell J.A., Jones E.B.G. and Vrijmoed L.L.P. (1999) Extracellular cellulolytic enzyme profiles of five lignicolous mangrove fungi. Mycological Research 103: 696–700.

Pointing, S.B., Vrijmoed, L.L.P. and Jones E.B.G. (1998) A qualitative assessment of lignocellulose degrading enzyme distribution in marine fungi. Botanica Marina 41: 293–298.

Pomponi, S.A. (1999) The bioprocess – technological potential of the sea. Journal of Biotechnology 70: 5–13.

Proksch, P., Edrada, R.A. and Ebel, R. (2002) Drugs from the seas – current status and microbiological implications. Applied Microbiology and Biotechnology 59: 125–134.

Qiu, X., Hong, H. and MacKenzie, S.L. (2001) Identification of a Δ4 fatty acid desaturase from *Thraustochytrium* sp. involved in the biosynthesis of docosahexanoic acid by heterologous expression in *Saccharomyces cerevisiae* and *Brassica juncea*. Journal of Biological Chemistry 276(34): 31561–31566.

Rahbæk, L., Sperry, S., Piper, J.E. and Crews, P. (1998) Deoxynortrichoharzin, a new polyketide from the saltwater culture of a sponge-derived *Paecilomyces* fungus. Journal of Natural Products 61: 1571–1573.

Renner, M.K., Jensen, P.R. and Fenical, W. (1998) Neomangicols: Structures and absolute stereochemistries of unprecedented halogenated sesterterpenes from a marine fungus of the genus *Fusarium*. Journal of Organic Chemistry 63: 8346–8354.

Renner, M.K., Jensen, P.R. and Fenical, W. (2000) Mangicols: Structures and biosynthesis of a new class of sesterterpene polyols from a marine fungus of the genus *Fusarium*. Journal of Organic Chemistry 65: 4843–4852.

Romero, F., Espliego, F., Baz, J.P., de Quesada, T.G., Gravalos, D., de la Calle, F. and Fernandez-Puentes, J.L. (1997) Thiocoraline, a new depsipeptide with antitumor activity produced by a marine *Micromonospora*. I. Taxonomy, fermentation, isolation, and biological activities. Journal of Antibiotics 50(9): 734–737.

Rowley, D.C., Kelly, S., Kauffman, C.A., Jensen, P.R. and Fenical, W. (2003) Halovirs A - E, new antiviral agents from a marine-derived fungus of the genus *Scytalidium*. Bioorganic and Medicinal Chemistry 11: 4263–4274.

Sakaki, H., Kaneno, H., Sumiya, Y., Tsushima, M., Miki, W., Kishimoto, N., Fujita, T., Matsumoto, S., Komemushi, S. and Sawabe, A. (2002) A new carotenoid glycosyl ester isolated from a marine microorganism, *Fusarium* strain T-1. Journal of Natural Products 65: 1683–1684.

Sallenave, C., Pouchus, Y.F., Bardouil, M., Lassus, P., Roquebert, M-F. and Verbist, J-F. (1999) Bioaccumulation of mycotoxins by shellfish: Contamination of mussels by metabolites of a *Trichoderma konigii* strain isolated in the marine environment. Toxicon 37: 77–83.

Sallenave-Namont, C., Pouchus, Y.F., Robiou du Pont, T., Lassus, P. and Verbist, J-F. (2000) Toxigenic saprophytic fungi in marine shellfish farming areas. Mycopathologia 149: 21–25.

Schlingmann, G., Milne, L. and Carter, G.T. (2002) Isolation and identification of antifungal polyesters from the marine fungus *Hypoxylon oceanicum* LL-15G256. Tetrahedron 58: 6825–6835.

Schulz, B., Römmert, A-K., Dammann, U., Aust, H-J. and Strack, D. (1999) The endophyte-host interaction: A balanced antagonism?, Mycological Research 103: 1275–1283.

Shao, Z., Lin, Y., Jiang, G., Zhou, S., Vrijmoed, L.L.P. and Jones, E.B.G. (1999) The novel compound with the skeleton of furanopyran from marine fungus *Penicillium* sp. from the South China Sea. Acta Scientiarum Naturalium Universitas Sunyatseni 38: 131–132.

Shigemori, H., Komatsu, K., Mikami, Y. and Kobayashi, J. (1999) Seragakinone A, a new pentacyclic metabolite from a marine-derived fungus. Tetrahedron 55: 14925–14930.

Smith, C.J., Abbanat, D., Bernan, V.S., Maiese, W.M., Greenstein, M., Jompa, J., Tahir, A. and Ireland, C.M. (2000) Novel polyketide metabolites from a species of marine fungi. Journal of Natural Products 63: 142–145.

Son, B.W., Choi, J.S., Kim, J.C., Nam, K.W., Kim, D.S., Chung, H.Y., Kang, J.S. and Choi, H.D. (2002) Parasitenone, a new epoxycyclohexenone related to gabosine from the marine-derived fungus *Aspergillus parasiticus*. Journal of Natural Products 65: 794–795.

Son, B.W., Jensen, P.R., Kauffman, C.A. and Fenical, W. (1999) New cytotoxic epidithiodioxopiperazines related to verticillin A from a marine isolate of the fungus *Penicillium*. Natural Product Letters 13(3): 213–222.

Sparidans, R.W., Henrar, R.E.C., Jimeno, J.M., Faircloth, G., Floriano, P. and Beijnen, J.H. (1999) Bioanalysis of thiocoraline, a new marine antitumoral depsipeptide, in plasma by high-performance liquid chromatography and fluorescence detection. Journal of Chromatography B, Biomedical Sciences and Applications 726: 255–260.

Sperry, S., Samuels, G.J. and Crews, P. (1998) Vertinoid polyketides from the saltwater culture of the fungus *Trichoderma longibrachiatum* separated from a *Haliclona* marine sponge. Journal of Organic Chemistry 63: 10011–10014.

Sponga, F., Cavaletti, L., Lazzarini, A., Borghi, A., Ciciliato, I., Losi, D. and Marinelli, F. (1999) Biodiversity and potentials of marine-derived microorganisms. Journal of Biotechnology 70: 65–69.

Strobel, G.A. (2003) Endophytes as sources of bioactive products, Microbes and Infection 5: 535–544.

Suenaga, K., Aoyama, S., Xi, W., Arimoto, H., Yamaguchi, K., Yamada, K., Tsuji, T., Yamada, A. and Uemura, D.

(2000) Isolation and structure of kasarin, a novel azetinone compound, isolated from a marine microorganism. Heterocycles 52(3): 1033–1036.

Suresh, P.V. and Chandrasekaran, M. (1999) Impact of parameters on chitinase production by an alkalophilic marine *Beauveria bassiana* in solid state fermentation. Process Biochemistry 34: 257–267.

Tabata, N., Ohyama, Y., Tomoda, H., Abe, T., Namikoshi, M. and Omura, S. (1999) Structural elucidation of roselipins, inhibitors of diacylglycerol acyltransferase produced by *Gliocladium roseum* KF-1040. Journal of Antibiotics 52(9): 815–826.

Tan, L.T., Cheng, X.C., Jensen, P.R. and Fenical, W. (2003) Scytalidamides A and B, new cytotoxic cyclic heptapeptides from a marine fungus of the genus *Scytalidium*. Journal of Organic Chemistry 68: 8767–8773.

Tepšič, K., Gunde-Cimerman, N. and Frisvad, J.C. (1997) Growth and mycotoxin production by *Aspergillus fumigatus* strains isolated from a saltern. FEMS Microbiology Letters 157: 9–12.

Tomoda, H., Ohyama, Y., Abe T., Tabata, N., Namikoshi, M., Yamaguchi, Y., Masuma, R. and Omura, S. (1999) Roselipins, inhibitors of diacylglycerol acyltransferase, produced by *Gliocladium roseum* KF-1040. Journal of Antibiotics 52(8): 689–694.

Toske, S.G., Jensen, P.R., Kauffmann, C.A. and Fenical, W. (1998) Aspergillamides A and B: Modified cytotoxic tripeptides produced by a marine fungus of the genus *Aspergillus*. Tetrahedron 54: 13459–13466.

Tsuda, M., Mugishima, T., Komatsu, K., Sone, T., Tanaka, M., Mikami, Y., Shiro, M., Hirai, M., Ohizumi, Y. and Kobayashi, J. (2003) Speradine A, a new pentacyclic oxindole alkaloid from a marine-derived fungus *Aspergillus tamarii*. Tetrahedron 59: 3227–3230.

Uchida, R., Tomoda, H., Arai, M. and Omura, S. (2001) Chlorogentisylquinone, a new neutral sphingomyelinase inhibitor, produced by a marine fungus. Journal of Antibiotics 54(11): 882–889.

Ui, H.K., Yamaguchi, Y., Masuma, R., Nagamitsu, T., Takano, D., Sunazuka, T., Namikoshi, M. and Omura, S. (2001) Nafuredin, a novel inhibitor of NADH-fumarate reductase, produced by *Aspergillus niger* FT-0554. Journal of Antibiotics 54(3): 234–238.

Usami, Y., Aoki, S., Hara, T. and Numata, A. (2002) New dioxopiperazine metabolites from a *Fusarium* species separated from a marine alga. Journal of Antibiotics 55(7): 655–659.

Usami, Y., Yamaguchi, J. and Numata, A. (2004) Gliocladins A-C and glioperazine; cytotoxic dioxo- or trioxopiperazine metabolites from a *Gliocladium* sp. separated from a sea hare. Heterocycles 63(5): 1123–1129.

Usui, T., Kondoh, M., Cui, C.B., Mayumi, T. and Osada, H. (1998) Tryprostatin A, a specific and novel inhibitor of microtubule assembly, Biochemical Journal 333: 543–548.

Vala, A.K., Vaidya, S.Y. and Dube, H.C. (2000) Siderophore production by facultative marine fungi. Indian Journal of Marine Sciences, 29: 339–340.

Varoglu, M. and Crews, P. (2000) Biosynthetically diverse compounds from a saltwater culture of sponge-derived *Aspergillus niger*. Journal of Natural Products 63: 41–43.

Verbist, J-F., Sallenave, C. and Pouchus, Y.F. (2000) Marine Fungal Substances. In: Studies in Natural Product Chemistry, (ed), Atta-Ur Rahman, Elsevier Science Publishers, Amsterdam. 24: 979–1092.

Wan, J., Lin, Y., Wu, X., Zhou, S. and Vrijmoed, L.L.P. (2001) Avicennin A, a new isocoumarine from mangrove entophytic fungus No. 2533 from the South China Sea. Acta Scientiarum Naturalium Universitas Sunyatseni 40(1): 127–128.

Wang, C-Y., Wang, B-G., Brauers, G., Guan, H-S., Proksch, P. and Ebel, R. (2002) Microsphaerones A and B, two novel γ-pyrone derivatives from the sponge-derived fungus *Microsphaeropsis* sp. Journal of Natural Products 65: 772–775.

Wang, G-Y-S., Abrell, L.M., Avelar, A., Borgeson, B.M. and Crews, P. (1998) New hirsutane based sesquiterpenes from salt water cultures of a marine sponge-derived fungus and the terrestrial fungus *Coriolus consors*. Tetrahedron 54: 7335–7342.

WHO World Health Reports for years 1996, 1997, and 1999.

Wright, A.D., Osterhage, C. and König, G.M. (2003) Epicoccamide, a novel secondary metabolite from a jellyfish-derived culture of *Epicoccum purpurascens*, Organic and Biomolecular Chemistry 1: 507–510.

Yamada, T., Iritani, M., Doi, M., Minoura, K., Ito, T. and Numata, A. (2001) Absolute stereostructures of cell-adhesion inhibitors, macrosphelides C, E - G and I, produced by a *Periconia* species separated from an *Aplysia* sea hare. Journal of the Chemical Society, Perkin Transactions 1, 3046–3053.

Yamada, T., Iritani, M., Minoura, K., Numata, A., Kobayashi, Y. and Wang, Y-G. (2002a) Absolute stereostructures of cell adhesion inhibitors, macrosphelides H and L, from *Periconia byssoides* OUPS-N133, Journal of Antibiotics 55: 147–154.

Yamada, T., Iwamoto, C., Yamagaki, N., Yamanouchi, T., Minoura, K., Hagishita, S. and Numata, A. (2004) Leptosins O - S, cytotoxic metabolites of a strain of *Leptosphaeria* sp. isolated from a marine alga. Heterocycles 63: 641–653.

Yamada, T., Iwamoto, C., Yamagaki, N., Yamanouchi, T., Minoura, K., Yamori, T., Uehara, Y., Andoh, T., Umemura, K. and Numata, A. (2002b) Leptosins M-N_1, cytotoxic metabolites from a *Leptosphaeria* species separated from a marine alga. Structure determination and biological activities, Tetrahedron 58: 479–487.

Yin, W., Lin, Y., Zhou, S. and Vrijmoed, L.L.P. (2002) The cyclopeptides from marine fungus 2516 from the South China Sea. Acta Scientiarum Naturalium Universitas Sunyatseni 41: 56–58.

14. Recent Advances in the Development of Mycoinsecticides in Australia

Richard J. Milner
Private Biopesticide Consultant, 81 Birriwa Road, Bungendore, NSW 2621, Australia

SUMMARY This article outlines some of the advances with development of mycoinsecticides made in Australia over the past 10 years or so and discusses some of the factors regarded as critical for commercial success. Most of the research has focussed on a hyphomycete *Metarhizium anisopliae* but many of the points considered will apply to related fungi such as *Beauveria* spp. and *Paecilomyces* spp. While these fungi are genetically diverse and have strains with the necessary characteristics for development as mycoinsecticides, each pest and cropping system needs to be considered independently and success is often not predictable. These points are illustrated with examples of two successful products, viz for control of scarab beetle larvae in sugarcane crops and for control of locusts and grasshoppers. Other products for pecan borers in orchard trees and for termite control in the urban environment, while being scientifically successful, have not yet been developed into commercial products. The success of BioCane™ and Green Guard® in the market place over several years has shown that mycoinsecticides can be effective in niche markets, through they are not yet able to compete directly with mainstream chemical insecticides. Inhibiting factors include the slow kill rate, market size, and the difficulty and cost of production scale-up. Profits may be slow to achieve and this can lead to manufacturers introducing products onto the market prematurely and so risking the reputation of mycoinsectides. It is essential that quality control and formulation factors are adequately addressed to ensure reliability in the real world.

Introduction

Many publications in recent years have outlined the promising features of mycoinsecticides for insect control and their value as crop protection agents for reducing the use of chemical pesticides, resistance management and in 'organic' production. Unfortunately the promise has only rarely resulted in commercial success. This paper will outline some of the advances recently made in Australia and discuss some of the features which are regarded as crucial to commercial success in the future. Most of our research has been with strains of the hyphomycete, *Metarhizium anisopliae*, but other hyphomycetes such as *Beauveria* and *Paecilomyces* have similar characteristics and the same factors will apply.

M. anisopliae is genetically highly diverse (Driver et al. 2000), and strains with appropriate virulence and host range can be found for many pests. However each pest and even crop system has to be investigated separately and success is rarely predictable. Research in the USA has found that whiteflies, despite being highly susceptible to *Beauveria bassiana,* are not effectively controlled on cotton even when the application is excellent. The reasons have not yet been elucidated (Wraight and Carruthers, 1999). In Australia, experiments over many years with the Childers grub, *Antitrogus parvulus*, a melolonthine scarab pest of sugarcane have shown that, while it is often killed in the field by isolates of *M. anisopliae*, no isolate has been found to be highly virulent and effective under field conditions. In contrast, locusts and grasshoppers are very rarely killed under natural conditions by strains of *M. anisopliae*, yet one naturally occurring strain

(FI-985) not only is highly virulent for our most serious locust pest, *Chortoicetes terminifera*, but will also control a wide range of acridid pests while not infecting non-acridids such as crickets and tettigonids.

Other advantages of *M. anisopliae* include the fact that it can be easily mass produced on cheap substrates, the conidia are amenable to being dried, formulated and stored. Perhaps because they are heavily pigmented, the infectious conidia are quite resistant to UV (Braga et al. 2001) and will persist on sprayed vegetation for over a week.

Despite these well-known advantages, there are still many challenges to be overcome before *M. anisopliae* is part of main-stream pest control. The current low-technology mass production system is effective for markets of up to say 100,000 ha per year for locust control but for major pests in broad acre agriculture, this production technology needs to be scaled up by a factor of 10-100 utilizing tonnes of substrate per batch rather than kilos as at present. The slow kill is a problem in many situations and present methods for overcoming this – mixing with low doses of chemicals and genetic engineering, would also not be acceptable to the major target markets. Improvements in formulation and application technology are being made and will enhance market acceptability by making products more reliable.

In developing alternatives to chemical pesticides, the problems caused by chemicals need to be highlighted. These include substantial adverse community reaction, problems of residues, the resistance of pests, the increasing loss of efficacy due to biodegradation and the lack of specificity. One factor often forgotten is the hazards presented by the large-scale manufacture and transport of dangerous chemicals.

Life-Cycle of *M. anisopliae*

The only infectious stage of *M. anisopliae* is the haploid asexual conidium. The conidia adhere to the cuticle of insects where they produce a germ tube penetrating into the haemocoel where the hyphae develop eventually killing the host. After death, if conditions are warm and moist, the fungus grows back out through the cuticle forming chains of new green conidia. The conidia do not disperse by wind or in the soil and so each killed cadaver forms a pool of conidia ready to attack and infect a suitable insect when it moves into the zone. It is this lack of a dispersal mechanism which makes *M. anisopliae* ineffective as a natural control agent. Thus man can intervene and provide dispersal by means of sprays or other formulations to develop the fungus into a mycoinsecticide. Many isolates are specific to particular groups while other isolates have a more general host range. Most isolates grow best between 20 and 30°C though some are classed as low temperature isolates growing at 10°C or below while others are high temperature isolates growing at 35°C or above. Most isolates have green conidia though other colours such as brown and yellow are known. The main reservoir of conidia is the soil where conidia can remain alive but dormant for many years. In my experience no saprophytic growth in soil is possible. *M. anisopliae* has been recorded from human infections (Revankar et al.1999) but it is extremely rare outside insects and the vast majority of 'safety testing' has shown that the fungus is not harmful to live forms other than insects.

Status of Metarhizium Based Mycoinsecticides in Australia

The first product registered in Australia was BioGreen™ for red-headed cockchafer, *Adoryphorus couloni*, in pasture and turf (Milner, 2000). Another isolate, FI-1045, was subsequently registered for use against the greyback canegrub, *Dermolepida albohirtum*, a serious pest of sugarcane in the tropical coast of north Queensland (Milner et al. 2002; Logan et al. 1999). Both products are produced by Bekker Underwood Pty Ltd in Somersby, near Sydney, New South Wales and consist of natural granules of rice on which the fungus has sporulated. A third product, Green Guard® based on FI-985 has recently been considered for registration and is for use against locusts and grasshoppers (Milner and Hunter, 2001). This is a completely different product, being an oil formulation of harvested dried conidia and is also now produced by Bekker Underwood Pty Ltd. Other products for use against other species of sugarcane scarab beetles, whiteflies, crickets and termites have been successfully field tested but have not yet been commercialised. The features which have enabled some products to be commercialised while others are still being considered will be discussed in this article.

Production of Hyphomycete Conidia for Use in Mycopesticides

A production technology has been developed which can be used for a wide range of entomopathogenic fungi. The basic process as used for production of the locust fungus, FI-985, has been outlined in a recent paper (Milner and Hunter, 2001). Isolates are stored as pure conidia powder in cryovials at –70 °C in order to minimize any genetic change and enables very long term (25 years so far) storage in a small space. Initial cultures are grown on Sabouraud's dextrose agar with 1% yeast to check for strain purity and sporulation capacity. Mycelium is then produced in bulk in a liquid medium (10 g yeast extract, 10 g peptone and 40 g dextrose in 1 L water) in shake flasks and this is used to inoculate moist sterile rice in special self-aerating plastic bags. Phase contrast microscopy is used to check the purity of the shake flask culture. Ten litre bags (Van Leer, France, or Unicorn, USA) are filled with 2.5 kg of rice and 250 ml of water added. The bags are then heat sealed and autoclaved for 1 hour. When cool, the rice in broken up and another litre or so of sterile water is added along with 100 ml of the mycelium medium under sterile conditions in a laminar flow cabinet. The yield can be markedly affected by moisture with FI-985, *M. anisopliae* var. *acridum,* being unusual in having a very high moisture optimum (Fig. 1). The bag is again sealed and the contents mixed thoroughly. The bags are then incubated on wire racks at 25°C for 14 days. During this time the bags are frequently monitored for contaminants and any contaminated bags discarded. Also, any clumps of rice are broken up and the rice/fungus mixture is well mixed. Initial sporulation is visible in 4-5 days while maximum sporulation may take 21 days or longer. Growth is stopped at 14 days because this is close to the maximum sporulation and further time risks an increase in contamination. However as experience is gained so the level of contamination has been reduced, so it is now close to zero. Causes of contamination are inadequate sealing of bags, improper sterilization and contamination in the initial shake flask culture.

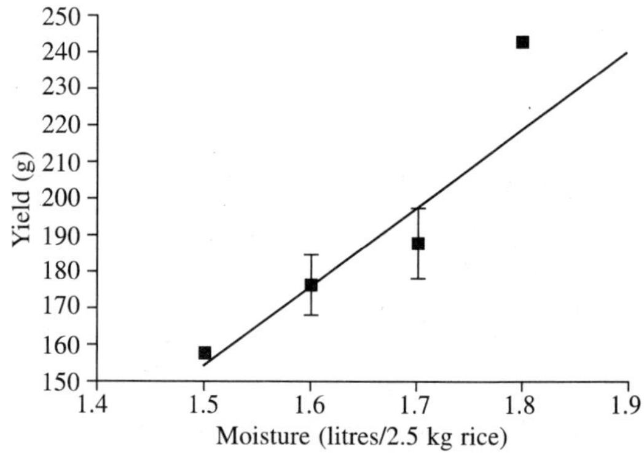

Fig. 1. Effect of moisture on yield of dry conidial powder of FI-985 in the bag production system.

After 14 days growth, the bags are inverted and the bottom plastic removed. The opened bags are then dried at 30°C and 20-30% relative humidity for about 10 days. Then the conidia are removed using an industrial electric sieve with 3 sieve plates of reducing aperture. The enables a very pure product to be harvested which has moisture content of 5-10%. The conidia can then be stored dry or formulated in oil.

Different isolates may have different moisture and temperature optima and often have very different yield potentials. The production of 3 isolates, one from each of the main 3 hyphomycete genera (*Paecilomyces, Beauveria* and *Metarhizium*) being developed for whitefly control in Australia was compared (Table 1). Overall yields were highest from the *Metarhizium* isolate and lowest from the *Paecilomyces*. All isolates produced well at the median temperature of 26°C, while only the *Metarhizium* produced good yields at 30°C. *Beauveria* and *Metarhizium* both produced well at high moisture levels though at the highest moisture the yield of *Beauveria* was reduced. In contrast, *Paecilomyces* yielded best at the two lowest moisture contents. These results highlight the need to optimise the production conditions for each isolate. Different

Table 1. Production of harvested conidia (g dry conidia/kg rice) of fungi being evaluated as mycoinsecticides against whiteflies in Australia using a sheep drench

A. Effect of temperature at 56.8% moisture			
Temperature	*Paecilomyces* (FI-1194)	*Beauveria* (FI-1378)	*Metarhizium* (FI-1248)
20°C	4.6 ± 1.2	66.7 ± 3.6	83.0 ± 3.5
26°C	13.0 ± 0.41	70.0 ± 3.2	99.7 ± 4.3
30°C	6.5 ± 1.5	12.0 ± 3.4	99.8 ± 5.3
B. Effect of moisture at 26°C			
Moisture (%)	*Paecilomyces* (FI-1194)	*Beauveria* (FI-1378)	*Metarhizium* (FI-1248)
38.6	43.7 ± 11.1	69.7 ± 8.2	47.7 ± 9.4
46.5	58.7 ± 7.4	71.7 ± 5.7	84.7 ± 6.4
56.8	13.0 ± 0.4	70.0 ± 3.2	99.7 ± 4.3
65.9	8.7 ± 1.8	76.5 ± 6.6	97.5 ± 6.7
75.0	7.2 ± 1.7	51.7 ± 4.8	91.7 ± 6.7

isolates of each species would also show individual preferences though probably not as dramatic as in this example. Nahar et al. (2003) presented some data on comparative yields using a very similar production system. Again *M. anisopliae* was the most productive yielding 40 g/kg substrate with *Beauveria* yielding 10 g/kg and *Nomuraea* just 6.5 g/kg. In developing a mycoinsecticide, the cost per hectare of the final product is a key element and the efficiency of the mass production is a key factor in the cost estimation. In field trials, it would be useful to compare doses on the basis of equivalent amounts of substrate for mass production, rather than conidial concentration as is normally the case.

Development of BioCane™

Sugarcane is a major export crop grown mainly on the tropical coastal plain of Queensland. The roots are attacked by at least 18 different species of scarabaeidae larvae and at times these may devastate the crop. Control is mainly by the prophylactic use of a slow release formulation of chlorpyriphos, but the reliance on this chemical pesticide, coupled with problems associated with its use, led the industry to sponsor a major research project in 1994 to develop naturally occurring strains of *M. anisopliae* as mycosinsecticides. The initial target was the greyback canegrub, *D. albohirtum*, which is the most destructive of the 18 species and is often the only pest present making control with a specific natural enemy attractive. It is also found in the alkaline soils of the irrigated Burdekin region where the alkalinity reduces the persistence of chlorpyriphos.

Initial laboratory bioassays demonstrated that FI-1045 was the most pathogenic of the isolates tested with some isolates obtained from field-killed greyback canegrub larvae surprisingly low in virulence. Isolates from other related species were not virulent and so FI-1045 was selected for further development (Milner et al. 2002). Small-scale field trials in 1996 confirmed the promise of this isolate and so mass production by BioCare Pty Ltd (now Bekker Underwood) commenced and, in 1997, the initial commercial trials were established. The commercial product BioCane™ was registered in 2000 and recent sales estimates are that about 40 tonnes were sold in a year.

The product is the natural rice granules with the conidia which have been produced on this rice. It is partially dried and has clay granules mixed with it to minimise clumping and ensure smooth flow through a granule applicator. The granules are placed about 20 cm below the soil surface during 'fill-in' a few months after the sugarcane sets are planted. Thus the conidia are close to the root system and will infect and kill any greyback canegrubs which come into contact with the granules. The control efficiency is dose dependant (Fig. 2) and a dose of 33 kg/ha was established as it gave reasonable control at a realistic price. At the recommended rate, the granules being quite well spread, the efficacy is dependant on the insects moving to contact with the granules. The conidia decline in viability over time but still have some activity 3.5 years after application (Milner et al. 2003). In the season of application, control has been found to be about 50% while the fungus will also give control in subsequent ratoon crops (Fig. 3). This level of control is adequate under most conditions but where the infestation is severe, crop damage will occur because too many insects survive and even those that die may do so slowly thus resulting in some feeding prior to death. For these situations, the use of a combination with chlorpyrifos is presently being evaluated.

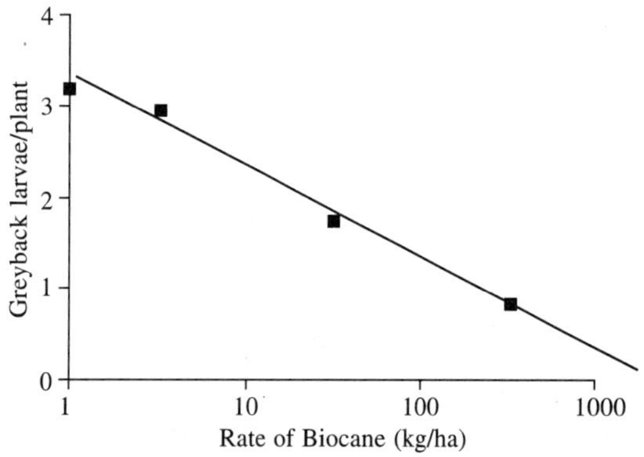

Fig. 2. Level of control achieved with different doses of BioCane™ in large-scale field trials (each point is the mean from 4 farms).

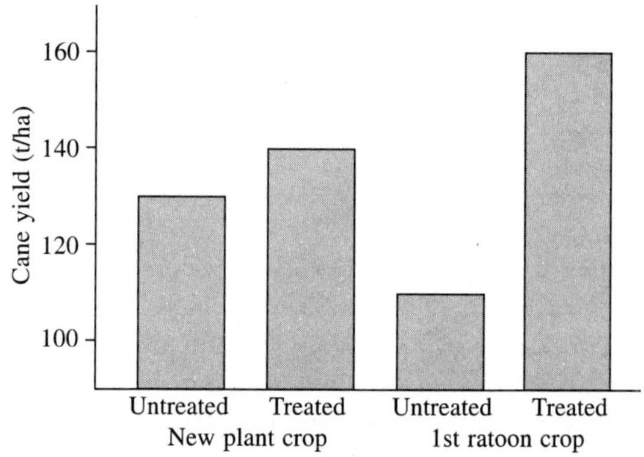

Fig. 3. Yield benefits from application of BioCane™ at the recommended rate on new plant crop as well as the next season's ratoon crop.

A key element of the development of this mycoinsecticide was the extensive development work done on farmers fields before the product was commercially released. This not only enabled the manufacturer to improve mass production, packaging, storage and transport but also solved key field application problems such as flow through the granular applicators. One serious factor was the variability in survival of the conidia after application. This was revealed through extensive monitoring of conidial concentrations in the soil. Soil cores were taken before and at various times after application. Sub-samples were sonicated in Tween 80 and plated out onto the selective CAD medium (Milner et al. 2003) which can detect concentrations down to 10^3 conidia/g. It was found that the time after application and before covering with soil was critical and a maximum time of 15 min. is now recommended.

Since various chemical fungicides and insecticides are often applied to sugarcane at planting, it was important to understand the effects of these materials when used in combination with the mycoinsecticide. Tests involving 8 fungicides and 3 insecticides revealed that while some of the fungicides had an adverse effect under laboratory conditions in culture media, none of them had any adverse effect under field conditions. The materials were applied using standard farmer equipment and the concentrations of viable conidia in the soil assessed comparing plots with and without potentially damaging fungicides. Even the most toxic fungicide (as revealed by plate tests) Shitan (methoxy ethyl mercuric chloride) did not have any detectable effect on colony counts over time from treated soil. Thus it was concluded that *M. anisopliae* was compatible with all the treated chemicals when used under commercial conditions (Samson et al. 2005).

A Product for Cerambycid Borers in Pecan Orchards

Pecans, and many other perennial tree crops, are attacked by large cerambycid beetles. The adults lay their eggs in crack in the bark and the larvae bore into the branches and trunks eating out substantial tunnels and galleries. The larvae feed in the tree for 4 or 5 years before pupating and emerging as adults. The feeding weakens the tree and they become vulnerable to storm damage. Control is very difficult with chemical insecticides as the insects are not inherently very susceptible and the material is difficult to apply. Parasites, particularly of the egg stage, are known and may be adversely affected by the use of chemicals. In addition, the major grower of pecans in Australia is an 'organic' grower and so was keen to develop a biological method for control. The larvae were known to succumb naturally to fungal diseases but these were very uncommon and not effective in reducing the pest status of the insect.

The initial stage of the project was to collect a large number of living larvae and keep them in pecan shavings in the hope of finding some naturally occurring fungal pathogens. Three isolates obtained were *Beauveria bassiana* (FI-1378), *M. anisopliae* (FI-1375) and a *Gliocladium* sp. The overall mortality due to fungi was found to be less than 5%. Very little is known about *Gliocladium* and so this isolate was not considered further. Laboratory bioassays were undertaken with the other two as well as a general isolate of *M. anisopliae*, FI-1248, isolated from termites. All three isolates caused over 80% mortality at doses of 10^7 conidia/ml and above (Table 2) after 8 weeks at 25°C. Further glasshouse experiments, simulating field conditions, also showed all 3 isolates to be effective with FI-1378 (*B. bassiana*) being the most effective. Since no isolate could be eliminated on the basis of these tests, all three isolates were field tested.

Table 2. Summary of laboratory bioassays of 2 isolates of *Metarhizium anisopliae* and 1 isolate of *Beauveria bassiana* against pecan borers *Agrianome* sp. (Cerambycidae). Percentage mortality and sporulation (in brackets) after 8 weeks incubation at 25°C

Dose (conidia/ml)	FI-1248 (*Metarhizium*)	FI-1375 (*Metarhizium*)	FI-1378 (*Beauveria*)
10^8	100 (96)	96 (96)	100 (96)
10^7	n.d.	80 (72)	80 (76)
10^6	60 (48)	64 (56)	28 (28)
10^5	n.d.	52 (28)	20 (16)
10^4	n.d.	44 (32)	8 (4)
10^3	n.d.	48 (16)	8 (8)
Control	24 (4)	21 (4)	12 (4)

n.d. = not done.

A field trial was undertaken on a large commercial pecan farm which had been treating the borers with an antifumigant chemical for several years. They had developed methods of mapping the incidence using frass deposits and used a 1 cm drill to detect and access the galleries. All positive holes were treated. A similar technique was therefore adopted with the mycoinsecticide. Holes were drilled adjacent to where frass was being expelled and the holes treated with an oil/water emulsion of conidia using a sheep drenching backpack (Fig. 4). Four months later, all treated branches were removed with a chain saw and destructively sampled to extract all live and fungus infected larvae. The best treatment was the *B. bassiana* strain FI-1378 which at a dose of 1g conidia/litre gave 84% mortality with almost all dead insects showing white outgrowth and sporulation. High kill rates, 74% and 69% were also recorded by the two *M. anisopliae* strains (FI-1375 and FI-1248) at a dose of 0.1g/litre. Most of the surviving larvae were found well away from treatment holes suggesting that almost all the detected insects were successfully controlled.

Fig. 4. Application of fungus to pecan borers using a sheep drenching apparatus.

Subsequent trials showed that spraying the conidia directly on the tree trunks gave a high level of control (up to 80%, unpublished). A problem in commercialising this type of product is the small size of the market for any product. Given the detailed research still needed to satisfy registration authorities, significant funding is required and this is unlikely given the market size. The rationale behind using a general isolate like FI-1248 was that if this was commercialised for other targets, then as a spin-off it may be used for pecan borer control. Favourable factors include the high value of the crop, the need for a non-chemical control method and the demonstrated high efficacy. The expertise of a single grower was a key factor in our successful field trial and so other potential users may find the product less effective because they do not have the same detailed knowledge of the borer infestation in their orchard.

Control of Termites with *M. anisopliae*

In 1990, a new project was established to build on the promising results obtained by Hanel and Watson (1983) using *M. anisopliae* for control of urban termites. They obtained a good initial effect but then the colony steadily recovered and they cited 'inhibitory factors' in the mound as being responsible. The new project aimed to identify these factors, obtain better adapted fungal strains from Australian termites (Hanel and Watson had used a strain collected from a homoptera in Mexico) and as a result develop strategies for using the fungus as a commercial termiticide.

The main potential inhibitory factors, anti-fungal termite secretions, humidity and temperature in the nest, were studied and found to be unimportant. Also no specifically termite adapted strain of the fungus could be found. It was concluded that behavioural factors were predominately responsible for the lack of transmission in the nest. These responses (referred to as 'repellancy') included walling off treated areas so that healthy termites did not enter, removal of infectious conidia from fellow termites by grooming and burying dead termites, often below the base of the nest, to minimise exposure to new conidia.

Following four strategies for using *M. anisopliae* for termite control have been tested:

1. The application of unformulated conidia to mound and tree nests for killing the colony;
2. Application of unformulated conidia to feeding galleries in houses to provide temporary control by killing and repelling termites;
3. The use of conidia formulated to mask the repellency in baits to kill colonies; and
4. The use of high concentrations of conidia in soil to act as a barrier to prevent termites gaining entry to the protected building.

While a number of different strains of *M. anisopliae* have been tested for these strategies, including a less repellent strain for use in baits, most successful has been the originally selected isolate (FI-610) (Milner et al. 1998) This isolate which is a typical clade 8 isolate (Driver et al. 2000) was obtained from a mound of *Coptotermes lacteus* in New South Wales and was found to be very virulent in a series of screening bioassays. This promise was confirmed in an experiment using laboratory jar colonies so that behaviour of the termites could be a factor in the assessment. The isolate is also productive in our rice mass production system, kills termites at the high temperatures often found in the nest and does not produce destruxins.

Field trials showed that colonies of *Coptotermes* spp. and *Nasutitermes exitisosus* could be killed in 3 to 12 months by blowing pure dry conidia into the centre of the nest using holes drilled into the mound (Table 3) (Milner, 2002). Treatment of feeding galleries in houses using a similar technique provided local control for 12 months or more but the fungus had no long term effect on the responsible colony. Application to soil as a barrier was effective but only at doses over 10^8/g which was not economically viable. The most attractive strategy is baiting and recent field trials using a special masking formulation have successfully killed mound colonies using remote baiting (Milner, 2002). However further work is needed on this strategy under more realistic conditions before it can be commercialised.

A large number of field trials over a period of over 10 years have concluded that the following factors are important in using *M. anisopliae* for termite control:

1. The virulence of the fungus isolate (the most virulent isolates are also the most repellent).
2. The dose of conidia applied.
3. The structure of the nest (diffuse nests are difficult to treat while some tree nests may have too dense a structure to enable good distribution of the fungal conidia).

Table 3. Summary of results from field trial of *Metarhizium anisopliae* against mound-nests of *Nasutitermes exitiosus*. Data are the number of nests with no living queen out of 6 replicates treated at different times of the year

Dose (g conidia)	April	June	August	February
0	0	1	0	1
1	3	6	6	4
5	6	6	6	5
10	4	6	6	6
20	5	6	6	6
30	5	5 (out of 5)	6	5

4. The species of termite.
5. The time of year (in temperate climates, termites are more easily controlled in winter).

It is unlikely that future studies will identify more suitable strains of the fungus, rather progress will come from improvements in formulation and application strategy, with some specific termite problems more amenable to this type of control.

Despite the large amount of research over a long period of time, the commercialisation of *M. anisopliae* for termite control in Australia has not been achieved. Control using fungus is still far from reliable and given the high value of houses, it is difficult to market a product which 'may' work. The demand for non-chemical methods is high but this can to some extent be satisfied using mechanical barriers such as stainless steel mesh. While damage done by termites is slow, home-owners, once an infestation has been detected, tend to want rapid control and are therefore reluctant to wait the many months for a mycoinsecticide baiting strategy to be effective.

Development of Green Guard®

Locusts have always been a major problem for agriculture in Australia with three main pest species, two of which only occur in Australia. Grasshoppers, particularly the wingless grasshopper, *Phaulacridium vittatum*, are a problem on the Tablelands of SE Australia as well as the SW part of Western Australia. Extended wet periods in the arid parts of Australia cause a population build up every few years and the resulting high numbers of the Australian plague locust, *Chortoicetes terminifera*, have to be treated to prevent major losses in the intensive irrigation areas of the Riverina and Murrumbidgee Irrigation Area. Control is largely the responsibility of a permanent Federal agency, The Australian Plague Locust Commission (APLC). This body continually monitors locust numbers even in recession years, and implements control measures as required to prevent populations building up to the extent that they threaten crops. This preventative control relies on early detection and treatment of isolated populations and so depends on high quality forecasting and monitoring.

Until recently, aerial spraying of fenitrothion from fixed wing aircraft was the sole means of control. Given concerns over the environmental safety of this chemical together with increasing restrictions on its use, for example near water, and finally the expansion of the 'organic' beef industry, the APLC were keen to develop alternatives to fenitrothion. Preliminary results with FI-985, *M. anisopliae* var. *acridum*, an Australian isolate from the spur-throated locust, *Austracris guttulosa*, were very promising and so, in 1997, a national committee was formed to further develop FI-985 as a commercial product. The committee included government representatives, APLC, the commercial partner and CSIRO. Through this committee, a major expansion in field testing and associated research resulted in Green Guard® being used operationally in 2000/2001 for control of the Australian Plague Locust (Milner and Hunter, 2001). Green Guard was fully registered in 2005 by the Australian Pesticides and Veterinary Medicines Authority.

Large scale field trials have enabled the following conclusions to be made:

1. Doses of 25 to 75 g/ha are effective (equivalent to 1×10^{12} to 3×10^{12} conidia) and will usually give over 90% control in 10 to 21 days (Fig. 5).

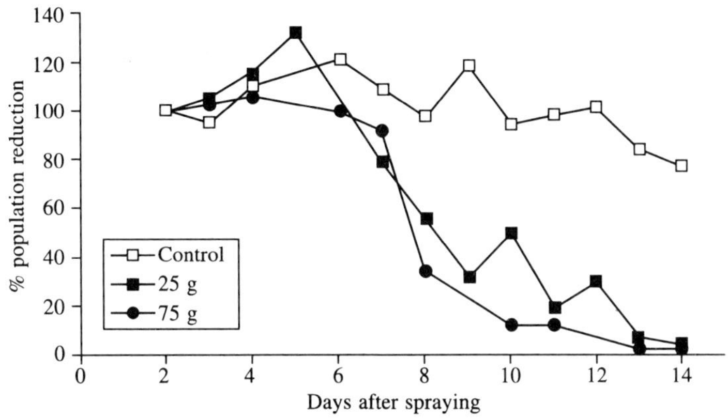

Fig. 5. Effect of two doses of Green Guard on Australian Plague Locust bands (January 2000).

2. Application is best by air using 500 ml of oil/ha but ground spraying using an oil/water emulsion is also effective.
3. The same isolate can be used against all the major locust and grasshopper pests in Australia with little effect seen on insects other than acridids (Table 4). Other Orthoptera such as crickets and tettigonids are not affected. If sprayed directly on foraging worker bees then a small proportion could be killed (< 20%).
4. The fungus is safe to spray on and near water (Milner et al. 2002).
5. The fungus persists for about 7 days after spraying at a sufficient level to control invading hopper bands but will not persist into the next season. Fungus-killed locusts are rapidly scavenged and no resporulation has been detected under field conditions.

Table 4. Host range of two isolates of *Metarhizium* var. *anisopliae* (FI-610) and var. *acridum* (FI-985). Combined data for % sporulation after 9 days incubation at 25°C from two doses (10^7 and 10^8 conidia/g substrate) tested against a range of insects

Order	Species	Common name	FI-985	FI-610
Isopoda	*Armadillidium vulgare*	Woodlouse	0.0	0.0
Coleoptera	*Tenebrio molitor*	Mealworm	7.5	92.5
	Chilocorus sp.	Ladybird	0.0	85.0
	Acalolepta vastator	Fig longicorn	0.0	30.0
Lepidoptera	*Galleria melonella*	Wax moth	32.5	87.5
	Mnesampela privata	Autumn gum moth	43.6	87.5
Hymenoptera	*Polistes* sp.	Paper wasp	2.5	50.0
	Iridomyrmex purpurius	Ant	8.1	82.3
Orthoptera	*Teleogryllus* spp.	Crickets (2 spp.)	0.0	24.0
Isoptera	*Coptotermes lacteus*	Termites	52.5	90.0

6. The infection does not spread outside the treated plot.
7. In the case of grasshoppers where reinvasion is a major problem, it has been found that fungus-treated plots are reinvaded more slowly than fenitrothion treated plots.
8. The fungus develops best when the locust's body temperature is between 20 and 30°C. Thus fungus is most effective under hot, cloudy, conditions as the basking of the locusts under sunny conditions increases the body temperature to over 35°C delaying mortality while cool conditions are also detrimental.

Two major problems were faced in taking this fungus from pilot field trials to a commercial product. The first was the formulation. The use of powder mixed *in situ* with oil was not considered for health and safety as well as consistency factors. Dilute oil formulations, ready to apply directly, were unsuitable. The conidia settled out during transport into a thick, hard to resuspend, cake and transporting the high volumes required was expensive. The current product is an oil-concentrate in corn oil. Corn oil was selected because of its relative high viscosity which minimises the settling problems. This concentrate is supplied in 25 litre pails which can be easily and quickly transported. The concentrate is then diluted in a ratio of 1:7 *in situ* with a light mineral oil for spraying. The spray oil diluent can be stored for prolonged periods of time at strategic places near the likely infestations. The second problem was the cost per hectare. Vast areas are treated for locust control in Australia and cost is a major factor. A reduction in cost down to the level which is directly comparable with chemical insecticides has been achieved by improvements in the mass production, development of a cheap effective drying method and a reduction of the dose used in the field. Recent trials have shown that even 12 g/ha is effective against the Australian plague locust (Milner and Hunter, 2001).

An important element in the success of Green Guard® was the selection of an isolate which is very virulent for all acridids over a wide range of temperatures and is very productive in mass production. The consumer demand due primarily to the expanding organic beef industry and the involvement of the APLC were essential. In addition, the project was helped by the involvement of an experienced and committed commercial partner at an early stage. Finally, the APLC was able to enter into a long term agreement with the commercial partner to buy a minimum amount of Green Guard® each year providing the incentive for the necessary capital expenditure.

Conclusions

Mycoinsecticides can be effective in small-scale high value horticulture as well as in broad-scale agriculture. The slow kill and the problems of scale-up of mass production are the main disincentives. Mycoinsecticides are more difficult to use, less reliable and more costly than chemicals, making farmer confidence a key factor in marketing. Too many products are released onto the market prematurely with problems of due to quality control, formulation and a lack of clear understanding of the most effective application method. Improvements are needed in formulation and application in the 'real world' to ensure products are more reliable.

Some of the factors discussed in this chapter include the choice of an effective isolate which is amenable to mass production and is effective over a range of conditions, particularly temperature. Unfortunately, the failings of a particular isolate may not be apparent until late in development by which time too much time and money has been expended to change isolate. An example of this is a very good isolate which is only found to be genetically unstable when production is expanded just prior to commercial release. For this reason, it is important that the choice of isolate for commercialisation be kept open for as long as possible and very large-scale field trials be undertaken early in development to test mass production capacity. Scientists may disregard as unimportant say a factor of two in the yield of conidia from two isolates, but if this results in one product being half the price of the other, there is a very clear outcome. Quality control at all stages of manufacture as well as transportation, storage and final use is vitally important. Thus, a product such as BioCane™ can be delivered to the farmer in excellent condition but if it is then left exposed to the sun for an hour during application, the product is rendered useless. Thus 'product stewardship' becomes the final and most important of all factors.

Experience in Australia suggests that non-scientific factors often not recognised by the scientists directly involved, may be critical. For example, expansion of organic beef farming was crucial in the development of Green Guard®. Also having a central organised market, as in the case of both locust control and sugarcane, was the key to having a commercially viable stable market for the product as well as for education of users.

References

Braga, G.U.L., Flint, S.D., Miller, C.D., Anderson, A.J. and Roberts, D.W. (2001) Variability in response to UV-B among species and strains of *Metarhizium* isolated from sites at latitudes from 61°N to 54°S. Journal of Invertebrate Pathology. 78: 98–108.

Driver, F., Milner, R.J. and Trueman, J.W.H. (2000) A taxonomic revision of *Metarhizium* based on sequence analysis of ribosomal DNA. Mycological Research 104: 135–151.

Hanel, H. and Watson, J.A.I. (1983) Preliminary field tests on the use of *Metarhizium anisopliae* for control of *Nasutitermes exitiosus* (Hill). Zeitschrift für Angewandte Entomologie. 94: 305–313.

Hunter, D.M., Milner, R.J. and Spurgin, P.A. (2001) Aerial treatment of the Australian plague locust, *Chortoicetes terminifera* (Orthoptera: Acrididae) with *Metarhizium anisopliae* (Deuteromycotina: Hyphomycetes). Bulletin of Entomological Research 91: 93–99.

Logan, D.P., Robertson, L.N. and Milner, R.J. (1999) Review of the development of *Metarhizium anisopliae* as a microbial insecticide, BioCane™, for the control of greyback canegrub *Dermolepida albohirtum* (Waterhouse) (Coleoptera: Scarabaeidae) in Queensland sugarcane. IOBC/WPRS Bull. 23:131–137.

Milner, R.J. (2000) Current status of *Metarhizium* for insect control in Australia. Biocontrol News and Information 21: 47N-50N.

Milner, R.J., Samson, P. and Bullard, G. (2002) FI-1045: A profile of a commercially useful isolate of *Metarhizium anisopliae* var. *anisopliae*. Biological Control Science and Technology 12: 43–58.

Milner, R.J., Staples, J.A. and Lutton, G.G. (1998) The selection of isolates of the hyphomycete fungus, *Metarhizium anisopliae,* for control of termites in Australia. Biological Control 11: 240–247.

Milner, Richard J. (2002) Application of biological control agents in mound building termites – experiences with *Metarhizium* in Australia. International Conference on Formosan Termite, New Orleans, May 2001. Sociobiology 40: 419–428.

Milner, Richard J. and Hunter, David. M. (2001) Recent developments in the use of fungi as a biopesticides against locusts and grasshoppers in Australia. Journal of Orthopteran Research, 10: 271–276.

Milner, Richard J., Lim, Richard P. and Hunter, David M. (2002) Risks to the aquatic ecosystem from the application of *Metarhizium anisopliae* for locust control in Australia. Pest Management Science 58: 718–723.

Milner, Richard J., Samson, Peter and Morton, Richard (2003) Persistence of conidia of *Metarhizium anisopliae* in sugarcane fields: effect of isolate and formulation on persistence over 3.5 years. Biocontrol Science and Technology. 13: 507–513.

Nahar, P., Kulye, M., Yadav, P., Hassani, M., Tuor, U., Keller, S. and Deshpande, M.V. (2003) Comparative evaluation of indigenous fungal isolates, *Metarhizium anisopliae* M34412, *Beauveria bassiana* B3301 and *Nomuraea rileyi* N812 for the Control of *Helicoverpa armigera* (Hub.) on chickpea. Indian Journal of Mycology and Plant Pathology. 33: 372–377. 013813.

Revankar, S.G., Sutton, D.A., Sanche, S.E., Rao, Jyothi, Zervos, M., Dashti, F., Rinaldi, M.G. and Rao, J. (1999) *Metarhizium anisopliae* as a cause of sinusitis in immunocompetent hosts. Journal of Clinical Microbiology 37: 195–198.

Samson, P.R., Milner, R.J., Sander, E.D., and Bullard, G.K. (2005) Effect of chemicals used during planting of sugarcane on viability of *Metarhizium anisopliae* and its efficacy against whitegrubs. BioControl 50: 151–163.

Wraight, S.P. and Carruthers, R.I. (1999) Production, delivery, and use of mycoinsecticides for control of insect pests on field crops, pp 233- 269. In "Methods in Biotechnology, vol. 5: Biopesticides: Use and delivery, (eds), Hall, F. R. and Menn, J.J. Humana Press, NJ, pp. 626.

15. Entomogenous Fungi and Development of Mycoinsecticides

T.S. Keshava Prasad and D.J. Bhat
Department of Botany, Goa University, Taleigao Plateau, Goa-403206, India

SUMMARY Fungi growing on insects have been often thought of as possible biocontrol agents for various pests. A glimpse of floristics, ecology and biology of entomogenous fungi belonging to various taxonomic groups studied over a period of time has been presented in this article. Efforts made to convert entomopathogenicity into field reality as mycopesticides have been identified and elaborated. Future researches needed in this direction have been discussed.

Introduction

Fungi are considered as the second largest group among living organisms, after insects. After careful and critical analysis, Hawksworth (2002) suggested that his earlier proposition of 1.5 million fungi on earth surface (Hawksworth, 1991) is an underestimate. Fungi live on a variety of substrates and habitats and one specialized group among them growing on insects is called 'entomogenous' or 'entomo-pathogenic fungi'. Though less known, this group, in recent days, has gained considerable importance in view of its potential as biocontrol agents of insect pests (Rossman, 1994).

Most subdivisions of Eumycota are represented by entomogenous fungi (Agarwal and Rajak, 1988). Over 700 species in 90 genera of fungi were reported to be pathogenic to insects and mites. Several reviews dealing with their taxonomy, biology, host range and economic importance are available (Burges, 1981; Roberts and Humber, 1981; Samson et al. 1988; Boucias and Pendland, 1991; Lacey and Goettel, 1995; Evans and Hywel-Jones, 1997; Tzean et al. 1997: Evans, 1999; Goettel et al. 2000; Roy and Pell, 2000; Lacey et al. 2001; Klingen et al. 2002a, b, c, d). Many species of fungi are known to grow on and regulate insect population (Butt et al. 2001; Inglis et al. 2001; Pell et al. 2001).

The potential of fungi as biocontrol agents in vector control programs has been recognized since the time of Pasteur. Although not used so far on a measurable scale, reports of mass production of fungal spores aiming at control of insects are available in the literature (Agarwal and Rajak, 1988). In this work, an attempt has been made to review the taxonomy and biology of some of the entomogenous fungi known for their bio-control potential.

The salient features of genera of hyphomycetes and coelomycetes are given in Table 1.

Coelomomyces Keilin

While describing *Coelomomyces*, Keilin in 1921 suggested its affinity to Chytridiales. In 1945, Couch, amending the genus, accommodated *Coelomomyces* in Blastocladiales. About 25 species are accommodated in *Coelomomyces* (Hawksworth et al. 1995).

The life cycle of members of *Coelomomyces* alternates between mosquito larvae and copepods or

Table 1. Salient features of genera of entomopathogenic hyphomycetes and coelomycetes

Genus	Condiophores	Conidiogenous cells	Conidia	Teleomorphs known
Akanthomyces	Synnematous, terminally tapering synnema	Arranged in a hymenial layer over synnema, denticulate	Catenate, aseptate, pleomorphic, ellipsoidal to elongated.	*Cordyceps*
Aspergillus	Mononematous with swollen terminal vesicle	Arranged in uni or biseriate over the vesicle	Catenate, aseptate, globose to subglobose.	*Emericella*
Beauveria	Mononematous	Clustered on lateral cells, sympodially denticulate	Solitary, aseptate, smooth, thin-walled, globose.	*Cordyceps*
Gibellula	Synnematous to mononematous, rough-walled and pigmented, conidiophores with a terminal vesicle	One-celled metulae and phialides over the vesicle	Catenate or solitary, aseptate, smooth-walled and ellipsoidal	*Torrubiella*
Hirsutella	Synnematous to mononematous, terminally tapering	Arranged laterally, swollen at the base and with a long slender neck, monophialidic	Solitary, aseptate, ellipsoidal to cylindrical conidia	*Cordyceps*
Hymenostilbe	Synnematous, terminally tapering	Arranged in a hymenial layer, denticulate	Solitary, aseptate, pleomorphic, ellipsoidal to elongated	Nectriaceae
Metarhizium	Mononematous, penicillate	One-celled metulae and monophialides	Catenate, aseptate, pale to olive green, ovoid to cylindrical,	*Cordyceps*
Nomuraea	Mononematous to synnematous	Verticillately arranged below the septum	Catenate to solitary, aseptate, hyaline to green, ovoid to cylindrical	–
Paecilomyces	Mononematous, penicillate or solitary	Terminal and intercalary	Catenate, aseptate, hyaline to green, globose to ellepsoidal	*Byssochlamys*
Verticillium	Mononematous, Verticillately branched	Verticillately arranged monophialides	Solitary, aseptate, hyaline, ovoid to ellipsoidal	*Hyphomyces* and *Torrubiella*
Tolypocladium	Mononematous, sparingly branched	Swollen phialides	Solitary in slimy balls, aseptate, globose to cylindrical	*Cordyceps* and *Hyphomyces*
Aschersonia	Pycnidia in the stroma, slender, branched simple pycinidiophores	Slender, awl-shaped	Hyaline, mostly fusoid, smooth, one-celled conidia	*Hypocrella*

ostracods (Whisler, 1985). In most species, the haploid gametes that emerge from copepods are isogametes of two mating types. Whisler (1979) confirmed meiosis in resting sporangia of *Coelomomyces* and presumed that segregation of different mating types occurs during meiosis. Isogametes fuse to form zygotes which eventually settle on the inter-segmental membranes of mosquito larva. Flagella are retracted in zygotes and adhesion vesicles attach it to the host. A cyst wall develops around the zygote. An appressorium from which a narrow penetration tube enters through a cuticle into the epidermal cells of the host. The fungus ramifies the epidermis, depletes nutrients in the haemocoel and thereby kills the mosquito larva. Resting sporangia are produced at swollen ends of hyphae. Federici and Chapmann (1977) estimated 10000 to 60000 resting sporangia in a single 4th instar larva depending on the species of host. The fungus over-winters as resting sporangia. Usually larvae die in the 4th instar stage and spores are released from the decomposing cadaver. Posteriorly uniflagellated meiozoospores emerging out from the resting spores, infect appropriate copepod or ostracod host and establish haploid heterothallic gametophytic stage which develops in the haemocoel.

In adult female *Aedes aegypti*, the infection is mostly localized in ovaries. During enlargement of ovaries, hyphae in the haemocoel get transferred to interstitial spaces of the ovaries and penetrate epithelial cells of the ovaries (Lucarotti, 1992). Following a sumptuous blood meal, fungal hyphae in the ovaries develop into resting sporangia (Lucarotti and Klein, 1988). Resting sporangia are laid down, in place of eggs, when female mosquitoes (infected with *Coelomomyces stegomyiae*) attempt to oviposit. Meiozoospores emerge out from the resting spores infect copepod host and complete the life cycle (Padua et al. 1986).

Coelomomyces spp. have a life cycle involving an intermediate crustacean host, which helps in maintaining a viable progeny. Development in an intermediate host provides the fungi a launching pad to invade mosquito larvae. The gametes are released simultaneously when copepods and mosquito larvae are both near the water surface (Apperson et al. 1992) which facilitates mating between two types of gametes and helps the resultant zygote to interact with mosquito larvae. Species of *Coelomomyces* have the potential for use in natural control because they can cause epizootics. These can cycle in the environment and infect adult female mosquitoes and thereby effect their potential dispersal (Lucarotti and Andreadis, 1995).

Lagenidium giganteum Couch

Lagenidium giganteum (Lagenidiales: Oomycetes) is a facultative parasite on mosquito larvae, first observed by Couch in 1935. Umphlett (1973) reisolated the fungus from the host and recognized its biocontrol potential. *L. giganteum* is the only fungus that attained operational stage in mosquito control. Biflagellate motile zoospores are the effective phase of the fungus that adheres to the larval cuticle. Infection is initiated by mechanical and enzymatic activity of encysted zoospores allowing entry of the fungus into larva and resulting in death of the latter within 24-72 hours. On maturation, the fungus reproduces either sexually or asexually. Asexual reproduction amplifies infection, with a new round of zoospore release occurring every 24-72 hours depending upon the mosquito host and environmental conditions. Sexual reproduction results in the formation of oospores that can survive desiccation, environmental extremes and mechanical abrasion. Oospores can be stored for at least 7 years in a viable condition in the laboratory or the field (Kerwin et al. 1994).

Three formulations of *Lagenidium giganteum* consisting of various combinations of sexual and asexual spores have been registered by the US Environmental Protection Agency (USEPA Regulation Nos. 56984-1, 56984-2 and 56984-3) with Department of Health Service as registrant (Kerwin et al. 1994). The fungus can be commercially grown on a large scale and applied by ground or air appliances (Kerwin and Washino, 1986, 1987, 1988). Its host specificity and safety has been evaluated using a variety of non-target organisms (Kerwin et al. 1988). Temperature intolerance (>3°C) of zoospores, absence of any toxin production, safety against mammals and birds makes the fungus very lucrative for pest control management (Siegel and Shadduck, 1987; Kerwin et al. 1990).

Entomophthorales

The Order Entomophthorales is placed in the Subdivision: Zygomycotina, Class: Zygomycetes. Six families are recognized in the Order. These include, Entomophthoraceae, Neozygitaceae, Completoriaceae, Ancylistaceae, Meristacraceae and Basidiobolaceae (Humber, 1989). Entomogenous fungi are accommodated in Entomophthoraceae and Neozygitaceae. About 200 species in Entomophthoraceae and 15 in Neozygitaceae are known (Keller, 1997). Several of the Entomophthorales regulate host population through epizootics. They have a narrow host range exhibiting close associations with foliar insect or mite hosts (Evans, 1989).

The members of Entomophthorales have complex life cycles with two or more types of spores. *Entomophthora muscae* is a typical example. To increase the efficiency of out-reaching insect hosts, many show deviation from the basic pattern of life cycle. Conidia are infective units in ideal conditions. Conidiophores of *E. muscae* arise through membranous regions of the host integument and large amounts of primary conidia are actively discharged (Soper et al. 1976; Samson et al. 1979). The mitospores are fragile, short-lived and germinate instantly. A primary conidium can produce and actively discharge a secondary conidium and the latter may produce and discharge a tertiary conidium. The size of successive conidia diminishes while the shape remains the same.

In species of *Neozygites*, primary conidia are not effective whereas secondary conidia produced are infective and differ in shape and size from the former. In *Zoophthora,* successive conidia are produced on fine capilli-conidiophores. The capilli-conidium is a sticky propagule borne at the tip of the conidiophore some distance above the surface and readily attaches to the host. Spherical appressoria are produced for host penetration although it is not a prerequisite (Brobyn and Wilding, 1977; Lambiase and Yendol, 1977). Entomophthorales grow as protoplasts lacking sugar rich cell walls within the haemocoel. Lack of cell wall is presumed to help in escaping detection by the host's immune system (Beauvais et al. 1989). Reduced feeding, negative geotaxis are some of the symptoms of infection. Physiological starvation and short-lived cells are factors reported as cause of host death (Hajek,1997b).

Entomophthorales survive hostile periods as resting spores. The zygospores or azygospores are thick-walled and remain dormant for several months before germination (Hajek, 1997a). These germinate any time when hosts are present in the field and produce one to several actively ejected infective germ conidia or capilli-conidia as in case of *Neozygites* spp.

Entomophthora muscae

E. muscae was the first Entomophthoralean fungus described (Cohn, 1855). The biology and biocontrol potential of this fungus against adult flies, especially housefly, *Musca domestica*, has been studied (MacLeod et al. 1976). The fungus is apparent as dead flies get attached to vegetation, walls, etc. by rhizoids emerging through the proboscis and by the legs of the dead flies. *E. muscae* is known from a range of dipteran hosts. Keller (1984, 1987) demonstrated that *E. muscae* is a complex of species. Recently *E. muscae* sensu stricto has been described (Keller et al. 1999).

E. muscae is not confined to one host species but may be transmitted to other dipteran species and epizootiology of the fungus involves several hosts (Kramer and Steinkraus, 1981; Mullens et al. 1987; Mullens, 1989; Eilenberg et al. 1990; Jensen and Eilenberg, 2000). The entire life cycle, including production of resting spores in the field population, has been studied (Wilding and Lauckner, 1974; Carruthers et al. 1985; Thomsen and Eilenberg, 2000). In other species, resting spores were observed occasionally (Steinkraus et al. 1993b). The natural epizootics caused by *E. muscae* are reported in the populations of *M. domestica, Delia* spp. and *C. rosae* (Carruthers and Haynes, 1985; Mullens and Rodriguez, 1985; Eilenberg, 1987a; Bellini et al. 1992; Moller, 1993; Six and Mullens, 1997).

Behavioural fever was observed in *E. muscae* infected *M. domestica* in which they prefer higher temperatures than uninfected flies, with the result that the fungus dies and the fly survives (Watson et al. 1993). Females of *C. rosae* infected with *E. schizophorae* do not recognize their hosts, though they are capable of depositing eggs (Eilenberg, 1987b). This behaviour influences the effect of *E. schizophorae* on *C. rosae* populations, since infected females do not contribute to population growth. Towards the end of the infection, flies attach themselves to the vegetation with their abdomen exposed. The fungus discharges primary conidia after death of the host. (Krasnoff et al. 1995). Recent works in host-pathogen interactions include that of Thomsen et al. (2001) and Thomsen and Jensen (2002).

Neozygites fresenii

Isolated from aphids, the fungus was originally recognized as *Empusa fresenii* Nowakowski (Nowakowski, 1883). Witlaczil (1885) described it as *Neozygites aphidis*. Thaxter (1888) placed *E. fresenii* in a new subgenus, *Triplosporium*. Giard (1888) noted that the genera *Triplosporium* and *Neozygites* were synonymous. Remaudiere and Keller (1980) replaced *Triplosporium* with *Neozygites*. About 15 *Neozygites* species are currently recognized (Keller, 1997). More undescribed species of *Neozygites* from Collembola have recently been discovered (Dromph et al. 2000; 2001). *N. fresenii* has a worldwide distribution with reports of infected aphids from Africa, Australia, Europe, India, Israel, the South Pacific and North America (Kuntz, 1925; Gustafsson, 1965; Ramaseshiah, 1968; Thoizon, 1970; Bitton et al. 1979; Milner and Holdom, 1986; Thaxter, 1888; Silvie and Papierok, 1991; Keller, 1997). The fungus is considered as the most common aphid pathogen of tropical regions (Remaudiere, 1977). *Neozygites floridana* was reported from Brazil, India, Israel, Poland, USA and the West Africa (Ramaseshiah, 1971; Kenneth et al. 1972; Brandenburg and Kennedy, 1981; Smitley et al. 1986; Mietkiewski et al. 1993; Yaninek et al. 1996).

Members of Neozygitaceae are specialized as pathogens of small arthropods such as mites, springtails, thrips and aphids (Keller, 1997). Because the hosts are of small size, *Neozygites* spp. may not have been well studied. *N. fresenii*, *N. cinarae* and *N. microlophii* attack only aphids, whereas *N. cucumeriformes* and *N. parvispora* are known only from Thysanoptera (Balazy, 1993) and *N. sminthuri* only from Collembola (Keller and Steenberg, 1997). *N. floridana* and *N. tetranychi* are restricted to mites in the Tetranychidae (Keller, 1997).

Neozygites spp. are recognized as important natural control agents of cassava green mite, *Mononychellus tanajoa*, and mites on groundnuts cotton and lima beans (Carner and Canerday, 1968; Brandenburg and Kennedy, 1983; Boykin et al. 1984; Keller, 1997; Elliot et al. 2000). A very short life cycle, ability to attack all stages except eggs of the hosts, production of a large number of primary conidia per host make them very effective as natural control agent (Steinkraus et al. 1993a). There are a number of published reports dealing with life cycle and epizootiology of *Neozygites* spp. (Bitton et al. 1979; Steinkraus et al. 1991, 1993a, 1995, 1996, 1999; Steinkraus and Slaymaker, 1994; Keller, 1997; McLeod et al. 1998).

Entomophaga maimaiga

Though said to be native to northern Asia, *E. maimaiga* has been reported from North America (Andreadis and Weseloh, 1990; Hajek et al. 1990). Molecular studies have disclosed that *E. aulicae* is a 'species complex', with *E. maimaiga* included in one of the four groups. Fungi in the other three groups currently retain the name *E. aulicae* (Walsh, 1996). *E. maimaiga* was initially differentiated from the *E. aulicae* complex based on the fact that it is the only member able to infect *Lymantria dispar* (Soper et al. 1988). Host range of this species has been extensively studied in view of its potential as biocontrol agent (Hajek et al. 1995a, b; Bidochka and Hajek, 1996, 1998).

Entomophaga grylli

E. grylli was first collected on a *Gryllus* sp. by Fresenius (1858) from Europe and has been referred as *Entomophthora grylli*, *Empusa grylli* and *Conidiobolus grylli* (Carruthers et al. 1997). *E. grylli* exists as a complex of pathotypes. *E. grylli* sensu stricto is distributed mainly in Europe (Carruthers et al. 1989; Humber, 1989). All members of the '*E. grylli* species complex' are pathogens of grasshoppers and locusts (MacLeod, 1963; Carruthers et al. 1989). North American pathotypes of the two species have a different host range (Pickford and Riegert, 1964; Carruthers and Soper, 1987; Ramoska et al. 1988; Streett and McGuire, 1990). Both pathotypes have resting spores which produce infective germ-tubes to initiate the infection cycle in their hosts. The pathotype 1, produces actively discharged conidia or resting spores in the host whereas pathotype 2 has only resting spores in all infected hosts (Carruthers et al. 1997).

Infection with *E. grylli* is commonly called 'summit disease' as infected and dying insects exhibit abnormal behaviour and climb to the top of vegetation, dying in a head up position, grasping the plant stems (Evans, 1989). Insect death occurs in the late afternoon and early evening, synchronizing sporulation and infection with optimal conditions of high humidity, cool temperatures and zero ultraviolet radiation during the night (Carruthers and Soper, 1987; Carruthers et al. 1988, 1992). *E. grylli* sporulates within hours of host death. If abiotic conditions are not favourable, the cadaver desiccates but the fungus remains viable for extended periods of time, with ability to rehydrate, sporulate and desiccate repeatedly (Sawyer et al. 1997).

Grasshoppers and locusts are highly mobile insects, and infection has little impact on their mobility during early stages. Though dispersing grasshoppers are likely to carry infection with them, at least over moderate distances. Carruthers et al. (1997) suggested that long range migration by hoppers may be a way of disease escape, particularly when the fungus can survive at a site for several seasons as resting spores. They also noted that some species leave favored feeding sites to lay eggs in more open areas, which may separate susceptible early instars from over-wintering sites of the fungus.

Erynia neoaphidis

E. neoaphidis has a wide distribution, being recorded from Europe, Asia, Africa, North and South America and Australia (Wilding and Brady, 1984; Glare and Milner, 1991; Hatting et al. 1999). Besides *Erynia* (Keller, 1991), the species was earlier assigned to *Pandora* (Humber, 1989) and *Zoopthora* (Balazy, 1993). *E. neoaphidis* has been recorded from more than 70 species of aphids on annual and perennial crops, weeds and wild flowers (Pell et al. 2001). Epizootiology of *Erynia neoaphidis* has been the subject of several publications (Milner et al. 1984; Morgan, 1994; Brown et al. 1995; Morgan et al. 1995; Roy, 1997; Dromph et al. 1997, 2001; Nielsen et al. 1998; McLeod et al. 1998; Hemmati, 1999; Roy et al. 1998, 1999, 2001).

Zoophthora radicans

Z. radicans was originally described from *Pieris brassicae* (Brefeld, 1870) as *Empusa radicans*. With a worldwide distribution, it has been recorded from members of Coleoptera, Diptera, Hemiptera, Hymenoptera, Lepidoptera, Orthoptera, Thysanoptera and Trichoptera (Glare and Milner, 1991). Individual isolates are better adapted to infect taxonomically related insect hosts (Papierok et al. 1984; Milner and Mahon, 1985; Mietkiewski et al. 1986; McGuire et al. 1987a; Magalhaes et al. 1988; Goettel et al. 1990). Infection may have an impact on host behaviour (Furlong et al. 1997; Reddy et al. 1998). Under certain circumstances, persistent resting spores (azygospores) are produced within the host in response to changing climatic factors, particularly low temperature and high humidity, high inoculum density, host age or inappropriate hosts (Ben Ze'ev and Uziel, 1979; Perry et al. 1982; McCabe et al. 1984; McGuire et al. 1987b; Glare et al. 1989). Resting spore production has not been observed in all isolates (Glare et al. 1989; Pell et al. 1993). Literature regarding effects of temperature and circadian rhythm on production and discharge of conidia (Milner and Lutton, 1983; Yamamoto and Aoki, 1983; Milner et al. 1984; van Roermund et al. 1984; Glare et al. 1987; Wraight et al. 1990; Pell et al. 1993; Sawyer et al. 1994; Leite et al. 1996b), effects of host or substrate (Wraight et al. 1990; Leite et al. 1996a, c), and effects of nutrients and pH (van Roermund et al. 1984; Magalhaes et al. 1990, 1991a, b) on germination and subsequent development are available.

Ascomycetes

Several entomogenous fungi reported to date belong to Ascomycetes. The genera such as *Cordyceps* (Fr.) Link, *Hypocrella* Sacc., and *Torrubiella* Bound. of Sphaeriales and members of Laboulbenniales are discussed.

Cordyceps Link

Typified by *C. militaris* (Fr.) Link, the genus *Cordyceps* accommodates more than 300 species (Mains, 1958). Host range comprises Coleoptera, Hymenoptera, Homoptera and Lepidoptera. *C. ophioglossoides* attacks deer truffle, *Elaphomyces* sp. The classic Japanese book on *Cordyceps* and related genera, by Shimizu and Kobayasi (ISBN 4-259-53866-7) contains colour paintings of these fungi. Tzean et al. (1997) described 6 species of *Cordyceps* from Taiwan. The genus is characterized by stalked stromatic fruit body with a fertile head. Those arising from insect larvae or pupae are known as 'vegetable caterpillars' and are used in traditional Chinese medicine for various ailments. Masses of hyphae usually fill the host body, emerging out in the form of stalks of hyphae through the exoskeleton. The perithecia are formed either superficially or immersed in the stalks. Thin-walled asci with 8 ascospores are typical of the genus with ascospores being long, filiform and sometimes end as part-spores at maturity (Evans and Hywel-Jones, 1997). The development of part-spores can be seen as adaptation to increase targeting-efficiency. Both shooting of spores into air and oozing into water films are reported. Usually each of the 8 long ascospores break into 128 part-spores, often while within the ascus. The number of part-spores in *Cordyceps* varies. Kendrick (2002) estimated that a single stroma of *Cordyceps* species can produce as many as 614,400,000 propagules. Anamorphic genera of *Cordyceps* belong to *Akanthomyces, Hirsutella, Hymenostilbe, Paecilomyces, Metarhizium* and *Verticillium. Tolypocladium niveum,* producer of cyclosporin, was recently discovered as the anamorph of *Cordyceps subsessilis* (Kendrick, 2002).

Torrubiella Boudier

The genus, typified by *Torrubiella arachnida* Boudier (Petch, 1923), accommodates about 54 species (Kobayasi and Shimizu, 1982). Fruiting body is distinguished from that of *Cordyceps* by absence of an erect stalk and the perithecium is borne directly on the host. Host range comprises of homoptera and spiders (Evans and Hywel-Jones, 1997). Peritheicial ascocarps, with wall covered by anamorphic conidial structures, long cylindrical asci with 8-spores, with thickened ascal apices penetrated by a fine canal, filiform multiseptate ascospores which break into several part-spores at maturity characterize the genus. Tzean et al. (1997) described 6 species of *Torrubiella* from Taiwan.

Hypocrella Saccardo

Typified by *Hypocrella discoidea* (Berk. & Br.) Sacc. (Petch, 1921), this genus accommodates about 38 species (Petch, 1921; Evans and Hywel-Jones, 1997). The genus *Hypocrella* is one of the most specific of entomopathogenic genera being restricted to the members of the Aleyrodidae and Coccidae (Evans and Hywel-Jones, 1997). Stroma exosclerotium, cushion-like, brightly coloured, irregular and undulating due to semi-erumpent perithecia. Asci are cylindrical to filiform, with prominent apical thickening, 8-spored. Ascospores are filiform and break into part-spores at maturity. Blackening of otherwise brightly coloured fruiting bodies is observed while drying herbarium collections. Anamorphs of *Hypocrella* belong to *Aschersonia.* Anamorph and teleomorph can be found on the same stroma adjacent to each other.

Laboulbenniales

Laboulbenniales are most interesting of all entomogenous fungi, named after French mycologist Alexandre Laboubene. The group contains around 2000 described species. The fungi usually are ectoparasitic, superficially penetrating the host cuticle and evidence for deleterious effects against host is wanting. Therefore, this group has received little attention as agents of biological control (Weir and Beakes, 1995).

The fungal body consists of a fixed number of cells and organs arranged in a definite pattern. The main axis supporting the reproductive organs is called a receptacle. Lower foot cell of the ascospore attaches to the host cuticle and gives rise to a primary receptacle, which further extends to a secondary receptacle to form secondary axis which may bear one or more peritheica. The structural complexity of the receptacle is the key criterion in the taxonomy of Laboulbenniales. Upper spore segment gives rise to primary appendages. Secondary appendages are also formed in some genera. Sometimes antheridia are produced in the septate columns of these appendages. It is assumed that spermatia are produced either exogenously like conidia (e.g. *Ceratomyces* sp.), endogenously in flask-shaped structures (e.g. *Laboulbenia* sp.) or compound antheridia in which antheridial cells discharge spermatia into a common antheridial cavity (e.g. *Eucantharomyces* sp.). Perithecial ascocarp is an outgrowth of receptacle comprising of stalk cells and a procarp which later develops into trichogyne, ascogonium and related cells. Arrangement of perithecial wall cells has importance in delineating families and subfamilies. Ascospores are usually 4 or rarely 8, elongated, spindle shaped and always two-celled.

The exact nature of host specificity, position specificity on the host body and sex specificity of the host is poorly understood. Transmission of the fungus seems to be by direct contact among host individuals. The fungus can be gathered round the year by collecting exact hosts. The fungus appears as short bristles either darkly pigmented or pale-coloured on the host body which can be seen microscopically.

Deuteromycotina

Hyphomycetes

Akanthomyces Lebert

The genus *Akanthomyces,* typified by *A. aculeata* Leb., was established by Lebert in 1858 (cited from Mains, 1950b). An anamorph of *Cordyceps* (Seifert, 1985), the genus is diagnosed by cylindrical, terminally tapering synnemata, conidiogenous cells in a hymenium and unicellular conidia in chains. The host range includes members of Lepidoptera, Coleoptera, Diptera, Orthoptera, and Arachnida (Mains, 1950b). Mains (1950b) described 4 species of *Akanthomyces*. Tzean and co-workers (1997) have described 13 species of *Akanthomyces* from Taiwan. Other useful references include Samson and Brady (1982), Vincent et al. (1988) and Hywel-Jones (1996).

Aspergillus Micheli

Species of *Aspergillus* such as *A. flavus, A. fumigatus, A. parasiticus* and *A. sclerotium* were reported from insects and spiders (Lisansky and Hall, 1983; Tzean et al. 1997; Evans and Hywel-Jones, 1997). Erect conidiophores with terminally swollen vesicles bearing uniseriate or biseriate phialides and 1-celled, catenate conidia characterize the genus *Aspergillus* (Gams and Samson, 1985). *A. fumigatus* is found to be infective to mammals and hence excluded from biocontrol studies (Lisansky and Hall, 1983).

Beauveria Vuillemin

The genus, typified by *Beauveria bassiana* (Bals.) Vuill., is important as a source of biocontrol agents and novel metabolites (Ferron, 1981). The type species is an insect pathogen and recorded from more than 500 insect hosts (Hall and Papierok, 1982; Ekesi et al. 1998; Hallsworth and Magan, 1999; Dromph, 2001; Dromph and Vestergaard, 2002; Bruck and Lewis, 2002; Klingen et al. 2002a,b). Altogether 49 species of *Beauveria* are so far recorded. Seven of them were found to be synonymous with *B. bassiana,* 2 with *B. brongniartii* and one with *B. felina.* Many of the recorded species have been transferred to and accommodated in related genera -5 in *Tolypocladium,* 2 each in *Acrodontium* and *Tritirachium* and 1 each in *Engyodontium, Nomurae* and *Pleurodesmospora.* Sympodial conidiophores with a zig-zag appearance; densely clustered conidiogenous cells, one-celled, hyaline, smooth, thin-walled, globose conidia characterize the genus (Tzean et al. 1997). According to Samson (1995), although there are considerable variations among isolates, 6 species can be recognized in the genus based on morphology and biochemical analysis. Though there are reports of allerginicity to *Beauveria bassiana* (Lisansky and Hall, 1983), several formulations of the fungus are available commercially from Mycotech, Troy Bioscience (USA), NPP (France), and Nitto Denko (Japan).

Gibellula Cavara

Though species of *Gibellula,* anamorphs of *Torrubiella,* are generally encountered as parasites of spiders, they have not been tested for their biocontrol efficacy. The genus is typified by *Gibellula pulchra* (Sacc.) Cavara. The species range from synnematous to mononematous with septate, rough-walled and pigmented conidiophores bearing terminal vesicular, hyaline, smooth-walled zone with one-celled metulae and phialides. Conidia are one-celled, smooth-walled, ellipsoidal and hyaline. About 35 species have been recorded so far (Petch,1932; Mains,1950a; Evans and Samson, 1987; Samson and Evans,1992; Tzean et al.,1997).

Hirsutella Pat.

The genus, typified by *H. entomophila* Pat., has been studied by Minter and Brady (1980). Altogether 87 records of species belonging to this genus can be found in literature. Several species have their perfect stages in *Cordyceps.* Synnematous or mononematous conidiophores, with monophialidic conidiogenous cells that are swollen at the base and with slender necks. One celled, ellipsoidal to cylindrical conidia covered with mucus characterize the genus. *H. thompsonii* attacks only mites and is well-known for its epizootics among eriophyid mites, particularly citrus rust mite (*Phyllocoptruta oleivora*), citrus bud mite (*Eriophyes sheldonii*) and coconut mite (*Eriophyes guerreronis*). It is known to have considerable potential as an acaricide (Mc Coy and Couch, 1978; Hall et al. 1980).

Hymenostilbe Petch

The genus is established by Petch in 1931 to accommodate *Hymenostilbe muscaria* Petch, an inhabitant on flies. Similar to *Akanthomyces, Hymenostilbe* produces synnema with phialides on lateral branches. *Hymenostilbe* however differs in having solitary conidia whereas *Akanthomyces* is characterized by catenate conidia (Mains, 1950b). In all, 25 species have been recorded under the genus.

Metarhizium Sorokin

The genus, typified by *M. anisopliae* (Metschn.) Sorokin, is characterized by mononematous and penicillate conidiophores, monophialidic conidiogenous cells, basipitally developing conidial chains and ovoid to cylindrical, aseptate, pale to olive green conidia in mass. *M. anisopliae* is one of the best studied entomopathogenic fungi. It was the first fungus to be produced in a large scale from a biological control perspective. Besides mosquitoes, *Metarhizium* spp. infect wide range of insects including members of Coleoptera, Collembola, Lepidoptera, Diptera and Homoptera (Prior, 1990; Hallsworth and Magan, 1999; Arthurs and Thomas, 2001a,b; Butt et al. 2001; Dromph, 2001; Ekesi et al. 2001; Hunter et al. 2001; Dromph and Vestergaard, 2002; Klingen et al. 2002a,b). Lately, *M. anisopliae* has been licensed for indoor control of cockroaches (Ward et al. 2000).

***Nomuraea* Maublanc**

Botrytis rileyi Farlow, the basionym of *Nomuraea rileyi* (Farl.) Samson, typifies the genus. Following is the diagnostic feature of the genus. Conidiophores mononematous to synnematous, with verticillately arranged cylindrical to oval conidiogenous cells, aseptate, hyaline to green, smooth, ovoid to cylindrical conidia (Samson, 1974). The members of the genus are well-known for their epizootics among Lepidoptera, though they were reported from Homoptera, Diptera and spiders. Currently 6 species are accommodated in the genus, viz. *Nomuraea anemonoides* A.D. Hocking, *N. atypicola* (Yasuda) Samson, *N. cylindrosporae* (Chen and Guo) Tzean, Hsieh, Chen and Wu, *N. prasina* Maubl., *N. rileyi* (Farl.) Samson and *N. viridula* Tzean, Hsieh, Chen and Wu. *Nomuraea prasina* was however recognized as a synonym of *N. rileyi.*

***Paecilomyces* Bainier**

Typified by *P. variotii* Bainier, the genus is cosmopolitan in distribution as saprophyte in soil and as pathogen on nematodes or insects. Although 119 species have been recorded, many are synonyms of *Acremonium, Acrophialophora, Gabarnaudia, Mariannaea, Penicillium, Sagenomella, Scopulariopsis, Septofusidium, Sesquicillium* and *Verticillium.* The species have their teleomorphs in the genus *Byssochlamys. Paecilomyces fumosoroseus* and *P. farinosus* were studied from the biocontrol utility point of view (Ekesi et al. 1998; Hallsworth and Magan, 1999).

***Verticillium* Nees**

The genus, typified by *V. tenerum* (Nees. Ex Pers.) Link, has erect, slender, branched, tapering conidiophores, verticillately branched phialides, 1-celled, smooth, hyaline, ovoid to ellipsoid, conidia which are solitary or sometimes in chains. The genus is heterogenous with species usually saprophytic or parasitic on plants but also mycoparasitic and entomopathogenic. Altogether 261 species have been recorded for the genus. Many have been later accommodated in genera such as *Acremonium, Acrostalagmus, Cladobotryum, Clonostachys, Gliocladium, Paecilomyces, Phaeostalagmus, Phialophora, Sesquicillium* and *Trichoderma.*

V. lecanii, originally described from scale insects, has a broad host range among Arthropoda. There is a great deal of variation among the isolates of the fungus from different hosts making *V. lecanii* a 'species complex' (Yun et al. 1991). As a consequence, the species of *Verticillium* has an impressive list of synonyms (Evans and Hywel-Jones, 1997). *V. lecanii* has its teleomorph associated with *Torrubiella,* although remaining species have it with *Hyphomyces.*

Verticillium lecanii has been used successfully to control aphids and white flies in glass houses (Lisansky and Hall, 1983). It was the first fungus to be registered for commercial use and produced in a controlled and standardized conditions. Three products of *V. lecanii,* Vertilac against whiteflies, Mycotal against aphids and Thriptal against thrips, produced by Microbial Resources Ltd. were in the market in early 1980's (Evans and Hywel-Jones, 1997). *V. chlamydosporium* is well-known for its efficacy against cyst nematode of cereals. Chlamydospores of this fungus are able to infect the target in soil, whereas the conidia are helpful in survival and dispersal (Lisansky and Hall, 1983).

***Tolypocladium* Gams**

The genus contains anamorphs of *Cordyceps* and *Hyphomyces. Tolypocladium* was established by Gams in 1971, typified by *Tolypocladium inflatum* W. Gams. Sparingly branched conidiophores, swollen phialides, and small, one-celled conidia borne in slimy phialides (Bisset, 1983) characterize the genus. *Tolypocladium niveum,* producer of cyclosporine, is the anamorph of *Cordyceps subsessilis* (Kendrick, 2002). *Tolypocladium* spp. are wide spread as soil fungi, some of which are pathogenic on insects (Bisset, 1983; Samson and Soares, 1984). Some strains of *T. cylindrosporum* are considered as potential candidates for control of mosquitoes and other insect pests (Lam et al. 1988; Samson and Soares, 1984; Klingen et al. 2002a, b). The fungus causes epizootics in mosquito larval population (Weiser and Pillai, 1981). The susceptible mosquitoes include species of *Aedes, Culex, Culiseta, Anopheles, Maorigoeldia, Opifex,* etc. (Pinnock et al. 1973; Soars et al. 1979 and 1985; Yu et al. 1980; Weiser and Pillai, 1981; Gardner et al. 1986; Goettel, 1987a,b). *Tolypocladium* spp. produce a wide range of metabolites including cyclosporins, efrapeptins, elvapeptins and the antibiotic LP237-F8 (Dreyfuss et al. 1976; Jackson et al. 1979; Krasnoff et al. 1991; Tsantrizos et al. 1996). However, the effect of these compounds during infection of mosquitoes is not known so far.

Laboratory studies have shown that relatively high dosages of *T. cylindrosporum* are required to elicit a response in mosquitoes (Pinnock et al. 1973; Goettel, 1987a). It is unlikely that such high doses occur in the field under natural conditions. Conidia are produced only on floating cadavers whereas blastospores are produced in the haemolymph of insects. Action of the fungus is said to be slow and possibilities of using the fungus as a biocontrol candidate is yet to be ascertained.

Coelomycetes

Aschersonia Montagne

The genus *Aschersonia*, typified by *A. taitensis* Montagne (Petch, 1921), is characterized by production of pycnidia containing conidiophores and/or paraphyses, formed in hemispherical or cushion-shaped centrum; conidiophores are slender, branched consisting of thin-walled usually awl-shaped conidiogenous cells with hyaline, mostly fusoid, smooth, one-celled conidia. Species of *Aschersonia* are parasitic on homopteran insects and have *Hypocrella* teleomorph (Tzean et al. 1997). Rolf and Fawcett (1913) first described their entomogenous nature. Reports on their biocontrol value are given by Ramakers and Samson (1984); Frasen et al. (1987), and Rombach and Gillespie (1988). Petch (1921) examined the herbarium specimens then available for *Aschersonia* and recognized 25 species. Mains (1959a,b, c) reviewed species of *Aschersonia* described from America. Recent works on ecology, *in vitro* growth and biocontrol aspects are reported by Hywel-Jones and Evans (1993), Evans (1994) and Meeks et al. (2002).

Mycoinsecticide Development

Development of mycoinsecticides is a complex process demanding enormous capital, technology, infrastructure, manpower and organizational inputs in order to make them environment-friendly principles. Promising fungi have not been researched much despite known advantages. A potential biocontrol organism can be realized only by obtaining fungi from different habitats and testing them for insecticidal activity. Lacey and Brooks (1997) advocated isolation of microfungi from insects and arachnids from as many geographical locations and climatic conditions as possible and maintenance in culture collections, as a priority area of research. Selection of a candidate requires evaluation for its ability to withstand adverse field conditions. Screening of fungi for insecticidal activity can be done by following relatively simple procedures (Kerwin and Petersen, 1997).

Conventional genetic crossing, mutation and molecular cloning can be of some help (Deshpande, 1999). Mass production of virulent and live propagules of fungi is another step in the process for which many standard methods are available (Jenkins and Goettel, 1997). Fungal entities need to be viable at the time of application (Bateman and Chapple, 2001; Wraight et al. 2001). Once cultivated on a suitable substrate, spores have to be collected and concentrated. Optimum drying rate and desired moisture level for each fungus has to be determined. Spore concentration will have to be formulated for stability during storage and application. Further, the formulation has to be stored for months to sustain agricultural market and pest control programs (Couch and Ignoffo, 1981). Desired physical characteristics need to be rendered by augmenting with wetters, stickers, dispersants, UV protectants etc. (Wraight et al. 2001). Identification of host pathogen recognition factors has major implications for commercial development of fungal principles.

Host specificity can be conferred to otherwise non-specific virulent isolates, by incorporating host specific antigens to fungal propagules. By incorporating different antigenic elements specific to different insect hosts, the host range of an isolate can be widened to a greater extent (Rath et al.,1996; St. Leger and Screen, 2001).

Ability of a biocontrol agent to persist on its host will have far-reaching implications in the efficacy of naturally occurring and introduced pathogens (Jacques, 1983; Meekes et al. 2000; Magan, 2001). The impact of rain on persistence of fungal propagules on hosts and their habitats has not been given much attention. Fate of fungal entities following field application can be investigated by variety of protocols that have been developed in recent times (Goettel et al. 2000; Bidochka, 2001).

The benefit of using a biological control agent on a target organism should always be linked with biosafety of non-target organisms (Goettel et al. 2001). Fungi have a wide spectra of host ranges among insects. Adverse effect on non-target organisms, competitive displacement of microorganisms and allergy to humans are matters of concern linked with utilization of microbes for pest control (Cook et al. 1996; Goettel and Jaronski, 1997; Hajek and Goettel, 2000; Goettel and Hajek, 2001; Holt and Hochberg, 2001; Waage et al. 2001; Strong and Pemberton, 2001; Lynch et al. 2001; Gassmann and Louda, 2001; Lonsdale et al. 2001; Neuenschwander and Markham, 2001). The risk of consumption of pathogens is addressed by standard maximum challenge tests which require acute mammalian oral toxicity/pathogenicity tests. Safety tests being observed for chemical pesticides are irrelevant for biopesticides. Stringent regulations have posed impediments to the development and use of microbial agents since these are required as components of IPM (Lacey and Goettel, 1995; Hajek and Goettel, 2000; Hopper 2001). Most regulatory authorities are following a 'tier' system of biosafety testing (Goettel and Jaronski, 1997). In the USA, fungi are tested in 5 tiers and those that clear the 1st tier are approved for sale. Some of the fungi cleared at first tier of tests include *Lagenidium giganteum* (Kerwin et al. 1994), *Hirsutella thompsonii* (Ignoffo et al. 1973), *Nomuraea rileyi* (Ignoffo, 1981) and *Verticillium lecanii* (Lisansky and Hall, 1983).

Although production of infective structures is not expensive, many additional downstream processes such as formulation, storage, packing and delivery substantially add to the cost (Wraight et al. 2001). Lately, bacterial *Bt* and *Bs* formulations, to some extent, are placed in a more lucrative position than representative fungi (Federici, 1995) with cost effective control in the field. Advantage of using fungi over *Bt* or *Bs* is that of effective development of resistant structures of fungi which can persist and recycle in the field. It appears that in many habitats only a single application is required per season. Large-scale field trials have not been conducted with fungi, so production costs and rates of application cannot be related to level and length of vector control.

Epilogue

Hitherto research has resulted in a treasure of knowledge on the occurrence, basic biology and biocontrol potential of numerous viruses, bacteria, fungi, protozoa and nematodes that are encountered as pests in nature. In case of fungi, the major advantage is their recycling ability. Search has to be intensified for isolation of newer strains that have better potential than already known fungal vector control agents. Newer combinations of existing candidates have to be tested through biotechnological means to overcome adverse field conditions and to increase the efficacy of these agents.

Efforts must be made to identify the fungal biocontrol agents correctly. Genera such as *Cordyceps, Aschersonia, Verticillium, Beauveria* and *Metarhizium* need urgent revision (Samson, 1995). The complete life cycle of species of these genera has to be elucidated as part of safety procedure, before they are released as mycoinsecticides. Destruction of habitats and ecosystems of these fungi has resulted in the disappearance of fungal germplasm of potential value. Therefore, the need to assess the biodiversity and creativity of these fungi on a priority basis cannot be overemphasized.

Acknowledgements
The authors sincerely thank Indian Council of Medical Research, New Delhi and the Ministry of Environment and Forests, Government of India, New Delhi, for financial support in the form of research grants.

References
Agarwal, G.P. and Rajak, R.C. (1988) Entomogenous Fungi in Retrospect and Prospect. In: Perspectives in Mycology and Plant Pathology. (eds.) Agnihotri, V.P., Sarabhoy, A.K. and Kumar, D. Malhotra Publishing House, New Delhi. 519–532.

Andreadis, T.G. and Weseloh, R.M. (1990) Discovery of *Entomophaga maimaiga* in North American gypsy moth *Lymantria dispar*. Proceedings of the National Academy of Sciences USA 67: 2461–2465.

Apperson, C.S., Federici, B.A., Stewart, W. and Tarver, F.R. (1992) Evidence for copepods *Acanthocyclops robbustus* and *Mesocyclops edax* as competent intermediate hosts for *Coelomomyces punctatus* during an epizootic in a larval population of the *Anopheles quadrimaculatus*. Journal of Invertebrate Pathology 60: 229–236.

Arthurs, S. and Thomas, M.B. (2001a) Effect of dose, pre-mortem host incubation temperature and thermal behaviour on host mortality, mycosis and sporulation of *Metarhizium anisopliae* var. *acridum* in *Schistocera gregaria*. Biocontrol Science and Technology 11: 411–420.

Arthurs, S. and Thomas, M.B. (2001b) Effect of temperature and relative humidity on sporulation of *Metarhizium anisopliae* var. *acridum* in mycosed cadavers of *Schistocera gregaria*. Journal of Invertebrate Pathology 78: 59–65.

Balazy, S. (1993) Flora of Poland. Fungi (Mycota) Vol. 24. Entomophthorales. Institute of Botany, Krakow, 356 pp.

Bateman, R. and Chapple, A. (2001) The Spray Application of Mycopesticide Formulations. In: Fungi as Biocontrol Agents: Progress, Problems and Potential. (eds), Butt, T.M., Jackson, C. and Magan, N. CAB International, London 289–309.

Beauvais, A., Latge, J.P., Vey, A. and Prevost, M.C. (1989) The role of surface components of the entomopathogenic fungus *Entomophaga aulicae* in the cellular immune response of *Galleria mellonella*. Journal of General Microbiology 135: 489–498.

Bellini, R., Mullens, B.A. and Jespersen, J.B. (1992) Infectivity of two members of the *Entomophthora muscae* complex (Zygomycetes: Entomophthorales) for *Musca domestica* (Dipt.: Muscidae). Entomophaga 37: 11–19.

Ben Ze'ev, I. and Uziel, A. (1979) *Moniellia costalis* (Fitch), a new host for *Zoophthora radicans* (Brefeld) Batko and *Entomophthora planchoniana* Cornu (Zygomycetes: Entomophthoraceae) in Israel. Phytoparasitica 7: 159–167.

Bidochka, M.J. (2001) Monitoring the fate of biocontrol fungi. In: *Fungi as Biocontrol Agents: Progress, Problems and Potential*. (eds.) Butt, T.M., Jackson, C. and Magan, N. CAB International, London 193–218.

Bidochka, M.J. and Hajek, A.E. (1996) Protoplast plasma membrane glycoproteins in two species of entomopathogenic fungi. Mycological Research 100: 1094–1098.

Bidochka, M.J., and Hajek, A.E. (1998) A nonpermissive Entomophthoralean fungal infection increases activation of insect prophenoloxidase. Journal of Invertebrate Pathology 72: 231–238.

Bisset, J. (1983) Notes on *Tolypocladium* and related genera. Canadian Journal of Botany 61: 1311–1329.

Bitton, S., Kenneth, R.G. and Ben Ze'ev, I. (1979) Zygospore overwintering and sporulative germination in *Triplosporium fresenii* (Entomophthoraceae) attacking *Aphis spiraecola* on citrus in Israel. Journal of Invertebrate Pathology 34: 295–302.

Boucias, D.G. and Pendland, J.C. (1991) Attachment of mycopathogens to cuticle—the initial event of mycoses in arthropod hosts. In: The Fungal Spore and Disease Initiation in Plants and Animals. (eds.) Cole, G.T. and Hoch, H.C. Plenum Press. New York and London 101–127.

Boykin, L.S., Campbell, W.V. and Beute, M.K. (1984) Effects of pesticides on *Neozygites floridana* (Entomophthorales: Entomophthoraceae) and arthropod predators attacking the twospotted spider mite (Acari: Tetranychidae) in North Carolina peanut fields. Journal of Economic Entomology 77: 969–975.

Brandenburg, R.L. and Kennedy, G.G. (1981) Overwintering of the pathogen *Entomophthora floridana* and its host, the twospotted spider mite. Journal of Economic Entomology 74: 428–431.

Brandenburg, R.L. and Kennedy, G.G. (1983) Interactive effects of selected pesticides on the two spotted spider mite and its fungal pathogen, *Neozygites floridana*. Entomologia Experimentalis et Applicata 34: 240–244.

Brefeld, O. (1870) Entwicklungsgeschichte der *Empusa muscae* und *Empusa radicans*. Botanische Zeitschrift 28: 161–166.

Brobyn, P.J. and Wilding, N. (1977) Invasive and developmental processes of Entomophthora species infecting aphids. Transactions of the British Mycological Society 69: 349–366.

Brown, G.C., Prochaska, G.L., Hildebrand, D.F., Nordin, G.L. and Jackson, D.M. (1995) Green leaf volatiles inhibit conidial germination of the entomopathogen *Pandora neoaphidis* (Entomophthorales: Entomophthoraceae). Environmental Entomology 24: 1637–1643.

Bruck, D.J. and Lewis, L.C. (2002) Rainfall and crop residue effects on soil dispersion and *Beauveria bassiana* spread to corn. Applied Soil Ecology 20: 183–190.

Burges, H.D. (1981) Safety, safety testing and quality control of microbial pesticides. In: Microbial Control of Pests and Plant Diseases. (ed) Burges, H.D. Academic press, New York. 737–767.

Butt, T.M., Jackson, C. and Magan, N. (2001) Introduction-Fungal Biological Control Agents: progress, problems and potential. In: Fungi as Biocontrol Agents: Progress, Problems and Potential. (eds), Butt, T.M., Jackson, C. and Magan. N. CAB International, London 1–8.

Carner, G.R. and Canerday, T.D. (1968) Field and laboratory investigations with *Entomophthora fresenii*, a pathogen of *Tetranychus* spp. Journal of Economic Entomology 61: 956–959.

Carruthers, R.I. and Haynes, D.L. (1985) Laboratory transmission in vivo of *Entomophthora muscae* (Entomophthorales: Entomophthoraceae) in the onion fly, *Delia antiqua* (Diptera: Anthomyiidae). Journal of Invertebrate Pathology 45: 282–287.

Carruthers, R.I. and Soper, R.S. (1987) Fungal diseases. In: Epizootiology of Insect Diseases. (eds), Fuxa, J.R. and Tanada, Y. Wiley Intersciences, New York 357 416.

Carruthers, R.I., Feng, Z., Ramos, M.E. and Soper, R.S. (1988) The effect of solar radiation on the survival of *Entomophaga grylli* conidia. Journal of Invertebrate Pathology 52: 154–162.

Carruthers, R.I., Haynes, D.L. and MacLeod, D.M. (1985) *Entomophthora muscae* (Entomophthorales: Entomophthoraceae) in the onion fly, *Delia antiqua* (Diptera: Anthomyiidae). Journal of Invertebrate Pathology 45: 81–93.

Carruthers, R.I., Humber, R.A. and Ramos, M.E. (1989) Biological control of rangeland grasshoppers using a fungal pathogen from Australia: a proposed field release of an exotic fungal pathogen. USDA APHIS Report, 23 pp.

Carruthers, R.I., Larkin, T.S., Firstencel, H. and Feng, Z. (1992) Influence of thermal ecology on the mycosis of a rangeland grasshopper. Ecology 73: 190-204.

Carruthers, R.I., Ramos, M.E., Larkin, T.S., Hostetter, D.L. and Soper, R.S. (1997) The *Entomophaga grylli* (Fresenius) Batko species complex: its biology, ecology and use for biological control of pest grasshoppers. Memoirs of the Entomological Society of Canada 171: 329–353.

Cohn, F. (1855) *Empusa muscae* und die Krankheit der Stubenfliegen. Hedwigia 1: 57–61.

Cook, R.J., Bruckart, W.L., Coulson, J.R., Goettel, M.S., Humber, R.A., Lumsden, R.D., Maddox, J.V., McManus, M.L., Moore, L., Meyer, S.F., Quimby, P.C., Stack, J.P. and Vaughn, J.L. (1996) Safety of microorganisms intended for pest control and plant disease control: a framework for scientific evaluation. Biological Control 7: 333–351.

Couch, J.N. (1935) A new saprophytic species of *Lagenidium,* with notes on other forms. Mycologia 27: 376–387.

Couch, T.L. and Ignoffo, C.M. (1981) Formulation of insect pathogens. In *Microbial Control of Pests and Plant Diseases 1970-1980.* (ed), Burges, H.D. Academic Press, London and New York 621–634.

Deshpande, M.V. (1999) Mycopesticide production by fermentation: potential and challenges. Critical Review of Microbiology 25: 229–243.

Dreyfuss, M., Harri, E., Hofmann, H., Pache, W. and Tscherter, H. (1976) Cyclosporin A and C, new metabolites from *Trichoderma polysporum*. European Journal of Applied Microbiolology 3: 123–135.

Dromph, K.M, Pell, J.K. and Eilenberg, J. (1997) Sporulation of *Erynia neoaphidis* from alate *Sitobion avenae*. In: Insect Pathogens and Insect Parasitic Nematodes. IOBC Bulletin 21: 91–94.

Dromph, K.M. (2001) Dispersal of entomopathogenic fungi by Collembolans. Soil Biology and Biochemistry 23: 2047–2051.

Dromph, K.M. and Vestergaard, S. (2002) Pathogenicity and attractiveness of entomopathogenic hyphomycete fungi to collembolans. Applied Soil Ecology 21: 197–210.

Dromph, K.M., Eilenberg, J. and Esbjerg, P. (2000) Natural occurrence of entomophthoralean fungi on Collembolans. IOBC/WPRS Bulletin 23: 147–149.

Dromph, K.M., Eilenberg, J. and Esbjerg, P. (2001) Natural occurrence of entomophthorale an fungi pathogenic to Collembolans. Journal of Invertebrate Pathology 78: 226–231.

Eilenberg, J. (1987a) The culture of *Entomophthora muscae* (C.) Fres. in the carrot fly (*Psila rosae* F.) and the effect of temperature on the pathology of the fungus. Entomophaga 32: 425–435.

Eilenberg, J. (1987b) Abnormal egg laying behaviour of female carrot flies (*Psila rosae*) induced by the fungus *Entomophthora muscae*. Entomologia Experimentalis et Applicata 52: 17–24.

Eilenberg, J., Bresciani, J. and Latge, J.P. (1990) Primary spore and resting spore formation in vitro of *Entomophthora schizophorae* and *E. muscae*, both members of the *E. muscae* complex. Cryptogamic Botany 1: 365–371.

Ekesi, S., Maniania, N.K. and Lohr, B. (1998) Pathogenicity of entomopathogenic fungi (Hyphomycetes) to the legume flower thrips, *Megalurothrips sjostedti* (Trybom) (Thysan., Thripidae). Journal of Applied Entomology 122: 629–634.

Ekesi, S., Maniania, N.K., Nyarko, K.A. and Akpa, A.D. (2001) Importance of timing of application of the entomopathogenic fungus, *Metarhizium anisopliae* for the control of the legume flower thrips, *Megalurothrips sjostedti* and its persistence on cowpea. Archiv Fur Phytopathologie Und Pflanzenschutz 33: 431–445.

Elliot, S.L., de Moraes, G.J., Delalibera I. Jr, da Silva, C.A.D., Tamai, M.A. and Mumford, J.D. (2000) Potential of the mite pathogenic fungus *Neozygites floridana* (Entomophthorales: Neozygitaceae) for control of the cassava green mite *Mononychellus tanajoa* (Acari: Tetranychidae). Bulletin of Entomological Research 90: 191–200.

Evans, H.C. (1989) Mycopathogens of insects of epigeal and aerial habitats. In: *Insect Fungus Interactions* (eds.) Wilding, N., Collins, N.M., Hammond, P.M. and Webber, J.F. Academic Press, London 205–238.

Evans, H.C. (1994) Spore germination in the entomopathogenic genus *Aschersonia*. Mycological Research 98: 165–168.

Evans, H.C. (1999) Biological control of weed and insect pests using fungal pathogens, with particular reference to Sri Lanka. Biocontrol News and Information 20: 63–68.

Evans, H.C. and Hywel-Jones, N.L. (1997) Entomopathogenic fungi. In: Soft Scale Insects—Their Biology, Natural Enemies and Control. (eds) Ben-Dov, Y. and Hodgson, C.J. Elsevier Science B.V. 3–27.

Evans, H.C. and Samson, R.A. (1987) Fungal pathogen of spiders. Mycologist 1: 152–159.

Federici, B.A. (1995) The future of microbial insecticides as vector control agents. Journal of American Mosquito Control Association 11: 260–268.

Federici, B.A. and Chapmann, H.C. (1977) *Coelomomyces dodgei*: establishment of an *in vivo* laboratory culture. Journal of Invertebrate Pathology 30: 288–297.

Ferron, P. (1981) Pest control by fungi *Beauveria* and *Metarhizium.* In: Microbial Control of Pests and Plant Diseases *1970-80.* (ed), Burges, H.D. Academic Press, London 465–482.

Frasen, J.J., Winkelman, K. and Lenteren, J.C. van. (1987) The different mortality at various life stages of the green house white fly, *Trialeurodes vaporariorum* (Hymenoptera: Aleyroridae), by infection with the fungus *Aschersonia aleyrodis* (Deuteromycotina: Coelomycetes). Journal of Invertebrate Pathology 50: 158–165.

Furlong, M.J., Pell, J.K. and Reddy, G.V.P. (1997) Premortality effects of *Zoophthora radicans* (Zygomycetes: Entomophthorales) infection on larval feeding rate and adult fecundity in *Plutella xylostella* (Lepidoptera: Yponomeutidae). Journal of Invertebrate Pathology 70: 214–220.

Gams, W. (1971) *Cephalosporium-artige* Schimmelpilze (Hyphomycetes). Gustav Fischer Verla, Stuttgart.

Gams, W. and Samson, R.A. (1985) Typification of *Aspergillus* and related teleomorph genera. In: *Advances in Penicillium and Aspergillus Systematics* (eds), Samson, R.A. and Pitt, J.I. Plenum Press. New York and London 23–29.

Gardner, J.M., Ram, R.C., Kumar, S. and Pillai, J.S. (1986) Field trials of *Tolypocladium cylindrosporum* against the larvae of *Aedes polynesiensis* breeding in crab holes in Fiji. Journal of American Mosquito Control Association 2: 292–295.

Gassmann, A. and Louda, S.M. (2001) *Rhinocyllus conicus*: initial evaluation and subsequent ecological impacts in North America. In: Evaluating Indirect Ecological Effects of Biological Control. (eds), Wajnberg, E. Scott, J.K. and Quimby, P.C. CAB International, London 147–183.

Giard, A. (1888) Note sur deux types remarquables d'Entomophthorees, *Empusa fresenii* Now. et *Basidiobolus ranarum* Eid. suivie de la description de quelques especes nouvelles. Compte Rendu des Seances de la Societe de Biologie Paris 40: 783–787.

Glare, T.R. and Milner, R.J. (1991) Ecology of entomopathogenic fungi. In: Handbook of Applied Mycology, Vol. 2: Humans, Animals, and Insects. (eds), Arora, D.K., Ajello, L. and Mukerji, K.G. Marcel Dekker, New York 547–612.

Glare, T.R., Milner, R.J. and Chilvers, G.A. (1989) Factors affecting the production of resting spores by *Zoophthora radicans* in the spotted alfalfa aphid *Therioaphis trifolii* f. *maculata*. Canadian Journal of Botany 67: 848–855.

Glare, T.R., Milner, R.J., Chilvers, G.A., Mahon, R.J. and Brown, W.V. (1987) Taxonomic implications of intraspecific variation amongst isolates of the aphid pathogenic fungi *Zoophthora radicans* Brefeld and *Z. phalloides* Batko (Zygomycetes: Entomophthorales). Australian Journal of Botany 35: 49–67.

Goettel, M.S. (1987a) Field incidence of mosquito pathogens and parasites in Central Alberta. Journal of American Mosquito Control Association 3: 231–238.

Goettel, M.S. (1987b) Preliminary field trials with the entomopathogenic Hyphomycete *Tolypocladium cylindrosporum* in Central Alberta. Journal of American Mosquito Control Association 3: 239–245.

Goettel, M.S. and Hajek, A.E. (2001) Evaluation of non-target effects of pathogens used for management of arthropods. In: Evaluating Indirect Ecological effects of Biological Control (eds), Wajnberg, E., Scott, J.K. and Quimbly, P.C. CAB international. London 81–97.

Goettel, M.S. and Jaronski, S.T. (1997) Safety and registration of microbial agents for control of grasshoppers and locusts. Memoirs of Entomological Society of Canada 171: 83–99.

Goettel, M.S. Hajek, A.E., Siegel, J.P. and Evans, H.C. (2001) Safety of fungal biocontrol agents. In: Fungi as Biocontrol Agents: Progress, Problems and Potential. (eds), Butt, T.M., Jackson, C. and Magan, N. CAB International, London 347–375.

Goettel, M.S., Inglis, G.D. and Wraight, S.P. (2000) Fungi. In: Field manual of techniques in Invertebrate pathology (eds), Lacey, L.A. and Kaya, H.K. Kluwer Academic Publishers. Netherlands 255–282.

Goettel, M.S., Poprawski, T.J., Vandenberg, J.D., Li, Z. and Roberts, W.D. (1990) Safety to nontarget invertebrates of fungal biocontrol agents. In: Safety of Microbial Insecticides. (eds), Laird, M. Lacey, L.A. and Davidson, E.W. CRC Press, Boca Raton, Florida 209–232.

Gustafsson, M. (1965) On species in the genus *Entomophthora* Fres. in Sweden. I. Classification and distribution. Lantbrukshogskolans Annaler 31: 103–212.

Hajek, A.E. (1997a) Ecology of terrestrial fungal entomopathogens. Advances in Microbial Ecology 15: 193 249.

Hajek, A.E. (1997b) *Entomophaga maimaiga* reproductive output is determined by the spore type initiating an infection. Mycological Research 101: 971–974.

Hajek, A.E. and Goettel, M.S. (2000) Guidelines for evaluating effects of entomopathogens on non-target organisms. In: Field Manual Techniques in Invertebrate Pathology. (eds), Lacey, L.A. and Kaya, H.K. Kluwer Academic Publishers. Netherlands 847–868.

Hajek, A.E., Butler, L. and Wheeler, M.W. (1995a) Laboratory bioassays testing the host range of the gypsy moth fungal pathogen *Entomophaga* maimaiga. Biological Control 5: 530–544.

Hajek, A.E., Humber, R.A. and Elkinton, J.S. (1995b) Mysterious origin of *Entomophaga maimaiga* in North America. American Entomologist 41: 31–42.

Hajek, A.E., Humber, R.A., Elkinton, J.S., May, B., Walsh, S.R.A. and Silver, J.C. (1990) Allozyme and RFLP analyses confirm *Entomophaga maimaiga* responsible for 1989 epizootics in North American gypsy moth populations. Proceedings of the National Academy of Sciences USA 87: 6979–6982.

Hall, R.A. and Papierok, B. (1982) Fungi as *Biological Control* agents of arthropods of agricultural and medicinal importance. Parasitology 84: 205–240.

Hall, R.A., Hussey, N.W. and Mariau, D. (1980) Results of survey of *Biological Control* agents of coconut mite, *Eriophyes guerreronis*. Oleagineux 35: 395–400.

Hallsworth, J.E. and Magan, N. (1999) Water and Temperature Relations of Growth of the Entomogenous Fungi *Beauveria bassiana, Metarhizium anisopliae* and *Paecilomyces farinosus*. Journal of Invertebrate Pathology 74: 261–266.

Hatting, J.L., Humber, R.A., Poprawski, T.J. and Miller, R.M. (1999) A survey of fungal pathogens of aphids from South Africa with special reference to cereal aphids. Biological Control 16: 1–12.

Hawksworth, D.L., Kirk, P.M., Sutton, B.C. and Pegler, D.N. (1995) Ainsworth & Bisby's Dictionary of Fungi. 8th edn. CAB international. Wallingford.

Hawksworth, D.L. (1991) The fungal dimension of biodiversity: magnitude, significance and conservation. Mycological Research 95: 641–655.

Hawksworth, D.L. (2002) Why Study Tropical Fungi? In: Tropical Mycology, Vol. 2, Micromycetes. (eds) Watling, R., Frankland, J.C., Ainsworth, A.M., Isaac, S. and Robinson, C.H.. CAB International. London 1–11.

Hemmati, F. (1999) Aerial dispersal of the entomopathogenic fungus *Erynia neoaphidis*. Ph.D. thesis, University of Reading, UK, 186 pp.

Holt, R.D. and Hochberg, M.E. (2001) Indirect interactions, community modules and biological control: a theoretical perspective. In: Evaluating Indirect Ecological Effects of Biological Control. (eds), Wajnberg, E. Scott, J.K. and Quimby, P.C. CAB International, London 13–37.

Hopper, K.R. (2001) Research needs concerning non-target impacts of Biological Control introductions. In: Evaluating Indirect Ecological Effects of Biological Control. (eds), Wajnberg, E., Scott, J.K. and Quimby, P.C. CAB International, London 38–56.

Humber, R.A. (1989) Synopsis of a revised classification for the Entomophthorales (Zygomycotina). Mycotaxon 34: 441–460.

Hunter, D.M., Milner, R.J., Spurgin, P.A. (2001) Aerial treatment of the Australian plague locust, *Chortoicetes terminifera* (Orthoptera: Acrididae) with *Metarhizium anisopliae* (Deuteromycotina: Hyphomycetes). Bulletin of Entomological Research 91: 93–99.

Hywel-Jones, N.L. (1996) *Cordyceps myrmecophila*-like fungi infecting ants in the leaf litter of tropical forest in Thailand. Mycological Research 100: 613–619.

Hywel-Jones, N.L. and Evans, H.C. (1993) Taxonomy and ecology *of Hypocrella discoidea* and its anamorph, *Aschersonia samoensis*. Mycological Research 97: 871–876.

Ignoffo, C.M. (1981) The fungus *Nomuraea rileyi* as a microbial insecticide. In: Microbial Control of Pests and Plant Diseases 1970 80. (ed). Burges, H.D. Academic Press, London 513–538.

Ignoffo, C.M., Barker, W.M. and McCoy, C.W. (1973) Lack of per os toxicity or pathogenicity in rats fed the fungus *Hirsutella thompsonii*. Entomophaga 18: 333–335.

Inglis, G.D., Goettel, M.S., Butt T.M. and Strasser, H. (2001) Use of Hyphomycetous Fungi for Managing Insect Pests. In: Fungi as Biocontrol Agents: Progress, Problems and Potential. (eds), Butt, T.M., Jackson, C. and Magan, N. CAB International, London 23–69.

Jackson, C.G., Linnett, P.E., Beechey, R.B. and Henderson, P.J.F. (1979) Purification and preliminary structural analysis of the efrapeptins, a group of antibiotics that inhibit the mitochondrial adenosine triphosphatase. Biochemical Society Transactions 59th Meeting, London pp: 224–226.

Jacques, R.P. (1983) The potential of pathogens for pest control. Agriculture, Ecosystems and Environment 10: 101–126.

Jenkins, N.E. and Goettel, M.S. (1997) Methods for mass production of microbial control agents of grasshoppers and locusts. Memoirs of the Entomological Society of Canada 171: 37–48.

Jensen, A.B. and Eilenberg, J. (2000) Genetic variation within the insect pathogenic fungus genus *Entomophthora*, focusing on the *E. muscae* complex, using PCR RFLP of the ITS II and the LSU rDNA. Mycological Research 105: 307–312.

Keilin, D. (1921). On a new type of fungus: *Coelomomyces stegomyiae*, n.g., n.sp., parasitic in the body cavity of the larva of *Stegomyia scutellaris* Walker (Diptera, Nematocera, Culicidae). Parasitology 13: 225–234.

Keller, S. (1984) *Entomophthora muscae* als Artenkomplex. Mitteilungen der Schweizerischen. Entomologischen Gesellschaft 57: 131–132.

Keller, S. (1987) Arthropod pathogenic Entomophthorales of Switzerland. I. *Conidiobolus*, *Entomophaga* and *Entomophthora*. Sydowia 40: 122–167.

Keller, S. (1991) Arthropod pathogenic Entomophthorales of Switzerland. II. *Erynia*, *Eryniopsis*, *Neozygites*, *Zoophthora* and *Tarichium*. Sydowia 43: 39–122.

Keller, S. (1997) The genus *Neozygites* (Zygomycetes, Entomophthorales) with special reference to species found in tropical regions. Sydowia 49: 118–146.

Keller, S. and Steenberg, T. (1997) *Neozygites sminthuri* sp. nov. (Zygomycetes: Entomophthorales) a pathogen of the springtail *Sminthurus viridis* L. (Collembola: Sminthuridae). Sydowia 49: 21–24.

Keller, S., Kalsbeek, V. and Eilenberg, J. (1999) Redescription of *Entomophthora muscae* (Cohn) Fresenius. Sydowia 51: 197–209.

Kendrick. B. (2002) The Fifth Kingdom. Mycologue Publications, Sidney, Canada.

Kenneth, R., Wallis, G., Gerson, U. and Plaut, H.N. (1972) Observations and experiments on *Triplosporium floridanum* (Entomophthorales) attacking spider mites in Israel. Journal of Invertebrate Pathology 19: 366–369.

Kerwin, J.L., Dritz, D.A., and Washino, R.K. (1988) Nonmammalean safety tests for *Lagenidium giganteum* (Oomycetes: Lagenidiales). Journal of Economic Entomology 81: 158–171.

Kerwin, J.L., Dritz, D.A., and Washino, R.K. (1994) Pilot scale production and application in wildlife ponds of *Lagenidium giganteum* (Oomycetes: Lagenidiales). Journal of American Mosquito Control Association 10: 451–455.

Kerwin, J.L. and Petersen, E.E. (1997) Fungi: Oomycetes and Chytridiomycetes. In: Manual Techniques in Insect Pathology (ed), Lacey, L.A. Academic Press, London 252–268.

Kerwin, J.L. and Washino, R.K. (1986) Ground and aerial application of sexual and asexual stages of *Lagenidium giganteum* (Oomycetes: Lagenidiales) for mosquito control. Journal of American Mosquito Control Association 2: 182–189.

Kerwin, J.L. and Washino, R.K. (1987) Ground and aerial application of asexual stages of *Lagenidium giganteum* for control of mosquitoes associated with rice culture in Central Valley of California. Journal of American Mosquito Control Association 3: 59–64.

Kerwin, J.L. and Washino, R.K. (1988) Field evaluation of *Lagenidium giganteum* (Oomycetes: Lagenidiales) and a description of a natural epizootic involving a new isolate of the fungus. Journal of Medical Entomology 25: 452–460.

Kerwin, J.L., Dritz, D.A. and Washino, R.K. (1990) Confirmation of the safety of *Lagenidium giganteum* (Oomycetes: Lagenidiales) to mammals. Journal of Economic Entomology 83: 374–376.

Klingen, I., Eilenberg, J. and Meadow, R. (2002d) Effects of farming system, field margins and bait insect on the occurrence of insect pathogenic fungi in soils. Agriculture, Ecosystems and Environment 91: 191–198.

Klingen, I., Hajek, A., Meadow, R. and Renwick, J.A.A. (2002c) Effect of brassicaceous plants on the survival and infectivity of insect pathogenic fungi. Biocontrol 47: 411–425.

Klingen, I., Meadow, R. and Aandal, T. (2002b) Mortality of *Delia floralis, Galleria mellonella* and *Mamestra brassicae* treated with insect pathogenic hyphomycetous fungi. Journal of Applied Entomology 126: 231–237.

Klingen, I., Salinas, S.H. and Meadow, R. (2002a) Checklist of naturally occuring pathogens of insects and mites in Norway. Norw. J. Entomol. 49: 23–28.

Kobayasi, Y. and Shimizu, D. (1982) Monograph of the genus *Torrubiella.* Bulletin of the National Science Museum, Tokyo. Series. B. (Botany) 8: 43–78.

Kramer, J.P. and Steinkraus, D.C. (1981) Culture of *Entomophthora muscae* in vivo and its infectivity for six species of muscoid flies. Mycopathologia 76: 139–143.

Krasnoff, S.B., Gupta, S., St. Leger, R.J., Renwick, J.A.A. and Roberts, D.W. (1991) Antifungal and insecticidal properties of the efrapeptins: metabolites of the fungus *Tolypocladium niveum.* Journal of Invertebrate Pathology 58: 180–188.

Krasnoff, S.B., Watson, D.W., Gibson, D.M. and Kwan, E.C. (1995) Behavioral effects of the entomopathogenic fungus, *Entomophthora muscae*, on its host *Musca domestica*: postural changes in dying hosts and gated pattern of mortality. Journal of Insect Physiology 41: 895 903.

Kuntz, W.A. (1925) Aphid diseases. In: Florida Agricultural Experimental Station report, Gainesville, Florida, pp. 46R 49R.

Lacey, L.A. and Brooks, W.M. (1997) Initial handling and diagnosis of diseased insects. In: Manual Techniques in Insect Pathology. (ed) Lacey, L.A., Academic Press, New York.

Lacey, L.A. and Goettel, M.S. (1995) Current developments in microbial control of insect pests and prospects for the early 21st century. Entomophaga 40: 3–27.

Lacey, L.A., Frutos, R., Kaya, H.K. and Vail, P. (2001) Insect pathogens as Biological Control agents: Do they have a future? Biological Control 21: 230–248.

Lam, T.N.C., Goettel, M.S. and Soares, G.G. Jr. (1988) Host records for the entomopathogenic hyphomycete *Tolypocladium cylindrosporum.* Florida Entomologist 71: 86–89.

Lambiase, J.T. and Yendol, W.G. (1977) The fine structure of *Entomophthora apiculata* and its penetration of *Trichoplusia ni.* Canadian Journal of Microbiology 23: 452–464.

Leite, L.G., Alves, S.B., Wraight, S.P., Galaini Wraight, S. and Roberts, D.W. (1996a) Comportamento de isolados do fungo *Zoophthora radicans* sobre *Empoasca kraemeri*. Ecosistema 21: 16–22.

Leite, L.G., Alves, S.B., Wraight, S.P., Galaini Wraight, S., Roberts, D.W. and Magalhaes, B.P. (1996b) Conidiogenese de *Zoophthora radicans* (Brefeld) Batko sobre Empoasca kraemeri (Ross and Moore) a diferentes temperaturas. Arquivos do Instituto Biologico, Sao Paulo 63: 47–53.

Leite, L.G., Alves, S.B., Wraight, S.P., Wraight, S.G. and Roberts, D.W. (1996c) Habilidade de infecca de isolados de *Zoophthora radicans* sobre *Empoasca kraemeri*. Scientia Agricola 53: 152–158.

Lisansky, S.B. and Hall, R.A. (1983) Fungal control of insects. In: The Filamentous Fungi Volume IV: Fungal Technology. (eds), Smith, J.E., Berry, D.R. and Kristiansen, B. Edward Arnold (Publishers) Limited. London. 327–345.

Lonsdale, W.M., Briese D.T. and Cullen, J.M. (2001) Risk analysis and weed biological control. In: Evaluating Indirect Ecological Effects of Biological Control. (eds), Wajnberg, E., Scott, J.K. and Quimby, P.C. CAB International, London 185–210.

Lucarotti, C.J. (1992) Invasion of *Aedes aegypti* ovaries by *Coelomomyces stegomyiae*. Journal of Invertebrate Pathology 60: 176–184.

Lucarotti, C.J. and Andreadis, T.G. (1995. Reproductive strategies and adaptations for survival among obligatory and fungal parasites of mosquitoes: A comparative analysis of *Amblyospora* and *Coelomomyces*. Journal of American Mosquito Control Association 11: 111–121.

Lucarotti, C.J. and Klein, M.B. (1988) Pathology of *Coelomomyces stegomyiae* in adult *Aedes aegypti* ovaries. Canadian Journal of Botany 66: 877–884.

Lynch, L.D., Hokkanen, H.M.T., Babendreier, D., Bigler, F., Burgio, G., Gao, Z.H., Kuske, S. Loomans, A., Hokkanen, M.I., Thomas, M.B., Tommasini, G., Waage, J.K., Lenteren, J.C.van and Zeng, Q.Q. (2001) Insect *Biological Control* and non-target effects: a european perspective. In: Evaluating Indirect Ecological Effects of Biological Control. (eds), Wajnberg, E., Scott, J.K. and Quimby, P.C. CAB International, London 99–125.

MacLeod, D.M. (1963) Entomophthorales infections. In: Insect Pathology: an Advanced Treatise, Vol. 2. (ed.) Steinhaus, E.A., Academic Press, New York, 189–232.

MacLeod, D.M., Muller Kogler, E. and Wilding, N. (1976). *Entomophthora* species with *E. muscae*-like conidia. Mycologia 68: 1–29.

Magalhaes, B.P., Butt, T.M., Humber, R.A., Shields, E.J. and Roberts, D.W. (1990) Formation of appressoria in vitro by the entomopathogenic fungus *Zoophthora radicans* (Zygomycetes: Entomophthorales). Journal of Invertebrate Pathology 55: 284–288.

Magalhaes, B.P., Humber, R.A., Shields, E.J. and Roberts, D.W. (1991a) Effects of environment and nutrition on conidium germination and appressorium formation by *Zoophthora radicans* (Zygomycetes: Entomophthorales): a pathogen of the potato leaf hopper (Homoptera: Cicadellidae). Environmental Entomology 20: 1460–1468.

Magalhaes, B.P., Lord, J.C., Wraight, S.P., Daoust, R.A. and Roberts, D.W. (1988) Pathogenicity of *Beauveria bassiana* and *Zoophthora radicans* to the coccinellid predators *Coleomegilla maculata* and *Eriopis connexa*. Journal of Invertebrate Pathology 52: 471–473.

Magalhaes, B.P., Wayne, R., Humber, R.A., Shields, E.J. and Roberts, D.W. (1991b) Calcium regulated appressorium formation of the entomopathogenic fungus *Zoophthora radicans*. Protoplasma 160: 77–88.

Magan, N. (2001) Physiological Approaches to Improving the Ecological Fitness of Fungal Bicontrol Agents. In: Fungi as Biocontrol Agents: Progress, Problems and Potential. (eds.) Butt, T.M. Jackson, C. and Magan, N. CAB International, London 239–251.

Mains, E.B. (1959a) North America species of *Aschersonia* parasitic on Aleyrodidae. Journal of Insect Pathology 1: 43–47.

Mains, E.B. (1959b) Species of *Aschersonia* (Sphaeropsidales). Lloydia 22: 215–221.

Mains, E.B. (1959c) Species of *Hypocrella*. Mycopathologia et Mycologia applicata 11: 311–326.

Mains, E.B. (1950a) The genus *Gibellula* on spiders in North America. Mycologia 42: 306–321.

Mains, E.B. (1950b) Entomogenous species of *Akanthomyces, Hymenostilbe* and *Insecticola* in North America. Mycologia 42: 566–589.

Mains, E.B. (1958) North american entomogenous species of *Cordyceps*. Mycologia 50: 169–222.

McCabe, D.E., Humber, R.A. and Soper, R.S. (1984) Observation and interpretation of nuclear reductions during maturation and germination of Entomophthoralean resting spores. Mycologia 76: 1104–1107.

McCoy, C.W. and Couch, T.L (1978) *Hirsutella thompsonii,* a potential mycoacaricide. Developments in Industrial Microbiology, New York 20: 89–96.

McGuire, M.R., Maddox, J.V. and Ambrust, E.J. (1987a) Host range studies of an *Erynia radicans* strain (Zygomycetes: Entomophthorales) isolated from *Empoasca fabae* (Homoptera Cicadellidae). Journal of Invertebrate Pathology 50: 75–77.

McGuire, M.R., Maddox, J.V. and Ambrust, E.J. (1987b) Effect of temperature on distribution and success of introduction of an *Empoasca fabae* (Homoptera: Cicadellidae) isolate of *Erynia radicans* (Zygomycetes: Entomophthoraceae). Journal of Invertebrate Pathology 50: 291–301.

McLeod, P.J., Steinkraus, D.C., Correll, J.C. and Morelock, T.E. (1998) Prevalence of *Erynia neoaphidis* (Entomophthorales: Entomophthoraceae) infections of green peach aphid (Homoptera: Aphididae) on spinach in the Arkansas River Valley. Environmental Entomologist 27: 796–800.

Meekes, E.T.M., Fransen, J.J. and Lenteren, J.C. van. (2002) Pathogenicity of *Aschersonia* spp. against whiteflies *Bemisia argentifolii* and *Trialeurodes vaporariorum*. Journal of Invertebrate Pathology 81: 1–11.

Meekes, E.T.M., Voorst, S. van, Joosten, N.N. and Fransen, J.J. (2000) Persistence of the fungal whitefly pathogen, *Aschersonia aleyrodis*, on different plant species. Mycological Research 104: 1234–1240.

Mietkiewski, R., Balazy, S. and van der Geest, L.P.S. (1993) Observations on a mycosis of spider mites (Acari: Tetranychidae) caused by *Neozygites floridana* in Poland. Journal of Invertebrate Pathology 61: 317–319.

Mietkiewski, R.T., van der Geest, L.P.S. and Balazy, S. (1986) Preliminary notes on the pathogenicity of some *Erynia* strains (Mycophyta: Entomophthoraceae) towards larvae of *Pieris brassicae*. Journal of Applied Entomology. 102: 499–504.

Milner, R.J. and Holdom, D.G. (1986) First record of *Neozygites fresenii* (Nowakowski) Batko, a fungal pathogen of aphids, in Australia. Journal of the Australian Entomological Society 25: 85–86.

Milner, R.J. and Lutton, G.G. (1983) Effect of temperature on *Zoophthora radicans* (Brefeld) Batko: an introduced microbial control agent of the spotted alfalfa aphid, *Therioaphis trifolii* (Monell) f. *maculata*. Journal of the Australian Entomological Society 22: 167–173.

Milner, R.J. and Mahon, R.J. (1985) Strain variation in *Zoophthora radicans*, a pathogen on a variety of insect hosts in Australia. Journal of the Australian Entomological Society 24: 195–198.

Milner, R.J., Holdom, D.G. and Glare, T.R. (1984) Diurnal patterns of mortality in aphids infected by Entomophthoralean fungi. Entomologia Experimentalis et Applicata 36: 37–42.

Minter, D.W. and Brady, B.L. (1980) Mononematous species of *Hirsutella*. Transactions of the British Mycological Society 74: 271–282.

Moller, A.P. (1993) A fungus infecting domestic flies manipulates sexual behaviour of its hosts. Behavioral Ecology and Sociobiology 33: 403–407.

Morgan, L.W. (1994) Survival, germination responses and infectivity of conidia of *Erynia neoaphidis* (Zygomycetes: Entomophthorales). Ph.D. thesis, University of Cardiff, 192 pp.

Morgan, L.W., Boddy, L., Clark, S.J. and Wilding, N. (1995) Influence of temperature on germination of primary and secondary conidia of *Erynia neoaphidis* (Zygomycetes: Entomophthorales). Journal of Invertebrate Pathology 65: 132–138.

Mullens, B.A. (1989) Cross transmission of *Entomophthora muscae* (Zygomycetes: Entomophthoraceae) among naturally infected muscoid fly (Diptera: Muscidae) hosts. Journal of Invertebrate Pathology 53: 272–275.

Mullens, B.A. and Rodriguez, J.L. (1985) Dynamics of *Entomophthora muscae* (Entomophthorales: Entomophthoraceae) conidial discharge from *Musca domestica* (Diptera: Muscidae) cadavers. Environmental Entomology 14: 317–322.

Mullens, B.A., Rodriguez, J.L. and Meyer, J.A. (1987) An epizoological study of *Entomophthora muscae* in muscoid fly populations on Southern California poultry facilities, with emphasis on *Musca domestica*. Hilgardia 55: 1–41.

Neuenschwander, P. and Markham, R. (2001) Biological Control in Africa and its possible effects on biodiversity. In: Evaluating Indirect Ecological Effects of Biological Control. (eds), Wajnberg, E., Scott, J.K. and Quimby, P.C. CAB International, London, pp: 127–146.

Nielsen, C., Hajek, A.E., Humber, R.A. and Eilenberg, J. (1998) Soil a natural source of entomophthoralean fungi infecting aphids. In: Insect Pathogens and Insect Parasitic Nematodes, IOBC Bulletin 21: 45–48.

Nowakowski, L. (1883) Entomophthoreae. Przyczynek do znajmosci pasorzytnych grzybkow sprawiajacych pomor owadow. Pamietn. Akad. Umiejetn. Krakowie, Wydz. Mat. Przyr. 8: 153–183.

Padua, L.E., Whisler. H.C., Gabriel. B.P. and Zeboid, S.L. (1986) In vivo culture and life cycle of Coelomomyces stegomyiae. Journal of Invertebrate Pathology 68: 284–288.
Papierok, B., Valadao, B., Torres, L. and Arnault, M. (1984) Contribution to the study of the specificity of the entomopathogenic fungus *Zoophthora radicans* (Zygomycetes: Entomophthorales). Entomophaga 29: 109–119.
Pell, J.K., Eilenberg, J., Hajek A.E. and Steinkraus, D.C. (2001) Biology, Ecology and Pest Management Potential of Entomophthorales. In: Fungi as Biocontrol Agents: Progress, Problems and Potential. (eds), Butt, T.M., Jackson, C. and Magan, N. CAB International, London 71–153.
Pell, J.K., Wilding, N., Player, A.L. and Clark, S.J. (1993) Selection of an isolate of *Zoophthora radicans* (Zygomycetes: Entomophthorales) for biocontrol of the diamondback moth *Plutella xylostella* (Lepidoptera: Yponomeutidae). Journal of Invertebrate Pathology 61: 75–80.
Perry, D.F., Tyrrell, D. and Delyzer, A.J. (1982) The mode of germination of *Zoophthora radicans* zygospores. Mycologia 74: 549–554.
Petch, T. (1921) Studies on entomogenous fungi: II. The genera of *Hypocrella* and *Aschersonia*. Annals of the Royal Botanic Gardens. Peradeniya 7: 167–278.
Petch, T. (1923) Studies in entomogenous fungi. III. *Torrubiella*. Transactions of the British Mycological Society 9: 108–128.
Petch, T. (1931) Notes on entomogenous fungi. Transactions of the British Mycological Society 16: 55–75.
Petch, T. (1932). Gibellula. Annals of Mycology 30: 386–393.
Pickford, R. and Riegert, P.W. (1964) The fungus disease caused by *Entomophthora grylli* Fres., and its effects on grasshopper populations in Saskatchewan in 1963. The Canadian Entomologist 96: 1158–1166.
Pinnock, D.E., Garcia, R. and Cubbin, C.M. (1973) *Beauveria tenella* as a control agent of mosquito larvae. Journal of Invertebrate Pathology 22: 143–147.
Prior, C. (1990) The biological basis for regulating the release of microorganisms, with particular reference to the use of fungi for pest control. Aspects of Applied Biology 24: 231–238.
Ramakers, P.M.J. and Samson, R.A. (1984) *Aschersonia aleyrodis*, a fungal pathogen of whitefly. II. Application as a biological insecticide in glasshouses. Journal of Applied Entomology 97: 1–8.
Ramaseshiah, G. (1968) *Entomophthora fresenii*, parasitic on aphids in India. Journal of Invertebrate Pathology 10: 439–441.
Ramaseshiah, G. (1971) Occurrence of an *Entomophthora* on tetranychid mites in India. Journal of Invertebrate Pathology 18: 421–424.
Ramoska, W.A., Hajek, A.E., Ramos, M.E. and Soper, R.S. (1988) Infection of grasshoppers by members of the *Entomophaga grylli* species complex. Journal of Invertebrate Pathology 52: 309–313.
Rath, A.C., Guy, P.L. and Webb, W.R. (1996) *Metarhizium* spore surface antigens are correlated with pathogenicity. Mycological Research 100: 57–62.
Reddy, G.V.P., Furlong, M.J., Poppy, G.M. and Pell, J.K. (1998). *Zoophthora radicans* infection inhibits the response to and the production of sex pheromone in the diamondback moth. Journal of Invertebrate Pathology 72: 167–169.
Remaudiere, G. (1977) Sur quelques Aphidoidea de la Polynesie Francaise. Bulletin de Societe Entmologique France 82: 151–155.
Remaudiere, G. and Keller, S. (1980) Revision systematique des genres d'Entomophthoraceae potentialite entomopathogene. Mycotaxon 11: 323–338.
Roberts, D.W. and Humber, R.A. (1981) Entomogenous fungi. In Biology Conidial Fungi, Volume 2. (eds), G.T. Cole and B. Kendrick, Academic Press, New York, 201–236.
Rolf, P.H. and Fawcett, H.S. (1913) Fungus diseases of scale insects and white fly. University of Florida. Agricultura Experiment Station (Gainesville) Bulletin 119: 71–82.
Rombach, M.C., Gillespie, A.T. (1988) Entomogenous Hyphomycetes for insect and mite control on greenhouse crops. Biocontl. News and Information 9: 7–18.
Rossman, A.Y. (1994) A strategy for an all-taxa inventory of fungal biodiversity. In: Biodiversity and terrestrial ecosystems. (eds), Peng, C.I. and Chou, C.H., Bull. Academia Sinica, Monograph series No. 14, Institute of Botany, Taipei, 169–194.
Roy, H.E., Pell, J.K. and Alderson, P.G. (2001) Targeted dispersal of the aphid pathogenic fungus *Erynia neoaphidis* by the ahid predator *Coccinella septempunctata*. Biocontrol Science and Technology 11: 99–110.

Roy, H.E., Pell, J.K., Clark, S.J. and Alderson, P.G. (1998) Implications of predator foraging on aphid pathogen dynamics. Journal of Invertebrate Pathology 71: 236–247.

Roy, H.E. (1997). Interactions between aphid predators and the entomopathogenic fungus *Erynia neoaphidis*. Ph.D. thesis, University of Nottingham, 206 pp.

Roy, H.E. and Pell, J.K. (2000) Interactions between entomopathogenic fungi and other natural enemies: Implications for biological control. Biocontrol Science and Technology 10: 737–752.

Roy, H.E., Pell, J.K. and Alderson, P.G. (1999) Effects of fungal infection on the alarm response of pea aphids. Journal of Invertebrate Pathology 74: 69–75.

Samson, R.A. (1974) *Paecilomyces* and some allied hyphomycetes. Studies in Mycology 6: 1–117.

Samson, R.A. (1995) Constraints associated with the taxonomy of biocontrol fungi. Canadian Journal of Botany 73(s): s83–s88.

Samson, R.A. and Brady, B.L. (1982) *Akanthomyces novoguineensis* sp. nov. Transactions of the British Mycological Society 79: 571–572.

Samson, R.A. and Evans, H.C. (1992) New species of *Gibellula* on spiders (Araneida) from South America. Mycologia 84: 300–314.

Samson, R.A. and Soares, G.G. Jr. (1984) Entomopathogenic species of Hyphomycete genus *Tolypocladium*. Journal of Invertebrate Pathology 43: 133–139.

Samson, R.A., Evans, H.C., Latge, J.P. (1988) Atlas of Entomopathogenic Fungi. Springer-Verlag, Berlin.

Samson, R.A., Ramakers, P.M.J. and Oswald, T. (1979) *Entomophthora thripidum*, a new fungal pathogen of *Thrips tabaci*. Canadian Journal of Botany 57: 1317–1323.

Sawyer, A.J., Griggs, M.H. and Wayne, R. (1994) Dimensions, density and settling velocity of Entomophthoralean conidia: implications for aerial dissemination of spores. Journal of Invertebrate Pathology 63: 43–55.

Sawyer, A.J., Ramos, M.E., Poprawski, T.J., Soper, R.S. and Carruthers, R.I. (1997) Seasonal patterns of cadaver persistence and sporulation by the fungal pathogen *Entomophaga grylli* (Fresenius) Batko (Entomophthoraceae) *infecting Camnula pellucida* (Scudder) (Orthoptera: Acrididae). Memoirs of the Entomological Society of Canada 171: 355–374.

Seifert, K.A. (1985) A monograph of *Stilbella* and allied hyphomycetes. Studies in Mycology 27: 1–235.

Siegel, J.P. and Shadduck, J.A. (1987) Some observation on the safety of entomopathogen *Lagenidium giganteum* to mammals. Journal of Economic Entomology 80: 994–997.

Silvie, P. and Papierok, B. (1991) Les ennemis naturels d'insectes du cotonnier au Tchad: premieres donnees sur les champignons de l'ordre des Entomophthorales. Coton et Fibres Tropicale 46: 293–308.

Six, D.L. and Mullens, B.A. (1997) Distance of conidial discharge of *Entomophthora muscae* and *Entomophthora schizophorae*. Journal of Invertebrate Pathology 67: 253–258.

Smitley, D.R., Kennedy, G.G. and Brooks, W.M. (1986) Role of the entomogenous fungus, *Neozygites floridana*, in population declines of the two-spotted spider mite, *Tetranychus urticae*, on field corn. Entomologia Experimentalis et Applicata 41: 255–264.

Soars, G.G. Jr., Pinnock, D.E. and Samson, R.A. (1979) *Tolypocladium*, a new fungal pathogen of mosquito larvae with the promise for use microbial control. Proceedings of the California Mosquito and Vector Control Association 47: 51–54.

Soars, G.G. Jr., Riba, G., Caudal, A. and Vincent, J.J. (1985) Comparative studies of eleven isolates of the fungal entomopathogen *Tolypocladium cylindrosporum* and two isolates of *Tolypocladium extiguens*. Journal of Invertebrate Pathology 46: 115–120.

Soper, R.S., Shimazu, M., Humber, R.A., Ramos, M.E. and Hajek, A.E. (1988) Isolation and characterization of *Entomophaga maimaiga* sp. nov., a fungal pathogen of gypsy moth, *Lymantria dispar*, from Japan. Journal of Invertebrate Pathology 51: 229–241.

Soper, R.S., Smith, L.F.R. and Delyzer, A.J. (1976) Epizootiology *of Massospora levispora* in an isolated population of *Okanagana rimosa*. Annals of the Entomological Society of America 69: 275–283.

St. Leger R.J. and Screen S. (2001) Prospects for Strain Improvement of Fungal Pathogens of Insects and Weeds. In: Fungi as Biocontrol Agents: Progress, Problems and Potential. (eds), Butt, T.M., Jackson, C. and Magan, N., CAB International, London, 219–237.

Steinkraus, D.C. and Slaymaker, P.H. (1994) Effect of temperature and humidity on formation, germination, and infectivity of conidia of *Neozygites fresenii* (Zygomycetes: Neozygitaceae) from *Aphis gossypii* (Hemiptera: Aphididae). Journal of Invertebrate Pathology 64: 130–137.

Steinkraus, D.C., Boys, G.O. and Slaymaker, P.H. (1993a) Culture, storage, and incubation period of *Neozygites fresenii* (Entomophthorales: Neozygitaceae), a pathogen of the cotton aphid. Southwestern Entomologist 18: 197–202.

Steinkraus, D.C., Geden, J.G. and Rutz, D.A. (1993b) Prevalence of *Entomophthora muscae* (Cohn) Fresenius (Zygomycetes: Entomophthoraceae) in house flies (Diptera: Muscidae) on dairy farms in New York and induction of epizootics. Biological Control 3: 93–100.

Steinkraus, D.C., Hollingsworth, R.G. and Boys, G.O. (1996) Aerial spores of *Neozygites fresenii* (Entomophthorales: Neozygitaceae): density, periodicity, and potential role in cotton aphid (Hemiptera: Aphididae) epizootics. Environmental Entomology 25: 48–57.

Steinkraus, D.C., Hollingsworth, R.G. and Slaymaker, P.H. (1995) Prevalence of *Neozygites fresenii* (Entomophthorales: Neozygitaceae) on cotton aphids (Homoptera: Aphididae) in Arkansas cotton. Environmental Entomology 24: 465–474.

Steinkraus, D.C., Howard, M.N., Hollingsworth, R.G. and Boys, G.O. (1999) Infection of sentinel cotton aphids (Homoptera: Aphididae) by aerial conidia of *Neozygites fresenii* (Entomophthorales: Neozygitaceae). Biological Control 14: 181–185.

Steinkraus, D.C., Kring, T.J. and Tugwell, N.P. (1991) *Neozygites fresenii* in *Aphis gossypii* on cotton. Southwestern Entomologist 16: 118–122.

Streett, D.A. and McGuire, M.R. (1990) Pathogenic diseases of grasshoppers. In: Biology of Grasshoppers. (eds) Chapman, R.F. and Joern, A. Wiley Interscience, New York, 483–516.

Strong, D.R. and Pemberton, R.W. (2001) Food webs, risks of alien enemies and reform of biological control. In: Evaluating Indirect Ecological Effects of Biological Control. (eds), Wajnberg, E., Scott, J.K. and Quimby, P.C. CAB International, London, 57–79.

Thaxter, R. (1888) The Entomophthoreae of the United States. Memoirs of the Boston Society of Natural History 4: 133–201.

Thoizon, G. (1970) Specificite du parasitisme des aphides par les Entomophthorales. Annales de Societe de Entomologie Francais (N.S.) 6: 517–562.

Thomsen, L. and Eilenberg, J. (2000) *Entomophthora muscae* resting spore formation in vivo in the host *Delia radicum*. Journal of Invertebrate Pathology 76: 127–130.

Thomsen, L. and Jensen, A.B. (2002) Application of nested-PCR technique to resting spores from the *Entomophthora muscae* complex: implications for analyses of host pathogen population interactions. Mycologia 94: 794–802.

Thomsen, L., Bresciani, J. and Eilenberg, J. (2001) Formation and germination of resting spores from different strains from the *Entomophthora muscae* complex produced in *Musca domestica*. Canadian Journal of Botany 79: 1076–1082.

Tsantrizos, Y.S., Pischos, S. and Sauriol, F. (1996) Structural assignment of the peptide antibiotic LP237-F8, a metabolite of *Tolypocladium geodes*. Journal of Organic chemistry 61: 2118–2121.

Tzean, S.S., Hsieh, L.S. and Wu, W.J. (1997) Atlas of Entomopathogenic Fungi from Taiwan. Council of Agriculture, Executive Yuan, Taiwan, R.O.C., 214 pp.

Umphlett, C.J. (1973) A note to identify a certain isolate of *Lagenidium* which kills mosquito larvae. Mycologia 65: 970–972.

van Roermund, H.J.W., Perry, D.F. and Tyrrell, D. (1984) Influence of temperature, light, nutrients and pH in determination of the mode of conidial germination in *Zoophthora radicans*. Transactions of the British Mycological Society 82: 31–38.

Vincent, M.A., Seifert, K.A. and Samson R.A. (1988) *Akanthomyces johnsonii,* a saprophytic synnematous Hyphomycete. Mycologia 80: 685–688.

Waage, J.K. (2001) Indirect ecological effects in biological control: the challenge and the opportunity. In: Evaluating Indirect Ecological Effects of Biological Control. (eds), Wajnberg, E., Scott, J.K. and Quimby, P.C. CAB International, London, 1–12.

Walsh, S.R.A. (1996) Development of molecular markers for the detection and differentiation of *Entomophaga* strains pathogenic for insects. Ph.D. dissertation, University of Toronto, Canada.

Ward, M.D.W., Madison, S.L., Andrews, D.L., Sailstad, D.M., Gavett, S.H. and Selgrade, M.J.K. (2000) Comparison of respiratory responses to *Metarhizium anisopliae* extract using two different sensitization protocols. Toxicology 147: 133–145.

Watson, D.W., Mullens, B.A. and Petersen, J.J. (1993) Behavioral fever response of *Musca domestica* (Diptera: Muscidae) to infection of *Entomophthora muscae* (Zygomycetes: Entomophthorales). Journal of Invertebrate Pathology 61: 10–16.

Weir, A. and Rossi, W. (1995) Laboulbeniales parasitic on British Diptera. Mycological Research. 99: 841–849.

Weiser, J. and Pillai, J.S. (1981) *Tolypocladium cylindrosporum* (Deuteromycetes, Moniliaceae) a new pathogen of mosquito larvae. Entomophaga 26: 357–361.

Whisler, H.C. (1985) Life history of species of Coelomomyces. In: The Genus Coelomomyces. (eds), Couch, J.N. and Bland, C.E., Academic Press Inc., Orlando, Florida 9–22.

Whisler, H.C. (1979) The fungi versus the arthropods. In: Insect-fungus Symbiosis: Nutrition, Mutualism and commensalism. (ed) Batra, L.R., Allanheld, Osmun and Co., Montclair, New Jersy, 1–32.

Wilding, N. and Brady, B.L. (1984) Descriptions of pathogenic fungi and bacteria, Set 82, Nos. 812, 814, 815, 817, 820. CMI, Kew, UK.

Wilding, N. and Lauckner, F.B. (1974) *Entomophthora* infecting wheat bulb fly at Rothamsted, Hertfordshire, 1967 1971. Annals of Applied Biology 76: 161–170.

Witlaczil, E. (1885) *Neozygites aphidis,* eine neue Gregarine. Archiven f r Mikrobiologie und Anatomie 24: 599–603.

Wraight, S.P., Butt, T.M., Galaini Wraight, S., Allee, L.L., Soper, R.S. and Roberts, D.W. (1990) Germination and infection processes of the Entomophthoralean fungus *Erynia radicans* on the potato leafhopper *Empoasca fabae*. Journal of Invertebrate Pathology 56: 157–174.

Wraight, S.P., Jackson M.A. and de Kock S.L. (2001) Production, stabilization and formulation of fungal biocontrol agents. In: Fungi as Biocontrol Agents: Progress, Problems and Potential. (eds) Butt, T.M., Jackson, C. and Magan, N. CAB International, London 253–287.

Yamamoto, M. and Aoki, J. (1983) Periodicity of conidial discharge of *Erynia radicans*. Transactions of the Mycological Society of Japan 24: 487–496.

Yaninek, J.S., Saizonou, S., Onzo, A., Zannou, I. and Gnanvossou, D. (1996) Seasonal and habitat variability in the fungal pathogens, *Neozygites* cf. *floridana* and *Hirsutella thompsonii*, associated with cassava mites in Benin, West Africa. Biocontrol Science and Technology 6: 23–33.

Yu, H., Cho, H.W. and Pillai, J.S. (1980) Infection studies of mosquito pathogen *Culicinomyces* sp. against *Aedes* and *Culex* larvae. Korean Journal of Entomology 10: 62–63.

Yun, Y., Bridge, P.D. and Evans, H.C. (1991) An integrated approach to the taxonomy of the genus *Verticillium*. Journal of General Microbiology 137: 1437–1444.

16. Advances in Immunochemical Methods for Mycotoxin/Mycotoxigenic Fungi Detection in the Agri-Food System

Elisabete Yurie Sataque Ono[1], Luciana Biazon[1], Simone Fujii[2], Mario Augusto Ono[3] and Elisa Yoko Hirooka[2]

Department of [1]Biochemistry, [2]Food and Drug Technology, [3]Pathological Sciences, State University of Londrina, P.O. Box 6001, 86051-990, Londrina, Paraná, Brazil

SUMMARY Mycotoxigenic fungi/mycotoxin detection in the agri-food system depends on efficient analytical methods at the quality control level. Conventional fungal identification methods (morphology, isoenzyme and secondary metabolite profiles) as well as the usual techniques for mycotoxin detection (thin-layer chromatography, high-performance liquid chromatography, gas-liquid chromatography and mass spectroscopy) are time-consuming, and the costs restrict their application in routine laboratory work. Immunochemical techniques are a promising alternative due to their high sensitivity, specificity, reproducibility and quickness. Taking into account the importance of prevention strategy for minimizing the human and animal hazard exposure, the immunodetection methods and advances in their application toward monitoring of mycotoxigenic fungi and mycotoxins (aflatoxins, fumonisins and ochratoxin A) in agri-food systems are reviewed.

Introduction

Toxigenic fungi produce hundreds of secondary metabolites, i.e. mycotoxins, that often contaminate agricultural staples along the food production chain from field, harvest, transport and storage.

Mycotoxins have diversified chemical structures, and are produced by field and storage fungi, including many species of *Fusarium*, *Penicillium,* and *Aspergillus* (Ayres, 1979). These compounds can cause acute and/or chronic intoxication symptoms in both humans and animals at mg/kg to µg/kg range.

The discovery and characterization of aflatoxin, and its involvement as a potential carcinogen in humans, may be regarded as the catalyst for triggering the modern mycotoxin research programs. Although the major effort was initially centered on aflatoxin, attention rapidly turned to the possible cause/effect implications of other mycotoxins in idiopathic diseases (Sydenham and Shephard, 1996).

In addition to the health hazard, the natural occurrence of mycotoxins in agricultural staples causes serious economic losses for commercial sectors such as crop, livestock and poultry producers, as well as for food and feed processors. Losses may result from poor grain quality, low crop yields, reduced animal performance, crop downgrading, and/or the necessity of detoxification processes in affected grains (Pestka et al. 1995).

The Food and Agricultural Organization (FAO) estimated the worldwide mycotoxin contaminated crop at 25%. The human health hazard associated with the natural occurrence of harmful mycotoxins compelled several countries to adopt regulatory guidelines, although the criteria to establish the permitted tolerance levels are inconsistent (Van Egmond, 1989).

Fungal growth and mycotoxin production depend on biological (susceptible crop) and environmental factors, with emphasis on regional climatic conditions during plant development and crop harvest (Pestka and Casale, 1990; Ono et al. 1999). The major means for mycotoxin elimination is efficient detection followed by the diversion of contaminated raw materials from the feed and food production chain. Analytical surveillance can also assist in identification of geographical regions where mycotoxins are a recurrent problem, and provide a database for human exposure in epidemiological studies (Pestka et al. 1995).

Taking into account that mycotoxigenic fungi/mycotoxin is a problem difficult to avoid, the most effective control measure depends on a rigorous program of monitoring in food and feed. Therefore, sensitive and accurate methods for mycotoxigenic fungi and/or mycotoxin detection are essential to minimize the exposure.

Current methods for fungal identification/detection include culturing on various media, microscopic examination, and chemical assays based on chitin, ergosterol or secondary metabolites (Thrane, 1989). Except for the last one, which includes mycotoxins, the low specificity restricts their use. Even though the identification of specific secondary metabolites provides useful confirmatory information, its use is limited due to the inconsistency of the induction/repression system in triggering metabolic pathways under different conditions (Abramson et al. 1998; Proctor et al. 1999).

Conventional mycotoxin detection techniques include thin-layer chromatography (TLC), high-performance liquid chromatography (HPLC), gas-liquid chromatography, and mass spectroscopy (Pestka et al. 1995). The Association of Official Analytical Chemists (AOAC) officially accepts TLC and HPLC, but the last one requires extensive clean-up steps to overcome the food matrix interferences, trained staff and a time-consuming procedure, in addition to high cost and use of harmful solvents (Barna-Vetró et al. 1996; Sydenham et al. 1996a, b).

Immunoassays provide an attractive and promising alternative for conventional mycotoxigenic fungi and/or mycotoxin analysis due to their high specificity, sensitivity, low cost, potential for automation and use in rapid quality control under field conditions (Hefle, 1995).

Basic principles of immunoassay formats and their applications for mycotoxigenic fungi and/or mycotoxin analysis (aflatoxins, fumonisins and ochratoxin A) are reviewed.

Basic Concepts in Immunology

Antibodies or immunoglobulins are glycoproteins produced by B lymphocytes in response to pathogens (bacteria, fungus, virus, parasites) and constitute an important host immune defense mechanism against infectious agents. The antibodies bind with high specificity and affinity to epitopes, the regions recognized on antigen molecules.

Antigens that induce an immune response are named immunogens and are, in general, compounds with high molecular weight. The haptens are low molecular weight compounds (e.g. mycotoxins) that are able to induce antibody production only when previously conjugated with high molecular weight proteins (bovine serum albumin, ovalbumin, keyhole limpet hemocyanin). A water-soluble carbodiimide or mixed anhydride method is commonly used to conjugate a mycotoxin onto a protein carrier (Chu, 1984). Aflatoxin B_1 (AFB_1) and AFM_1, which contain a carbonyl group, have been conjugated with proteins after the formation of a carboxymethyloxime derivative. Fumonisin B_1 has been conjugated with a protein carrier using its free amine group via glutaraldehyde (Azcona Olivera et al. 1992a). Ochratoxin A has been directly conjugated with a protein carrier by its carboxyl group, but without any derivatization (Kawamura et al. 1989; Chu, 1984; Xiao et al. 1995).

Polyclonal antibodies resulting from the activation of several B lymphocytes clones are produced by antigen inoculation in animals (mouse, rabbit, goat, etc.). Although the procedure for polyclonal antibody production is relatively simple and inexpensive, they may cross-react with other antigens. Furthermore, the scale up is limited, due to the animal variability, requirement of high purity antigen for immunization, and maintenance of biotery (Wyatt, 1992).

Monoclonal antibodies are produced by hybridomas, that are cells resulting from the fusion of myeloma cells with B lymphocytes clones. In contrast to polyclonal antibodies, the monoclonal antibodies are homogenous reagents with high specificity and constant affinity, which is essential for antigen identification in a complex matrix (Gazzaz et al. 1992; Dietrich et al. 1995).

Recently a new technique for antibody production has been introduced, using recombinant antibodies that are antibody fragments expressed in bacteria, phage or fungus. The advantage of this technique is low cost and the possibility of antibody DNA manipulation to improve the specificity and affinity characteristics (Yau et al. 2003). Moghaddam et al. (2001) selected a single chain antibody from phage display recombinant antibody libraries that could bind to aflatoxin B_1 with high affinity (6×10^{-9}).

Therefore, antibodies are very useful tools for the detection of different substances in many biological systems (e.g. hormones, antibiotics, pesticides, mycotoxins), allowing development of several immunoassays with detection limit at nanogram levels. All immunoassays are based on high specificity and sensitivity of antigen-antibody interactions.

RIA (radioimmunoassay) was initially used in mycotoxin analysis, but the disadvantages such as short half-life and radioactive hazards led to its replacement by enzyme-linked immunosorbent assay (ELISA) (Hefle, 1995). ELISA is an analytical procedure based on the conversion of a substrate (chromogenic, fluorescent or chemiluminescent) by an enzyme labeled antigen or antibody as a means of detecting/ quantifying the analyte (Gazzaz et al. 1992).

ELISAs can be classified as homogeneous and heterogeneous assays. In the first type the enzyme activity is modified by antibody binding and does not require the separation of free and bound reagents, while in the latter, the enzyme activity is not altered and requires a solid phase, such as a microtitre plate, for separation of free and bound reagents (Chu, 1984).

ELISAs commonly used in mycotoxin detection are direct competitive-ELISA (dcELISA) and indirect competitive-ELISA (icELISA). In dcELISA (Fig. 1) the mycotoxin in a food sample competes with enzyme-labeled mycotoxin for immobilized antibody binding in the solid phase, while in icELISA (Fig. 2) the sample mycotoxin competes with immobilized mycotoxin for antibody binding (Hefle, 1995).

Sandwich-ELISA (Fig. 3) is an assay format that can be used for fungus detection in food. In this technique, an antibody specific to the fungus, immobilized on solid phase will capture the fungal antigen that will be recognized by an enzyme-labeled antibody (Hefle, 1995).

Immunoassays in Mycotoxin Detection

Several immunochemical methods have been applied for mycotoxin detection, with emphasis on ELISA, immunoaffinity columns (IACs), membrane based methods, biosensors and immunofiltration-based assays.

The ELISA techniques are widely used for analysis of batch samples and they can provide qualitative results based on color development by an enzymatic reaction with a chromogenic substrate, or semi-quantitative/quantitative results by spectrophotometric determination. The disadvantage of ELISA is matrix interference due to the non-specific interaction of food components (proteins, fatty acids) with the antibody, which causes false-positive reaction or overestimation of the mycotoxin concentration (Hefle, 1995; Barna-Vetró et al. 1996). Matrix interference can be minimized by diluting the sample before assay, or by including a prior clean-up step (Pestka et al. 1994; Barna-Vetró et al. 1996; Ono et al. 2000).

Recently AOAC has validated several analytical methods based on food extract clean-up by an immunoaffinity column prior to HPLC analysis (Burdaspal et al. 2001; Dragacci et al. 2001; Entwisle et al. 2001; Visconti et al. 2001). Although these methods apply HPLC as a separation-analytical technique, the use of TLC coupled to imunoaffinity columns has been demonstrated as a promising procedure comparable

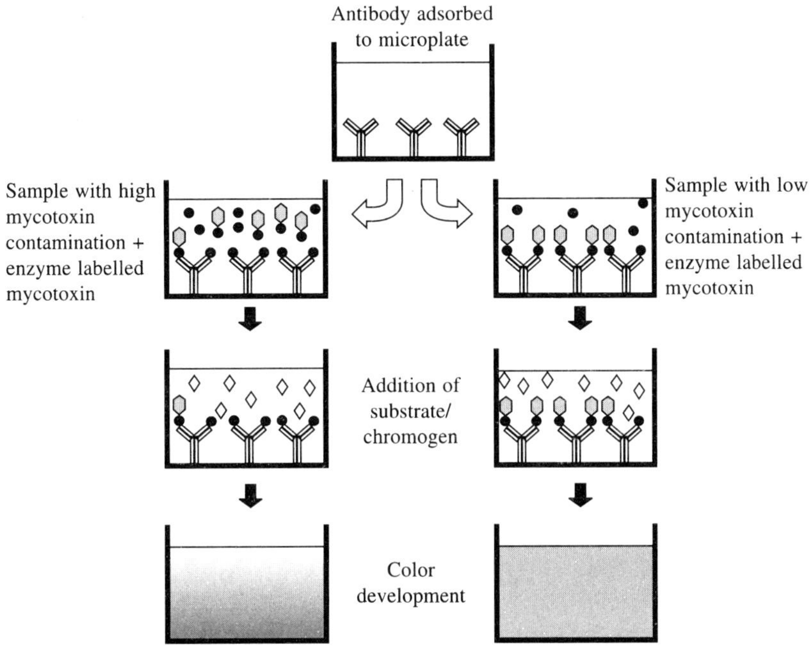

Fig. 1. Direct competitive ELISA (dcELISA).

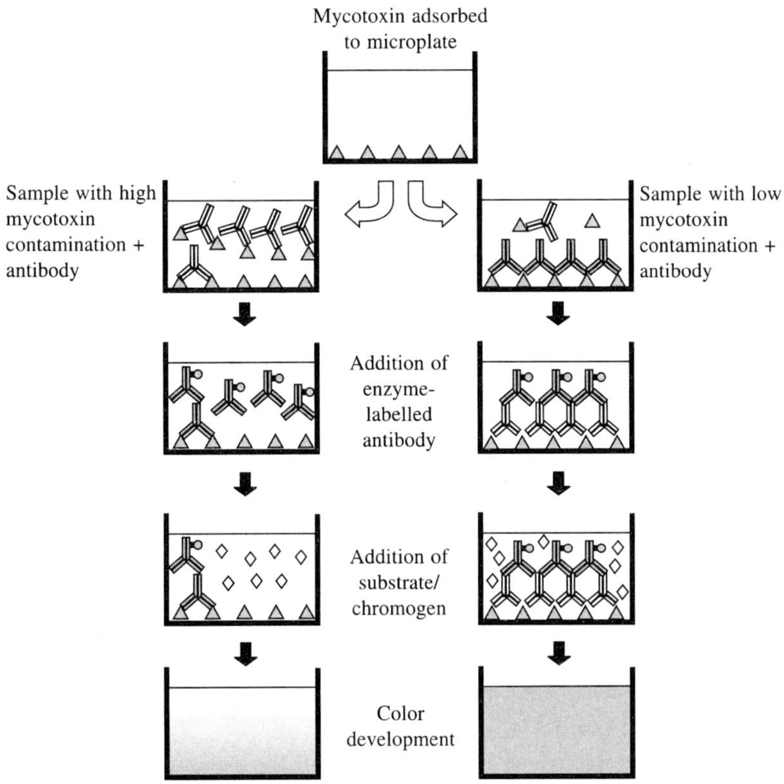

Fig. 2. Indirect competitive ELISA (Ic-ELISA).

to HPLC (Stroka and Anklam, 2002), specially if small-particle-size silica-gel (high-performance TLC-HPTLC) is used (Valenta, 1998). Therefore, HPTLC could be an alternative for quality control in developing countries. The disadvantage regarding the high cost of commercial IACs tends to be minimized by the column regeneration process (Scott and Trucksess, 1997; Fazekas and Tar, 2001; Watanabe et al. 2001; Kondo et al. 2002).

Regulatory guidelines in foodstuff importing countries have imposed more and more demanding thresholds for mycotoxin contamination, increasing the demand for rapid and reliable methods for field conditions. Numerous membrane-based methods (dip-sticks, immunochromatographic test strips, flow-through devices) have been developed for field application (De Saeger and Van Peteghem, 1996; Sibanda et al. 2000; Ho and Wauchope, 2002). However, most of them had low sensitivity, and were inadequate for batch analysis of samples. Lately, substantial progress has also occurred in biosensors such as surface plasmon resonance, fiber optic probes and microbead-based assays for onsite applications (Carlson et al. 2000; Daly et al. 2000; Maragos, 2002; Nasir and Jolley, 2002), but only one sample can be assayed at a time. The major limitations of these methods are the low reusability of the immunosurface and high consumption of reagent (Gonzalez-Martinez et al. 1999). Recently, an immunofiltration-based assay was developed for rapid analysis of batch samples under field conditions, providing reliable semi-quantitative analysis without instrumentation (Pal and Dhar, 2004).

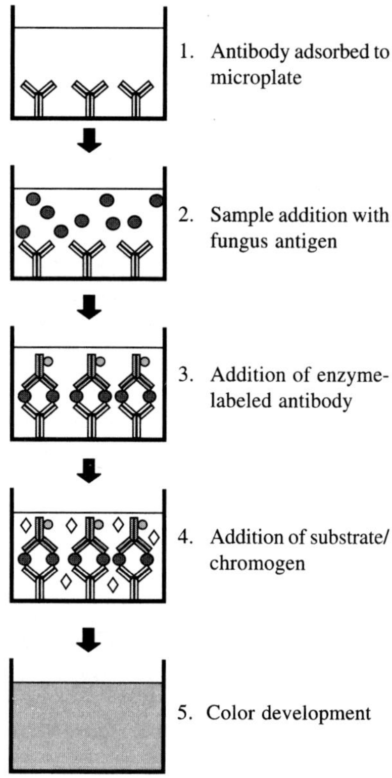

Fig. 3. Sandwich ELISA.

Aflatoxins

Aflatoxins are a group of toxic secondary metabolites produced by *Aspergillus flavus* and *A. parasiticus* on several agricultural commodities. These heterocyclic mycotoxins are composed of a dihydrodifuran or tetrahydrodifuran moiety fused to a substituted coumarin. The major members are the four naturally occurring toxins AFB_1, AFB_2, AFG_1 and AFG_2 (Fig. 4), which are normally converted to the derived products as AFM_1, AFM_2, AFP_1, AFQ_1 and aflatoxicol by microbial or animal metabolic pathways (Bradburn et al. 1993).

Aflatoxin B_1 (AFB_1) is a powerful teratogen, mutagen and hepatocarcinogen, through its hepatic microsomal biotransformation to the epoxide form which can covalently bind to DNA. The International Agency for Research on Cancer (IARC) classified AFB_1 as a Group 1 carcinogen, due to its association with primary liver cancer in humans (IARC, 2002). A stringent regulatory guideline has been imposed on the presence of AFB_1 in agricultural commodities. Several countries have set the limit for aflatoxin levels ranging from non-detectable to 50 μg/kg, where regulations specify only the AFB_1 level, with the most frequent specified limit at 5 μg/kg (Van Egmond, 1989; FAO, 1997).

AFM_1, a hydroxylated derivative of AFB_1, is produced in the milk of lactating animals consuming

AFB_1-contaminated feed (Dragacci and Fremy, 1996). The IARC (2002) classified it as Group 2B carcinogen (possibly carcinogenic to humans) and the maximum limit set by most countries is either 0.05 μg/kg or 0.5 μg/kg (Van Egmond, 1989; FAO, 1997).

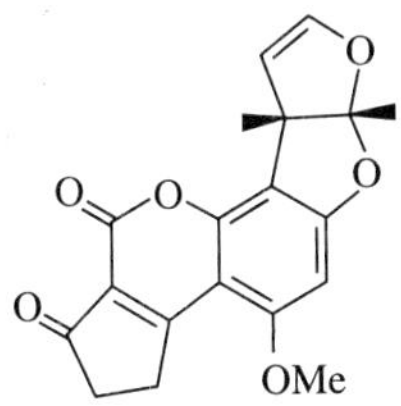

Aflatoxin B_1
Aflatoxin B_2: 15, 16-Dihydro

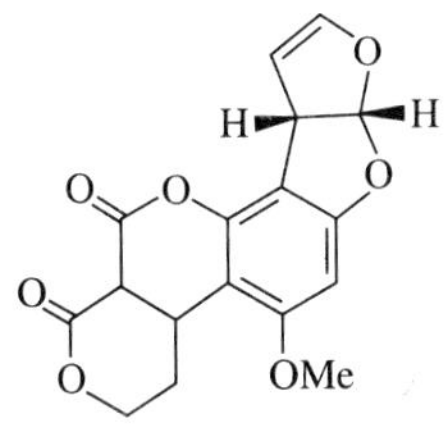

Aflatoxin G_1
Aflatoxin G_2: 15, 16-Dihydro

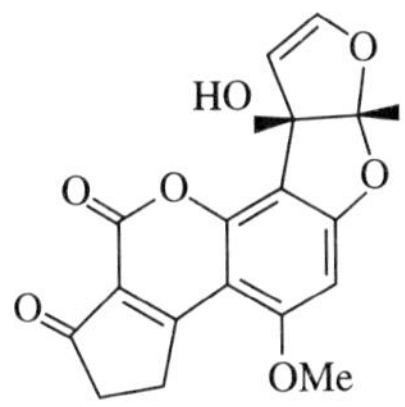

Aflatoxin M_1

Fig. 4. Chemical structures of aflatoxins B_1, B_2, G_1, G_2, M_1.

Polyclonal and monoclonal antibodies to aflatoxins have been raised for application in dc-ELISA or ic-ELISA (Morgan et al. 1983; Chu et al. 1987; Kawamura et al. 1988; Li et al. 1994; Aldao et al. 1995, Devi et al. 1999; Thirumala-Devi et al. 2002; Lipigorngoson et al. 2003).

Lipigorngoson et al. (2003) developed a highly sensitive in-house dc-ELISA for AFB_1 detection based on a monoclonal antibody with a detection limit of 4 μg/kg and a recovery range from 88.1 to 99.5%. Correlation coefficients with commercial ELISA kits and thin-layer chromatography were 0.912 and 0.802 for corn, 0.941 and 0.832 for peanuts, respectively ($p < 0.05$). The screening for contamination in Thai corn and peanuts with this assay showed mean AFB_1 levels (per cent positive) of 73 μg/kg (85.7%) in corn and 102 μg/kg (67.9%) in peanuts.

A number of studies on the performance of commercial kits for aflatoxin determination have been carried out (Dorner and Cole, 1989; Koeltzow and Tanner, 1990; Beaver et al. 1991). Hongyo et al. (1992) compared maize extract analysis by monoclonal antibody based dc-ELISA with TLC and HPLC, and found correlation coefficients of 0.93 and 0.95, respectively, when the contamination level was up to 200 μg/kg.

Several collaborative studies on commercial ELISA kits and immunoaffinity columns have been reported since 1988 (Mortimer et al. 1988; Park et al. 1989; Trucksess et al. 1989; Patey et al. 1992; Trucksess and Stack, 1994; Patey et al. 1991; Trucksess et al. 1990, 1991; Barmark and Larsson, 1994). The Aflatest-P affinity column coupled to fluorimetry or HPLC has been adopted as the official method by AOAC for total aflatoxin in corn, raw peanuts and peanut butter at concentrations ≥ 10 μg/kg (Trucksess et al. 1991).

Aflatoxins spiked at 6.5, 13.0, and 19.5 μg/kg levels in tahini, sesame butter, were analyzed by HPLC, fluorometry and ELISA. An immunoaffinity column was used to clean-up extracts prior to HPLC and fluorometry. Fluorometry-HPLC correlation was 0.978, and both methods showed high recovery and low variance, while ELISA was found applicable only as a screening method (Nilüfer and Boyacyoglu, 2002).

Gathumbi et al. (2003) developed a highly sensitive immunochemical method for rapid detection of AFB_1 in chicken liver, and the effectiveness of immunoaffinity chromatography (IAC) prior to dc-ELISA analysis was evaluated. The detection limit ranged from 15 to 17 pg/mL, with AFB_1 recovery of 54.3 to 65.5% in artificially contaminated tissues at 1 to 5 ng/g, respectively. The sensitivity and specificity in spiked samples (1 ng/g AFB_1) were 100% for IAC processed samples, while without IAC they were 91.7 and 100%, respectively.

A rapid colorimetric sequential injection immunoassay (SIIA) system using a jet ring flow cell packed with polymethylmethacrylate beads as solid phase has been developed for AFB_1 (Garden and Strachan, 2001). The detection limit of 4 ng/mL in spiked pistachio nut was comparable to ELISA, but the automated SIIA system required less than 10 min to conclude a measurement, whereas ELISA took 2.5 h.

Pal and Dhar (2004) developed a simple analytical device for performing non-instrumental immunofiltration-based assay, which consisted of membrane strips with antibody-immobilized zones on a polyethylene card. It detected very low AFB_1 concentration in food by the improved catalyzed reporter deposition (Super-CARD) method of signal amplification, involving biotinylated tyramine (B-T), avidin-horseradish peroxidase conjugate and 4-chloro-1-naphthol as substrate. Quantitative estimation was possible by densitometry (detection limit 0.25 pg/spot, 0.01 ng/mL), with mean recoveries from different foods between 91 and 104%, and high correlation with HPLC ($r^2 = 0.99$) in spiked corn and peanut samples. A batch of 12 samples was analyzed on a single card in 12 min.

A plastic-backed nitrocellulose strip was the basis for AFB_1 detection at < 20 ng by AFB_1-tagged liposomes encapsulating a visible dye, and could serve as a rapid visual screening or an inexpensive quantitative assay by densitometry (Ho and Wauchope, 2002). In this method, the sample migrated along the strip by capillary action into a zone with immobilized antibodies; then AFB_1-tagged, dye-containing liposomes migrated into the same area with subsequent filling of the remaining antibody sites, resulting in an intense purplish pink color.

An ic-ELISA based on polyclonal antibodies has been developed to estimate AFM_1 in milk and derived products. The assay showed accurate results at AFM_1 concentration exceeding 0.25 ng/mL, and the relative cross-reactivity with AFB_1 was 5%, but much less for AFB_2, AFG_1, AFG_2, and no reaction with ochratoxin A (Thirumala-Devi et al. 2002).

Sibanda et al. (1999) developed a membrane-based flow-through enzyme immunoassay for AFM_1 detection in liquid and powered milk with sensitivity of 0.05 ng/mL, and total assay time of 30 min including an IAC clean-up step. The assay consisted of nylon membrane sensitized with anti-mouse IgG mounted in a plastic snap-fit device with cotton wool absorbent, mouse anti-AFM_1 monoclonal antibody and AFB_1-horseradish peroxidase conjugate.

Fumonisins

Fumonisins are mycotoxins produced mainly by *Fusarium verticillioides* and *F. proliferatum*, which are worldwide corn pathogens. Since their isolation and characterization in 1988, fumonisins have been extensively studied due to their toxic effects in animals and humans (Marasas, 1996).

Although several analogues have been identified and characterized, fumonisin B_1 (FB_1), FB_2, and FB_3 occur at significant levels in naturally contaminated corn and corn-based products (Rheeder et al. 2002). Fumonisins are diesters of propane-1,2,3-tricarboxylic acid (tricarballylic acid), and various 2-amino-12,16-dimethyl polyhydroxy-eicosanes, in which the hydroxyl groups on C_{14} and C_{15} are esterified with the terminal carboxyl group of tricarballylic acid (Fig. 5).

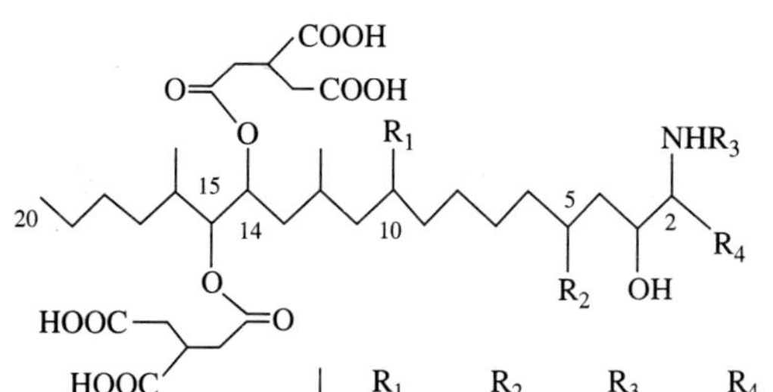

	R_1	R_2	R_3	R_4
FB_1	OH	OH	H	CH_3
FB_2	H	OH	H	CH_3
FB_3	OH	H	H	CH_3

Fig. 5. Chemical structures of major fumonisins.

FB_1 causes equine leukoencephalomalacia in horses (Kellerman et al. 1990), pulmonary edema in pigs (Harrison et al. 1990), and it has been associated with high incidences of human esophageal cancer in Transkei, Southern Africa, and China (Rheeder et al. 1992; Ueno et al. 1997). The International Agency for Research on Cancer classified the *F. verticillioides* toxins as possible carcinogens for humans (IARC, 2002). Maximum advisory levels for FB_1 have been set at 5, 10 and 50 μg/g for equine, swine, and poultry feed, respectively (Munkvold and Desjardins, 1997), and at 1 μg/g in corn products for human consumption in Switzerland (Visconti and Boenke, 1995).

Polyclonal and monoclonal antibodies to fumonisins have been raised for incorporation in dc-ELISA and ic-ELISA (Azcona-Olivera et al. 1992a, b; Usleber et al. 1994; Yu and Chu, 1996; Iijima et al. 1996; Yeung et al. 1996; Christensen et al. 2000; Barna-Vetró et al. 2000). A dc-ELISA based on monoclonal antibodies developed for FB_1 detection in cereals (Barna-Vetró et al. 2000) showed detection limit of 7.6 ng/g FB_1 and 100, 91.8, and 209% mean cross-reactivity for FB_1, FB_2 and FB_3, respectively, but no reactivity with hydrolyzed fumonisin. The toxin recovery from spiked cereals at 50-200 ng/g FB_1 ranged from 61 to 84%, and comparative analysis of naturally contaminated corn samples proved its suitability for rapid screening of food and feed.

Several comparative studies on competitive ELISAs with chemical methods (HPLC, GC-MS) were carried out, some of them showing good correlations, although overestimation by ELISA could occur (Pestka et al. 1994; Sydenham et al. 1996a, b; Usleber et al. 1994; Iijima et al. 1996; Sutikno et al. 1996; Yu and Chu, 1996; Scott et al. 1997; Ono et al. 2000; Kim et al. 2002).

Danielsen and Funck Jensen (1998) evaluated 35 maize samples from three geographic regions of Costa Rica, and detected fumonisin at 4 to 16,000 ng/g range, with an average of 2500 ng/g, using a commercial dc-ELISA based on FB_1-monoclonal antibody (RIDASCREEN Fumonisin Fast, R-Biopharm GmbH, Germany). The antibodies cross-reacted with FB_2 (40%) and FB_3 (100%), resulting in an average overestimate of < 20% for the FB_1 readings. The recovery rate of this assay was 60%, and the detection limit 3.4 ng/g.

Desjardins et al. (2000) evaluated fumonisin occurrence in Nepalese maize by dc-ELISA based on monoclonal antibody with a detection limit of 1 μg/g, and fumonisins were >1 μg/g in 22% of 74 samples.

The performance of a monoclonal-antibody based ic-ELISA for fumonisin detection was evaluated in post-harvest corn of the 1995 and 1996 crop (150 samples, 62 hybrids) from three regions of Paraná state, Brazil (Ono et al. 2000; 2001). Fumonisins were detected in 147 (98%) samples at a concentration range of 0.096 to 22.6 μg/g. The assay showed a detection limit of 93 ng/g and correlation coefficient of 0.94 with HPLC. Therefore, it could be effectively applied for fumonisin screening in corn, with advantages of simplicity, sensitivity and easiness of sample preparation.

Kim et al. (2002) evaluated the FB_1 occurrence in seventy-six corn-based food samples from Seoul, Korea using dc-ELISA and HPLC. Average recoveries from corn flake samples spiked at 5-1000 ng/g FB_1 were 104% by dc-ELISA (detection limit of 5 ng/g), and 82% by HPLC (detection limit of 20 ng/g), with good correlation ($r^2 = 0.992$).

A collaborative study involving 13 laboratories in the United States was carried out using dc-ELISA for total fumonisin (B_1, B_2, and B_3) determination in corn fields with naturally contaminated sites at three levels, and artificially contaminated sites spiked at 1.0, 3.0, and 5.0 mg/kg. The average recovery of total fumonisin was 120, 100 and 90%, respectively, showing acceptable intra- and interlaboratory precision (Bird et al. 2002).

A comparative study was also performed to evaluate the FB_1 extraction efficiency in deionized water, PBS, PBS-Tween, 70% aqueous methanol and 50% aqueous acetonitrile using 16 field corn samples, by polyclonal antibody based dc-ELISA. Aqueous phosphate buffer was suitable for FB_1 extraction, avoiding the use of organic solvents. HPLC and ELISA were well correlated in FB_1 spiked samples at concentrations between 1 and 50 μg/mL of extract (Kulisek and Hazebroek, 2000).

Ho and Durst (2000) developed a flow-injection liposome immunoanalysis (FILIA) system for FB_1 quantification, based on competition between FB_1 and sulforhodamine B dye encapsulating, FB_1-tagged liposomes for a limited number of antibody binding sites. The antibodies were immobilized via protein-A in a capillary immunoreactor column, and 35% methanol was used for regeneration of antibody binding sites after each assay, which allowed usefulness for up to 70 sequential sample injections. The relevant features of FILIA when compared to HPLC are lower detection limit (ca. 25 × lower) and less complex and rapid sample preparation (Ho and Durst, 2003).

Maragos et al. (2001) developed a rapid portable fluorescence polarization-based assay, where the competition of unlabeled fumonisin was compared with a fluorescently tagged fumonisin (FB_1-FL) for specific monoclonal antibody. The fluorescence polarization (FP) of tagged fumonisin increased on binding with the antibody, while when free toxin was present, less FB_1-FL was bound and decreased the polarization signal. The assay was very simple to perform, requiring < 2 min per sample, and showed a detection limit of 0.5 μg of FB_1/g in spiked corn and a mean recovery of 94.3 ± 13.8% over 0.5-20 μg/g range. The correlation with HPLC was 0.85-0.88 (Maragos et al. 2001).

A surface plasmon resonance (SPR) immunosensor for direct assay of FB_1, with a detection limit of 50 ng/mL and analysis time of less than 10 min was developed using polyclonal antibodies adsorbed onto a thin gold film substrate, coupled to a glass prism (Mullett et al. 1998). The output beam of a planar light-emitting diode was focused through the prism to excite SPR at gold film surface. When a sample with FB_1 was added to a cell on the outside of the gold film, the angular profile of reflected light intensity was shifted proportionally to FB_1 concentration.

Ochratoxin A

Ochratoxins, a mycotoxin group with dihydroisocoumarin linked to phenylalanine (Fig. 6) are produced by *Aspergillus* sp. and *Penicillium* sp. Ochratoxin A is the major analogue detected in naturally contaminated agricultural commodities (maize, wheat, barley, coffee beans). *A. ochraceus* and *P. verrucosum* are the predominant toxigenic species (Petzinger and Weidenbach, 2002; Batista et al. 2003), but recently new species, mainly belonging to the *Aspergillus* group, have been reported, due to the advance in molecular taxonomy of toxigenic fungi (Mouyna et al. 2002; Sweeney and Dobson, 1999).

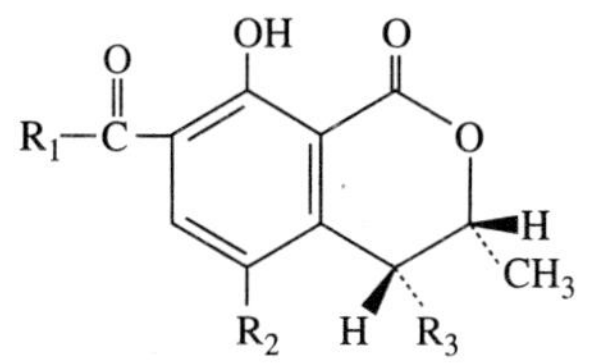

Ochratoxin	R_1	R_2	R_3
Ochratoxin A	$C_6H_5CH_2COOHNH$	CI	H
Ochratoxin B	$C_6H_5CH_2COOHNH$	H	H
Ochratoxin C	$C_6H_5CH_2COO(C_2H_5)NH$	CI	H
(4R)-4-Hydroxi-ochratoxin A	$C_6H_5CH_2COOHNH$	CI	OH
Ochratoxin α	H	CI	H

Fig. 6. Chemical structures of ochratoxins.

Ochratoxin A (OA) causes several adverse effects in laboratory animals including nephrotoxicity, teratogenicity, cytotoxicity, immunotoxicity, and genotoxicity (Hussein and Brasel, 2001; Grosso et al. 2003). It has been classified as a Group 2B carcinogen (possibly carcinogenic to humans) due to its probable association with primary liver cancer in humans (IARC, 2002). The maximum limit for OA has been set in several countries, with values ranging from 0.5 to 5 μg/kg in infant foods, 2 to 50 μg/kg in cereals and 5 to 300 μg/kg in animal feeds (Van Egmond, 1996; Beretta et al. 2002).

Several commercial immunoaffinity columns are available for OA analysis: Ochratest (Vicam®); Ochraprep and Ochrascan Easi-Extract ochratoxin column (Rohne Diagnostics®); Rida ochratoxin A column (R-Biopharm®) (Scott and Trucksess, 1997; Castellari et al. 2000). Ochratoxin quantification can be improved by immunoaffinity clean up of the sample before HPLC analysis (Castellari et al. 2000, Visconti et al. 2000, Jornet et al. 2000).

Solfrizzo et al. (1998) developed a rapid and reliable method for OA detection in cereals, based on sample extraction with acidic chloroform, defatting with chloroform followed by HPLC fluorimetric detection. The poor clean-up of this procedure limited its application for trout feed and rye analysis, but the use of the Ochratest column overcame this problem by eliminating false positive results and reaching a detection limit of 0.2 μg/kg.

Castellari et al. (2000) evaluated different clean-up procedures for OA detection in wine samples by HPLC. Comparing the clean-up performance of Ochraprep, Ochratest column and chloroform extraction, the quantification limit ranging from 0.020 to 0.045 μg/L indicated no significant difference among these procedures. Nevertheless, Visconti et al. (2000) evaluated ochratoxin contamination in beer using the Ochratest clean-up step prior to HPLC analysis and reached a detection limit of 0.01 ng/ml, with ochratoxin levels in 50% of samples ranging from < 0.01 to 0.135 ng/ml.

ELISAs based on polyclonal and monoclonal antibodies have been developed for OA detection in several agricultural commodities (Morgan et al. 1983; Lee and Chu, 1984; Candlish et al. 1988; Kawamura et al. 1989; Barna-Vetró et al. 1996; Thirumala-Devi et al. 2002). Candlish et al. (1988) developed an ic-ELISA based on OA-monoclonal antibody in barley with a detection limit of 5 μg/kg, and cross reactivity of 8.2% with OC. Kawamura et al. (1989) developed an ic-ELISA based on monoclonal antibodies with a detection limit of 50 pg/mL. Monoclonal antibodies OA 1, 3, 4, 5 and 7 showed cross reactivity with OA (100%) and OC (63.1-99.5%), but reacted poorly with the analogues OB (0.63-1.14%), Oα (0.04-0.07%) and 4-hydroxi OA (1.19-1.51%), confirming the high specificity of these antibodies.

A dc-ELISA based on monoclonal antibodies with a detection limit of 0.5 ng/g was developed for OA detection in different cereals. The toxin recovery from cereals spiked with 5-100 ng/mL OA ranged from 90 to 130% (Barna-Vetró et al. 1996).

An ic-ELISA based on polyclonal antibodies for OA determination in chili samples showed a detection limit of 0.1 ng/mL, and mean recovery of 90-110% from chilies spiked with 1 to 100 μg of OA per kg. These antibodies did not cross-react either with ochratoxin B, coumarin, 4-hydroxycoumarin, L-phenylalanine or aflatoxin B_1 (Thirumala-Devi et al. 2000).

A flow-through enzyme immunoassay was developed for screening green coffee beans for OA with sensitivity of 8 ng/g, and a recovery range from 70 to 89% (spiked samples) and from 86 to 95% (naturally contaminated samples) (Sibanda et al. 2001).

A manual vacuum manifold and an automated solid phase extraction (ASPEC) system were applied to cleanup OA in wheat, rye, barley, and oat samples with immunoaffinity columns, followed by HPLC separation and fluorescence detection. The detection limit of both extraction systems was 0.1 μg/kg, and mean recoveries in spiked cereals at 0.6 and 2.5 μg/kg levels ranged from 68 to 106% (Eskola et al. 2002).

A collaborative study to validate a novel field immunoassay kit for rapid OA detection in cereals (wheat, rye, maize and barley) was carried out and compared to HPLC. The detection limit was 4 μg/kg and the percentages of false positive and false negative results were 2% and 4%, respectively. This study has shown that the OA kit can be used to screen OA in cereals (De Saeger et al. 2002).

Immunoassays for Mycotoxigenic Fungi

The estimation of fungal contamination in food and feed by immunochemical techniques has been the subject of extensive research since 1985.

Most fungal antigens have been obtained from the soluble component of cells or the culture broths (Cousin, 1990).

Notermans et al. (1986, 1987) developed a highly sensitive genus-specific assay for several fungi using polyclonal antibodies against heat-stable, water-soluble extracellular polysaccharide (EPS) antigens, and they showed its applicability for fungi detection and quantification in processed and fresh foods. Otherwise, most of the assays based on EPS antigens until now have shown low specificity (Li et al. 2000).

Exoantigens, the soluble immunogenic molecules produced by fungi, can be readily detected in culture broths or aqueous extracts of slant cultures (Kaufman and Standard, 1987). Many fungi produce exoantigens that are unique to a specific genus and/or species, therefore these can be used in identification (Kaufman and Standard, 1987).

Kaufman and Standard (1987) produced antibodies against exoantigens of medically important fungi, and assayed them by immunodiffusion. Although highly specific, this method was relatively insensitive, time-consuming and required a larger quantity of antibody than other immunoassay formats, such as ELISAs, which have been developed for detection/identification of mold species with variable success (Banks et al. 1993; Dewey et al. 1990; Fuhrmann et al. 1990, 1992; Notermans et al. 1986; Tsai and Cousin, 1990; Wong et al. 1988). Some ELISAs have been developed for detection and quantification of major storage (Dewey et al. 1990; Lu et al. 1994, 1995a, b) or field fungi (Wong et al. 1988).

Lu et al. (1994) developed a polyclonal antibody based ic-ELISA against *Penicillium aurantiogriseum* exoantigens, showing good sensitivity and low or no cross-reactivity with extracts from grain and fungi, including several *Penicillium*, *Aspergillus*, *Fusarium*, *Mucor*, and *Alternaria* species. Lu et al. (1995b) also developed a polyclonal antibody based ic-ELISA for detection/quantification of *A. ochraceus* in wheat grain. Immunoblotting was used to characterize the antibodies' reactivity towards exoantigens of four *Aspergillus*, seven *Penicillium*, three *Fusarium*, one *Mucor* and one *Alternaria* species. The antibodies were specific for *A. ochraceus*, and did not react with other fungi or water-soluble grain components.

Immunoblotting studies (Lu et al. 1995a) have also demonstrated that *A. ochraceus* exoantigens were present in both liquid and solid cultures, but the type of culture influenced the relative concentrations. Therefore such variation could reduce the predicting accuracy of mycelial dry weight based on exoantigen content. However, these shortcomings are probably no worse than those observed in other mold assays such as chitin (glucosamine), ergosterol, and fungal propagules count (Cousin, 1996; Gourama and Bullerman, 1995; Pitt and Hocking, 1997). The chitin assay can also yield false-positive results due to insect contamination (Cousin, 1996).

Neucere et al. (1992) used surface proteins from *A. flavus* and *A. parasiticus* to develop an immunoassay to differentiate the two species, but there were persistent cross-reactions. Shapira et al. (1997) produced polyclonal antibodies to *A. parasiticus* exoantigens and chimeric proteins expressed in *Escherichia coli*. Although antibodies to exoantigens showed cross-reactions with *Aspergillus* and *Penicillium* species, the antibodies elicited against chimeric proteins were highly specific.

A double-sandwich ELISA was performed for detection of *A. parasiticus* in artificially contaminated corn, rice, wheat and peanut, as well as for detection of naturally occurring aflatoxigenic molds in the same cereals (Tsai and Yu, 1999). Although *A. parasiticus*/*A. flavus* were detected in 5 out of 40 cereal samples by standard plate counting, the ELISA was not able to detect any positive sample. The method showed positive results only when the aflatoxigenic mold count exceeded 10^4 CFU/g.

Chen et al. (2000) immunized laying hens with exoantigens from *A. flavus*, *A. ochraceus* and *A. versicolor*, and developed a polyclonal antibody based ELISA. Antibodies showed no cross-reaction with exoantigens from a further 17 *Aspergillus* species and 10 other genera species, nor with grain extracts. The antibodies against *A. ochraceus* exoantigens were species-specific, whereas the antibodies against *A. flavus* and *A. versicolor* cross-reacted to variable degrees with the antigens from other *Aspergillus* species.

Yong and Cousin (2001) developed a double-sandwich ELISA based on polyclonal antibodies against an extracellular antigen from *A. parasiticus*. These antibodies cross-reacted with *A. flavus*, *A. sojae* and *A. oryzae*, phenotypically similar species to *A. parasiticus*. Low and high aflatoxin levels corresponded to low and high mould antigens (ELISA readings) in naturally contaminated corn samples. Therefore, this ELISA showed potential as an early detection method for aflatoxigenic fungi.

Gan et al. (1997) immunized chickens with mycelia-soluble antigens and the exoantigens of *F. sporotrichioides*, *F. poae* and *F. graminearum*, and developed a polyclonal antibody-based ELISA to determine their concentration in grain. These antigens were characterized by ELISA, immunoblotting and sodium dodecyl sulphate polyacrylamide gel electrophoresis. Antibodies to the exoantigens and mycelia-soluble antigens from 11 species of *Fusarium* and 11 species of fungi from other genera did not cross-react with grain extracts. The antibodies against mycelia-soluble antigens tended to cross-react with other genera antigens, although antibodies raised to exoantigens were genus specific. Antibodies to exoantigens from *F. poae* were species specific, demonstrating the feasibility of exoantigen immunoassays for *Fusarium* sp. detection in grain and food. This ELISA (Gan et al. 1997) was also used to estimate *Fusarium* exoantigen levels in 40 samples of *Fusarium* head-blight-infected hard wheat, and in 50 samples of infected soft wheat. These samples were also assayed for deoxynivalenol and ergosterol (Abramson et al. 1998). *F. sporotrichioides* antisera showed linear correlation between the exoantigen level and deoxynivalenol content with coefficients of 0.80 (hard wheat) and 0.76 (soft wheat). The same antisera also indicated linear correlations between the exoantigen level and total ergosterol concentration with coefficients of 0.66 (hard wheat) and 0.81 (soft wheat). This assay could be a useful tool for detection of *Fusarium* species, as preharvest confirmation, providing an early warning about deoxynivalenol production in cereals (Abramson et al. 1998).

A sandwich ELISA based on monoclonal antibodies specific to *Fusarium* extracellular polysaccharide (EPS) was developed for the rapid detection of *Fusarium* species, with a detection limit of 0.001 μg/ml against *F. verticillioides* EPS. Among the 59 strains tested, high reactivity occurred with most of *Fusarium* species, but no reactivity was observed with other genera, except *A. versicolor* and *Trichoderma viride*. The mean recoveries for *F. verticillioides* EPS spiked in foods from 1 to 100 μg/g were 151% (rice), 84% (potato), and 94% (mandarine orange). The correlation coefficient between the EPS levels (sandwich ELISA) and dry mycelial weight (liquid culture of *F. verticillioides*) was 0.97, while the correlation between the EPS and colony forming unit in solid culture was 0.91 (Kwak et al. 2003).

More studies are needed on antigens from mycotoxigenic fungi for further development of biosensors with potential in reducing the time required to easily detect and identify molds in food and feedstuff.

Conclusion

The introduction of immunoassays as reliable analytical methods in the early detection of mycotoxins and/or toxigenic fungi is an unquestionable fact, in order to comply with the trade requirements. The current immunoassays fulfill the features of screening methods providing rapid diagnosis with high sensitivity and accuracy to avoid exposure of humans and animals to mycotoxin hazards. ELISA is currently available for analysis of major mycotoxins, although false positive reactions still hinder its application as an alternative method recognized by the AOAC, and confirmation should be performed by chemical methods. The immunoaffinity chromatography provided specificity and efficiency for the clean-up and concentration steps prior to HPLC analysis, and has been widely employed in mycotoxin determination. In addition, ELISA based on antibodies to fungal exoantigens can contribute to semi-quantitative detection of fungi in food. New approaches for rapid methods are based on biosensors, dipstick kits, liposome immunoanalysis as well as surface plasmon resonance, which can be used in the rapid analysis of large number of samples. These immunochemical methods show the trend for screening tests as the main approaches of a reliable and cost-effective determination and prevention strategy. Therefore the production of antibodies with high specificity and affinity is a challenge to immunoassay development for food safety assurance.

Acknowledgements

The authors would like to thank National Council of Scientific and Technological Development (CNPq, Brazil) and Araucaria Foundation for financial support.

References

Abramson, D., Gan, Z., Clear, R.M., Gilbert, J. and Marquardt, R.R. (1998) Relationships among deoxynivalenol, ergosterol and *Fusarium* exoantigens in Canadian hard and soft wheat. International Journal of Microbiology 45: 217–224.

Aldao, M.A.J., Carpinella, M.C., Corelli, M. and Herrero, G.G. (1995) Competitive ELISA for quantifying small amounts of aflatoxin B_1. Food and Agricultural Immunology 7(4): 307–314.

Ayres, J.C. (1979) Significance of food mycology - an overview. In: Food mycology, (ed), Rhodes M.E. Massachusetts: G.K. Hall & Co., 3–10.

Azcona-Oliveira, J.I., Abouzied, M.M., Plattner, R.D., Norred, W.P., Pestka, J.J. (1992a) Generation of antibodies reactive with fumonisins B1, B2 and B3 by using cholera toxin as the carrier-adjuvant. Applied and Environmental Microbiology 58(1): 169–173.

Azcona-Olivera, J.I., Abouzied, M.M., Plattner, R.D., Pestka, J.J. (1992b) Production of monoclonal antibodies to the mycotoxins fumonisins B_1, B_2 and B_3. Journal of Agricultural and Food Chemistry 40(3): 531–534.

Banks, J.N., Cox, S.J. and Northway, B.J. (1993) Polyclonal and monoclonal antibodies to field and storage fungi. International Biodeterioration and Biodegradation 32: 137–144.

Barmark, A.-L. and Larsson, K. (1994) Immunoaffinity column clean-up/liquid chromatography determination of aflatoxins: An interlaboratory study. Journal of AOAC International 77: 46–53.

Barna-Vetró, I., Solti, L., Téren, J., Gyöngyösi, A., Szabó, E. and Wölfling, A. (1996) Sensitive ELISA Test for determination of ochratoxin A. Journal of Agricultural and Food Chemistry 44: 4071–4074.

Barna-Vetró, I., Szabo, E., Fazekas, B. and Solti, L. (2000) Development of a sensitive ELISA for the determination of fumonisin B_1 in cereals. Journal of Agricultural and Food Chemistry 48: 2821–2825.

Batista, L.R., Chalfoun, S.M., Prado, G., Schwan, R.F. and Wheals, A.E. (2003) Toxigenic fungi associated with processed (green) coffee beans (*Coffea arabica* L.). International Journal of Food Microbiology 85(3): 293–300.

Beaver, R.W., James, M.A. and Lin, T.Y. (1991) Comparison of an ELISA-based screening test with liquid chromatography for the determination of aflatoxins in corn. Journal of the Association of Official Analytical Chemists 74: 827–829.

Beretta, B., Domenico, R., Gaiaschi, A., Ballabio, C., Galli, C.L., Gigliotti, C. and Restani, P. (2002) Ochratoxin A in cereal-based baby foods: occurrence and safety evaluation. Food Additives and Contaminants 19(1): 70–75.

Bird, C.B., Malone, B., Rice, L.G., Ross, P.F., Eppley, R. and Abouzied, M.M. (2002) Determination of total fumonisins in corn by competitive direct enzyme-linked immunosorbent assay: collaborative study. Journal of AOAC International 85(2): 404–410.

Bradburn, N., Blunden, G., Coker, R.D. and Jewers, K. (1993) Aflatoxin contamination of maize. Tropical Science 33: 418–428.

Burdaspal, P., Ma Legarda, T. and Gilbert, J. (2001) Immunoaffinity column clean-up with liquid chromatography for the determination of ochratoxin A in baby food: collaborative study. Journal of AOAC International 84: 1445–1452.

Candlish, A.A.G., Stimson, W.H. and Smith, J.E. (1988) Determination of ochratoxin A by monoclonal antibody-based enzyme immunoassay. Journal of the Association of Official Analytical Chemists International 71(5): 961–964.

Carlson, M.A., Bargeron, C.B., Benson, R.C., Fraser, A.B., Phillips, T.E., Velky, J.T., Groopman, J.D., Strickland, P.T. and Ko, H.W. (2000) Na automated, handheld biosensor for aflatoxin. Biosensors and Bioelectronics 14: 841–848.

Castellari, M., Fabri, S., Fabiani, A., Amati, A. and Galassi, S. (2000) Comparation of different immunoaffinity clean-up procedures for high performance liquid chromatographic analysis of ochratoxin A in wines. Journal Chromatography A 888(1–2): 129–136.

Chen, F.S., Chen, J.W., Zhao, S., Gan, Z.B, Luo, X.C. and Zhou, Q. (2000) Characterization of chicken antisera raised against *Aspergillus* spp. by enzyme linked immunosorbent assay. Nahrung 44: 323–327.

Christensen, H.R., Yu, F-Y. and Chu, F.S. (2000) Development of a polyclonal antibody-based sensitive enzyme-linked immunosorbent assay for fumonisin B_4. Journal of Agricultural and Food Chemistry 48: 1977–1984.

Chu, F.S. (1984) Immunoassays for analysis of mycotoxins. Journal of Food Protection 47: 562–569.

Chu, F.S., Fan, T.S.L., Zhang, G.S., Xu, Y.C., Faust, S. and McMahon, P.L. (1987) Improved enzyme-linked immunosorbent assay for aflatoxin B_1 in agricultural commodities. Journal of the Association of Official Analytical Chemists 70: 854–857.

Cousin, M.A. (1990) Development of the enzyme-linked immunosorbent assay for detection of molds in foods: a review. Developments in Industrial Microbiology 31: 157–163.

Cousin, M.A. (1996) Chitin as a measure of mold contamination of agricultural commodities and foods. Journal of Food Protection 59: 73–81.

Daly, S.J., Keating, G.J., Dillon, P.P., Manning, B.M., O'Kennedy, R., Lee, H.A. and Morgan, M.R. (2000) Development of surface plasmon resonance-based immunoassay for aflatoxin B1. Journal of Agricultural and Food Chemistry 48(11): 5097–5104.

Danielsen, S. and Funck Jensen, D. (1998) Relationships between seed germination, fumonisin content, and *Fusarium verticillioides* infection in selected maize samples from different regions of Costa Rica. Plant Pathology 47: 609–614.

De Saeger, S. and Van Peteghem, C. (1996) Dipstick Enzyme immunoassay to detect *Fusarium* T-2 toxin in wheat. Applied and Environmental Microbiology 62: 1880–1884.

De Saeger, S., Sibanda, L., Desmet, A. and Van Peteghem, C. (2002) A collaborative study to validate novel field immunoassay kits for rapid mycotoxin detection. International Journal of Food Microbiology 75: 135–142.

Desjardins, A.E., Manandhar, G., Plattner, R.D., Maragos, C.M., Shrestha, K., and McCormick, S.P. (2000) Occurrence of *Fusarium* species and mycotoxins in Nepalese maize and wheat and the effect of traditional processing methods on mycotoxin levels. Journal of Agricultural and Food Chemistry 48: 1377–1383.

Devi, K.T., Mayo, M.A., Reddy, K.L.N., Delfosse, P., Reddy, G., Reddy, S.V. and Reddy, D.V.R. (1999) Production and characterization of monoclonal antibodies for aflatoxin B_1. Letters in Applied Microbiology 29: 284–288.

Dewey, F.M., MacDonald, M.M., Phillip, S.I. and Priestley, R.A. (1990) Development of monoclonal-antibody-DIP-STICK immunoassays for *Penicillium islandicum* in rice grains. Journal of General Microbiology 136: 753–760.

Dietrich, R., Schneider, E., Usleber, E. and Martlbauer, E. (1995) Use of monoclonal antibodies for the analysis of mycotoxins. Natural Toxins 3: 288–293.

Dorner, J.W. and Cole, R.J. (1989) Comparison of two ELISA screening tests with liquid chromatography for determination of aflatoxins in raw peanuts. Journal of the Association of Official Analytical Chemists 72(6): 962–964.

Dragacci, S. and Fremy, J.M. (1996) Application of immunoaffinity column cleanup to aflatoxin M_1 determination and survey in cheese. Journal of Food Protection 59: 1011–1013.

Dragacci, S., Grosso, F. and Gilbert, J. (2001) Immunoafinity column clean-up with liquid chromatography for determination of aflatoxin M_1 in liquid milk: collaborative study. Journal of AOAC International 84: 437–443.

Entwisle, A.C., Williams, A.C., Mann P.J., Russel, J., Slack, P.T. and Gilbert, J. (2001) Combined phenyl silane and immunoaffinity column clean-up with liquid chromatography for the determination of ochratoxin A in roasted coffee: collaborative study. Journal of AOAC International 84: 444–450.

Eskola, M., Kokkonen, M. and Rizzo, A. (2002) Application of Manual and Automated Systems for Purification of Ochratoxin A and Zearalenone in Cereals with Immunoaffinity Columns. Journal of Agricultural and Food Chemistry 50: 41–47.

Food and Agriculture Organization of the United Nations (FAO). (1997) Application of risk management to food safety matters. Report of the joint FAO/WHO consultation, FAO Food and Nutrition paper no. 65, Rome: FAO.

Fazekas, B. and Tar, A. (2001) Determination of zearalenone content in cereals and feedstuffs by immunoaffinity column coupled with liquid chromatography, Journal of the Association of Official Analytical Chemists International 84(5): 1453–1459.

Fuhrmann, B., Lebreton, V., Van Hoegaerden, M., Kamphuis, H.J. and Strosberg, A.D. (1992) A monoclonal antibody specific for conidia and mycelium wall layer of *Penicillium* and *Aspergillus*. Microbiology and Immunology 36: 1–12.

Fuhrmann, B., Roquebert, M.F., Lebreton, V. and Van Hoegaerden, M. (1990) Immunological differentiation between *Penicillium* and *Aspergillus* taxa. In: Modern Concepts in *Penicillium* and *Aspergillus* Classification, (eds), Samson, R.A. and Pitt, J.I. Plenum Press, New York, 423–432.

Gan, Z., Marquardt, R.R., Abramson, D. and Randy, M.C. (1997) The characterization of chicken antibodies raised against *Fusarium* spp. by enzyme-linked immunosorbent assay and immunoblotting. International Journal of Food Microbiology 38: 191–200.

Garden, S.R. and Strachan, N.J.C. (2001) Novel colorimetric immunoassay for the detection of aflatoxin B_1. Analytica Chimica Acta 444: 187–191.

Gathumbi, J.K., Usleber, E., Ngatia, T.A., Kangethe, E.K., Martlbauer, E. (2003) Application of immunoaffinity chromatography and enzyme immunoassay in rapid detection of aflatoxin B1 in chicken liver tissues. Poultry Science 82: 585–590.

Gazzaz, S.S., Rasco, B.A. and Dong, F.M. (1992) Application of immunochemical assays to food analysis. Critical Reviews in Food Science and Nutrition 32(3): 197–229.

Gonzalez-Martinez, M.A., Puchades, R. and Maquieira, A. (1999) Online immunoanalysis for environmental pollutants: from batch assays to automated sensors. Trends in Analytical Chemistry 18: 204–218.

Gourama, H. and Bullerman, L.B. (1995) Detection of molds in foods and feeds: potential rapid and selective methods. Journal of Food Protection 58: 1389–1394.

Grosso, F., Saïd, S., Mabrouk, I., Fremy, J.M., Castegnaro, M., Jemmali, M., Dragacci, S. (2003) New data on the occurrence of ochratoxin A in human sera from patients affected or not by renal diseases in Tunisia. Food and Chemical Toxicology 41(8): 1133–1140.

Harrison, L.R., Colvin, B.M., Green, J.T., Newman, L.E. and Cole, J.R. (1990) Pulmonary edema and hydrothorax in swine produced by fumonisin B_1, a toxic metabolite of *Fusarium moniliforme*. Journal of Veterinary Diagnostic Investigation 2: 217–221.

Hefle, S.L. (1995) Immunoassay fundamentals. *Food Technology* 49(2): 102–107.

Ho, J.-A. A and Durst, R.A. (2000) Development of a flow-injection liposome immunoanalysis system for fumonisin B_1. Analytica Chimica Acta 414: 61–69.

Ho, J.-A.A. and Durst, R.A. (2003) Detection of fumonisin B_1: comparison of flow-injection liposome immunoanalysis with high-performance liquid chromatography. Analytical Biochemistry 312: 7–13.

Ho, J.-A.A. and Wauchope, R.D. (2002) A strip liposome immunoassay for aflatoxin B_1. Analytical Chemistry 74: 1493–1496.

Hongyo, K., Itoh, Y., Hifumi, E. and Takeyasu, A. (1992) Comparison of monoclonal antibody-based enzyme-linked immunosorbent assay with thin-layer chromatography and liquid chromatography for aflatoxin B_1 determination in naturally contaminated corn and mixed feed. Journal of AOAC International 75: 307–312.

Hussein, S.H. and Brasel, J.M. (2001) Toxicity, metabolism and impact of mycotoxins on humans and animals. Toxicology 167 (2): 101–134.

Iijima, K., Kawamura, O, Wang, D.-S., Manabe, M., Tanaka, K., Chen, G., Yu, S.Z. and Ueno, W. (1996) Development of highly sensitive immunosorbent assay for fumonisins and its application for contaminated corn. Mycotoxins 42: 63–66.

International Agency for Research on Cancer. IARC Monographs on the Evaluation of Carcinogenic Risks to Humans and Their Supplements: A Complete List. Vol. 82, Some Traditional Herbal Medicines, Some Mycotoxins, Naphthalene and Styrene; International Agency for Research on Cancer, World Health Organization: Lyon, France, 2002; http://monographs.iarc.fr/htdocs/ monographs/vol82/82-05.html.

Jornet, D., Busto, O. and Guash, J. (2000) Solid phase extraction applied to the determination of ochratoxin A in wines by reversed-phase high-performance liquid chromatography. Journal of Chromatography A 882 (1-2): 29–35.

Kaufman, L. and Standard, P.G. (1987) Specific and rapid identification of medically important fungi by exoantigens detection. Annual Review of Microbiology 41: 209–225.

Kawamura, O., Nagayama, S., Sato, S., Ohtani, K., Ueno, I. and Ueno, Y. (1988) A monoclonal antibody-based enzyme-linked immunosorbent assay for aflatoxin B_1 in peanut products. Mycotoxin Research 4: 75–88.

Kawamura, O., Sato, S., Kajii, H., Nagayama, S., Ohtani, K., Chiba, J. and Ueno, Y. (1989) A sensitive Enzyme-Linked Immunosorbent assay of Ochratoxin A based on Monoclonal Antibodies. Toxicon 27(8): 887–897.

Kellerman, T.S., Marasas, W.F.O., Thiel, P.G., Gelderblom, W.C.A., Cawood, M. and Coetzer, J.A.W. (1990) Leukoencephalomalacia in two horses induced by oral dosing of fumonisin B1. Onderstepoort Journal Veterinary Research 57: 269–275.

Kim, E.K., Shon, D.H., Chung, S.H. and Kim, Y.B. (2002) Survey for fumonisin B_1 in Korean corn-based food products. Food Additives Contaminants 19(5): 459–464.

Koeltzow, D.E. and Tanner, S.N. (1990) Comparative evaluation of commercially available aflatoxin test methods. Journal of the Association of Official Analytical Chemists 73: 584–589.

Kondo, F., Ito, Y., Yamada, S., Tsuji, K., Imokawa, M., Niimi, Y., Harada, K.-I., Ueno, Y. and Miyazaki, Y. (2002) Determination of microcystins in lake water using reusable immunoaffinity column. Toxicon 40: 893–899.

Kulisek, E.S. and Hazebroek, J.P. (2000) Comparison of extraction buffers for the detection of fumonisin B_1 in corn by immunoassay and high-performance liquid chromatography. Journal of Agricultural and Food Chemistry 48: 65–69.

Kwak, B.Y., Kwon, B.J., Kweon, C.H. and Shon, D.H. (2003) Detection of *Fusarium* species by enzyme-linked immunosorbent assay using monoclonal antibody. Journal of Microbiology and Biotechnology 13(5): 794–799.

Lee, S.C. and Chu, F.S. (1984) Enzyme-linked immunosorbent assay of ochratoxin A in wheat. Journal of the Association of Official Analytical Chemists 67: 45–49.

Li, S., Marquardt, R.R. and Abramson, D. (2000) Immunochemical detection of molds: A review. Journal of Food Protection 63: 281–291.

Li, S., Marquardt, R.R., Frohlich, A.A., Xiao, H. and Clarke, J.R. (1994) The development of a quantitative ELISA for aflatoxin B_1 in ground peanuts using antibodies isolated from the yolk of a laying hen. Journal of Food Protection 57: 1022–1024.

Lipigorngoson, S., Limtrakul, P., Suttajit, M. and Yoshizawa, T. (2003) In-house direct cELISA for determining aflatoxin B_1 in Thai corn and peanuts. Food Additives Contaminants 20(9): 838–845.

Lu, P., Marquardt, R.R., Frohlich, A.A. and Mills, J.T. (1994) Detection of *Penicillium aurantiogriseum* by ELISA utilizing antibodies produced against its exoantigens. Microbiology 140: 3267–3276.

Lu, P., Marquardt, R.R., Frohlich, A.A. and Mills, J.T. (1995a) Development of solid phase immunoassays for the detection and quantification of *A. ochraceus* in wheat grain. Food and Agriculture Immunology 7: 81–93.

Lu, P., Marquardt, R.R. and Kierek-Jaszczuk. (1995b) Immunochemical identification of fungi using polyclonal antibodies raised in rabbits to exoantigens from *Aspergillus ochraceus*. Letters in Applied Microbiology 20: 41–45.

Maragos, C.M, Jolley, M.E., Plattner, R.D. and Nasir, M.S. (2001) Fluorescence polarization as a means for determination of fumonisins in maize. Journal of Agricultural and Food Chemistry 49: 596–602.

Maragos, C.M. (2002) Novel assays and sensor platforms for the detection of aflatoxins. Advances in Experimental Medicine and Biology 504: 85–93.

Marasas, W.F.O. (1996) Fumonisins: History, world-wide occurrence and impact. In: Fumonisins in Food, (eds), Jackson L.S., De Vries J.W. and Bullerman L.B. Plenum Press, New York, 3–18.

Moghaddam, A., Lobersli, I., Gebhardt, K., Braunagel, M., Mervik, O.J. (2001) Selection and characterization of recombinant single-chain antibodies to the hapten Aflatoxin-B1 from naïve recombinant antibody libraries. Journal of Immunological Methods 254: 169–181.

Morgan, M.R.A., McNerney, R. and Chan, H.W.S. (1983) Enzyme-linked immunosorbent assay of ochratoxin A in barley. Journal of the Association of Official Analytical Chemists 66: 1481–1484.

Mortimer, D.N., Shepherd, M.J., Gilbert, J. and Clark, C. (1988) Enzyme-linked immunosorbent (ELISA) determination of aflatoxin B_1 in peanut butter: collaborative trial. Food Additives and Contaminants 5: 601–608.

Mouyna, I., Sarfati, J., Recco, P., Fontaine, T., Henrissat, B. and Latge, J.P. (2002) Molecular characterization of a cell wall-associated beta (1-3) endoglucanase of *Aspergillus fumigatus*. Medical Mycology 40(5): 455–464.

Mullett, W., Lai, E.P.C., Yeung, J.M. (1998) Immunoassay of fumonisins by a surface plasmon resonance biosensor. Analytical Biochemistry 258: 161–167.

Munkvold, G.P. and Desjadins, A.E. (1997) Fumonisin in maize: can we reduce their occurrence?. Plant Disease 81: 556–565.

Nasir, M.S. and Jolley, M.E. (2002) Development of a Fluorescence Polarization Assay for the Determination of Aflatoxins in Grains. Journal of Agricultural and Food Chemistry 50: 3116–3121.

Neucere, J.N., Ullah, A.H.J. and Cleveland, T.E. (1992) Surface proteins of two aflatoxin-producing isolates of *A. flavus* and *A. parasiticus* mycelia: 1. A comparative immunochemical profile. Journal of Agricultural and Food Chemistry 40: 1610–1612.

Nilufer, D. and Boyacioglu, D. (2002) Comparative study of three different methods for the determination of aflatoxins in tahini. Journal of Agricultural and Food Chemistry 50(12): 3375–3379.

Notermans, S., Heuvelman, C., Van Egmond, H.P., Paulsch, W.E. and Besling, J.R. (1986) Detection of mold in food by enzyme-linked immunosorbent assay. Journal of Food Protection 49: 786–791.

Notermans, S., Wieten, G., Engel, H.W.B., Rombouts, F.M., Hoogerhout, P. and van Boom, J.H. (1987) Purification and properties of extracellular polysaccharide (EPS) antigens produced by different mould species. Journal of Applied Bacteriology 62: 157–166.

Ono, E.Y.S., Kawamura, O., Ono, M.A., Ueno, Y. and Hirooka, E.Y. (2000) A comparative study of indirect competitive ELISA and HPLC for fumonisin detection in corn of the State of Paraná, Brazil. Food and Agricultural Immunology 12(1): 5–14.

Ono, E.Y.S., Ono, M.A., Funo, F.Y., Medina, A.E., Oliveira, T.C.R.M., Kawamura, O., Ueno, Y. and Hirooka, E.Y. (2001) Evaluation of fumonisin-aflatoxin co-occurrence in Brazilian corn hybrids by ELISA. Food Additives and Contaminants 18: 719–729.

Ono, E.Y.S., Sugiura, Y., Homechin, M., Kamogae, M., Vizzoni, E., Ueno, Y. and Hirooka, E.Y. (1999) Effect of climatic conditions on natural mycoflora and fumonisins in freshly harvested corn of the State of Paraná, Brazil. Mycopathologia 147: 139–148.

Pal, A. and Dhar, T.K. (2004) An analytical device for on-site immunoassay: Demonstration of its applicability in semiquantitative detection of aflatoxin B_1 in a batch of samples with ultrahigh sensitivity. Analytical Chemistry 76: 98–104.

Park, D.L., Miller, B.M., Nesheim, S., Trucksess, M.W., Vekich, A., Bigigare, B., McVey, J.L. and Brown, L.H. (1989) Visual and semiquantitative spectrophotometric ELISA screening method for aflatoxin B_1 in corn and peanut products: follow-up collaborative study. Journal of the Association of Official Analytical Chemists 72(4): 638–643.

Patey, A.L., Sharman, M. and Gilbert, J. (1991) Liquid chromatographic determination of aflatoxin levels in peanut butters using Na immunoaffinity column cleanup method: international collaborative trial. Journal of the Association of Official Analytical Chemists International 74: 76–81.

Patey, A. L., Sharman, M. and Gilbert, J. (1992) Determination of total aflatoxin levels in peanut butter by enzyme-linked immunosorbent assay: collaborative study. Journal of Association of Official Analytical Chemists International 75: 693–697.

Pestka, J.J. and Casale, W.L. (1990) Naturally occurring fungal toxins. In: Food Contamination from Environmental Sources, (eds), Simmons M.S. and Nriagy, J. John Wiley and Sons Ltd., New York, 613–638.

Pestka, J.J., Abouzied, M.N. and Sutikno. (1995) Immunological assays for mycotoxin detection. Food Technology 2: 120–128.

Pestka, J.J., Azcona-Oliveira, J.I., Plattner, R.D., Minervini, F., Doko, M.B. and Visconti, A. (1994) Comparative assessment of fumonisin in grain based foods by ELISA, GC-MS, and HPLC. Journal of Food Protection 57(2): 169–172.

Petzinger, E. and Weidenbach, A. (2002) Mycotoxins in the food chain: the role of ochratoxins. Livestock Production Science 76: 245–250.

Pitt, J.I. and Hocking, A.D. (1997) Fungi and food spoilage, 2nd ed., Blackie Academic & Professional, London.

Proctor, R.H., Desjardins, A.E., Plattner, R.D. and Holn, T.H. (1999) A polyketide synthase gene required for biosynthesis of fumonisin mycotoxins in *Gibberella fujikuroi* mating population A. Fungal Genetics and Biology 27: 100–112.

Rheeder, J.P., Marasas, W.F.O. and Vismer, H.F. (2002) Production of fumonisin analogs by *Fusarium* species. Applied and Environmental. Microbiology 68: 2101–2105.

Rheeder, J.P., Marasas, W.F.O, Thiel, P.G., Sydenham, E.W., Shephard, G.S. and Van Schalkwyk, D.J. (1992) *Fusarium moniliforme* and fumonisins in corn in relation to human esophageal cancer in Transkei. Phytopathology 82: 353–357.

Scott, P.M. and Trucksess, M.W. (1997) Application of Immunoaffinity Columns to Mycotoxin Analysis. Journal of the Association of Official Analytical Chemists International 80(5): 941–949.

Scott, P.M., Yeung, J.M., Lawrence, G.A. and Prelusky, D.B. (1997) Evaluation of enzyme-linked immunosorbent assay for analysis of beer for fumonisins. Food Additives and Contaminants 14(5): 445–450.

Shapira, R., Paster, N., Menasherov, M., Eyal, O., Mett, A., Meiron, T., Kuttin, E. and Salomon, R. (1997) Development of polyclonal antibodies for detection of aflatoxigenic moulds involving culture filtrate and chimeric proteins expressed in *Escherichia coli*. Applied and Environmental Microbiology 63: 990–995.

Sibanda, L., De Saeger, S. and Van Peteghem, C. (1999) Development of a portable field immunoassay for the detection of aflatoxin M_1 in milk. International Journal of Food Microbiology 48: 203–209.

Sibanda, L., De Saeger, S., Bauters, T.G.M., Nelis, H.J. and Van Peteghem, C. (2001) Development of a flow-through enzyme immunoassay and application on screening green coffee samples for ochratoxin A with confirmation by high-performance liquid chromatograpy. Journal of Food Protection 64: 1597–1602.

Sibanda, L., De Saeger, S., Van Peteghem, C., Grabarkiewicz-Szczesna, J. and Tomczak, M. (2000) Detection of T-2 Toxin in Different Cereals by Flow-Through Enzyme Immunoassay with a Simultaneous Internal Reference. Journal of Agricultural and Food Chemistry 48: 5864–5867.

Solfrizzo, M., Avantaggiato, G. and Visconti, A. (1998) Use of various clean-up procedures for the analyses of ochratoxin A in cereals. Journal of Chromatography A 815: 67–73.

Stroka, J. and Anklam, E. (2002) New strategies for screening and determination of aflatoxins and for detection of aflatoxin producing mould in food and feed. Trends in Analytical Chemistry 21: 90–95.

Sutikno Abouzied, M.M., Azcona-Olivera, J.I., Hart, L.P. and Pestka, J. (1996) Detection of fumonisins in *Fusarium* cultures, corn, and corn products by polyclonal antibody-based ELISA: relation to fumonisin B_1 detection by liquid chromatography. Journal of Food Protection 59(6): 645–651.

Sweeney, M.J. and Dobson, A.D.W. (1999) Molecular biology of mycotoxin biosynthesis, FEMS Microbiology Letters 175(2): 149–163.

Sydenham, E.W. and Shephard, G.S. (1996) Chromatographic and allied methods of analysis for selected mycotoxins. In: Progress in Food Contaminants Analysis, (ed), Gilbert, J. Blackie Academic & Professional Press, London, 65–146.

Sydenham, E.W., Shephard, G.S., Thiel, P.G., Bird, C. and Miller, B.M. (1996a) Determination of fumonisins in corn: evaluation of competitive immunoassay and HPLC techniques. Journal of Agricultural and Food Chemistry 44(1): 159–164.

Sydenham, E.W., Stockenstrom, S., Thiel, P.G., Rheeder, J.P., Doko, M.B., Bird, C. and Miller, B.M. (1996b) Polyclonal antibody-based ELISA and HPLC methods for the determination of fumonisins in corn: a comparative study. Journal of Food Protection 59(8): 893–897.

Thirumala-Devi, K., Mayo, M.A., Hall, A.J., Craufurd, P.Q., Wheeler, T.R., Waliyar, F., Subrahmanyam, A. and Reddy, D.V.R. (2002) Development and application of an indirect competitive enzyme-linked immunoassay for aflatoxin M_1 in milk and milk-based confectionery. Journal of Agricultural and Food Chemistry 50: 933–937.

Thirumala-Devi, K., Mayo, M.A., Reddy, G., Reddy, S.V., Delfosse, P. and Reddy, D.V. (2000) Production of polyclonal antibodies against ochratoxin A and its detection in chilies by ELISA. Journal of Agricultural and Food Chemistry 48(10): 5079–82.

Thrane, U. (1989) *Fusarium* species and their specific profile of secondary metabolites. In: *Fusarium*: Mycotoxin, Taxonomy and Pathogenicity, (ed), Chelkowski J. Elsevier, Amsterdam, 199–225.

Trucksess, M.W. and Stack, M.E. (1994) Enzyme-linked immunosorbent assay of total aflatoxins B1, B2 and G1 in corn: follow-up collaborative study. Journal of Association of Official Analytical Chemists International 77(3): 655–658.

Trucksess, M.W., Stack, M.E., Nesheim, S., Page, S.W., Albert, R.H., Hansen, T.J. and Donahue, K.F. (1991) Immunoaffinity column coupled with solution fluorometry or liquid chromatography postcolumn derivatization for determination of aflatoxins in corn, peanuts, and peanut butter: collaborative study. Journal of the Association of Official Analytical Chemists International 74(1): 81–88.

Trucksess, M.W., Stack, M.E., Nesheim, S., Park, D.L. and Pohland, A.E. (1989) Enzyme-linked immunosorbent assay of aflatoxins B_1, B_2, and G_1 in corn, cottonseed, peanuts, peanut butter, and poultry feed: collaborative study. Association of Official Analytical Chemists International 72(6): 957–962.

Trucksess, M.W., Young, K., Donahue, K.F., Morris, D.K. and Lewis, E. (1990) Comparison of two immunochemical methods with thin-layer chromatographic methods for determination of aflatoxins. Association of Official Analytical Chemists International 73(3): 425–428.

Tsai, G.J. and Cousin, M.A. (1990) Enzyme-linked immunosorbent assay for detection of molds in cheese and yogurt. Journal Dairy Science 73: 3366–3378.

Tsai, G.-J. and Yu, S.-C. (1999) Detecting *Aspergillus parasiticus* in cereals by an enzyme-linked immunosorbent assay. International Journal of Food Microbiology 50: 181–189.

Ueno, Y., Iijima, K., Wang, S.-D., Sugiura, Y., Sekijima, M., Tanaka, T., Chen, C. and Yu, S.-Z. (1997) Fumonisins as a possible contributory risk factor for primary liver cancer: A 3-year study of corn harvested in Haimen, China, by HPLC and ELISA. Food Chemistry. Toxicology 5: 1143–1150.

Usleber, E., Straka, M. and Terplan, G. (1994) Enzyme immunoassay for fumonisin B_1 applied to corn-based food. Journal of Agricultural and Food Chemistry 42(6): 1392–1396.

Valenta, H. (1998) Chromatographic methods for the determination of ochratoxin A in animal and human tissues and fluids. Journal of Chromatography A, 815: 75–92.

Van Egmond, H.P. (1989) Current situation on regulations for mycotoxins. Overview of tolerances and status for standard methods of sampling and analysis. Food Additives and Contaminants 6: 139–188.

Van Egmond, H.P. (1996) Analytical Methodology and regulations for ochratoxin A. Food Additives and Contaminants 13: 11–13.

Visconti, A. and Boenke, A. (1995) Improvement of the determination of fumonisins (FB_1 and FB_2) in maize and maize-based feeds, European Commission BCR Information—Chemical Analysis. Brussels, Belgium.

Visconti, A., Pascale, M. and Centonze, G. (2000) Determination of ochratoxin A in domestic and imported beers in Italy by immunoaffinity clean-up and liquid chromatography. Journal of Chromatography A 888: 321–326.

Visconti, A., Solfrizzo, M. and De Girolamo, A. (2001) Determination of fumonisin B_1 and B_2 in corn and corn flakes by high performance liquid chromatography and immunoaffinity column clean-up: collaborative study. Journal of AOAC International 84(6): 1828–1837.

Watanabe, E., Yosimura, Y., Yuasa, Y. and Nakazawa, H. (2001) Immunoaffinity column clean-up for the determination of imazalil in citrus fruits. Analytica Chimica Acta 433: 199–206.

Wong, W.C., White, M. and Wright, I.G. (1988) Production of monoclonal antibodies to *Fusarium oxysporum* f.sp. cubense race 4. Letters in Applied Microbiology 6: 39–42.

Wyatt, G.M. (1992) Immunoassays for food poisoning bacteria and bacterial toxins, Chapman & Hall Press, London.

Xiao, H., Clarke, J., Marquardt, R.R. and Frohlich, A.A. (1995) Improved methods for conjugating selected mycotoxins to carrier proteins and dextran for immunoassays. Journal of Agricultural and Food Chemistry 43: 2092–2097.

Yau, K.Y.F., Lee, H. and Hall, J.C. (2003) Emerging trends in the synthesis and improvement of hapten-specific recombinant antibodies. Biotechnology Advances 21: 599–637.

Yeung, J.M., Prelusky, D.B., Savard, M.E., Dang, B.D.M. and Robinson, L.A. (1996) Sensitive immunoassay for FB_1 in corn. Journal of Agricultural and Food Chemistry 44: 3582–3586.

Yong, R.K. and Cousin, M.A. (2001) Detection of moulds producing aflatoxins in maize and peanuts by an immunoassay. International Journal of Food Microbiology 65: 27–38.

Yu, F.-Y. and Chu, F.S. (1996) Production and characterization of antibodies against fumonisin $B_{1.}$ Journal of Food Protection 59: 992–997.

17. Biosynthesis of Silver Nanoparticles Using Fungi

Absar Ahmad[1], Satyajyoti Senapati[2], M. Islam Khan[1], Rajiv Kumar[2] and Murali Sastry[3]
[1]Biochemical Sciences, [2]Catalysis and [3]Materials Chemistry Divisions, National Chemical Laboratory, Pune-411 008, India

SUMMARY The field of nanotechnology has generated great excitement and expectation in recent years among materials scientists. An important aspect of nanotechnology concerns the development of reliable experimental procedures for the synthesis of nanomaterials of controlled size, shape, monodispersity and chemical composition. Furthermore, the current need is to develop clean and eco-friendly synthetic protocols for nanomaterial synthesis. An attractive possibility is the use of micro-organisms in the biosynthesis of nanomaterials. In this chapter, we provide a brief overview of the global research efforts on the biosynthesis of nanomaterials using micro-organisms with particular emphasis on the recent and exciting results obtained at National Chemical Laboratory, Pune on the biosynthesis of inorganic nanomaterials using fungi. Some of the challenging aspects of these efforts are highlighted.

Introduction

The essence of nanotechnology is the fabrication and utilization of materials and devices at the atomic and molecular level and the use of their unique properties and phenomena of matter at the nanoscale (1-100 nm) level. This exciting field of nanotechnology embraces the areas of research as diverse as physics, chemistry, material science, microbiology, biochemistry and molecular biology. The fascinating properties shown by nanoparticles arise due to quantum confinement (Brus, 1984; Fendler, 1992; Henglein, 1993; Weller, 1993; Colvin et al. 1994; Alivisatos, 1996), and are of interest because of their applications in catalysis (Schmid, 1992; Hoffman et al. 1992), photocatalysis (Brugger et al. 1981), optoelectronics (Brus, 1984; Wang and Herron, 1991), single-electron transistors (SETs), light emitters (Klein et al. 1997; Weller, 1998), nonlinear optical devices (Wang, 1991) etc. Recently, magnetic nanoparticles are being viewed with a considerable interest (Shi et al. 1996) because of their application in drug delivery, tissue engineering, magnetic separation of biomolecules, hyperthermia treatment for malignant cells (Pankhurst et al. 2003; Pedro Tartaj et al. 2003; Berry and Curtis 2003; Salata, 2004; Shivshankar et al. 2004), magnetic memory storage devices, magnetic resonance imaging (MRI) enhancement (Mornet et al. 2004) and magnetic refrigeration (Shull et al. 1993). The ability to tune the optical properties of semiconductor nanoparticles (also termed as quantum dots) during UV-light irradiation by simply changing the nanoparticle size is particularly attractive in diverse areas of application such as cell labeling (Wu et al. 2003), cell tracking (Parak et al. 2002), *in vivo* imaging (Dubertret et al. 2002), diagnostics and DNA detection (Taylor et al. 2000; Xu et al. 2003). More recently, nanomaterials have been viewed with interest for their potential application in nanoelectronics and nanocomputers, synthesis of new hybrid materials, energy storage devices, electronic and optical displays and biomedical devices (Fendler and Meldrum, 1995). Recognizing the wide scope of nanomaterials in new future technologies, major initiatives have been taken by many countries for understanding the fundamental principles involved and amalgamation of the same with other established fields for further development and applications.

Recent Development in Synthesis Protocols of Bio-Based Nanomaterials

The synthesis of nanoparticles of different chemical compositions, sizes/shapes and controlled monodispersity is essential for the development of nanotechnology. Generally, a number of chemical methods exist in literature (Handley, 1989) that use toxic chemicals in synthesis protocols, which raises a great concern for environmental reasons. Consequently, researchers in the field of nanoscale materials synthesis and assembly have been eagerly looking at biological systems for inspiration. Recently, the utilization of biological systems has emerged as a novel method for the synthesis of nanoparticles. It is well known that many organisms, both unicellular and multicellular, are known to produce inorganic materials either intra-or extracellularly (Simkiss and Wilbur, 1989; Mann, 1996). Some examples of bio-organisms synthesizing inorganic materials include magnetotactic bacteria (which synthesize magnetite nanoparticles), (Lovley et al. 1987; Dickson 1999), diatoms (which synthesize siliceous materials) (Mann, 1993; Oliver et al., 1995; Kroger et al., 1999) and S-layer bacteria (which produce gypsum and calcium carbonate layers) (Pum and Sleytr 1999; Sleytr et al. 1999). The secrets gleaned from nature have lead to the development of biomimetic approaches for the growth of advanced nanomaterials.

Although many biotechnological applications such as remediation of toxic metals employ micro-organisms such as bacteria (Stephen and Maenaughton, 1999) and yeast (Mehra and Winge, 1991) (the detoxification often occurring via reduction of the metal ions/formation of metal sulfides), it is only relatively recently that materials scientists have been viewing with interest such micro-organisms as possible eco-friendly nanofactories (Beveridge and Murray, 1980; Southam and Beveridge, 1996). Beveridge and co-workers have demonstrated that nanoscale gold particles may be readily precipitated within bacterial cells by incubation of the cells with Au^{3+} ions (Beveridge and Murray, 1980; Southam and Beveridge, 1996; Fortin and Beveridge, 2000). Klaus-Joerger and co-workers have shown that the bacterium *Pseudomonas stutzeri* AG259 isolated from a silver mine, when placed in a concentrated aqueous solution of $AgNO_3$, resulted in the reduction of the Ag^+ ions and formation of silver nanoparticles of well-defined size and distinct morphology within the periplasmic space of the bacteria (Klaus et al. 1999; Joerger et al. 2000; Klaus-Joerger et al. 2001). Extending this approach a step forward, they demonstrated a new microbiologically based production method, by thermal treatment of biocomposites of nanocrystalline silver and bacteria to yield a carbonaceous (cermet) nanomaterial which shows interesting optical properties for potential application in thin-flim coatings (Joerger et al. 2000). Recently, Jose-Yacaman and co-workers have reported the growth of gold and silver nanoparticles inside live alfalfa plants by gold and silver uptake from solid media (Gardea-Torresdey et al. 2001, 2003). Nair and Pradeep (2002) have shown that nanocrystals of gold, silver and their alloys can be synthesized within cells of lactic acid bacteria, *Lactobacillus* strains, present in buttermilk, when exposed to the appropriate metal ions. Apart from metal nanoparticles, there is considerable current interest in the synthesis of semiconductor nanoparticles such as CdS for application in quantum-dot (QD) fluorescent biomarkers in cell labeling (Chan et al. 2002). The extremely high photostability of quantum dots and their ability to change color by simply tuning the particle size during UV-light irradiation allows real-time monitoring of intracellular processes over a long period of time (minutes to hours) and simultaneous imaging of multiple targets inside living cells or on the cell surface (Cunningham and Lundie, 1993). Recently, bacteria and yeast have been used successfully in the synthesis of CdS and PbS nanoparticles (Dameron, et al. 1989; Holmes et al. 1995; Smith, et al. 1998; Kowshik et al. 2002a, b). Holmes et al. (1995) demonstrated that exposure of Cd^{2+} ions to the bacterium, *Klebsiella aerogenes* results in the intracellular formation of CdS nanopaticles in the size range 20-200 nm. They also showed that the tolerance of *K. aerogenes* to cadmium ions and the formation of cadmium sulfide crystallites were dependent on the buffer composition of the growth medium. In a classical study, Dameron et al. (1989) for the first time demonstrated that yeasts such as *Schizosaccharomyces pombe* and *Candida glabrata* produced intracellular CdS nanocrystals when challenged with cadmium salt in solution. Recently, Kowshik et al. (2002a) identified the yeast, *Torulopsis* sp., as being able to synthesize intracellular PbS nanocrystals when exposed to aqueous lead salt. These crystallites were recovered from the biomass by freeze-thawing and analyzed by a variety of techniques. Recognizing the importance of competency with chemically synthesized protocols in terms of performance in the application of new devices, Kowshik et al. (2002b) also showed that the CdS nanocystals in the size range of 1-1.5 nm synthesized intracellularly in *Schizosaccharomyces pombe* yeast cells exhibit ideal diode characteristics. These nanocrystals were used in the fabrication of a heterojunction with poly (p-phenylenevinylene) and exhibited diode characteristics of ~75 mA/cm^2 current at 10 V in the forward bias mode while break down occurred at ~15 V in the reversed bias mode. Thus, the use of micro-organisms in the deliberate synthesis of nanoparticles is a relatively new and exciting area of research with considerable potential for development.

Developing on the strategy to widen the scope of bio-organisms, which has immensely relied on the use of prokaryotes such as bacteria in the intracellular synthesis of nanoparticles, we have recently embarked on an interdisciplinary programme aimed at investigating whether eukaryotic organisms such as fungi would be used to grow nanoparticles of different chemical compositions and sizes. A number of different genera of fungi (close to 200 different genera of fungi) have been investigated in this effort and it has been observed that fungi are extremely good candidates in the synthesis of metal (Mukherjee, et al. 2001a,b, 2002; Ahmad et al. 2003a) and metal sulfide nanoparticles (Ahmad et al. 2002). From this extensive study, we observed that two different genera of fungi, viz. *Verticillium* (Mukherjee et al. 2001a,b) and *Fusarium* (Mukherjee et al. 2002; Ahmad et al. 2003a) can reduce the metal ions of aqueous $AuCl_4^-$ and Ag^+ ions. In the case of *Verticillium* sp., the metal ions were reduced intracellularly leading to the formation of gold and silver nanoparticles in the size range 2-20 nm (Mukherjee et al. 2001a,b). However, *Fusarium oxysporum* behaved completely differently. The reduction of metal ions occurred extracellulary leading to the formation of highly stable gold and silver nanoparticles of 5-50 nm dimensions (Mukherjee et al. 2002; Ahmad et al. 2003). Further, the versatility of this approach can be illustrated by the fact that when *Fusarium oxysporum* was challenged with aqueous solution of $CdSO_4$, highly stable quantum dots of technologically important CdS nanocrystals were formed extracellularly by a purely enzymatic process (Ahmad et al. 2002). The CdS quantum dots are formed by reaction of Cd^{2+} ions with sulphide ions that are formed by the enzymatic reduction of sulphate ions where the enzymes are released *in situ* by *Fusarium oxysporum*. Thus the use of fungi is potentially exciting since they secrete large amounts of enzymes and are simpler to deal with in the laboratory. In a bid to achieve better size and shape, we have recently enlarged the scope of our studies and have screened a number of actinomycetes in the synthesis of metal nanoparticles (Ahmad et al. 2003b,c). We observed that the exposure of an alkalotolerant actinomycete, *Rhodococcus* sp. (Ahmad et al. 2003b) to aqueous $AuCl_4^-$ ions resulted in the intracellular synthesis of gold nanoparticles while the exposure of an extremophilic actinomycete, *Thermomonospora* sp. (Ahmad et al. 2003c) led to the formation of gold nanoparticles extracellularly with good monodispersity. In this article, we briefly illustrate the methods of nanoparticle biosynthesis developed by us through two examples: intra-and extracellular growth of silver nanoparticles in *Verticillium* sp. and *Fusarium oxysporum* respectively.

Intracellular Synthesis of Silver Nanoparticles Using Fungus *Verticillium* sp.

As briefly mentioned above, the fungus *Verticillium* sp. when exposed to metal ions ($AuCl_4^-$ and Ag^+ ions) causes the reduction of the metal ions and formation of nanoparticles intracellularly. The acidophilic fungus, *Verticillium* sp., was isolated from the *Taxus* plant and maintained on potato-dextrose agar (PDA) slants at 25°C. For the synthesis of silver nanoparticles, the fungus was grown in 500 mL Erlenmeyer flasks each containing 100 mL of MGYP medium containing malt extract (0.3%), glucose (1.0%), yeast extract (0.3%), and peptone (0.5%) at 25-28°C under shaking (200 rpm) for 96 h. The mycelial mass was separated from the culture broth by centrifugation (5000 rpm) at 10°C for 20 min and washed thrice with sterile distilled water. The washed mycelial mass (10 g wet weight) was then resuspended in 100 mL of 10^{-4} M aqueous $AgNO_3$ solution in a 500 mL Erlenmeyer flask and incubated on shaker at 28°C (200 rpm) for 72 h. The biotransformation was monitored by visual inspection of the biomass as well as measurement of the UV-Vis spectra of the fungal cells.

Thin sections of the *Verticillium* cells, after reaction with Ag^+ ions, were prepared for Transmission Electron Microscopy (TEM) analysis as follows. Approximately 1 mm^3 bits of the Ag nano-*Verticillium* biomass were taken and fixed in 2.5% gluteraldehyde in distilled water for 2 h at room temperature. After fixation, the cells were sedimented (1500 rpm, 10 min) and washed three times with distilled water. Without postfixation, the pellet was subjected to dehydration with 30, 50, 70, and 90% ethanol for 15 min at each concentration followed by two changes in absolute ethanol. Since ethanol is poorly miscible with epoxy resins, propylene oxide was used as a linking agent. The dehydrated pellet was kept in propylene oxide for 15 min following which the infiltration of the resin was carried out by placing the pellet in a 1:1 mixture of propylene oxide and Epon 812 overnight at room temperature. Embedding was carried out by using a mixture of the resin (Epon 812) and hardeners (DDSA: dodecynyl succinic anhydride and MNA: methyl nadic anhydride) in the ratio 1:1.5. To this, two drops of tridimethylaminomethyl phenol (DMP30) was added to accelerate the polymerization process which was carried out using this mixture at 60 °C for 3 days. Ultrathin sections were cut using an ultramicrotome (Leica Ultracut UCT) and were taken on copper TEM grids. TEM measurements were carried out on a JEOL Model 1200EX instrument operated at an accelerating voltage of 60 kV. A low operating voltage was used to minimize the damage to the thin section by electron beam heating. Films of the fungal cells (both before and after exposure to Ag^+ ions for 72 h) for UV/Vis spectroscopy, X-ray diffraction, and SEM studies were prepared by solution casting the fungal cells onto quartz and Si(111) substrates and thoroughly drying the films in a flow of N_2.

A picture of the *Verticillium* fungal cells before and after exposure to aqueous silver nitrate solution for 72 h is shown in Fig. 1. The pale yellow color of the fungal cells can clearly be seen in Fig. 1A. The presence of the dark brown color in the fungal cells after reaction with $AgNO_3$ solution (Fig. 1B) is a clear indicator of the reduction of Ag^+ ions and formation of silver nanoparticles in the fungal biomass. Silver nanoparticles exhibit striking colours (light yellow to brown) due to excitation of surface plasmon resonance in the particles (Henglein, 1993) and thus provide a convenient means of visually determining their presence in the fungal biomass. Fig. 1B shows that the aqueous medium is colorless, thereby strongly indicating that extracellular reduction of Ag^+ ions had not occurred. Fig. 1C shows the UV-vis spectra recorded from a film of the fungal cells before (curve 1) and after immersion in 10^{-4} M $AgNO_3$ solution for 72 h (curve 2). While there is no evidence of absorption in the spectral window 400-800 nm in the case of the As-harvested fungal cells, the fungal cells exposed to Ag^+ ions show a distinct and fairly broad absorption band centered at ca. 450 nm. The presence of the broad resonance indicates an aggregated structure of the silver particles in the film. It is pertinent to mention here that since the films of the bionanocomposite were rough, the results are not quantitative and have been used merely to detect the presence of silver nanoparticles in the biomaterial.

Fig. 2A shows a scanning electron microscopy (SEM) picture of the *Verticillium* cells after exposure to 10^{-4} M aqueous $AgNO_3$ solution for 72 h. The presence of uniformly distributed silver nanoparticles on the surface of the fungal cells is observed, indicating that the nanoparticles formed by the reduction of Ag^+ ions are bound to the surface of the cells. The silver nanoparticles seen outside the mycelia may be due to weakly bound silver nanoparticles dislodged from the biomass during preparation of the films for SEM investigation which, it may be recollected, involves washing the biomass thrice with distilled water. It is also possible that the cell walls of a small fraction of the mycelia rupture during the sample preparation. Fig. 2B shows the EDAX (Energy dispersive analysis of X-rays) spectrum recorded in the spot profile mode from one of the densely populated silver nanoparticle regions on the surface of the fungal cells. Strong signals from the silver atoms in the nanoparticles are observed, while weaker signals from C, O, S, P, Mg and Na atoms were also recorded. The C, O, S, P, Mg, and Na signals are likely to be due to X-ray emission from proteins/enzymes present in the cell wall of the biomass.

(A) B (C)

Fig. 1. Conical flasks containing *Verticillium* biomass before (A) and after (B) exposure to 10^{-4} M aqueous $AgNO_3$ solution for 72 h of reaction. (C) UV-Vis spectra recorded from films of the fungal cells before (curve 1) and after immersion in 10^{-4} M aqueous $AgNO_3$ solution for 72 h (curve 2).

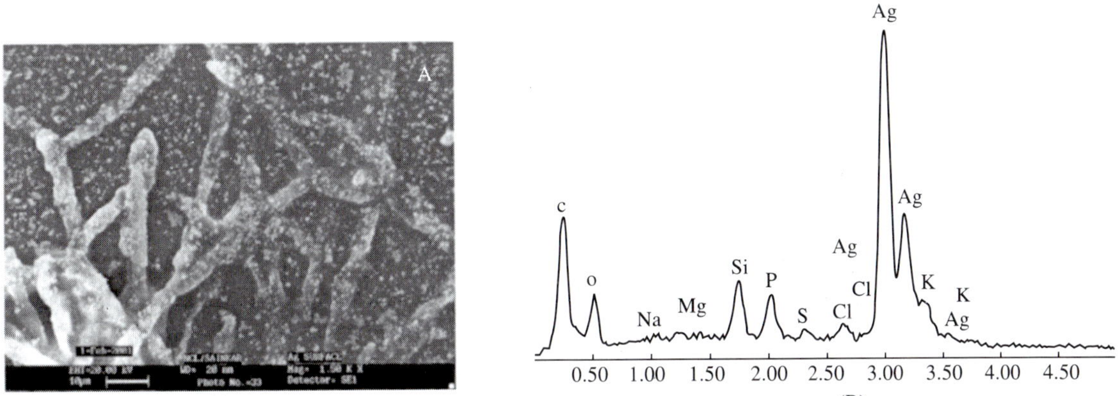

(A) (B)

Fig. 2. (A) SEM image of the *Verticillium* fungal cells after immersion in 10^{-4} M aqueous $AgNO_3$ solution for 72 h. (B) EDAX spectrum recorded from a film of fungal cells after formation of silver nanoparticles. Different X-ray emission peaks are labelled. (Fig. 2B reprinted with permission from reference Mukherjee, et al. (2001b), © 2001 American Chemical Society.)

In order to understand the location of reduction of silver ions relative to the fungal cells, TEM analysis of thin sections of the Ag nano-*Verticillium* cells were performed. Figure 3 shows representative TEM images at various magnifications of the Ag nano-*Verticillium* cells. At lower magnification, a number of mycelia can be observed, which on closer examination reveal assemblies of *Verticillium* cells within the mycelia (Fig. 3A). This image shows small particles of silver organized on the mycelial cell walls as well as some larger particles within the cells. Fig. 3B shows a slightly higher magnification image of the junction between two mycelia, wherein the individual cells are more clearly resolved. A number of silver particles can be seen on the mycelia wall surface. The mycelia in the upper right corner also show an individual *Verticillium* cell with silver particles clearly bound to the surface of the cytoplasmic membrane. Fig. 3C shows a single *Verticillium* cell with silver nanoparticles both on the cell wall (external boundary) and on the cytoplasmic membrane (inner boundary). Occasionally, large silver particles can be observed within the cytoplasm of the *Verticillium* cells (Fig. 3C). Analysis of the size of the silver nanoparticles from many different TEM images of the Ag nano-*Verticillium* cells yielded a particle diameter of 25 ±12 nm. Fig. 3D shows a higher magnification TEM image of the cell wall of one of the *Verticillium* cells after Ag nanoparticle formation. A number of silver particles are observed that are essentially spherical in morphology with fairly good monodispersity. In this respect, our results differ from those of Klaus et al. who observed that silver nanoparticles formed within the periplasmic space of silver resistant bacteria showed morphologies ranging from spherical to triangular to hexagonal (Klaus et al. 1999).

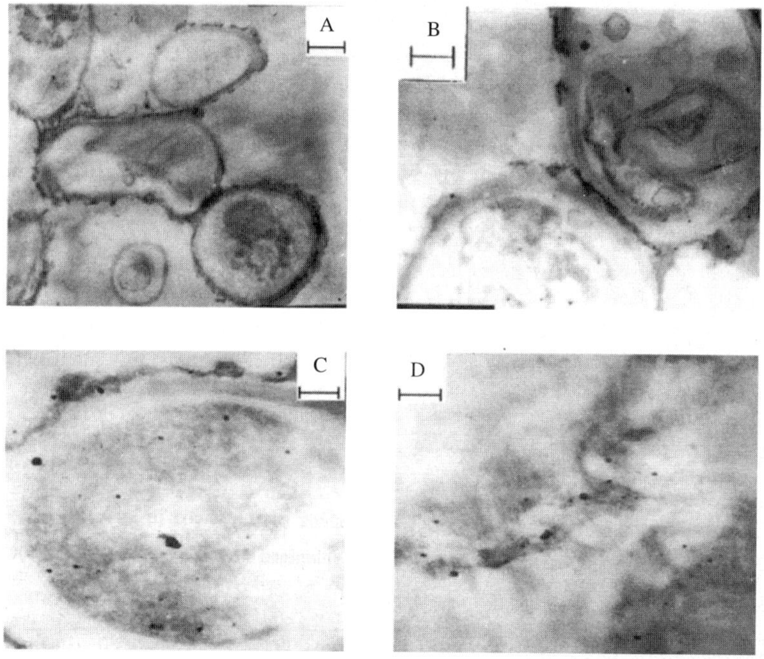

Fig. 3. TEM images of thin sections of *Verticillium* cells after reaction with Ag^{+} ions for 72 h at different magnifications (A-D). The scale bars in A-D correspond to 1 μm, 500, 200 and 100 nm, respectively (Fig. 3 (A and B) reprinted with permission from reference Mukherjee, et al. (2001b), © 2001 American Chemical Society).

The exact mechanism leading to the formation of silver nanoparticles by the fungal cells is not fully understood at the moment. We speculate that since the nanoparticles are formed on the surface of the mycelia and not in solution, the first step involves trapping of the Ag^{+} ions on the surface of the fungal cells possiblely via electrostatic interaction between the Ag^{+} and negatively charged carboxylate groups in enzymes present in the cell wall of the mycelia. Thereafter, the silver ions are reduced by the enzymes leading to the formation of silver nuclei, which subsequently grow by further reduction of Ag^{+} ions and accumulation on these nuclei. The TEM results (Fig. 3) showing the presence of some silver nanoparticles on the cytoplasmic membrane as well as within the cytoplasm suggested that some Ag^{+} ions diffuse through the cell wall and are reduced by enzymes present on the cytoplasmic membrane and within the cytoplasm. It may also be possible that some of the smaller silver nanoparticles diffuse across the cell wall to be trapped within the cytoplasm. A more detailed study is in progress to clarify this hypothesis and extend this to other metals such as Pt, Pd and alloys/bimetallics of Au-Pt, Au-Pd, Au-Ag etc.

Extracellular Synthesis of Silver Nanoparticles Using *Fusarium oxysporum*

From the application point of view, it would be more useful if the fungus reduces the metal ions extracellularly directly in the aqueous medium. Towards this objective, we screened a number of species of fungus belonging to different genera and observed the extracellular synthesis of silver nanoparticles by the fungus *Fusarium oxysporum*.

Fusarium oxysporum was maintained on potato dextrose agar slants at 25°C. Stock cultures were maintained by subculturing at monthly intervals. After growing the fungus at pH 7.0 and 25°C for 5-6 days, the slants were preserved at 15°C. For experimentation, subcultures were made on fresh slants and after 5-6 days incubation at pH 7.0 and 25°C, used as the starting material for the fermentation. *Fusarium oxysporum* was grown in 500 mL Erlenmeyer flasks each containing 100 mL of MGYP medium at 25-28°C under shaking (200 rpm) for 96 h. After 96 h of fermentation, mycelia were separated from the culture broth by centrifugation (5000 rpm) at 10°C for 20 min and the settled mycelia were washed thrice with sterile distilled water. The harvested mycelial mass (10 g wet weight of mycelia) was then resuspended in 100 mL of 10^{-3} M aqueous $AgNO_3$ solution in 500 ml Erlenmeyer flask and incubated on a shaker at 28°C (200 rpm) for 72 h in the dark. Periodically, aliquots of the reaction mixture were removed and subjected to UV-Vis and fluorescence spectroscopic measurements. The UV-Vis spectroscopy measurements were performed on a Shimadzu dual-beam spectrophotometer (model UV-1601 PC) operated at a resolution of 1 nm, while fluorescence measurements were carried out on a Perkin-Elmer LS 50B luminescence spectrophotometer. The excitation wavelength was 260 nm chosen to maximize optical transitions in tryptophan and tyrosine residues in proteins released into the solution by the fungus (Sastry et al. 1997). On completion of the reaction of the Ag^+ ions with the fungal biomass, films of the silver nanoparticles were formed on Si (111) substrates by drop-coating the nanoparticle solution. The films on Si wafers were subjected to Fourier transform infrared spectroscopic (FTIR) studies, which were carried out in a Shimadzu FTIR-8201 PC instrument in the diffuse reflectance mode at a resolution of 4 cm^{-1}. Similar films on Si (111) wafers were analyzed by X-ray diffraction (XRD), which was carried out in the transmission mode on a Philips PW 1830 instrument operating at 40 kV and a current of 30 mA with Cu K_α radiation. The silver nanoparticle films were also formed on carbon coated copper TEM grids and analyzed by transmission electron microscopy (TEM) on a JEOL 1200EX instrument at a voltage of 80 kV.

Fig. 4 shows two conical flasks with the *Fusarium oxysporum* biomass before (A) and after reaction with Ag^+ ions for 72 h (B). It is observed that the biomass has a pale yellow color before reaction with the silver ions (A), which changes to a brownish color on completion of the reaction (B). The appearance of a yellowish-brown color in solution containing the biomass is a clear indication of the formation of silver nanoparticles in the reaction mixture and is due to the excitation of surface plasmon vibrations in the nanoparticles (Henglein, 1993). Upon filtration, it was observed that the biomass was still pale yellow and that the aqueous solution contained the silver nanoparticles, characterized by an intense yellow color. This indicated that the reduction of the Ag^+ ions takes place extracellularly and is an important observation that we will return to subsequently. The UV-Vis spectra recorded from the *Fusarium oxysporum* reaction vessel at different times of reaction are plotted in Fig. 5. The time at which the aliquots were removed for analysis is indicated next to the respective curves. The strong surface plasmon resonance centered at ca. 413 nm clearly increased in intensity with time, stabilizing after ca. 72 h of reaction. Quite interestingly, the solution was extremely stable, with no evidence of flocculation of the particles even a month after reaction.

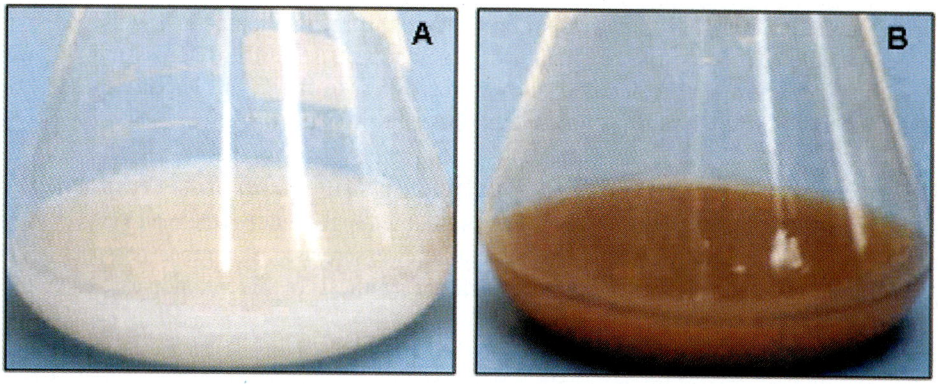

Fig. 4. Picture of conical flasks containing the *Fusarium oxysporum* biomass in aqueous solution of 10^{-3} M $AgNO_3$ at the beginning of the reaction (A) and after 72 h of reaction (B).

The plasmon resonance is sharp and indicates little aggregation of the particles in solution. The inset in Fig. 5 shows the UV-Vis spectrum in low wavelength region recorded from the reaction medium 72 h after reaction. An absorption band at ca. 270 nm is clearly visible and is attributed to electronic excitations in tryptophan and tyrosine residues in the proteins (Gole et al. 2001). This observation indicates the release of proteins into solution by *Fusarium oxysporum* and suggests a possible mechanism for the reduction of the metal ions present in the solution. In order to demonstrate that the reduction of the silver ions does indeed take place extracellularly, 10 g of the biomass were immersed in 100 ml of water for 72 h, following which the aqueous component was separated by filtration. To this solution, $AgNO_3$ was added, to yield an overall Ag^+ concentration in a solution of 10^{-3} M. It was observed that this initially colorless aqueous solution changed to a pale yellowish-brown within 24 h (data not shown), clearly indicating that the reduction of the ions occurs extracellularly through reducing agents released into the solution by *Fusarium oxysporum*. To identify the number of reducing agents present in the reaction mixture, a preliminary gel

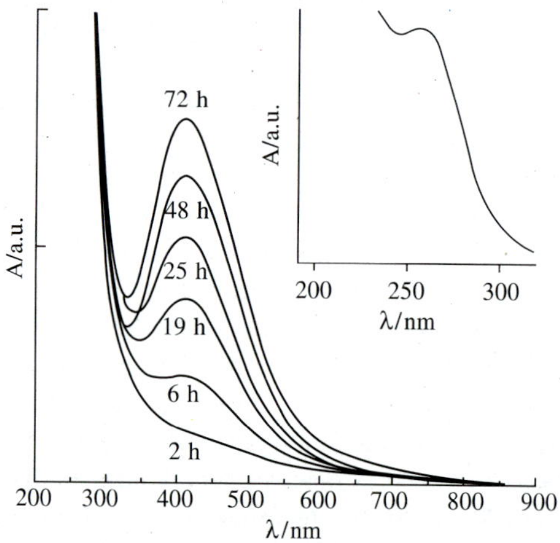

Fig. 5. UV-Vis spectra recorded as a function of time of reaction of an aqueous solution of 10^{-3} M $AgNO_3$ with the fungal biomass. Time of reaction is indicated next to the respective curves. The inset shows the UV-Vis absorption spectrum in the low wavelength region recorded from the reaction medium 72 h after commencement of the reaction (reprinted with permission from reference Ahmad et al. (2003a), © 2003 Elsevier).

electrophoresis was performed. It indicated that a minimum of four proteins of molecular masses between 66 kDa and 10 kDa were released by the fungal biomass (Laemmli, 1970). The protein extract for gel electrophoresis was obtained by suspending the *Fusarium oxysporum* mycelial biomass for 48 h and was concentrated by ultrafiltration through a YM-3 ultrafiltration membrane and then dialyzed against distilled water by using a 3K cut-off dialysis bag. This process removes the low molecular weight components in the extract such as co-factors. The protein mixture obtained after dialysis failed to reduce Ag^+ ions. However, on addition of stoichiometric amounts of NADH to the protein extract, the reduction of Ag^+ ions occurred quite readily. This clearly indicates the reduction of Ag^+ ions by NADH dependent reductases in the extract.

The FTIR spectrum shows the presence of three bands at 1650 cm^{-1} (1), 1540 cm^{-1} (2) and 1450 cm^{-1} (Fig. 6). The bands at 1650 and 1540 cm^{-1} are assigned to amide I and II bands and arise due to —CO stretch and —NH stretch vibrations in amide bonds present in the proteins, respectively (Macdonald and Smith, 1996; Sastry et al. 1997, 1998; Gole et al. 2001). The positions of these bands are close to those reported for native proteins. The FTIR results thus indicate that the secondary structure of the proteins is not affected as a consequence of reaction with the Ag^+ ions or binding with the silver nanoparticles. The band at ca. 1450 cm^{-1} is due to methylene scissoring vibrations from the proteins in the solution.

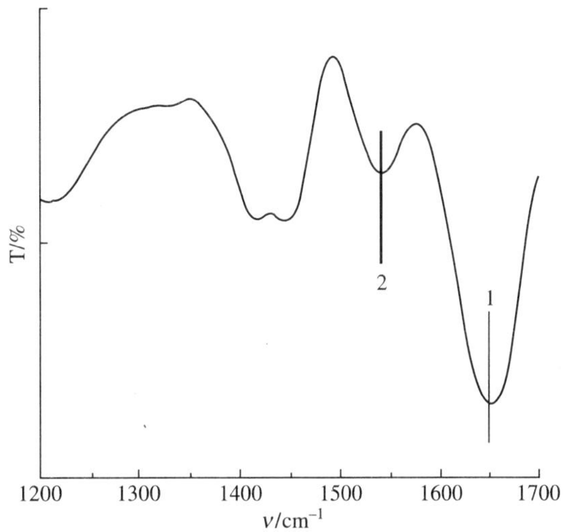

Fig. 6. FTIR spectrum recorded from a drop-coated film of an aqueous solution incubated with *Fusarium oxysporum* and reacted with Ag^+ ions for 72 h. The amide I and II bands are identified in the figure (reprinted with permission from reference Ahmad et al. (2003a), © 2003 Elsevier).

A representative TEM picture recorded from the silver nanoparticle film deposited on a carbon-coated copper TEM grid in Fig. 7 shows individual silver particles as well as a number of aggregates. The

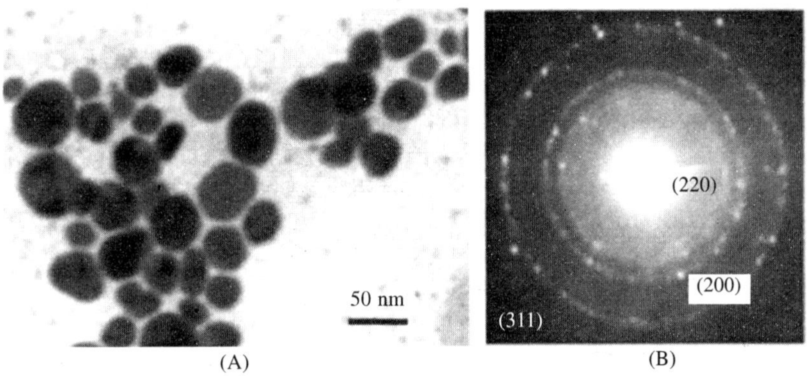

Fig. 7. (A) TEM micrograph recorded from a drop-coated film of an aqueous solution incubated with *Fusarium oxysporum* and reacted with Ag^+ ions for 72 h. The scale bar corresponds to 50 nm. (B) Selected area of electron diffraction pattern recorded from one of the silver nanoparticles shown in (A). The diffraction rings have been indexed with reference to fcc silver ((B) reprinted with permission from Ahmad et al. (2003a), © 2003 Elsevier).

morphology of the nanoparticles is highly variable, with spherical and occasionally triangular nanoparticles observed in the micrograph. Under observation of this image in an optical microscope, these assemblies are found to be aggregates of silver nanoparticles in the size range 5-50 nm. The nanoparticles are not in direct contact even within the aggregates, indicating stabilization of the nanoparticles by a capping agent. The silver particles are crystalline, as can be seen from the selected area diffraction pattern recorded from one of the nanoparticles in the aggregates (Fig. 7B). As discussed earlier, the silver nanoparticle solution, synthesized by the reaction of Ag^+ ions with *Fusarium oxysporum* is exceptionally stable. The stability is likely to be due to capping with proteins secreted by the fungus. The separation between the silver nanoparticles seen in the TEM image could be due to capping by proteins and would explain the UV-Vis spectroscopy measurements, which is characteristic of well dispersed silver nanoparticles. The use of specific enzymes secreted by organisms such as fungi in the extracellular synthesis of nanoparticles is exciting for the following reasons. The synthesis of nanoparticles in solution would be of importance in homogeneous catalysis and other applications such as non-linear optics. The nanoparticles may be immobilized in different matrices or in thin film form for optoelectronic applications, which is impossible to achieve if the nanoparticles are bound to the biomass. We believe that the biggest advantage of this protocol based on fungal enzymes is the possibility of developing a rational approach for the biosynthesis of nanomaterials over a range of chemical compositions.

Conclusions and Future Perspectives

A brief survey on the use of microorganisms such as bacteria, yeasts, algae, actinomycetes and fungi in the biosynthesis of inorganic nanomaterials has been made. The ability of the microorganisms to reduce the inorganic metals has opened up a new exciting green chemistry approach towards the development of natural 'nano-factories'. However, a number of issues have to be addressed from the nanotechnology and microbiology point of view before such biosynthetic approaches can compete with the existing physical and chemical procedures. The use of enzymes secreted by fungi in the specific reactions leading to inorganic nanoparticles is a new and rational biosynthesis strategy. The use of fungi in the synthesis of nanoparticles is exciting because of its ability to secrete large quantities of enzymes in a relatively pure state, free from other cellular proteins associated with the organism and can be easily separated by filtration and used for nanoparticles synthesis. The identification of these enzymes can definitely lead to the development of approaches to the synthesis of nanoparticles of different chemical compositions and sizes/shapes of controlled monodispersity. Understanding the surface chemistry of the biogenic nanoparticles (i.e. nature or number of capping proteins/peptides) is equally important. This would lead to the possibility of genetically engineering the microbes to overexpress specific reducing molecules and capping agents and thereby control not only the size of the nanoparticles but also their shape. The fungal mediated biosynthetic approach has a number of advantages over other microbial approaches such as ease with which the process can be scaled up, economic viability, possibility of easily covering large surface areas by suitable growth of the mycelia, etc. The bacterial fermentation involves the use of sophisticated equipment for getting clear filtrates from the colloidal broths while the fungal broths can be separated by simple filtration, thus saving considerable investment costs for equipment. Fungi have been known to be extremely efficient secretors of soluble proteins under optimized conditions of fermentations, e.g. mutant strains secrete up to 30 g per litre of extracellular protein. Compared to bacteria, this is a much higher amount and make the product more viable.

Acknowledgement

The authors thank Ms. Renu Pasricha and Dr. S.R. Sainkar, Materials Chemistry Division, NCL, Pune, for assistance in TEM and SEM measurements, respectively. Financial support from Department of Biotechnology (DBT), New Delhi, India is gratefully acknowledged.

References

Ahmad, A., Mukherjee, P., Mandal, D., Senapati, S., Khan, M.I., Kumar, R. and Sastry, M., (2002) Enzyme mediated extracellular synthesis of CdS nanoparticles by the fungus, *Fusarium oxysporum*. Journal of the American Chemical Society 124 (41): 12108–12109.

Ahmad, A., Mukherjee, P., Senapati, S., Mandal, D., Khan, M.I., Kumar, R. and Sastry, M., (2003a) Extracellular biosynthesis of silver nanoparticles by the fungus *Fusarium oxysporum*. Colloids and Surfaces B: Biointerfaces 28 (4): 313–318.

Ahmad, A., Senapati, S., Khan, M.I., Kumar, R., Ramani, R., Srinivas, V., and Sastry, M., (2003b) Intracellular synthesis of gold nanoparticles by a novel alkalotolerant actinomycete. *Rhodococcus* species. Nanotechnology 14 (7): 824–828.

Ahmad, A., Senapati, S., Khan, M.I., Kumar, R. and Sastry, M., (2003c) Extracellular biosynthesis of monodispersed gold nanoparticles by a novel extremophilic actinomycete, *Thermomonospora* sp. Langmuir. 19 (8): 3550–3553.

Alivisatos, A.P. (1996) Semiconductor clusters, nanocrystals, and quantum dots Science 271 (5251): 933–937.

Berry, C.C. and Curtis, A.S.G. (2003) Functionalisation of magnetic nanoparticles for applications in biomedicine. Journal of Physics D: Applied physics. 36: R198-R206.

Beveridge, T.J. and Murray, R.G.E. (1980) Sites of metal deposition in *Bacillus subtilis*. Journal of Bacteriology 141 (2): 876–887.

Brugger, P.A., Cuendet, P. and Graetzel, M. (1981) Ultrafine and specific catalysts affording efficient hydrogen evolution from water under visible light illumination. Journal of the American Chemical Society 103 (11): 2923–2927.

Brus, L.E. (1984) Electron–electron and electron-hole interactions in small semiconductor crystallites: The size dependence of the lowest excited electronic state. Journal of Chemical Physics 80 (9): 4403–4409.

Chan, W.C.W., Maxwell, D.J., Gao, X., Bailey, R.E., Han, M. and Nie, S. (2002) Luminescent quantum dots for multiplexed biological detection and imaging. Current Opinion in Biotechnology 13 (1): 40–46.

Colvin, V.L., Schlamp, M.C. and Alivisatos, A.P. (1994) Light-emitting diodes made from cadmium selenide nanocrystals and semiconducting polymer. Nature 370 (6488): 354–357.

Cunningham, D.P. and Lundie, L.L. (1993) Precipitation of cadmium by *Clostridium thermoaceticum*. Applied and Environmental Microbiology 59 (1), 7–14.

Dameron, C.T., Reese, R.N., Mehra, R.K., Kortan, A.R., Carroll, P.J., Steigerwald, M.L., Brus, L.E. and Winge, D.R. (1989) Biosynthesis of cadmium sulphide quantum semiconductor crystallites. Nature. 338 (6210): 596–598.

Dickson, D.P.E. (1999) Nanostructured magnetism in living systems. Journal of Magnetism and Magnetic Materials 203 (1-3): 46–49.

Dubertret, B., Skourides, P., Norris, D.J., Noireaux, V., Brivanlou, A.H. and Libchaber, A. (2002) In Vivo Imaging of Quantum Dots Encapsulated in Phospholipid Micelles. Science 298 (5599): 1759–1762.

Fendler, J.H. (1992) Membrane Mimetic Chemistry Approach to Advanced Materials. Springer-Verlag, Berlin.

Fendler, J.H. and Meldrum, F. (1995) The colloid chemical approach to nanostructured materials. Advanced Materials 7 (7): 607–632.

Fortin, D. and Beveridge, T.J. (2000). In Biomineralization. From Biology to Biotechnology and Medical Applications (ed. Baeuerien, E.), Wiley-VCH, Weinheim, p. 7.

Gardea-Torresdey, J.L., Parsons, J.G., Gomez, E., Peralta-Videa, J., Troiani, H.E., Santiago, P. and Yacaman, M.J. (2001) Formation and growth of Au nanoparticles inside live alfalfa plants. Nano Letters 2 (4): 397–401.

Gardea-Torresdey, J.L., Gomez, E., Peralta-Videa, J.R., Parsons, J.G., Troiani, H. and Jose-Yacaman, M. (2003) Alfalfa sprouts: a natural source for the synthesis of silver nanoparticles. Langmuir 19 (4): 1357–1361.

Gole, A., Dash, C., Ramachandran, V., Mandale, A.B., Sainkar, S.R., Rao, M. and Sastry, M. (2001) Pepsin-gold colloid conjugates: preparation, characterization, and enzymatic activity. Langmuir 17 (5): 1674–1679.

Handley D.A. (1989) Colloidal gold: Principles, Methods, and Applications. vol 1, chapter 2 (ed), Hayat M.A. Academic press, San Diego, CA.

Henglein, A. (1993) Physicochemical properties of small metal particles in solution: "microelectrode" reactions, chemisorption, composite metal particles, and the atom-to-metal transition. Journal of Physical Chemistry 97(21): 5457–5471.

Hoffman, A.J., Mills, G., Yee, H. and Hoffman, M.R. (1992) Q-sized cadmium sulfide: synthesis, characterization, and efficiency of photoinitiation of polymerization of several vinylic monomers. Journal of Physical Chemistry 96(13): 5546–5552.

Holmes, J.D., Smith, P.R., Evans-Gowing, R., Richardson, D.J., Russell, D.A. and Sodeau, J.R. (1995) Energy-dispersive X-ray analysis of the extracellular cadmium sulfide crystallites of Klebsiella aerogenes. Archives of Microbiology 163 (2): 143–147.

Joerger, R., Klaus, T. and Granqvist, C.G. (2000) Biologically produced silver-carbon composite materials for optically functional thin-film coatings. Advanced Materials 12 (6): 407–409.

Klaus, T., Joerger, R., Olsson, E. and Granqvist, C.G. (1999) Silver-based crystalline nanoparticles, microbially fabricated. Proceedings of the National Academy of Sciences, USA. 96 (24): 13611–13614.

Klaus-Joerger, T., Joerger, R., Olsson, E. and Granqvist, C.G. (2001) Bacteria as workers in the living factory: metal-accumulating bacteria and their potential for materials science. Trends in Biotechnology 19 (1): 15–20.

Klein, D.L., Roth, R., Lim, A.K.L., Alivisatos, A.P. and McEven, P.L., (1997) A single-electron transistor made from a cadmium selenide nanocrystals. Nature 389 (6652): 699–701.

Kowshik, M., Vogel, W., Urban, J., Kulkarni, S.K. and Paknikar, K.M. (2002a) Microbial Synthesis of Semiconductor PbS Nanocrystallites. Advanced Materials 14 (11): 815–818.

Kowshik, M., Deshmukh, N., Vogel, W., Urban, J., Kulkarni, S.K. and Paknikar, K.M. (2002b) Microbial synthesis of semiconductor CdS nanoparticles, their characterization, and their use in the fabrication of an ideal diode. Biotechnology and Bioengineering 78 (5): 583–588.

Kroger, N., Deutzmann, R. and Sumper, M. (1999) Polycationic peptides from diatom biosilica that direct silica nanosphere formation. Science 286 (5442): 1129–1132.

Laemmli, U.K. (1970) Cleavage of structural proteins during the assembly of the head of bacteriophage T4. Nature 227: 680–685.

Lovley, D.R., Stolz, J.F., Nord, G.L. and Phillips, E.J.P. (1987) Anaerobic production of magnetite by a dissimilatory iron-reducing microorganism. Nature 330 (6145): 252–254.

Macdonald, I.D.G. and Smith, W.E. (1996) Orientation of cytochrome *c* adsorbed on a citrate-reduced silver colloid surface. Langmuir 12 (3): 706–713.

Mann, S., (1993) Molecular tectonics in biomineralization and biomimetic materials chemistry. Nature 365 (6446): 499–505.

Mann, S. (ed.) (1996) Biomimetic Materials Chemistry, VCH Publishers. New York. 33.

Mehra, R.K. and Winge, D.R. (1991) Metal ion resistance in fungi: Molecular mechanisms and their regulated expression. Journal of Cellular Biochemistry 45 (1): 30–40.

Mornet, S., Vasseur, S., Grasset, F. and Duguet, E. (2004) Magnetic nanoparticle design for medical diagnosis and therapy 14 (14): 2161–2175.

Mukherjee, P., Ahmad, A., Mandal, D., Senapati, S., Sainkar, S.R., Khan, M.I., Ramani, R., Parischa, R., Ajaykumar, P.V., Alam, M., Sastry, M. and Kumar, R. (2001a) Bioreduction of $AuCl_4^-$ ions by the fungus, *Verticillium* sp. and surface trapping of the gold nanoparticles. Angewandte Chemie International Edition in English 40 (19): 3585–3588.

Mukherjee, P., Ahmad, A., Mandal, D., Senapati, S., Sainkar, S.R., Khan, M.I., Parischa, R., Ajaykumar, P.V., Alam, M., Kumar, R. and Sastry, M. (2001b) Fungus-mediated synthesis of silver nanoparticles and their immobilization in the mycelial matrix: A novel biological approach to nanoparticle synthesis. Nano Letters 1(10): 515–519.

Mukherjee, P., Senapati, S., Mandal, D., Ahmad, A., Khan, M.I., Kumar, R. and Sastry, M., (2002) Extracellular biosynthesis of gold nanoparticles by the fungus *Fusarium oxysporum*. ChemBioChem 3(5): 461–463.

Nair, B. and Pradeep, T. (2002) Coalescence of nanoclusters and formation of submicron crystallites assisted by *Lactobacillus* strains. Crystal Growth Design 2 (4): 293–298.

O'Brien, E., Stubbins, M., Asquith, S., Spurr, N.K., Lai, E.H., Xu, H., Sha, M.Y., Wong, E.Y., Uphoff, J., Xu, Y., Treadway, J.A., Truong, A., Mahoney, W. (2003) Multiplexed SNP genotyping using the Qbead™ System: a quantum dot-encoded microsphere-based assay. Nucleic Acids Research 31: 43.

Oliver, S., Kupermann, A., Coombs, N., Lough, A. and Ozin, G.A. (1995) Lamellar aluminophosphates with surface patterns that mimic diatom and radiolarian microskeletons. Nature 378 (6552): 47–50.

Pankhurst, Q.A., Connolly, J., Jones. S.K. and Dobson, J. (2003) Application of magnetic nanoparticles in biomedicine. Journal of Physics D: Applied Physics 36: R167-R181.

Parak, W.J., Boudreau, R., Gros, M.L., Gerion, D., Zanchet, D. Micheel, C.M., Williams, S.C., Alivisatos, A.P. and Larabell, C. (2002) Cell Motility and Metastatic Potential Studies Based on Quantum Dot Imaging of Phagokinetic Tracks. Advanced Materials 14 (12): 882–885.

Pedro Tartaj, P., Morales, M.D.P., Veintemillas-Verdaguer, S., Gonzalez-Carreno, T and Serna, C.J. (2003) The preparation of magnetic nanoparticles for applications in biomedicine. Journal of Physics D: Applied Physics. 36: R182-R197.

Pum, D. and Sleytr, U.B. (1999) The application of bacterial S-layers in molecular nanotechnology. Trends in Biotechnology 17 (1): 8–12.

Salata, O.V. (2004). Application of nanoparticles in biology and medicine. Journal of Nanobiotechnology. 2 (3): 1–6.

Sastry, M., Bandyopadhyay, K. and Mayya, K.S. (1997) pH dependent changes in the optical properties of carboxylic acid derivatized silver colloidal particles. Colloids and Surfaces A: Physicochemical and Engineering Aspect 127 (1-3): 221–228.

Sastry, M., Patil, V. and Sainkar, S.R. (1998) Electrostatically controlled diffusion of carboxylic acid derivatized silver colloidal particles in thermally evaporated fatty amine films, Journal of Physical Chemistry B 102 (8): 1404–1410.

Schmid, G. (1992) Large clusters and colloids. Metals in the embryonic state, Chemical Review 92 (8), 1709–1727.

Shi, J., Gider, S., Babcock, K. and Awschalom, D.D. (1996) Magnetic clusters in molecular beams, metals, and semiconductors, Science 271 (5251): 937–941.

Shivshankar, S., Rai, A., Ankamwar, B., Singh, A., Ahmad, A. and Sastry, M. (2004) Biological synthesis of triangular gold nanoprisms. Nature Materials 3: 482–488.

Shull, R.D., McMichael, R.D. and Ritter, J.J. (1993) Magnetic nanocomposites for magnetic refrigeration. Nanostructured Material 2 (2): 205–211.

Simkiss, K. and Wilbur, K.M. (1989) Biomineralization, Academic Press, New York.

Sleytr, U.B., Messner, P., Pum, D. and Sara, M. (1999) Crystalline bacterial cell surface layers (S Layers): From supramolecular cell structure to biomimetics and nanotechnology. Angewandte Chemie International Edition in English 38 (8): 1034–1054.

Smith, P.R., Holmes, J.D., Richardson, D.J., Russell, D.A. and Sodeau, J.R. (1998) Photophysical and photochemical characterisation of bacterial semiconductor cadmium sulfide particles, Journal of the Chemical Society, Faraday Transaction 94 (9): 1235–1241.

Southam, G. and Beveridge, T.J. (1996) The occurrence of sulfur and phosphorus within bacterially derived crystalline and pseudocrystalline octahedral gold formed in vitro. Geochimica et Cosmochimica Acta 60 (22): 4369–4376.

Stephen, J.R. and Macnaughton, S.J. (1999) Developments in terrestrial bacterial remediation of metals. Current Opinion in Biotechnology 10 (3): 230–233.

Taylor, J.R., Fang, M.M. and Nie, S. (2000) Probing specific sequences on single DNA molecules with bioconjugated fluorescent nanoparticles. Analytical Chemistry 72 (9): 1979–1986.

Wang, Y. (1991) Nonlinear optical properties of nanometer-sized semiconductor clusters. Accounts of Chemical Research 24 (5): 133–139.

Wang, Y. and Herron, N. (1991) Nanometer-sized semiconductor clusters: materials synthesis, quantum size effects, and photophysical properties. Journal of Physical Chemistry 95 (2): 525–532.

Weller, H. (1993) Colloidal semiconductor Q-particles: chemistry in the transition region between solid state and molecules. Angewandte Chemie International Edition in English 32 (1): 41–53.

Weller, H. (1998) Transistors and light emitters from single nanoclusters. Angewandte Chemie International Edition in English 37 (12): 1658–1659.

Wu, X., Liu, H., Liu, J., Haley, K.N., Treadway, J.A., Larson, J.P. Ge, N., Peale, F. and Bruchez, M.P. (2003). Immunofluorescent labeling of cancer marker Her2 and other cellular targets with semiconductor quantum dots Nature Biotechnology. 21 (1): 41–46.

18. Can Agricultural Practices and Land Use Patterns Affect Arbuscular Mycorrhizal Fungal Population and Diversity?

R. Lakshmipathy, D.J. Bagyaraj and A.N. Balakrishna
Department of Agricultural Microbiology, University of Agricultural Sciences, GKVK Campus, Bangalore-560065, India

SUMMARY Arbuscular mycorrhizal (AM) fungi belong to the class Zygomycetes and order Glomales. These fungi form mutualistic association with 90% of the terrestrial plant species. AM fungi form the main component of soil microbiota in most agro-ecosystems and enhance plant growth mainly by enhanced P uptake, and also enhance tolerance to pathogens and abiotic stresses such as drought and metal toxicity. These fungi have strong influence on plant species diversity in natural ecosystem. In turn, plant species determine the population and diversity of AM fungi. Apart from plant species, human activities also affect AM fungi.

Agricultural practices like application of fertilizers, herbicides, pesticides and cropping systems affect these fungi. Diversity of AM fungi decrease as the land use intensity increases. Crop rotation, mixed cropping and organic farming enhance the AM fungal activity compared to mono-cropping with intensive agriculture. Minimum tillage and non-weeding enhance AM fungal population than intensive tillage. Modern intensive agricultural practices are evidently a threat for AM fungi.

Introduction

Frank (1985), the German botanist first coined the term "mycorrhiza" which literally means "fungus root" to describe the mutualistic association between roots of higher plants and certain fungi. These associations are grouped, based on morphological and anatomical characters, as ectomycorrhizae and endomycorrhizae. Ectomycorrhizae are common among temperate forest tree species and can be cultured on laboratory media. Endomycorrhizae, include arbutoid, monotropoid, ericoid, orchid and arbuscular mycorrhizal (AM) forms. Arbuscular mycorrhizae are the most common and widely occurring of all the mycorrhizal associations and have great economic significance. They cannot be cultured on laboratory media.

AM fungi are said to establish a mutualistic relationship with 80% of vascular plants (Harley and Smith, 1983; Trappe, 1977; Gianinazzi and Gianinazzi Pearson, 1986). Plants that rarely form AM fungal association includes Caryophyllaceae, Brassicaceae, Chenopodiaceae and Cyperaceae (Hirrel et al. 1978). In addition to their widespread distribution throughout the plant kingdom, arbuscular mycorrhizae are ubiquitous and occur in plants grown in arctic, temperate and tropical regions (Mosse et al. 1981). They have been reported to be associated with plants grown in sand dunes, coal mines (Khan, 1978) and aquatic environments (Bagyaraj et al. 1978). Blaszkowski (1994) observed variations in AM fungal diversity with the changes in plant species. Plants of a particular family are colonized by specific types of AM fungi and a few AM fungal genera were found only in the plants of a particular family.

AM fungi have the widest host range and distribution of all the mycorrhizal associations. It is estimated that about 90% of vascular plants normally establish mutualistic relationships with VAM fungi. Arbuscular mycorrhizae have been observed in 1000 genera of plants representing some 200 families. There are at least 300,000 receptive hosts in the world flora (Kendrick and Berch, 1985) and there are about 150 species of AM fungi (Schenck and Perez, 1990). If the hosts are divided up evenly among the fungi, with no overlap in host range, each fungus would have more than 2500 potential partners. We know that the host range overlaps extensively, suggesting that some individual AM fungi may well have access to thousands of hosts (Kendrick and Berch, 1985).

Arbuscular Mycorrhizal Fungi: Classification, Importance and Occurrence

Arbuscular mycorrhizal fungi are obligate symbionts and cannot be cultured on synthetic media. Their penetration takes place through root hairs or epidermal cells, grow intercellularly or intracellularly in the root cortex, ultimately developing short haustoria like structures called arbuscules within the cortical cells. These arbuscules function as sites of nutrient exchanges between the fungus and host roots. Vesicles are formed in the cortical cells, which are thin walled structures of various sizes and shapes and function as storage organs. The presence of vesicles and arbuscules is the criteria for identifying AM fungus in the roots.

Arbuscular mycorrhizal fungi come under the class Zygomycetes, under the order Glomales. The order Glomales is divided into 2 sub-orders Glomineae and Gigasporineae. The sub-order Glomineae has four families namely Glomaceae, Acaulosporaceae, Paraglomaceae and Archeosporaceae. Further, Glomaceae has two genera, viz. *Glomus* and *Sclerocystis*, Acaulosporaceae has two genera, viz. *Acaulospora* and *Entrophospora,* Archaeosporaceae has only one genus *Archaeospora* and Paraglomaceae has one genus *Paraglomus*. The suborder Gigasporineae has only one family Gigasporaceae. Gigasporaceae includes the genera *Gigaspora* and *Scutellospora*.

Improved plant growth due to inoculation of soil with AM fungi has been demonstrated especially under P deficient conditions (Mosse, 1977a). The growth improvement is mainly because of enhanced P uptake. AM fungi can also enhance tolerance or resistance to root pathogens (Borowicz, 2001) and abiotic stresses such as drought and metal toxicity (Meharg and Cairney, 2000). AM fungi play a role in the formation of stable soil aggregates, build up a macroporous structure of soil that allows penetration of water and air and prevents erosion (Miller and Jastrow, 1992). There is well documented evidence that VA mycorrhizae have important effects on plant P uptake. Greater soil exploration by mycorrhizal roots as a means of increasing phosphate uptake is well established. In phosphate deficient soils immobile phosphate ions develop a phosphate depletion zone around the roots. The hyphae spread beyond this zone and directly translocate nutrients from the soil to the root cortex (Hayman, 1983). Experiments with ^{32}P labeled phosphate indicate that VA mycorrhizal hyphae obtain their extra phosphate from the labile pool rather than by accessing dissolving soluble phosphate (Raj et al. 1981). Sparingly soluble rock phosphate is better utilized by the hyphae by closer physical contact with the ions dissociating at the particle surface.

The increased growth of plants inoculated with VA mycorrhizal fungi is not only attributed to improved phosphate uptake but also to better availability of other elements like Zn, Cu, K, Al, Mn, Fe etc. AM fungi affect the levels of plant hormones. Allen et al. (1982) measured levels of plant hormones like cytokinins and gibberellin-like substances. VA mycorrhizal fungi can tolerate a wide range of soil water regimes and also improve water relationships of many plants. It is still unclear whether the observed effects are directly due to the fungus itself or indirectly due to some alteration in host physiology as a result of improved P nutrition. Anatomical and other physiological studies have brought out that mycorrhizal plants have increased rates of respiration, photosynthesis and increased amounts of sugars, amino acids, RNA etc. and larger and/ or more number of chloroplasts, mitochondria, xylem vessels, motor cells, etc. Changes in the root exudations and altered rhizosphere microorganisms (which also affect plant growth) may result because of colonization of roots by VAM fungi (Mada and Bagyaraj, 1993; Caroline Machado and Bagyaraj, 1995).

Mycorrhizal colonization may also allow introduced populations of beneficial soil organisms like *Azotobacter, Azospirillum* and phosphate solubilizing bacteria to be maintained in high numbers than around non-mycorrhizal plants and to exert synergistic effects on plant growth. It is apparent from the investigations on AM fungi-plant pathogen interactions that AM fungi can usually (though not always) deter or reduce the severity of disease caused by soil-borne pathogens. All these studies bring out that AM fungi help the host plant in more than one way and that AM fungal inoculation helps plant growth.

AM fungi, in addition to their widespread distribution throughout the plant kingdom, are also geographically ubiquitous and occur in plants growing in arctic, temperate and tropical regions (Mosse et al. 1981). As an explanation for their remarkably widespread distribution, Trappe (1977) proposed that AM fungi were disseminated intercontinentally prior to the continental drift. The super continent Gondwanaland is thought to have begun to break apart and drift north about 125 million years ago. Fossil records of plants containing AM like structures have occurred as Trappe suggested. In general, AM population is more in cultivated soil compared to virgin soil (Mosse and Bowen, 1968). They are mostly seen in the top 15-30 cm of soil, and their numbers decrease markedly below the top 15 cm (Redhead, 1977). They are normally not found in depths beyond the normal root range of plants (Mosse et al. 1981). Although AM fungi are ubiquitous in soils, the patterns or distribution of individual species have not been fully understood. Studies on the distribution of species have either covered large geographical areas (Mosse and Bowen, 1968; Hall, 1977) or have been limited to smaller regions (Abbott and Robson, 1977). The distribution of species of AM fungi varies with climatic and edaphic environment as well as with land use. For example, *Acaulospora laevis* is common in western Australia (Abbott and Robson, 1977) and New Zealand (Mosse and Bowen, 1968) but occurs less frequently in soils of eastern Australia (Mosse and Bowen, 1968). *Glomus* spp. appear to have the widest distribution. *Gigaspora* and *Sclerocystis* spp. are more common in tropical soils. *Acaulospora* seems to be better adapted to soils with pH < 5.0. In fact, certain AM fungi have been linked to particular kinds of soil: *Glomus mosseae* with fine textured, fertile high pH soils; *Acaulospora laevis* with coarse textured, acid soils; and *Gigaspora* species with sand dune soils (Kendrick and Berch, 1985).

A fundamental problem in studies on the distribution of AM fungi lies in identifying the fungi. It relies almost entirely on spore morphology (Hall, 1977), which may change with spore age (Hall, 1977; Abbott and Robson, 1979). Description of many of the taxa has been based on single collections of uncertain age. This, coupled with the possibility of finding undescribed taxa, can make the identification of spores collected from field samples very difficult. Another difficulty in determining the distribution of AM fungi is that spores of all species are not equally easy to recover from soil. Soils often contains spores of more than one species of AM fungus (Abbott and Robson, 1977). Moreover, the same root may become colonized by more than one species of AM fungi (Cooper, 1976; Mosse, 1977b). Some AM fungi develop characteristic morphological features within plant roots (Cooper, 1976, Abbott and Robson, 1979). This can allow quantitative estimates to be made on the total amounts of coarse or fine AM fungal hyphae within plant roots (Sparling and Tinker, 1978). It is also possible to assess the extent of infection formed by different species of AM fungi within coarse hyphae occurring in the same soil (Abbott and Robson, 1981).

Ecosystems and AM Fungi

Arbuscular mycorrhizal fungi form the main component of soil microbiota in most agro-ecosystems. Hence AM fungi have been shown to have a strong influence on plant species diversity. Since these fungi are obligate symbionts, their population and diversity are determined by the plant species present in the given ecosystem. Apart from plant species, human activities also affect these fungi. These modify the structure and functioning of plant communities in a complex and unpredictable way (Grime et al. 1987). AM fungi, by forming an extended, intricate hyphal network, can absorb mineral nutrients from the soil and deliver them to their host plants in exchange for carbohydrates. Facilitated nutrient uptake, particularly with respect to immobile nutrients such as phosphorus, is believed to be the main benefit of this symbiosis for plants (George et al. 1995; Miller and Jastrow, 1992). Apart from this, they can also enhance drought resistance, resistance against root pathogens and tolerance to heavy metal toxicity in plants (Borowrcz, 2001; Meharg and Cairney, 2000). AM fungi also play a role in the formation of stable soil aggregates, building up of macroporous structures of soils that allow penetration of water and air, and prevent erosion (Miller and Jastrow, 1992). In a given ecosystem these fungi play an important role in carbon allocation, nutrient cycling and maintenance of diversified ecosystems (Doss and Bagyaraj, 2001). The presence of these fungi and their genetic and functional diversities are important for both plant community and ecosystem productivity (Bever et al. 2001; Bidartondo et al. 2002; Klironomos et al. 2000).

The composition of the plant community may also affect mycorrhizal fungi causing differential reproduction and survival which will definitely act as a selective force on the composition of AM population in soil (Janos, 1980; Sanders and Fitter, 1992). It is important to develop AM fungal management strategies for sustainable low input but reasonably productive and ecologically sound agriculture (Bethlenfalvay, 1993; Muthukumar and Udaiyan, 2002; Tilman, 1999). Hence AM fungi are very important for the functioning of terrestrial ecosystems, environmental conservation and sustainable agriculture.

Human activities like application of fertilizer, crop rotation and soil management may also alter the population and diversity of arbuscular mycorrhizal fungi (Giovannetti and Gianinazzi-Pearson, 1994). Modern intensive farming practices are evidently a threat to AM fungi, as indicated by studies on these fungi (Douds and Millner, 1999; Mader et al. 2000). However, little is known about the effect of management practices on the species diversity of these fungi. Oehl et al. (2003) observed decrease in species richness of AM fungi with increased land use intensity in Central Europe. In recent days, AM fungal population and diversity are declining because of agricultural/land use intensification.

Ecosystems and Diversity of AM Fungi

Mycorrhizal associations are potential factors in determining the diversity in ecosystems. They can probably modify the structure and functioning of a plant community in a complex and unpredictable way (Grime et al. 1987). Any shift in the mycorrhizal fungal population could have consequences for the composition of plant communities (survival, competition, floristic diversity) causing changes in the biology of natural ecosystems (Molina et al. 1992). The composition of plant communities may also affect that of fungal communities. Any factor (e.g. host-symbiont combinations) causing differential reproduction and survival of arbuscular mycorrhizal fungi (e.g. sporulation rates) will operate as a selective force on the composition of the soil populations (Janos, 1980; Sanders and Fitter, 1992). Further, different abiotic factors like CO_2, pollutants may also provoke modifications in native plant communities thus modifying the mycorrhizal status of the ecosystems. Human activities like application of fertilizers, crop rotation and soil management may also alter the dynamics and diversity of the fungal community (Giovannetti and Gianinazzi-Pearson, 1994) and therefore, these mycorrhizal associations being dominant in agricultural and horticultural cropping systems have a lot of significance. Thus a knowledge of different factors influencing the population biology of arbuscular mycorrhizal fungi is essential in any attempt to use them in environmental conservation (Allen, 1991), biotechnology (Mulongoy et al. 1992) or in sustainable agriculture (Bethlenfalay and Linderman, 1992).

Glomalean fungi have following unique biological and ecological properties:

(i) Original is an ancient obligate symbiosis that has existed from at least 250 million years ago that has been conserved concomitant with the evolution of land plants (Malloch et al. 1980; Morton, 1990).

(ii) Most of the fungi appear to be obligately asexual, with only one unconfirmed exception reported

(Tommerup and Sivasithamparam, 1990).

(iii) Sub-cellular morphological diversity of spores is unparalleled among fungi in the kingdom Mycota in organization, complexity and stability (Morton and Benny, 1990).

(iv) Organisms of a species are so widely dispersed that they can be found in all major continents (Morton and Bebtivenga, 1994).

(v) Taxonomic groups do not correlate with ecological niches (Morton, 1990).

Effect of Agricultural Practices on AM Fungal Population and Diversity

Modern agricultural practices to enhance food production to meet the needs of increasing human population are posing problems to AM fungi. The agricultural intensification declined the AM fungal abundance and effectiveness with respect to root colonization and plant growth promotion. The important agricultural practices commonly followed are cropping pattern, crop rotation, tillage operations, organic amendments, season, fertilizer application, weeding, crop protection etc.

Monocropping with a particular crop results in the development of a predominant AM fungus in soil. Continuous cultivation of maize in Philadelphia, USA, for three years resulted in the development of *Gigaspora gigantia* in a soil (Schenck and Kinloch, 1980). Mixed cropping is common practice in the tropics. In an experiment, the effect of mono and mixed cropping with soybean and maize on AM fungal population in soil showed that mixed cropping stimulated the proliferation of AM fungi, compared with monocropping with maize or soybean (Harinikumar et al. 1990). Soybean being a legume, possibly provides nitrogen to maize through AM fungi. Sieverding and Leihner (1984) found that mycorrhizal root infection of cassava increased by intercropping with legumes. One reason for the higher propagule density under mixed cropping may be the more intensively rooted soil in the mixed system. Additionally, through higher plant density, nutrients are extracted faster from the soil, thereby stimulating AM fungal reproduction. Mixed cropping is a common practice in the tropics.

Graminaceous and leguminous crops are generally believed to increase AM fungal population, while non-mycotrophic plants decrease the population of mycorrhizal fungi (Sieverding and Leihner, 1984). Taking non-mycorrhizal hosts like mustard or leaving the land follow will reduce the propagules of AM fungi in soil (Harinikumar and Bagyaraj, 1988b). In contrast, use of a crop which is strongly mycorrhizal will increase their numbers. An experiment was carried out in a P deficient soil, where in first season, finger millet was grown in all the plots and in the second season, a mycorrhizal host (cowpea) was grown in two thirds of the plot and rest was left fallow. In the third season, cowpea was grown in all the plots. The results showed reduction in the mycorrhizal propagules in the soil left fallow (Harinikumar and Bagyaraj, 1988b). Growing a non-mycorrhizal host significantly reduced the native mycorrhiza, but the reduction was not as bad as when the land was left fallow. A mycorrhizal host taken in the third season resulted in a slow build up of mycorrhizal population in the soil. However, at the end of the third season it did not reach the same level as in plots cropped continuously with mycorrhizal host. This suggests the reduction in mycorrhizal population caused by leaving the land fallow or growing a non-mycorrhizal host. It may take at least two cropping seasons with a mycorrhizal host to rebuild the reduced mycorrhizal population to the original level. Therefore, it is best to grow a variety of crops in rotation. Further, some plants do not become colonized by AM fungi and therefore will depress populations of these fungi.

Weeds can act as a kind of an instantaneous crop rotation. Since the diversity of the AM fungus community can be proportional to the associated plant community, strict and complete weed control decreases the diversity and efficacy of the indigenous community of AM fungi. Earlier studies have shown that less tillage of soil is better for the buildup of mycorrhizal populations (McGonigle and Miller, 1993). Less disturbance of the soil with ridge tillage resulted in more mycorrhizal population, compared with moldboard ploughing. Studies have shown that less tillage of soil is better for the fill up of mycorrhizal population (McGonigle and Miller, 1993). Reduced disturbance of soil with ridge tillage resulted in more mycorrhizal population compared with mould board ploughing. Further, a cover crop of hairy vetch planted after harvest of winter wheat became significantly more colonized by AM fungi in non tilled soil where hyphal networks were intact than in plots subjected to mould board tillage. Further, a cover crop of hairy vetch planted after harvest of winter wheat became significantly more colonized by AM fungi in untilled soil, where hyphal networks were intact, than in plots subjected to mold board plough tillage. Untilled soils may have more mycorrhizal spores in the top soil, while tilled soils may have more at the 8-15 cm depth (Abbott and Robson, 1991). The mycorrhizal hyphae in the soil act as the nutrient-absorbing organ of the mycorrhiza and the way in which new roots are colonized. Tillage disrupts both of these functions. On the other hand, seedlings grown in untilled soils become colonized by AM fungi more rapidly and have greater phosphorus status than those grown in tilled soils.

Another practice that has negative impacts on AM fungi is over winter bare fallow. This removes potential host roots, from which the fungi can receive sugar during mild fall and spring weather, thereby decreasing viability and ability of the fungi to colonize the next crop. An over winter cover crop may not only be useful for the mycorrhizal fungi but will also boost the amount of AM fungi in the soil. Studies conducted in temperate countries reveal that AM fungi sporulate during summer with higher temperature and longer day length (Hayman, 1974; Furlan and Fortin, 1977). An experiment conducted by Harinikumar and Bagyaraj (1988a) and Mallesh and Bagyaraj (1991) revealed that AM fungi sporulate during winter in the tropics. This is probably because the optimum soil temperature for sporulation occurs during winter in the tropics while it occurs during summer in the temperate countries. The optimum temperature for sporulation by mycorrhizal fungi appears to be around 25°C.

In tropical soils, application of organic matter either in the form of FYM, compost or organic amendments stimulates proliferation of AM fungi (Harinikumar and Bagyaraj, 1989). This is probably because of the low organic matter content in tropical soils. Harinikumar and Bagyaraj (1988a) found that addition of organic amendments such as paddy straw, maize straw and pongamia leaf increased the mycorrhizal activity. Of the three amendments studied, the addition of pongamia leaf encouraged AM fungi to the maximum, followed by maize straw.

For modern agriculture, fertilizer application is an essential and often the most promising method to increase crop production in infertile soils. In tropical soils, phosphorus is one of the most limiting elements for crop production. High P availability is reported to be negatively correlated with AM fungal activity (Krishna and Bagyaraj, 1982). Apparently, the internal P content of plants regulates AM fungal infection and reproduction (Menge et al. 1978). Tissue P concentration is also not always a good estimate for mycorrhizal colonization, because the mycorrhizae themselves influence the factor. It is likely that P influences AM colonization by affecting concentrations of root carbohydrates or the amount of root exudates. The percentage of P in plants at the time of AM colonization is the best indicator to identify a soil, which provides good AM colonization (Jasper et al. 1979). The rock phosphate applied at 100 ppm P level resulted in more infective propagules of *Glomus fasciculatum* (Sreenivasa and Bagyaraj, 1989). Clarke and Mosse (1981) observed that rock phosphate encourages better proliferation of AM fungi, compared to bone meal and super phosphate. Research findings show that years of P fertilization can lead to very high soil P levels. Plants that are able to absorb sufficient P via their roots alone in high nutrient soils inhibit the spread of colonization by the fungus. This reduces the flow of sugars to the fungus, which lessens the amount of AM fungi in the soil. Hence, low or no P fertilization is necessary in such soils. Addition of phosphatic fertilizer decreased the AM fungal formation compared to no added P control and P supplied through organic matter. Even disturbance of soil pre-established with AM fungi affected its further establishment (Boddington and Dodd, 2000).

Application of heavy doses of nitrogen fertilizers (188 kg N per hectare per year) can have a large negative effect on AM population (Hayman, 1975) and nitrate as a source of nitrogen has been shown to be more inhibitory to AM development than ammonium salts (Menge, 1984: Sreenivasa and Bagyaraj, 1988). Menge (1984) noted that regular fertilization of citrus with more than 100 ppm N as a mixture of NO_3 and NH_4 retarded mycorrhizal development. Among the salts, calcium nitrate, urea and calcium ammonium nitrate at different levels were compared and it was found that calcium ammonium nitrate applied at 80 ppm N level to soilrite : perlite mix substrate with negligible N produced maximum number of infective propagules of *Glomus fasciculatum,* in association with Rhodes grass (Sreenivasa and Bagyaraj, 1988). Levels greater than 80 ppm decreased the number of infective propagules. These results indicate that nitrogen content in soils could greatly influence the distribution and abundance of AM fungi. Potassium content of acidic tropical soils is generally low, and sustainable crop yields depend on K application. Cassava has a very high K demand. About 5.8 kg has to be applied for each tonne of cassava yield (Howeler, 1980). It was found that increasing K application levels up to 200 kg K/ha increased mycorrhizal root infection ratings as well as tuber yield in Columbia.

Most pesticides inhibit colonization and development of AM fungi in plants (Ocampo and Hayman, 1980). Although the majority of pesticides tested adversely affect the symbiosis (Menge, 1982) others do not appear to damage mycorrhizal fungi. Some may even increase mycorrhizal colonization (Trappe et al. 1984). Fumigation of soil with biocides such as methyl bromide, chloropicrin, etc. effectively kills endophytes. However, AM fungi can reinvade most fumigated soils within several years (Trappe et al. 1984). The systemic fungicides like thiobendazole, benomyl and triademefon are most toxic to these fungi (Nemec, 1980). Pentachloronitrobenzene which is not systemic is also toxic (Nemec and Tucker, 1983). Fungicides like Captan and Rilon applied at half the recommended levels had no adverse effect on AM fungi (Praveen Kumar and Bagyaraj, 1999). The insecticide metasystox and aldrin differ in their activity on AM fungi, the former being less toxic whilst most nematicides exhibit slight inhibitory effects on AM fungi. Interestingly some nematicides such as 1,2-dibromo-3-chloropropane (DBCP) can stimulate root infection in host plants (Menge et al. 1978). This response could be due to control of competitive and pathogenic microflora and possible stimulation of root exudates of host plants or other factors (Trappe et al. 1984).

Effect of Land Use Patterns on the Arbuscular Mycorrhizal Fungal Population and Diversity

Land use pattern is the way/purpose/extent to which a particular area of land is being utilized. Land use patterns are mainly classified into 2 types viz., cultivated and uncultivated. In cultivated type different land use patterns are forests, grasslands, plantations, agricultural lands etc. In uncultivated type different land use patterns are human dwellings, industrial areas, roads, barren lands etc. Studies have been carried out in cultivated areas. Many studies were done to know AMF population and diversity because of change in land use intensities/patterns, i.e. conversion of forest to agricultural lands or to pasture, pastures to agriculture lands, barren lands to forest plantations etc. Studies were also done regarding comparison of AMF population diversity in different land use intensities or land use patterns, e.g. between natural forests, plantations, grasslands and within the plantations between different crop etc.

Oehl et al. (2003) studied the impact of land use intensity on the diversity of arbuscular mycorrhizal fungi at eight different sites; three sites with low input grasslands, two sites with low to moderate input farming with 7 year crop rotation and three sites with high input continuous maize monocropping. They observed a decrease in AM fungal diversity from low input grasslands to high input monocropping farming system. As the land use intensification increased, the AM fungal species diversity also decreased. Apart from this they also observed more spore abundance in low input grasslands than in low to medium input farming system and high input farming system. Even the AMF species diversity index was more in grasslands. Among two different farming systems, AMF species diversity index was more in sites with crop rotation rather than sites with monocropping. Galvez et al. (2001) studied the populations of spores of AM fungi, mycorrhiza formation and nutrient utilization of maize in moldboard ploughed, chisel-disked or untilled soil under conventional and low input agricultural systems. Soils under low input management had higher AM fungal spore populations than soils under conventional management. Spore load and colonization of maize roots by AM fungi were higher in non-tilled than in moldboard ploughed and chisel disked soil.

Glomalin, an arbusular mycorrhizal fungal soil protein playing an important role in soil aggregation was significantly affected by land use pattern. Glomalin concentrations were highest in native forest soils, moderate in afforested soils and lowest in agricultural lands. Soil C and N were highly correlated with glomalin across all soils and within each land use type, indicating that some glomalin may be under similar controls as soil C. These results also show that glomalin may be useful as an indicator of land use change (Rilling et al. 2003).

Carpenter et al. (2001) studied the spore density and diversity of AM fungi in different land uses and across different gradients of erosion. They found that the diversity and composition of AM fungi changed due to change in land use types and AM fungal diversity decreased due to erosion. Picone (2000) compared spore abundance and diversity of AM fungi in soils of lowland evergreen forests and pastures in Nicaragua. Species composition, dominance-diversity curves and Simpson's diversity indices were similar for both forest and pasture soils. Of 28 distinct fungus morphospecies, 11 produced more spores in pasture, while only 1 produced more spores in forest. According to species-accumulation curves, local AM fungal species richness did not significantly decline following conversion of forest to pasture. Because pastures contained a surprising abundance and diversity of AM fungal spores compared to native forest, a lack of mycorrhizal fungi is unlikely to limit plant succession, restoration, or reforestation in the pastures studied. In addition to these trends in diversity, species that sporulated more in pasture tended to have small spores, while the species that sporulated more in forest had the largest spores. Similarly, only large-spored fungi (300 μm) showed seasonal variation in spore abundance, being more common in the wet season. Wolfe (2002) compared species diversity of AM fungi in three pairs of forest fragment and adjacent pasture sites in a coffee farming community in Coto Brus, Southern Costa Rica. They determined species diversity of AM fungi by both direct assessment of field-collected soil samples and estimates from trap cultures in the greenhouse. Results suggested that conversion of forest to pasture produced shifts in abundance of many AM fungi species rather than general declines. Species richness of AM fungi was similar in pasture and forest sites, despite the dipartite nature of the pasture plant community relative to that of the forest. While some AM fungi species were common in both forest and pasture sites, others were abundant in one vegetation type and rare or absent in the other. These results suggested that pasture plants support AM fungi community that may be as species-rich as that of forested sites, but differing in its composition.

Lovelock et al. (2003) assessed the spatial and temporal distribution of the AM fungal community in a wet tropical rainforest in Costa Rica. Host tree species differed in their associated AM fungal communities, but differences in the AM community between the hosts could not be generalized over life history groupings of the hosts. Changes in the relative abundance of a few common AM fungal species were the cause of differences in AM fungal communities for different host tree species instead of differences in the presence and absence of AM fungal species. Thus, AM fungal communities were spatially distinguishable in the forest, even though all species were widespread. Soil fertility ranging between 5 and 9 mg/ha phosphorus did not affect composition of AM fungal communities, although sporulation was more abundant in lower fertility soils. Sampling soils over seasons revealed that some AM fungal species sporulated profusely in the dry season compared to the rainy season. On one host tree species sampled at two sites with vastly different rainfall, relative abundance of spores from *Acaulospora* was lower and that of *Glomus* was relatively higher at the site with lower and more seasonal rainfall. Johnson and Wedin (1997) found that diversity of mycorrhizal spore communities (measured by Sorenson's similarity index) was lower in the grassland plots than in the forest plots, indicating that grass invasion had caused some convergence. However, total spore density and alpha diversity of mycorrhizal spore communities (measured by species richness and Simpson's diversity index) were not altered by wildfires and grass invasion. These results suggest that persistence and regeneration of forest plant species in the grasslands may not be constrained to a significant degree by the lack of mycorrhizal symbionts. Stutz (2003) studied species richness, composition, spore density and diversity of arbuscular mycorrhizal fungi in four different land use types, viz. urban residential, urban non-residential, agriculture and desert. The agricultural sites were associated with decreased spore densities and decreased species richness. Spores of *G. microaggregatum* were most abundant in urban sites, while those of *G. eburneum* were most abundant in desert and agricultural sites.

Conclusion

AM fungal population and diversity are influenced by the composition of plant species, organic matter, organic carbon, and pH and P levels in soil of a given ecosystem. Further, these fungi are also affected by agricultural practices in the agro-ecosystem. Studies related to the effect of land use intensity and agricultural practices on AM fungi give an idea of managing these fungi for improving crop growth.

References

Abbott, L.K. and Robson, A.D. (1991) Field management of mycorrhizal fungi. In: The Rhizosphere and Plant Growth. (eds), Kelster, D.L. and Cregan, P.B., Kluwer Academic Publishers, Dordrecht, the Netherlands, 355–362.

Abbott, L.K. and Robson, A.D. (1977) The distribution and abundance of vesicular endophytes in some western Australian soils. Australian Journal of Botany 25: 515–522.

Abbott, L.K. and Robson, A.D. (1979) A quantitative study of the spores and anatomy of mycorrhizas formed of *Glomus*, with reference to its taxonomy. Australian Journal of Botany 27: 363–375.

Abbott, L.K. and Robson, A.D. (1981) Infectivity and effectiveness of five endomycorrhizal fungi: Competition with indigenous fungi in field soils. Australian Journal of Botany 32: 621–630.

Allen, M.F. (1991) The ecology of mycorrhizae. Cambridge University Press, Cambridge, U.K.

Allen, M.F., Moore, T.S. and Christensen, M. (1982) Phytohormone changes in *Boutelina gracilis* infected by vesicular arbuscular mycorrhizae. II . Altered levels of gibberillin-like substances and abscisic acid in the host plant. Canadian Journal of Botany 60: 468–471.

Bagyaraj, D.J. Manjunath, A. and Patil, R.B. (1978) Occurrence of vesicular arbuscular mycorrhiza in some tropical aquatic plants. Transactions of the British Mycological Society 72: 164–167.

Bethlenfalvay, G.J. and Linderman, R.G. (1992) *Mycorrhizae in Sustainable Agriculture*. ASA special publication, No, 54, Madison, WI, U.S.A.

Bethlenfalvay, G.J. (1993) Mycorrhizae in the agricultural plant-soil system. Symbiosis 14: 413–425.

Bever, J.D., Schultz, P.A., Pringlr, A. and Morton, J.B., (2001) Arbuscular mycorrhizal fungi: more diverse than meets the eye, and the ecological tale of why. Bioscience 51: 923–931.

Bidartondo, M.L., Redecker, D., Hijri, L., Wiemken, A., Bruns, T.D., Dominguez, L., Sersic, A., Leake, J.R. and Read, D.J. (2002) Epiparasitic plants specialized on arbuscular mycorrhizal fungi. Nature. 419: 389–392.

Blaszkowski, J. (1994) Comparative studies on the occurrence of arbuscular fungi and mycorrhizae (Glomales) in cultivated and uncultivated soils of Poland. Acta Mycologica 28: 93–140.

Boddington, C.L. and Dodd, J.C. (2000) The effect of agricultural practices on the development of indigenous arbuscular mycorrhizal fungi, II. Studies in experimental microcosms. Plant and Soil. 218: 145–157.

Borowicz, V.A. (2001) Do arbuscular mycorrhizal fungi alter plant-pathogen relations? Ecology 82: 3057–3068.

Caroline Machado and Bagyaraj, D.J. (1995) Mycorrhization bacteria and its influence on growth of cowpea. In: Mycorrhizae, Biofertilizers for the future (eds), Adholeya, Alok and Sujan Singh, Tata Energy Research Institute Pub. New Delhi, 192–196.

Carpenter, F.L., Palacioss S., Gonzalez , E. and Schroeder, M. (2001) Land-use and erosion of a Costa Rican Ultisol affects soil chemistry, mycorrhizal fungi and early regeneration. Forest Ecology and Management 144: 1–17.

Clarke, C. and Mosse, B. (1981) Plant growth response to vesicular-arbuscular mycorrhiza. XII. Field inoculation

responses of barley at two soil P levels. New Phytologist 87: 695–703.
Cooper, K.M. (1976) A field survey of mycorrhizas in New Zealand ferns. New Zealand Journal of Botany 14: 169–181.
Doss, D.D. and Bagyaraj, D.J. (2001) Ecosystem dynamics of mycorrhizae.In: Innovative Approaches in Microbiology (eds), Maheswari D.K. and Dubey, R.C. Bishen Singh and Mahendra Pal Singh, Dehradun, India. 115–129.
Douds, D.D. and Millner, P. (1999) Biodiversity of arbuscular mycorrhizal fungi in agroecosystems. Agriculture, Ecosystems and Environment 74: 77–93.
Frank, A.B. (1985) Uber die out Wurzetsymbiose berohende ernahruing gewisser Baume durch utnerimdinche. Berdent. Bot. Gessel. 3: 128–145.
Furlan,V. and Fortin, J.A. (1977) Effects of light intensity on the formation of vesicular-arbuscular endomycorrhizas on *Allium cepa* by *Giaspora calospora*. New Phytologist 79: 335–340.
Galvez, L., Douds, D.D. Jr, Drinkwater, L.E. and Wagoner, P. (2001) Effect of tillage and farming system upon VAM fungus populations and mycorrhizas and nutrient uptake of maize, Plant and Soil 228: 299–308.
George, E., Marshner, H. and Jakobsen, I. (1995) Role of arbuscular mycorrhizal fungi in uptake of phosphorus and nitrogen from soil. Critical Reviews in Biotechnology 15: 253–270.
Gianinazzi, S. and Gianinazzi-Pearson, V. (1986) Progress and headaches in endomycorrhiza biotechnology. Symbiosis 2: 139–149.
Giovannetti, M. and Gianinnazzi-Pearson, V. (1994) Biodiversity in arbuscular mycorrhizal fungi. Mycological Research 98: 705–715.
Grime, J.P., Mackey, J.M.L., Hillier, S.H. and Read, D.J. (1987) Floristic diversity in a model system using experimental microcosms. Nature. 328: 420–422.
Hall, I.R.S. (1977) Species and mycorrhizal infections of New Zealand Endogonaceae. Transactions of the British Mycological Society. 68: 341–356.
Harinikumar, K.M. and Bagyaraj, D.J. (1988a) The effect of season on VA mycorrhiza of leucaena and mango in semi-arid tropic. Arid Soil Research Rehabilitation 7: 139–143.
Harinikumar, K.M. and Bagyaraj, D.J. (1988b) Effect of crop rotation on native VA mycorrhizal propagules in soil. Plant and Soil 110: 77–80.
Harinikumar, K.M. and Bagyaraj, D.J. (1989) Effect of cropping sequences, fertilizers and farm yard manure on VA mycorrhizal fungi. Biology and Fertility of Soils 7: 173–175.
Harinikumar, K.M., Bagyaraj, D.J. and Mallesha B.C. (1990) Effect of intercropping and organic soil amendments on native VA mycorrhiza in semi-arid tropics. Arid Soil Research Rehabilitation 4: 193–197.
Harley, J.L. and Smith, S.E. (1983) *Mycorrhizal symbiosis*. Academic Press, London.
Hayman, D.S. (1974) Plant growth responses to vesicular-arbuscular mycorrhiza VI. Effect of light and temperature. New Phytologist 73: 71–80.
Hayman, D.S. (1975) The occurrence of mycorrhiza in crops as affected by soil fertility. In: Endomycorrhizas (eds) Sanders, F.E., Mosse, B., and Tinker, P.S. Academic Press, London, pp. 495–509.
Hayman, D.S. (1983) The physiology of vesicular arbuscular endomycorrhizal symbiosis. Canadian Journal of Botany 61: 944–963.
Hirrel, M.C., Mehravaran, H. and Gerdemann, J.W. (1978) Vesicular arbuscular mycorrhizae in Chenopodiaceae and Cruciferaceae: do they occur? Canadian Journal of Botany 56: 2813–2817.
Howeler, R.H. (1980) Soil related cultural practices for cassava. In: Cassava Cultural Practices. Proceedings of a workshop held in Salvador, Bahia, Brazil, 18-21 March (eds) Webes, E.J., Toro, M.J.C. and Graham, M. CAB International, pp. 159–169.
Janos, D.P. (1980) Vesicular-arbuscular mycorrhizae affect lowland tropical rain forest plant growth. Ecology 61: 151–162.
Jasper, P.A., Robson, A.D. and Abbott, L.B. (1979) Phosphorus and the formation of vesicular-arbuscular mycorrhizae on growth and metabolism of sweet orange. New Phytologist 90: 665–670.
Johnson, N.C., Wedin, D.A. (1997) Soil carbon, nutrients, and mycorrhizae during conversion of dry tropical forest to grassland. Ecological Applications 7: 171–182.
Kendrick, B. and Berch, S. (1985) Mycorrhizae: Applications in agriculture and forestry. In: *Comprehensive Biotechnology*, Vol. 4 (ed) Robinson, C.W., Pergamon Press, Oxford, 109–150.
Khan, A.G. (1978) Vesicular-arbuscular mycorrhizas in plants colonizing black wastes from bituminous coal mining in

the Illawarra region of New South Wales. New Phytologist 81: 53–63.

Klironomos, J.N., McCune, J., Hart, M. and Neville, J. (2000) The influence of arbuscular mycorrhizae on the relationship between plant diversity and productivity. Ecology Letters 3: 137–141.

Krishna, K.R. and Bagyaraj, D.J. (1982) Effect of vesicular-arbuscular mycorrhiza and soluble phosphate on *Abelmoscus esculentus* (L.) Moench. Plant and Soil 64: 209–213.

La-Rue, J.W. and Mcllellen, W.A. and Peacock, W.L. (1975) Mycorrhizal fungi and peach nutrition. California agriculture 29: 6–7.

Lovelock, C.E., Andersen, K. and Morton, J.B. (2003) Arbuscular mycorrhizal communities in tropical forests are affected by host tree species and environment. Oecologia 132: 268–279.

Mada, R.J. and Bagyaraj, D.J. (1993) Root exudations from *Leuceana leucocephala* in relation to mycorrhizal colonization. World Journal of Microbiology and Biotechnology 9: 342.

Mader, P., Edenhofer, S., Boller, T., Wiemken, A. and Niggli, U. (2000) Arbuscular mycorrhizae in a long term field trial comparing low input (organic biological) and high input (conventional) farming systems in a crop rotation. Biology and Fertility of Soils 31: 150–156.

Mallesha, B.C. and Bagyaraj, D.J. (1991) Season favouring sporulation of VA mycorrhizal fungi in cardamom plantations. Soil Biology and Ecology 11: 75–78.

Malloch, D.W., Pirozynski, K.A. and Raven, P.H. (1980) Ecological and evolutionary significance of mycorrhizal symbioses in vascular plants. Proceedings of the National Academy of Sciences, USA 77. 2113–2118.

McGonigle, T.P. and Miller, M.H. (1993) Mycorrhizal development and phosphorus adsorption in maize under conventional and reduced tillage. Soil Science Society of America Journal 57: 1002–1006.

Meharg, A.A. and Cairney, J.W.G. (2000) Co-evolution of mycorrhizal symbionts and their hosts to metal-contaminated environments. Advance in Ecological Research 30: 69–112.

Menge, J.A. (1982) Effect of soil fumigants and fungicides on vesicular-arbuscular fungi. Phytopathology. 72: 1125–1132.

Menge, J.A. (1984) Inoculum production. In: VA mycorrhiza. (eds) Powell, C.L., and Bagyaraj, D.J. CRC Press Inc., Boca Raton, Florida, 188–189.

Menge, J.A., Steirle, D., Bagyaraj, D.J., Johnson, E.L.V. and Leonard, T.T. (1978) Phosphorus concentrations in plants responsible for inhibition of mycorrhizal infection. New Phytologist 80: 575–578.

Miller, R.M. and Jastrow, J.D. (1992) The application of VA mycorrhizae to ecosystem restoration. In: Mycorrhizal Functioning (ed.) Allen, M.F., Chapman and Hall Ltd., London, England.

Molina, R., Massocotte, H. and Trapre, J.M. (1992) Specificity phenomena in mycorrhizal symbioses: community-ecological consequences and potential implications. In: Mycorrhizal functioning (ed), Allen, M.F. Chapman and Hall: New York, U.S.A. 357–372.

Morton, J.B. (1990) Species and clones of arbuscular mycorrhizal fungi (glomales, Zygomycetes): their role in macro and micro evolutionary process. Mycotaxon 37: 493–515.

Morton, J.B. and Benny, G.L. (1990) Revised classification of arbuscular mycorrhizal fungi (zygomycetes): a new order, Glomales, two new suborders, Glomineae and Gigasporineae and two new families, Acaulosporaceae and Gigasporacea, with an emendation of Glomaceae. Mycotaxon 37: 471–491.

Morton, J.B. and Bentivenga, S.P. (1994) Levels of diversity in endomycorrhizal fungi (Glomales, Zygomycetes) and their role in defining taxonomic and non-taxonomic groups. Plant and Soil 159: 47–59.

Mosse, B. (1977a) Plant growth responses to vesicular arbuscular mycorrhiza-IV. In soil given additional phosphorus. New Phytologist 72: 127–136.

Mosse, B. (1977b) Plant growth responses to vesicular–arbuscular mycorrhiza. X. Responses of Styloxanthus and maize to inoculation in unsterile soils. New Phytologist 78: 277–288.

Mosse, B. and Bowen, G.D. (1968) The distribution of *Endogone* spores in some Australian and New Zealand soils and in the experimental field soil at Rothamsted. Transactions of the British Mycological Society 51: 485–492.

Mosse, B., Stribley, D.P. and Le Tacon, F. (1981) Ecology of mycorrhizae and mycorrhizal fungi. Advances in Microbial Ecology 5: 137–209.

Mulongoy, K., Gianinazzi, S., Roger, P.A. and Dommergues, Y. (1992) Biofertilizers: Agronomic and environmental impacts and economics. In: Biotechnology, Economic and social aspects: Issues for developing countries. (eds) De Silva, E.J., Rutledger, C. and Sasson, A., UNESCO, Cambridge University Press, Cambridge, U.K. 55–69.

Muthukumar, T. and Udaiyan, K. (2002) Growth and yield of cowpea as influenced by changes in arbuscular mycorrhiza

in response to organic manuring. Journal of Agronomy and Crop Science 188: 123–132.
Nemec, S. (1980) Effects of eleven fungicides on endomycorrhizal development in sour oranges. Canadian Journal of Botany 58: 522–527.
Nemec, S. and Tucker, D. (1983) Effects of herbicides on endomycorrhizal fungi in Florida citrus (Citrus spp.) soils. Weed Science. 31: 417–431.
Ocampo, I.A. and Hayman, D.S. (1980) Effects of pesticides on mycorrhiza in field grown barley, maize and potatoes Transactions of the British Mycological Society 74: 413–416.
Oehl, F., Sieverding, E., Ineichen, L., Mader, P., Boller, T. and Wienmken, A. (2003) Impact of land use intensity on the species diversity of arbuscular mycorrhizal fungi in agroecosystems of central Europe. Applied and Environmental Microbiology 69: 2816–2824.
Picone, C.M. (2000) Diversity and abundance of arbuscular mycorrhizal fungus spores in tropical forest and pasture. Biotropica 32: 734–750.
Praveen Kumar, K.A. and Bagyaraj, D.J. (1999) Mass production of arbuscular mycorrhiza as influenced by some agrochemicals. Proceeding of National Academy of Sciences India 69: 61–66.
Raj, J., Bagyaraj, D.J. and Manjunath, A. (1981) Influence of soil inoculation with vesicular arbuscular mycorrhiza and a phosphate dissolving bacterium on plant growth and ^{32}P-uptake. Soil Biology and Biochemistry 13: 105–108.
Redhead, J.F. (1977) Endotrophic mycorrhizas in Nigeria: Species of the Endogonaceae and their distribution. Transactions of British Mycological Society 69: 275–280.
Rilling, M.C., Ramsey, P.W., Morris, S. and Paul, E.A. (2003) Glomalin, an arbuscular mycorrhizal fungal soil protein, responds to land use changes, Plant and Soil 253: 293–299.
Sanders, I.R. and Fitters, A.H. (1992) Evidence for differential responses between host-fungus combinations of vesicular arbuscular mycorrhizas from a grassland. Mycological Research 415–419.
Schenck, N.C. and Kinloch, R.A. (1980) Incidence of mycorrizal fungi on six field crops in monoculture on a newly cleared woodland site. Mycologia 72: 445–455.
Schenck, N.C. and Perez, Y. (1990) Manual for the Identification of VA Mycorrhizal Fungi, University of Florida, Gainesville, FL, pp. 245
Sieverding, E. and Leihner, D.E. (1984) Influence of crop rotation and intercropping of cassava with legumes on VA mycorrhizal symbiosis of cassava. Plant and Soil 80: 143–146.
Sparling, G.P. and Tinker, P.B. (1978) Mycorrhizal infection in Penning grassland. II. Effects of mycorrhizal infection on growth of some upland grasses on irradiated soil. Journal of Applied Ecology. 15: 951–958.
Sreenivasa, M.N. and Bagyaraj, D.J. (1988) *Chloris gayana* (Rhodes grass) a better host for mass production of *Glomus fasiculatum* inoculum. Plant and Soil 106: 109–112.
Sreenivasa, M.N. and Bagyaraj, D.J. (1989) Suitable form and level of phosphorus for mass production of the VA mycorrhizal fungus, *Glomus fasiculatum*. Zentralbl Mikrobiol., 144: 34–36.
Stutz, J.C. (2003) Preliminary assessment of arbuscular mycorrhizal fungal diversity and community structure in an urban ecosystem. Mycorrhiza 13: 319–326.
Tilman, D. (1999) Global environmental impacts of agricultural expansion: the need for sustainable and efficient practices. Proceedings of the National Academy of Sciences, USA 96: 5995–6000.
Tommerup, I.C. and Sivasithamparam, K. (1990) Zygospores and asexual spores of *Gigaspora decipiens*, an arbuscular mycorrhizal fungus. Mycological Research 94: 897–900.
Trappe, J.M. (1977) Biogeography of hypogeneous fungi: Trees, mammals, and continental drift, Abstr. 2nd Int. Mycol.Congr. (eds) Bigelow, H.E. and Simmons, E.G, University of South Florida, Tampa, pp. 675.
Trappe, J.M., Molina, R. and Castellano, M. (1984) Reactions of mycorrhizal fungi and mycorrhiza formation to pesticides. Annual Review of Phytopathology 22: 331–359.
Wolfe, A.L. (2002) Species diversity and community composition of arbuscular mycorrhizal fungi in tropical forest fragments and adjacent pastures, 87th Annual meeting of the Ecological Society of America and Annual International conference of the Society for Restoration, Tuscon, AR, USA, August 04-09, 2002, pp. 62.

Index